DICTIONARY OF

PURE
AND
APPLIED
PHYSICS

COMPREHENSIVE DICTIONARY
OF PHYSICS

Dipak Basu
Editor-in-Chief

FORTHCOMING AND PUBLISHED VOLUMES

Dictionary of Pure and Applied Physics
Dipak Basu

Dictionary of Material Science
and High Energy Physics
Dipak Basu

Dictionary of Geophysics, Astrophysics,
and Astronomy
Richard A. Matzner

A VOLUME IN THE
COMPREHENSIVE DICTIONARY
OF PHYSICS

DICTIONARY OF

PURE
AND
APPLIED
PHYSICS

Edited by

Dipak Basu

CRC Press

Boca Raton   London   New York   Washington, D.C.

Library of Congress Cataloging-in-Publication Data

Dictionary of pure and applid physics / edited by Dipak Basu.
    p.   cm. — (Comprehensive dictionary of physics)
    ISBN 0-8493-2890-X (alk. paper)
    1. Physics—Dictionaries. I. Basu, Dipak. II. Series.

QC5 .D485 2000 2001
530¢.03—dc21
                                                                00-052884

Preface

The *Dictionary of Pure and Applied Physics* (DPAP) is one of three physics dictionaries being published by CRC Press LLC, the other two being the *Dictionary of Material Science and High Energy Physics* and *Dictionary of Geophysics, Astrophysics and Astronomy.* Each of these dictionaries is entirely self-contained.

The aim of the DPAP is to provide students, researchers, academics and professionals in general, with definitions in a very clear and concise form. The presentation is such that readers will not have any difficulty finding any term being looked for. Each definition is written in detail as informative as possible supported by suitable diagrams, equations, and formulae whenever necessary.

With more than 3000 terms, the fields covered in the DPAP are acoustics, biophysics and medical physics, communication, electricity, electronics, geometrical optics, low temperature physics, magnetism, and physical optics.

Like most other branches of science, physics has grown rapidly over the last decade. As such, many of the terms used in older books have become obsolete and new terms have appeared in scientific and technical literature. Care has been taken to ensure that old terms are not included in the DPAP and new terminologies are not missed. Some of the terms are related to other fields, e.g., engineering fields (mostly electrical and mechanical), mathematics, chemistry, biology.

Authors are eminent scientists at research institutions and university professors from around the world. Readership includes physicists and engineers in all fields, teachers and students in physics and engineering at university, college and high school levels, technical writers and professionals in general.

The uniqueness of the DPAP lies in the fact that it is an extremely useful source of information in the form of meanings of scientific terms presented in a very clear language and concise form written by authoritative persons in the field. It would be a great aid to students in understanding textbooks, help academics and researchers fully appreciate research papers in professional scientific journals, provide authors in the field with assistance to clarify their writings and, in general, benefit the enhancement of literacy in physics by presenting scientists and engineers with meaningful and workable definitions.

Dipak Basu

CONTRIBUTORS

Barry I. Barker
Stanford University
Stanford, California

Dipak Basu
Carleton University
Ottawa, Ontario, Canada

Christopher Boswell
The Johns Hopkins University
Baltimore, Maryland

H.R. Chandrasekhar
University of Missouri
Columbia, Missouri

David Cheeke
Concordia University
Montreal, Quebec, Canada

Shenlin Chen
Boise, Idaho

Lee Chow
University of Central Florida
Orlando, Florida

T.C. Choy
University of Melbourne
Parkville, Victoria, Australia

J.M. Collins
Marquette University
Milwaukee, Wisconsin

Luis Cruz-Cruz
Boston University
Boston, Massachusetts

Robert T. Deck
University of Toledo
Toledo, Ohio

Vijai Dixit
St. Louis University
St. Louis, Missouri

Douglas M. Gingrich
University of Alberta
Edmonton, Alberta, Canada

Arthur A. Grossman
University of California, San Diego
San Diego, California

Shirin Haque-Copilah
University of the West Indies
St. Augustine, Trinidad, West Indies

Takafumi Hayashi
The University of Aizu
Fukushima, Japan

Cila Herman
The Johns Hopkins University
Baltimore, Maryland

Stanley Jeffers
York University
Toronto, Ontario, Canada

Joe Khachan
University of Sydney
Sydney, NSW, Australia

Vasudevan Lakshminarayanan
University of Missouri, St. Louis
St. Louis, Missouri

Scott A. Lee
University of Toledo
Toledo, Ohio

Mirko Mirkov
Cynosure, Inc.
Chelmsford, Massachusetts

Michael J. O'Shea
Kansas State University
Manhattan, Kansas

Vladimir Ostashev
NOAA
Boulder, Colorado

A.G. Unil Perera
Georgia State University
Atlanta, Georgia

Edward Rothwell
Michigan State University
East Lansing, Michigan

Kenneth Trantham
Arkansas Technology University
Russellville, Arkansas

Kainam Thomas Wong
The Chinese University of Hong Kong
Shatin, NT, Hong Kong

Editorial Advisor

Stan Gibilisco

Acknowledgments

The following figures have been reproduced by kind permissions as mentioned.

Kellner Schmidt optical system: Fundamentals of Optics (4th Edition), by F. Jenkins and H. White, McGraw Hill, New York, 1976.

Tangential and sagittal focal lines: Mirrors, Prisms and Lenses, by J. P. Southall, Dover Publishers, New York, 1964.

Lummer-Gehrke plate: Fundamentals of Optics (4th Edition), by F. Jenkins and H. White, McGraw Hill, New York, 1976.

Moira fringes: McGraw Hill Dictionary of Scientific and Technical terms (5th Edition), by S. P. Parker (Editor-in-Chief), McGraw Hill, New York, 1994.

Heat capacity for 4He. The Lambda phenomenon: John Lipa and Joel Nisser of Stanford University.

A

rays from the object and the image, h_1 and h_2 the object and image heights, n_1 and n_2 the refractive indices, then the *Abbe's sine condition* is

$$n_1 h_1 \sin \theta_1 + n_2 h_2 \sin \theta_2 = 0 .$$

Abbe number   Dispersion or separation of neighboring wavelengths by transparent material can be characterized by the Abbe number. If for a given material n_D, n_F, and n_C refer to the refractive indices for the Fraunhöfer D (589 nm), F (489 nm), and C (656 nm), then the chromatism of the material is characterized by its v-value or constringence

$$v = \frac{n_D - 1}{n_F - n_C} .$$

This is called the refractive efficiency, v-value or *Abbe number.* The larger the constringence, the lesser the chromatism. The inverse of the Abbe number is the dispersive power.

Abbe sine condition   (**1**) For an optical system, if one considers the object and image planes perpendicular to the optical axis, and θ to be the angle relative to the axis made by a ray from an axial object point in object space of index n and θ' to be that in image space of index n', then the transverse magnification, m:

$$m = \frac{n \sin \theta}{n' \sin \theta'} \quad \text{for all } \theta \text{ and } \theta' .$$

For an object at infinity:

$$\sin \ \theta' = -\frac{y}{f'}$$

where y is the height of a ray parallel to the axis and f' is the back (or secondary) focal distance. The condition is valid for a lens free of coma and the relationships hold to a good approximation in most lenses, but not where discontinuities appear in ray behavior (such as Fresnel lenses or zone plates). Deviations from this relationship are called the "offense against the same condition" and are associated with coma.

(**2**) The condition satisfied by rays refracted by spherical interfaces of an optical system (e.g., microscope) to form an image free of coma; if θ_1 and θ_2 are the angles made by non-paraxial

ABCD law   If q_{in} and q_{out} are the complex radii of curvature of the input and output beams of an optical system, they are related by the elements of the ABCD matrix according to *ABCD law:* $q_{out} = \frac{A\, q_{in} + B}{C\, q_{in} + D}$. *See* ABCD matrix.

ABCD matrix   The height (y) and angle (α) of a paraxial ray (measured with respect to the optical axis) can be described by a simple 2×2 composite system matrix as it passes through an optical system with several elements (lenses, mirrors, etc.). The elements of the matrix (M) are referred to as A, B, C, and D. In the equation

$$\begin{bmatrix} y_f \\ \alpha_f \end{bmatrix} = M \begin{bmatrix} y_i \\ \alpha_i \end{bmatrix} ; M = M_n M_{n-1} \cdots$$

$$M_1 = \begin{bmatrix} A & B \\ C & D \end{bmatrix} ,$$

the subscripts i and f refer to the initial (input) and final (output) rays and the system matrix M is a product of the ray matrices of each element of the optical system. The optical element closest to the initial ray has the matrix M_1 and the one closest to the final ray has M_n. The physical significance of the elements of the system matrix is the following: The input (output) plane corresponds to the first (second) focal plane when $D(A)$ is equal to zero. The output plane is the image plane conjugate to the input plane when $B = 0$ with A being the linear magnification. When $C = 0$, a parallel bundle of input rays emerges as a parallel bundle of output rays; D corresponds to the angular magnification. *See* cardinal points.

aberration   Optical systems form distortion-free images when the rays entering the system are parallel to the axis and are close to it (paraxial rays). However, this restriction leads to low throughput. Off-axis and non-paraxial rays lead to distortions which can be classified as (a) chromatic and (b) monochromatic aberrations. *See* aberration, chromatic and aberration, monochromatic.

aberration, chromatic   The refractive index of a material is dependent on wavelength. By Snell's Law, light rays of different wavelengths will be refracted at different angles, since index is not a constant. Because the index of refraction is higher for shorter wavelengths, these are focused closer to a lens when compared to longer wavelengths, when polychromatic or white light is incident on it. *Longitudinal chromatic aberration* is defined to be the axial distance from the nearest to the farthest focal points.

aberration, lateral   The inability of a system to focus rays from a source on its axis to a point on the axis, but instead focus at a point shifted perpendicularly or laterally to the axis.

aberration, longitudinal   When a system focuses the obliquely incident rays on the optical axis but at a point that is not the focal point for the paraxial rays.

aberration of optical systems   Any error in imaging, for example due to dispersion (chromatic), curvature of the surface (spherical), coma, astigmatism, distortion, etc., resulting in inability of an optical system to bring a broad beam to focus at a unique point.

aberrations, monochromatic   The aberrations in an optical system even for a monochromatic light, thus arising totally from the geometry of the system, in contrast to chromatic aberration (*see* chromatic aberration) due to dispersion of the medium (different refractive indices for different wavelengths of the incident light).

aberration, spherical   This distortion is caused by marginal or non-paraxial rays (i.e., rays that are parallel to the axis but are further from the axis). The image has a diffuse halo surrounding the sharp image formed by the paraxial rays. The marginal rays come to a focus at different points on the axis compared to paraxial rays and diverge at steeper angles. (*See* diagram.) This leads to longitudinal and transverse spherical aberration. The position at which the image blur is minimum is called the *circle of least confusion*. The spherical aberration is zero for an optical system if the object and image are arranged to be at the *conjugate points* of one another. (*See* conjugate points.) For a thin lens, if the shape is such that the radius of curvature of the first convex surface is about six times that of the second concave surface, then the *spherical aberration* is minimized. The meniscus shape of eyeglasses is chosen for this reason. Optical systems such as cameras can reduce this defect by using smaller apertures (large f-numbers) to block off the outer rays.

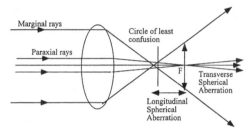

Aberration, spherical.

aberrations, Seidel   Monochromatic aberrations named after the German mathematician Ludwig von Seidel who studied aberration theory keeping higher order terms for parameters describing the departure of the incident beam from a paraxial beam. The five Seidel aberrations are astigmatism, coma, curvature of field, distortion, and spherical aberration.

abrupt junction   A diode junction in which the dopant level or material type changes abruptly at the junction.

A diode is formed at the junction of two materials with different Fermi levels (that is, each with a different chemical potential, the energy level which is the highest filled at a temperature of absolute zero). In semiconductors, one way to alter the Fermi level of the material is to add dopants that are either electron donors or acceptors. A diode constructed of the same basic semiconductor material can be realized with different dopants on each side of the junction. The dopant level can be changed gradually across the device, or change abruptly at the junction. In real devices, the dopant level changes in a region small compared to the depletion layer, the region of non-uniform carrier density about the junction.

Such abrupt junction diodes have the benefit of simplified numerical analysis. For instance, the depletion region depth depends on the square root of the voltage difference (which includes the intrinsic diode voltage and the applied reverse bias voltage).

A junction that is not abrupt is sometimes called a gradual or graded junction, and will have different properties depending on the nature of the dopant profile. *See* diode junction.

absolute zero   The absolute minimum in temperature. Any system in thermal equilibrium at *absolute zero* has its minimum energy and minimum entropy.   Absolute zero occurs at $-273.15°C$ ($-459.67°F$). The absolute temperature scale has an arbitrariness in the size of the degree, chosen such that the triple point of water is exactly $273.16°C$. It is not possible for a system to reach the absolute zero of temperature. This can be seen through the behavior of entropy as the temperature approaches zero. The third law of thermodynamics states that the entropy of a system at $T = 0K$ is a constant. If this is true, then the specific heat is zero at absolute zero:

$$\frac{C_X}{T} = \left(\frac{\partial S}{\partial T}\right)_X$$

where X describes the method of temperature changes (e.g., V for constant volume). Since the third law states S approaches a constant at absolute zero, the slope of the entropy vs. temperature curve must become zero at absolute zero, resulting in a zero heat capacity. It can be shown that this also results in the thermal expansion approaching zero at zero temperature and that

$$\frac{V\alpha}{C_X} \underset{T\to 0}{\to} \text{constant}$$

where V is the volume, α is the expansion coefficient, and C_X is the heat capacity. For an adiabatic change in some variable X, we find

$$dT = \frac{V\alpha}{C_X}T\,dX$$

and this shows us that the change in X needed to produce a temperature change dT grows without bound as $T \to 0$. This completes the argu-

ment that absolute zero is unattainable. *See also* thermodynamics, third law of.

absolute zero, unattainablility of   *See* absolute zero.

absorbance   *See* absorption, Lambert's law of.

absorptance   *See* absorbance.

absorption   If a beam of light with the intensity I_0 passes through a homogeneous substance of thickness x (unit: m) and emerges with a lower intensity I, according to the exponential law of absorption $I = I_0\exp(-\alpha x)$. α is the absorption (or extinction) coefficient (unit: m^{-1}). α depends on the material and is a function of the wavelength of light passing through it. For example, α for glass can be as low as 10^{-7} m^{-1} in the visible while it can be several orders of magnitude higher for the infrared and ultraviolet wavelengths of light. While deriving this law, effects due to scattering are ignored.

absorption, acoustic   Nonreversible transformation of acoustic energy into other forms of energy (for instance, heat). Two main kinds of this process are absorption of sound in its reflection from a surface and absorption of sound in its propagation through gaseous, fluid, or solid media. The latter phenomenon results in exponential decrease of an amplitude and intensity of a sound wave with distance of propagation, e.g., $A = A_0e^{-\alpha x}$. Here, A and A_0 are the amplitudes of sound pressure in a sound wave at two fixed points, x is a length of a sound path between these points, and α is the *absorption coefficient*.

absorption, anomalous, acoustic   Absorption of sound by *relaxation* processes. The term, *anomalous absorption,* had been used until the middle of the twentieth century. However, it is rarely used nowadays.

absorption bands/lines   The absorption spectra of atoms at small concentrations consist of discrete energies at which the absorption is high. The electronic energy levels of each atomic species are discrete and unique. An atom in one state of energy can be excited to a higher state by absorption of a photon whose energy is equal to the difference in energy between the two states of the atom. A spectrograph which displays the intensity of white light passing through the atoms of interest vs. wavelength will show dark lines due to the missing photons of specific energies absorbed by the atoms. Molecules, on the other hand, have vibrational and rotational energy levels superimposed on the electronic energy levels giving rise to a large number of closely spaced lines which smear and appear as bands. Bands also appear at higher concentration of atoms due to broadening of spectral lines. Energy levels of solids (crystalline and amorphous) also exhibit bands.

absorption, Beer's law of   For solutions with concentration C (unit: gm per liter) in a cell of path length b (unit: cm), the logarithmic form of Beer's law for the absorbance, A, can be written as $A = \log(\frac{I_0}{I}) = abC = \varepsilon bC$. Here a is called the *absorptivity* (unit: liters per gram-cm). If C is in units of moles per liter, ε is called the *molar absorptivity* (unit: liter per mole-cm).

absorption coefficient   *See* absorption.

absorption coefficient, acoustic   (**1**) A ratio of absorbed by a surface energy of a sound wave incident on this surface to the energy of the wave before interacting with the surface.

(**2**) An inverse distance at which the amplitude of pressure in a sound wave propagating in a medium decreases by a factor $e \approx 2.7183$. The absorption coefficient, α, can also be determined by $\alpha = x^{-1}\ln(A_0/A)$, where x is a length of a sound path between two fixed points, and A and A_0 are the amplitudes of sound pressure at these points.

absorption edge   Refers to the threshold energy below which photons are transparent in a substance.

1. *Absorption edge* in x-rays: The probability of absorption of photons in elements increases with wavelength until an absorption edge is reached when a sharp drop occurs. The sharp drops correspond to the binding energies of electrons in inner shells.

2. *Absorption edge* in semiconductors: The electronic energy states consist of completely filled bands of energies (Valence Band) followed by unoccupied bands (Conduction Band) separated by a *band gap*. The absorption of photons shows a sharp increase at and above the band gap energy due to the promotion of electrons from the filled Valence Band to the empty Conduction Band. The direct band gap semiconductors (e.g., gallium arsenide) have a sharp edge unlike the indirect gap (e.g., silicon) semiconductors in which the absorption is assisted by lattice excitations (phonons). The edge is also very sensitive to temperature. At low temperatures (when thermal energy is much less than the band gap energy) the edge is much sharper. At higher temperatures, the phonons and impurity states smear the edge.

absorption, Lambert's law of   The exponential law of absorption converted to logarithms to base 10 can be written as $A = \log(\frac{I_0}{I}) = \frac{\alpha x}{2.303}$. Here A is called the *absorbance* of the substance, I_0 and I are the incident and transmitted intensities through a substance of thickness x (unit: m) and absorption coefficient α (units: m^{-1}). *See* absorption.

absorptivity   *See* absorption, Beer's law of.

absorptivity, molar   *See* absorption, Beer's law of.

acceptor   An impurity introduced into a semiconductor crystal lattice to accept an electron, allowing a bond missing an electron (a positive hole) to migrate and thus carry charge. For example, in a silicon lattice, an atom of boron can be substituted for one of the silicon atoms. Silicon atoms have four valence electrons, whereas boron atoms have only three. However, the boron atom completes the covalent bonds with the neighboring silicon atoms by accepting an electron (and becoming negatively charged). The crystal remains electrically neutral, so a hole (seen as a positively charged lack of an electron) can be a charge carrier through the lattice.

Acceptor impurities (for silicon or germanium) include boron, aluminum, gallium and indium. Semiconductors with more acceptor than donor impurities have conductivity dominated by diffusion of holes (positive charges) and are labelled *p*-type. *See also* donor.

access coupler   An *access coupler* transfers optical signal from one conductor to another in an optical transmission system. The coupling coefficient, measuring the efficiency of a coupler, is defined as the ratio of optical power on the incident side to the optical power on the other side of the interface.

access line   An *access line* embodies the circuitry linking a user and a switching center (where message packets are routed onwards to their eventual destination for addresses served by other centers or are delivered to local addresses served by this center). An access line may be specifically adjusted to provide amplitude and phase equalization to the signal passing through it; this is labeled a special-grade access line. An access line connecting an individual user or a group of users to a communication network is called a network access line.

access time   *Access time* in a communication system refers to the time lag from the beginning of an access attempt to an access success. Access time is defined only for successful access attempts. Access time in a computer system refers to the time lapse from when data are delivered to when data storage starts.

accidents, laser   Lasers are dangerous due to the high intensity of laser light. The laser beam causes localized burning resulting in skin lesions, retinal damage, damage to mucous membranes, etc. The extent of damage is a function of laser wavelength and intensity and may be alleviated by wearing safety glasses and opaque clothing. Care must yet be taken to prevent damage to the safety glasses and starting fire to the opaque clothing.

accommodation   (**1**) To observe objects closely and far away, the lens in the eye accommodates, i.e., it adjusts its dioptric power to attain maximal sharpness of retinal imagery. When accommodating to a near object, the ciliary body contracts and hence the curvature of the lens decreases, increasing the power of the lens. The front surface of the lens moves forward slightly. For a distant object, the ciliary body relaxes and the lens assumes a flatter shape, thus increasing its radius of curvature and consequently increasing its focal length.

(**2**) *Optical:* the ciliary muscles tense or relax causing the lens to either (a) bulge resulting in a small radius of curvature and a concomitant shorter focal length allowing the viewer to focus on close objects, or (b) relax resulting in a larger radius of curvature and a concomitant longer focal length allowing the viewer to focus on more distant objects.

(**3**) *Sensory:* with constant, or nearly constant stimulation, sensory receptors may adapt to the stimulus and fail to respond. That is, a slowly rising depolarization at the molecular level may not initiate an action potential (*See* action potential); hence no stimulation.

(**4**) *Evolutionary:* a change in an organism that allows it to better interact with its environment, sometimes referred to as *selective evolution.*

accumulation time   The amount of time a signal (radiation, x-ray, light, etc.) is actually being recorded. Every recording instrument has dead time; the fraction of running time required for a signal to be processed before the next signal can be processed. Hence, the accumulation time is the running (or clock) time less the dead time.

accumulator   An electrochemical cell that can be electrically recharged repeatedly is called an *accumulator.* It is also called secondary cell. *See* secondary cell.

achromatic colors   Colors in which the attribute hue is lacking. Colors ranging from black (absence of light) through various shades of gray to dazzling white are called achromatic. The attribute that distinguishes them is brightness. *See* color.

achromatic doublet   A common way to eliminate chromatic aberration is to use an *achromatic doublet* that consists of a convex and a concave lens of different glasses cemented together. The focal lengths and refractive powers of the lenses differ (by shaping their surface curvatures) producing a net power, and the dispersion powers of the components are chosen such that they are in inverse proportion to these powers. The result is a compound lens that has a net focal length but reduced dispersion over a major portion of the visible spectrum. If the doublet, made up of 2 lenses of individual powers $F_1 + F_2$, has a net power of F diopters, then the conditions defining the doublet are:

$$F = F_1 + F_2 \, ,$$

and

$$\frac{F_1}{v_1} = -\frac{F_2}{v_2} \, .$$

The powers (F, F_1 and F_2) are measured for the sodium D-line (589 nm) and v_1 and v_2 are the Abbe numbers of the materials of the two lenses. Achromatic prisms are formed by combining 2 prisms with equal and opposite dispersions.

achromatism   An optical element or system is said to exhibit *achromatism* if chromatic aberration has been eliminated in the element, e.g., by use of an achromatic doublet.

achromatopsia   Total color blindness. The optic system responds to all frequencies of visible light; however, the person "sees" only shades of gray.

acoustic bridge   Device that is used to measure acoustic impedance of a substance. *Acoustic bridge* is closely analogous to electric bridge. For example, in one of possible arrangements of an acoustic bridge, two tubes (one filled with a substance whose impedance is known and the other with a substance of unknown impedance) form the ratio arms.

acoustic capacitance   The ratio of the volume displacement in an acoustic system to the pressure applied to the system.

acoustic density   Deviation of the density in a medium from its *ambient* value caused by passage of a sound wave. *See also* acoustic pressure; displacement, acoustic.

acoustic efficiency   The ratio of the acoustic energy radiated by a transducer to the energy (for example, electrical energy) supplied to this transducer. *Acoustic efficiency* is a dimensionless quantity, and, according to the principle of energy conservation, it is always less than one. It is customary to express acoustic efficiency in percent.

acoustic grating   A periodic structure or surface that affects an incident sound wave in such a way that, after transmission or reflection, a number of diffraction maxima and minima occur. *Acoustic grating* is similar to optic grating. An example of *acoustic grating* is a number of rods with width h placed in a row at a distance l between each other. If a plane sound wave is incident normally on this acoustic grating, the far field of the transmitted wave has maxima that are determined by the formula $\sin \alpha = n\lambda / (h + l)$. Here α is the angle between the normal to the grating and the direction of the diffracted wave, $n = 0, \pm 1, \pm 2, \ldots$, and λ is the sound wavelength.

acoustic lens   Material of special shape and kind that is used to focus sound waves. This focusing is based on the phenomenon of sound refraction. Acoustical lenses are similar to optical lenses. Acoustic lenses can be made from gaseous, fluid or solid substance. In the first two cases the substance is encased.

acoustic loss   A decrease in amplitude and intensity of a sound wave in its reflection from a surface or in its propagation through gaseous, fluid and solid media due to attenuation, geometrical spreading and other mechanisms. For example: excitation of a surface wave and acoustics barriers.

acoustic pressure   Deviation of a pressure in a medium from its *ambient* value, caused by passage of a sound wave. At a given point, acoustic pressure p oscillates in time t with the frequency of the sound wave. The root-mean-squared acoustic pressure $\overline{p} = (\frac{1}{T} \int_0^T p^2(t) \, dt)^{1/2}$ does not depend on t and is often used in practice. Here, T is a time interval of averaging, which should be properly chosen. Another quantitative measure of acoustic pressure that is often used is the sound pressure level $L = 20 \lg \frac{\overline{p}}{p_0}$, where p_0 represents a reference pressure. In atmospheric acoustics, $p_0 = 2 \cdot 10^{-5}$ Pa, while in underwater acoustics $p_0 = 10^{-6}$ Pa.

acoustics in moving media   The branch of acoustics that studies the effects of medium motion on sound propagation, and the effects of source and receiver motion on the emitted and received sound. Medium motion significantly affects sound propagation in the atmosphere where wind velocity and its fluctuations are relatively large: in the ocean if a sound wave crosses a region with strong currents, in ducts with mean gas flow, etc. Mean flow in a medium and its regular and random inhomogeneities cause bending in the path of a sound wave, phase change, diffraction and scattering of this wave. Source motion always results in *Doppler effect* and also generates a sonic boom if the source speed is supersonic.

acoustics of buildings   A part of architectural acoustics that deals with noise propagation through a building or its parts. Building acoustics treats outdoor noise transmission through walls of a building and the level of this noise in rooms, and considers indoor noise transmission from one room to another. It also designs buildings to reduce transmission of outdoor noise through walls, for example, by locating windows of a building away from a highway.

acoustics of rooms   A part of architectural acoustics that deals with sound propagation inside a room (auditorium, concert hall, studio, etc.). In a room, a listener hears direct sound from a source plus a series of its *echoes* due to reflection and scattering by walls and objects. The latter sound is called reverberant sound. Reverberant sound exponentially decreases in the course of time because of absorption by walls, objects and air. A time interval for which the level of the reverberant sound is decreased by 60 dB is known as reverberation time. Reverberation time is the main characteristic of acoustic quality of a room. If reverberation time is small, sound is toneless. If reverberation time is large, parts of speech or music overlap and are difficult to be heard.

acousto electronics   The discipline that deals with conversion of radiosignals by means of ultrasound devices. Acousto electronic devices allow one to amplify an amplitude of a radiosignal and to modulate this amplitude, change a phase of a radiosignal and its spectrum, delay a radio impulse and to change its lengths, integrate, decode and encode radiosignals, and to do other conversions. The frequencies of ultrasound waves used in acousto electronics are 10 MHz and higher.

acousto-optic deflector   A beam of light can be deflected by an angle θ by Bragg diffraction from a refractive index grating produced by acoustic waves launched in a crystal. *See* acousto-optic effect.

acousto-optic effect   Interaction of optical and acoustic waves in a crystal. By applying a periodic mechanical stress in a crystal (by attaching a piezo-electric transducer, for example) longitudinal acoustic waves can be launched in it. This leads to a moving refractive index grating in the crystal with the spacing, $d = \frac{v}{f}$, where v is the speed of the sound wave in the crystal and f is the frequency of the transducer. This grating can scatter light according to *Bragg's law* leading to diffracted beams.

acousto-optic modulator   A device that can modulate a beam of light by acousto-optic effect. The RF signal used to launch acoustic waves in the crystal can be (a) amplitude modulated which results in the modulation of the intensity of the undiffracted beam, (b) frequency modulated resulting in the modulation of the frequency shift of the first order beam or (c) modulated by changing the direction of the sound wave in the crystal. *See* acousto-optic effect.

acousto optics   The discipline that deals with the effects of sound waves in solids and fluids on light or electromagnetic wave propagating through these media. A progressive or standing sound wave in a medium periodically changes its dielectric permittivity in space and/or time. These changes cause diffraction and scattering of a light beam or an electromagnetic wave propagating through such a medium. This phenomenon, also known as light diffraction by ultrasound, is the basis of many acousto optical devices that are widely used to control the direction of light propagation, and its polarization, amplitude and spectrum. Among such devices are acousto optical filters, scanners and deflectors. Acousto optical devices allow one to control a light beam by changing amplitude and/or frequency of a sound wave.

actin A proteinaceous filament constituent of muscle tissue and noncontractile tissue causing motility, usually in association with myosin. Actin is a globular protein of molecular weight 42,000 Daltons. *See* molecular weight.

actinic radiation The portion of wavelengths of the total radiation at which a specific sensor is excited within the pass band of an optical system. This is a term used in the context of metrology.

action potential A membrane electric potential exists between the exterior and the interior of a living cell; usually of value 50 mV to 100 mV, with the zero potential reference in the outside medium and the inside of the cell negative polarity. This resting potential is due to an uneven distribution of ion species (co- and counter ions) inside and outside the cell. Whenever the membrane potential rapidly or suddenly changes, as for example during muscle movement, it is referred to as an action potential.

action potential, sodium conductance Sodium conductance is a measure of the ability of sodium ions, Na^+, to diffuse across a cell membrane. It is observed that depolarization of a cell membrane leads to an increase in sodium conductance across the membrane. The action potential (*see* action potential) causes a transient depolarization of the cell membrane; hence, action potentials create a transient increase in sodium conductance.

activation analysis An atom or a nucleus that absorbs energy may be transformed into either an excited state atom or nucleus, or, if the energy absorption is sufficient, the nucleus may be transformed in a nucleus of higher atomic mass. Many of these excitations are metastable and the decay may be observed and recorded yielding a signature of the original atom or nucleus present.

activation analysis, neutron In neutron activation analysis, a beam of neutrons (sometimes slow (thermal), sometimes fast) is incident on a sample of material for which the constituents are desired to be identified. Some of the nuclei in the sample will capture a neutron, thus increasing the atomic mass by one. Often the new, or daughter, nucleus will decay via beta emission: the form and energy of emission identifying the isotope originally present in the sample. Different nuclei have disparate capture efficiencies (cross sections) for neutrons of various energy (speed), hence the proper neutron energies must be selected for the anticipated sample nuclei composition.

activation analysis, photon When high energy photons are incident on atoms within a substance, some of the atoms will absorb the photon if either the photon energy matches one of the allowed molecular, atomic, or nuclear energy level differences, or if the photon energy is at or above the ionization energy of the atom. The atom is then placed in a metastable, excited state that decays to its ground state. This yields a characteristic frequency signature without altering the original atom present, unlike neutron activation, which permanently alters the original atom present.

activation energy (**1**) *Chemical:* the minimum amount of energy that must be provided to start a self-sustaining chemical reaction, that is, on the molecular level, the amount of energy per mole to break the requisite chemical bonds so the atoms are free to recombine.

(**2**) *General:* the amount of energy (mechanical, heat, light, etc.) that must be supplied to a system for the system to begin functioning in a desired way.

active device A device that introduces energy into the primary circuit, instead of being a purely storage or dissipative device. Some examples of active devices are transistors, amplifiers and mixers. *See also* passive device.

active voltage In a circuit element operated with alternating current, the voltage waveform can be viewed as the sum of two waveforms, one in phase with the current, one out of phase. The in-phase component is sometimes termed the active voltage. Energy dissipation in the device is proportional to the active voltage times the current. *See* phasor.

acuity, stereoscopic   This is the smallest difference in distance of an object perceivable by stereoscopic cues and is usually specified by angle of stereopsis. This is a measure of depth perception. The angle of stereopsis is the difference between the angle subtended at the centers of the entrance pupils of the two eyes by the two points at different distances from the eyes.

acuity, visual   Visual acuity depends on the eyes' ability to differentiate details at different distances from the eye. It is a measure of the resolving power of the eye. There are 3 types of resolution possible:

1. *Minimum separable resolution:* the ability to discriminate between two closely spaced points.

2. *Minimum visible resolution:* the ability to see the smallest resolvable angle subtended by a black bar on a white background

3. *Minimum legible resolution:* the ability to read the smallest angle subtended by block letters on a test chart such as a Snellen chart which is commonly used.

Acuity is denoted in terms of $6/x$ (known as a Snellen fraction) or in decimal form. In the Snellen chart, a standard test distance of 6 meters (20 feet) is chosen. At this distance, in the row of letters labeled $6/x$, each letter will subtend 5 minutes of arc and each individual feature in the letter (i.e., the gap in the letter C) will subtend 1 arc minute. In general the numerator of the Snellen fraction indicates the fixed distance or test distance and the denominator denotes the distance at which each letter in a given Snellen row subtends 5 arc minutes.

Visual acuity varies with region of retina stimulated, state of light adaptation of the eye, general illumination, background contrast, size and color of objects, refractive state of the eye, character of retinal image and time of exposure (viewing time).

adaptation of the eye   When the eye (or the visual system) is presented with a stimulus for a period of time, the system will "adapt" to the stimulus and will be less responsive or sensitive. This can be formally defined as the change in sensitivity due to continuous or repeated sensory stimulation. Examples of adaptation include:

1. dark adaptation, wherein the eye adjusts and becomes more sensitive to light as illumination is reduced (such as being in a dark room), and its converse, light adaptation (walking from a dark room to a well lit room); and

2. chromatic adaptation which is an altered sensitivity to color that results in apparent changes in hue and saturation due to prolonged viewing of a specific color.

ADC   (*Analog to digital converter*) A circuit or device that quantizes an incoming analog signal and puts out a digital representation of the input. A necessary precurser for digital signal processing, ADCs must be designed with the appropriate dynamic range, linearity and stability for the signal expected.

adder, analog   An amplifier circuit that provides the output proportional to the sum of the inputs. Also called a *summer.* Requires at least two inputs and one output.

adder, cascade   A device constructed of cascaded half adders and the appropriate logic circuits to allow the true addition of binary numbers. A simple cascade adder is also known as a *full adder. See* half–adder.

adiabatic nuclear demagnetization   A cooling method that uses the properties of nuclear paramagnets in high magnetic fields to reach temperatures below 1 mK. If a system of non-interacting nuclear spins, magnetic moment μ, is in thermal equilibrium at temperature T in a magnetic field B, its entropy is a function of $\mu B / k_B T$. If the system is then thermally isolated, and the magnetic field is decreased isentropically, the system temperature will decrease linearly with the magnetic field. $T_f = \frac{B_f}{B_i} T_i$. The most common simple nuclear paramagnet used in nuclear demagnetization cryostats is high purity copper. It is possible to cool a Cu demagnetization stage to a few μK by starting at a magnetic field of roughly $8\ T$ and a temperature of 5 to 10 mK . When higher initial temperatures and/or lower initial magnetic fields are to be used, $PrNi_5$ is a good choice of nuclear refrigerant. The 4f electrons in Pr form a nonmagnetic singlet ground state in zero magnetic field. In a large magnetic field, some of these electrons will become aligned with the magnetic field, enhancing the local magnetic field seen by the nuclei. This hyperfine enhancement is very large for $PrNi_5$ – the field produced by the electrons at the nucleus is roughly ten times the external field. The concomitant entropy decrease allows use of $PrNi_5$ at higher temperatures and lower fields than those used with Cu demagnetization cryostats. As a result of the large internal field, the minimum temperature attainable with $PrNi_5$ is approximately 0.4 mK as the interactions lead to magnetic ordering. *See also* cooling, magnetic.

adiabatic process in sound propagation
Process with no heat flow in a medium due to sound propagation. Usually, adiabatic process is mathematically formulated as the following condition:

$$(\partial/\partial t + v \cdot \nabla) S = 0 .$$

Here, $S(R, t)$ is the entropy per unit mass, $R = (x, y, z)$ are the Cartesian coordinates, t is time, v is the medium velocity vector, and $\nabla = (\partial/\partial x, \partial/\partial y, \partial/\partial z)$. Propagation of sound waves in air and water can be considered to be accurate as an adiabatic process if the sound frequencies are much less than 10^9 Hz and 10^{12} Hz, respectively.

adjacent channel   An adjacent channel refers to a channel adjoining another channel in the frequency domain, in the time domain, or in the spatial domain. For instance, in a frequency-division multiplexing communication system, the channel with carrier frequency just above or just below the carrier frequency of channel A embodies an adjacent channel to channel A. In the case of time-division multiplexing, the channel at the time slot right before or right after the time slot of channel A is an adjacent channel of channel A. Adjacent channel interference may occur if signal power from adjacent channels spills into the desired channel, say, due to frequency drift in a frequency-multiplexed system or due to mis-synchronization in a time-division-multiplexed system. The capability of a receiver to differentiate signals in the desired

channel from its adjacent channels is termed *adjacent channel selectivity*.

admittance   The inverse of impedance. If the voltage in a circuit element is V (the phasor) and the current is I (also a phasor), the admittance Y is the ratio of current "admitted" to applied voltage: $Y = \frac{I}{V}$. This is in general a complex value, and may depend upon the frequency ω. Some examples of admittance are given here for a pure resistor ($Y_R = \frac{1}{R}$, a real number), a pure capacitor ($Y_C = \sqrt{-1}\omega C$) and a pure inductor ($Y_L = \frac{\sqrt{-1}}{\omega L}$), both imaginary. The admittance of elements in parallel is the sum of the individual component admittances. Sometimes the admittance is broken into the length of the phasor and the phase shift angle. For AC devices, the admittance is the change in current over the change in voltage $Y = \frac{\partial I}{\partial V}$.

admittance, acoustic   Reciprocal of the *acoustic impedance*.

admittance, input   The admittance calculated by dividing the current driven into a device by the applied voltage. For antenna design or other small signal inputs, it is crucial to tune the input impedance to match that of the transmission line, or a large portion of the signal will be reflected back instead of being processed by the device.

admittance, output   The admittance calculated by dividing the current output of a device by the voltage signal output. Inverse of the output impedance.

adsorption pump   A pump which uses cryopumping to achieve low pressures. *See also* cryopumping.

aerodynamic sound (noise)   Sound (noise) generated (1.) by interaction of a flow with a solid surface, (2.) by low speed flow. Examples of aerodynamic sound of the first kind are eolian sounds produced by twigs and wires in the wind. Sources of this aerodynamic sound are associated with aerodynamic forces acting on a solid surface; they have a dipole character. Sources of the aerodynamic sound of the second kind are velocity fluctuations in a flow, which have a quadrupole character. The latter aerodynamic sound can be mathematically treated by using Lighthill's acoustic analogy: sound generation in a non-moving homogeneous medium by the turbulent velocities in a flow. *See also* air jet noise.

afocal system   A telescope or a beam expander is often referred to as an *afocal system* i.e., one without a focal length. Both the object and the image are located at infinity. The angular magnification is the ratio of the apparent angular size of the image to that of the object. When the object is centered on the axis the magnification, M, is $M = \frac{\tan \alpha_2}{\tan \alpha_1}$ where α_1 and α_2 are half-field angles in the object and image space, respectively.

air jet noise   Noise (sound) generated by subsonic and supersonic air jets. Noise produced by a subsonic turbulent jet can be approximately treated by using Lighthill's acoustic analogy (*see* aerodynamic sound) which yields that the noise power is proportional to the eighth power of the jet speed. Noise generation by a supersonic jet is completely different from that by a subsonic jet or a low speed flow, for example; shock waves may play an important role in noise generation. As a result, the noise power of a supersonic jet is proportional to the jet speed to the power of no greater than three.

air liquefier, Claude-Heylandt   The Claude-Heylandt liquefier produces liquid air by a combination of isentropic expansion of a gas and Joule-Thompson cooling. Incoming gas is pressurized to the order of 200 bars, then allowed into an expansion engine, where the pressure decreases to 1 bar. This gas exchanges heat with an incoming stream of air and then returns to the compressor. The incoming stream of gas, after being cooled in a series of heat exchangers, passes through a Joule-Thompson valve where some fraction of the air is liquefied. The remaining gas returns to the compressor cooling the incoming gas on the way. *See also* heat exchangers.

air, liquid   Air, composed of 78% nitrogen, 18% oxygen, and 1% trace gases, becomes a liquid at 78.8 K. Liquid air is a colorless, odorless liquid. The liquid oxygen is actually a pale blue, but is normally not visible in liquid air. Due to the liquid oxygen present, liquid air is a flammable liquid, and, as with all cryogenic liquids, is a severe frostbite hazard.

Airy disk   The pattern of diffraction for circular aperture has rotational symmetry about the axis. The central maximum is a circle of light that corresponds to the zeroth order of diffraction and is known as the *Airy disk*. The near field angular radius of the Airy disk is given by $\frac{1.22\lambda}{D}$ where D is the diameter of the aperture and λ is the wavelength of light used to illuminate the disk.

Airy function   In the context of the transmittance of a Fabry-Perot interferometer, a function of the form $\frac{1}{1+F\sin^2\left(\frac{\delta}{2}\right)}$ is called the *Airy function*. F is the coefficient of finesse and is a measure of the sharpness of fringes and δ is the phase difference between successive interfering rays. The figure below displays the Airy function for different values of F.

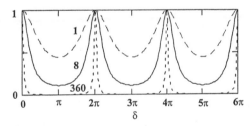

Airy function.

Airy's disc   The Fraunhofer diffraction pattern due to a circular aperture contains a central bright spot with concentric rings. The central spot is called the Airy's disc. The angular radius of the Airy's disk is $\frac{1.22\lambda}{D}$ where D is the diameter of the aperture and λ is the wavelength of light.

Alfven waves   Transverse hydromagnetic waves propagating along magnetic field lines in plasma or in a conducting fluid. Alfven waves were theoretically predicted in 1942 by H. Alfvén, Swedish physicist. The speed of the Alfven waves does not depend on frequency and is given by $v_a = H/\sqrt{4\pi\varrho}$. Here, H is the magnetic field strength, and ϱ is the density of a medium. The Alfven waves play an important role in astrophysics.

algorithm   A sequence of precisely defined procedures, often mathematical or logical operations, to perform a particular task possibly on a given set of data. This sequence of steps has a singular starting point and terminates either with the task accomplished or an indication that the specified task is impossible. The actual path taken through the defined procedures would depend on the input data set and on the prior process step. A practical algorithm accomplishes the given task within the time constraints of the problem and the memory limitations of the computer with sufficient robustness to contamination in the data set. An algorithm may also be embodied in the hardware architecture of the computer processor as a special-purpose dedicated machine.

aliasing   If signals are sampled at certain rates during measurements, a well-known phenomenon in information theory called *aliasing* places a limit on the highest frequency that is unambiguously processed. If Δt is the time interval between successive measurements, the Nyquist frequency, f_N, defined as $f_N = \frac{1}{2\Delta t}$ determines that the signal and noise present at a frequency f will be folded back at frequencies $2nf_N - f$ and $2mf_N + f$. Here n is a positive integer and m a negative integer chosen such that the aliasing is in the range of 0 to f_N. If the signals are band limited up to a maximum frequency, f_{max}, the sampling should be such that $f_N \leq f_{max}$. Stated differently, for a proper measurement of a spectrum the highest frequency component should be sampled at least twice. This is also referred as the *sampling theorem*. The diagram below illustrates this principle. A function due to a superposition of three harmonic waves of frequencies 0.3, 0.8 and 1.1 Hz is shown in the top part of figure. If this data is sampled at intervals of $\Delta t = 1$ sec, the Nyquist frequency, f_N, is 0.5 Hz. The signal components at 0.8 and 1.1 Hz do not satisfy the sampling theorem while the 0.3 Hz does. The 0.8 and 1.1 Hz will be aliased to 0.2 Hz (corresponding to $n = 1$) and 0.1 Hz (for $m = -1$),

respectively, as shown in (b) of the lower figure. The panel (a) shows the true spectrum where the sampling theorem is satisfied for the highest frequency component present in the data.

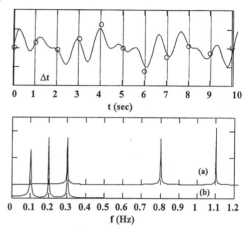

Aliasing. t (sec), f (Hz).

These ideas have applications in digital signal processing and Fourier transform spectroscopy, to name a few. Any function with two variables related by Fourier transforms (e.g., time and frequency, distance and spatial frequency) can be processed by the above criteria.

alnico   Manmade magnetic material exhibiting high coercivity (*see* coercivity), comprising approximately 8% Al, 13% Ni, 24% Co and 3% Cu.

alternating current   Alternating current (AC) is current that is alternating in the direction of the current flow. The typical alternating current is sinusoidal in shape. Alternating current has an advantage over direct current (DC) in that its voltage magnitude can be changed easily through a transformer.

alternator   An electrical device for generating alternating current.

ambient   Pertaining to a medium where a sound wave can propagate. If there is no sound wave propagating in a medium, the pressure, density and fluid velocity in the medium are called the ambient pressure, density and fluid velocity. If there is a sound wave in the medium and the approximation of linear acoustics holds, the pressure, density and fluid velocity in the medium are sums of the ambient and acoustic pressure, densities and fluid velocities.

ammeter An instrument used for the measurement of direct or alternating electrical current. It consists of resistors in series with a galvanometer.

amount of charge that flows in an electrolytic cell The rules are:

1. The mass of an element eroded at an electrode is directly proportional to the amount of electric charge Q passed through the electrode.

2. If the same charge Q is passed through several electrodes, the mass lost at each electrode is directly proportional to the atomic mass of the element, and to the number of moles of electrons required to erode one mole of electrode elemental material.

ampere Unit of electric current in the Systeme International (S.I.). One ampere of current is defined as the current when flow in each of two infinitely long parallel wires of negligible radius separated by a distance of one meter in a vacuum causes a transverse force per unit length of $2\mathrm{x}10^{-7}$ newton/meter to act between the wires. Equal to one coulomb of charge passing through a circuit element per second.

ampere balance An apparatus for measuring current, balancing the torque from forces between coils on the balance beam and stator coils (in the same circuit) with the torque from forces on the balance beam due to weights. If properly calibrated, the current through the coils can be read off from the position of the weights.

ampere hour The integrated amount of charge passing through a circuit element in an hour. Therefore 1 ampere-hour = 1 coulomb/second times 3600 seconds/hour, or 3600 coulombs.

Ampere's law This theorem states that the line integral of a magnetic field around a closed path is equal to the current enclosed by the path.

ampere turn In the case of a coil of N turns carrying current I the line integral of the field around a closed loop yields NI, the magnetomotive force. The units of magnetomotive force are ampere turns.

Amperian path An arbitrary closed path used in the definition of Ampere's law. In Ampere's law, a closed line integral of the magnetic field is equal to the total current enclosed in the path times the permitivity of the free space, μ_o. This arbitrary closed path in Ampere's law is sometimes called Amperian path.

amplifier A circuit or device designed to produce an output proportional to the input signal, or possibly some other function of the input. Often this proportionality is a single multiplication by a gain factor (g) so that the output voltage V_{out} is related to the input voltage V_{in} by the relationship $V_{\mathrm{out}} = gV_{\mathrm{in}}$. However, sometimes the output current (or voltage) is a function of the input voltage (or current). In general the gain may be complex.

amplifier, AC coupled An amplifier circuit constructed with a simple resistor-capacitor network to isolate a DC input voltage (or intermediate voltage in a multi-staged device) from the input directly to the amplifier (or following stage amplifier). Alternating current passes through the capacitor, while any DC offset voltage is not amplified. This allows operation of any components previous to the amplifier at operational voltages, or can be used from controlling temperature effects. Also called an RC-coupled amplifier (for the resistor-capacitor isolation circuit used).

amplifier, antilog An amplifier with the output level an exponential function of the input; thus the input is a logarithmic function of the output.

amplifier, audio-frequency An amplifier designed to operate at frequencies similar to the frequencies of audible sound, 100 Hz to 3 kHz. More generally, the term *audio amplifier* is used to imply a DC signal will not be amplified. In this usage, the design frequency may be much higher than sound.

amplifier, bipolar An amplifier that is based on a bipolar transistor. A bipolar transistor is a device containing two diode junctions. Since each junction has a polarity, the name bipolar is an apt descriptor of the device. The transistor has three active leads, labelled emitter, base and collector. *See* transistor, junction; bipolar code.

amplifier, broad band An amplifier circuit capable of operating over a wide frequency range with a nearly constant gain.

amplifier, cascaded An amplifier where two or more amplifier stages are connected in series (cascaded). This is done to provide increased gain. Mixing between stages may allow shifting of the frequency to allow filtering or optimal amplification in the intermediate stages.

Many radio receiver designs are staged in such a way, with filtering done in the intermediate frequency stages.

amplifier, cascode A low noise amplifier constructed of two amplifiers in a special series arrangement. If bipolar amplifiers are used, the first stage is a common emitter circuit, with input at the base and output to the collector. This goes into the second stage, a common base amplifier. The signal is input to the second stage emitter, with output to the second stage collector.

A cascode circuit need not be of bipolar devices. Any amplifiers, if arranged in an analogous fashion (stage 1: common source or common cathode, stage 2: common gate or common grid) can make up a cascode amplifier.

The high output impedance of the cascode circuit allows its use to drive circuits while keeping the effects of the amplified output away from the input. The rest of the parameters of the circuit behave much like the first stage would alone.

amplifier, classes A, B, AB, C Amplifier classes are defined by what fraction of the input waveform cycle results in amplified output. This is changed by changing the relationship of the amplifier response to the input signal.

Class A amplifiers are those for which the output signal follows the input at all times. Consider the amplifier response curve that shows the relationship between input voltage V_i and output voltage V_o (or current or other signal). Since the

response of the amplifier is well behaved for all V_i, the output signal includes the entire waveform and makes this a class A amplifier.

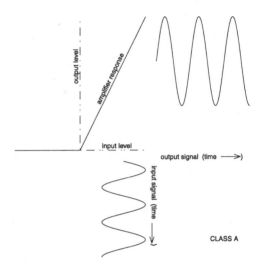

Class A amplifier operation.

Class B amplifiers are those that provide an amplified output for half of the input waveform cycle. This is due to the response cutoff value falling right at the input cycle average voltage.

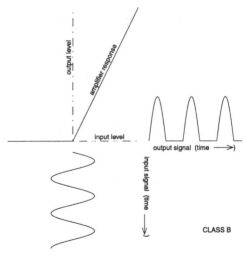

Class B amplifier operation.

Class AB amplifiers are those that provide output signal for less than the full input cycle, but more than half the cycle.

Class C amplifiers provide an output signal for less than half the cycle.

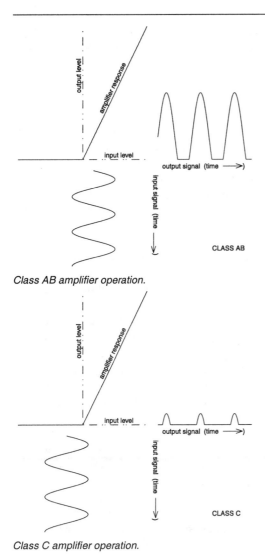

Class AB amplifier operation.

Class C amplifier operation.

amplifier, common base   A bipolar amplifier with the base itself connected to the common voltage. Thus, the input signal is the emitter to common, with output collector to common. The base to common connection generally includes an impedance circuit chosen for stability.

This type of amplifier circuit has a low input impedance and high output impedance. The output will be nearly in phase with the input.

amplifier, common collector   A bipolar amplifier with the collector connected to the common voltage. The input signal is connected to the base, while the output circuit connects through the emitter. The circuit has a low output impedance, with higher input impedance. The output is nearly in phase with the input. Of-

ten used as an impedance matching stage in a multiple-stage amplifier circuit. *See* amplifier, cascode.

amplifier, common drain   An amplifier circuit based on a field effect transistor, in which the drain is connected to the common. The input is to the gate, while the output is to the source. Analogous to the common collector amplifier.

amplifier, common emitter   A bipolar amplifier designed to amplify the input base to emitter signal to an output collector to emitter circuit. Since the emitter is connected to a common "local ground" voltage, the amplifier is termed *common emitter.* In most uses, the connection from emitter to common includes a resistor for stability. So the signal is base to common, while the output is collector to common. The circuit has a high output impedance, but a lower input impedance. The output will be out of phase with the input signal by 180°.

The common emitter amplifier circuit is the most common bipolar amplifier for single stage amplification. Most useful circuits, however, have multiple stages.

amplifier, DC   An amplifier constructed such that it is capable of the amplification of a direct current signal (DC offset). Since this means that the frequency of zero is within the bandwidth of the amplifier, this is the antonym of audio amplifier.

amplifier, difference   A circuit that produces an amplified output proportional to the difference of two input signals. This is especially useful with some signal transmission lines that may pick up induced voltages due to the electromagnetic environment. If these induced voltages will be nearly identical (common mode noise), the output will not be affected.

amplifier, feedback   An operational amplifier circuit in which there is a return path from the output to one of the inputs. For resistive return paths this feedback may be negative (tending to mitigate the input) or positive (tending to reinforce the input).

However, in general the return path may contain any impedance, introducing a magnitude attenuation and phase lag. For instance, an integrator circuit uses a capacitor in the feedback circuit.

amplifier, high gain   An amplifier circuit designed with high gain, producing a substantial output level with a very low signal input level.

amplifier, ideal   A mathematical construct for the representation of an amplifier circuit in calculations. The input impedance is modeled as a high resistance between inputs, and the output voltage (with respect to ground) is a perfect linear function of the input.

amplifier instability   Variations in the response of an amplifier circuit. Instabilities arise from many causes including thermal effects, coupling to other parts of the circuit (crosstalk), and environmental effects. Stable amplifier circuits provide reliable gain for a broad range of operating parameters, but must be optimized for the intended use. For example, the space environment poses special problems for the elimination of circuit instabilities.

The most pernicious types of instability are based on the device itself, as when a capacitive path back through a device can allow the circuit to meet the requirements for oscillation (*see* Barkhausen criterion). This may occur either by a feedback path or by causing parametric amplification. Often remedied by changing the impedance of the bias or input circuit paths.

amplifier, linear   An amplifier that produces an output that is linearly dependent on the input. The region over which this linearity holds is the usual operating range. However, amplifiers are sometimes deliberately operated outside this range, as in an oscillating circuit.

amplifier, logarithmic   An amplifier producing an output that is a logarithmic function of the input.

amplifier, narrow band   An amplifier circuit in which the gain drops appreciably for frequencies off the operating frequency. The width of the region with higher gain is narrow enough to allow the use in oscillators or filtering applications. Also termed a tuned amplifier.

amplifier, negative resistance   An amplifier in which the real part of the input impedance is negative. What this means is that when an increase in voltage is applied to the input, the circuit will allow less positive current (or more negative current) to flow. The simplest way to construct this is in a positive feedback circuit. Useful in latching and memory applications.

amplifier, nonlinear   An amplifier circuit that provides an output that is not a linear function of the input. For example, a logarithmic amplifier is a type of nonlinear amplifier.

amplifier, overdriven   An amplifier circuit operated outside of the normal design range of the amplifier. For a linear amplifier, if the input signal level is above some threshold, the output response will no longer be a linear function of the input due to the saturation of the amplifier. This operation of the circuit outside the amplification region is sometimes useful, for example in oscillator circuits. Most often, however, an overdriven amplifier is the cause of unintended performance problems.

amplifier, parametric   An amplifier that relies on a nonlinearity and a varying parameter of a device to perform amplification. For instance, a variable capacitor may be the parameter varied at some frequency (called the pump frequency). Suppose an input signal at some different (possibly lower) frequency traversed the circuit with the variable capacitor. A mixing occurs between the frequencies, introducing sum and difference frequencies into the circuit. Nonsymmetric treatment of the sum and difference (for instance filtering the difference) may allow the sum to be output, containing signal information and with a potentially high gain relative to the original signal input level.

amplifier, power   An amplifier device or circuit designed to handle higher power levels, i.e., higher currents, than most devices. Attention is paid in the design to thermal management as well as electronic characteristics.

amplifier, push-pull   An amplifier circuit constructed of two (often class B) amplifiers, one for amplifying each polarity of the input signal. Thus the output of the push-pull circuit reproduces the entire waveform, while the operating voltages may be lower than needed for a single (possibly class A) circuit, with better efficiency.

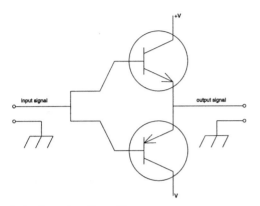

A simple push-pull amplifier.

amplifier, radio frequency   An amplifier circuit optimized for use in the radio frequency (RF) bands. Usually, RF amplifiers are multistage devices with frequency mixing used to place the desired signal in a shifted frequency region in different stages of the device. Such an intermediate frequency (IF) allows the construction of more complex filtering and amplification stages that are tuned for use at the IF. Some radio amplifiers (especially those used above 30 MHz) have multiple IFs.

In addition to designating the operating frequency range, the term RF amplifier allows distinction from the rest of the receiver circuit, which may include amplifiers that handle the audio portion of the signal. Filtering circuits may exist here as well, but do not in general replace the filtering in the IF stages of the circuit.

amplifier, video   A high-bandwidth amplifier circuit ($\sim$ 100 MHz) designed to amplify without signal distortion, even in the higher frequencies of use. Named for its utility in video and CRT applications.

amplitudes of waves, acoustic   Maximum absolute values in oscillations of pressure, fluid velocity, and density in a medium, caused by passage of a sound wave. For a plane monochromatic wave, these oscillations are given by $\xi = \xi_0 \exp(i\mathbf{k} \cdot \mathbf{R} - i\omega t)$. Here, ξ stands for acoustic pressure, fluid velocity, or density, ξ_0 is the amplitude of the sound wave, $\mathbf{k}$ is the wavevector, $\mathbf{R} = (x, y, z)$ are the Cartesian coordinates, ω is the angular frequency, and t is time.

analog operations   Procedures acting on analog signals that provide analog results. For instance, it is easy to construct a circuit to provide the analog sum of two inputs. This sum has been synthesized from the continuous response of the circuit, so any small change in the inputs may provide the correct smoothly varying output.

Other analog operations besides the analog sum include the difference, integration, differentiation, multiplication, etc. Analog circuits may be used in the construction of analog computers, for which digital calculations may be ill suited. Therefore the term *analog operations* is used as an antonym of digital operations.

analyzer   The state of polarization of light that has been passed through a dichoric polarizer can be tested by a second dichoric polarizer, which can then function as an analyzer. When the transmission axis of the analyzer is oriented at 90° relative to the transmission axis of the polarizer, the light is effectively extinguished. As the analyzer is rotated the light transmitted by the pair increases reaching a maximum when their transmission axes are aligned. The transmitted intensity is given by Malus' Law which states that the irradiance for any relative angle θ between the transmission axes is given by:

$$I = I_o \cos^2 \theta ,$$

where I_o is maximum transmitted intensity.

anamorphic system   An optical system with different powers (or magnifications) in different meridians. Such systems are used to correct astigmatism of the eye which arises from uneven curvature of the cornea.

anastigmatism   In an anastigmatic lens, both astigmatism and curvature of field are corrected. Such lenses must contain negative lenses. Typical anastigmatic lenses include:
 1. the Celor (or Gauss) type,
 2. the Cooke triplet and

3. the Dager type.

The Cooke triplet in particular, consists of a negative lens at the aperture stop with two positive lenses: one in front, the other in back. If the last positive lens is a cemented doublet, it is called a Pepan lens and a Heliar if both positive lenses are cemented. The Celor type is made up of two airspaced achromatic doublets, one on each side of the stop of the system. The Dager type consists of two lens systems (each having three or more lenses) placed symmetrically with respect to the stop.

AND   A logic operation that is true if and only if all the inputs are true. For a binary (2 input) digital logic circuit where true is 1 and false is 0, all cases are as follows:

input A	0	0	1	1
input B	0	1	0	1
output	0	0	0	1

Anderson bridge   A bridge used to measure inductance values over a wide range while only requiring a fixed capacitance of a moderate value. A schematic diagram of Anderson bridge is shown below. The resistors are chosen such that $R_1 R_3 = R_2 R_x$. Adjustments of r and R_x are used to balance the circuit. At balance, the inductance is given by

$$L = C[R_1 R_3 + (R_3 + R_x)r] \ .$$

Some circuit element values that give optimum sensitivity are:

$$R_1 = R_2, \quad R_3 = 2R_1, \quad R_x \cong 2R_1 \cong \sqrt{\frac{2L}{C}} \ .$$

Andronikashvili's experiment   An experiment, first performed by Andronikashvili, which measures the density of the normal fluid in a superfluid. Thin disks, closely spaced, are placed in a superfluid helium bath at the end of a torsional fiber. This set of disks is set into oscillatory motion and the period measured. Any normal fluid is locked to the disks due to the viscosity of the normal fluid and thereby contributes to the system's moment of inertia. As normal fluid is converted into superfluid, the moment

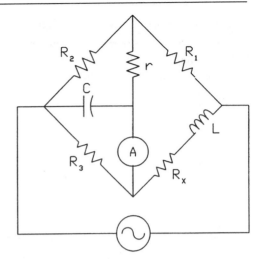

Anderson bridge.

of inertia decreases and the period decreases. Thus, the measurement of period provides a direct measurement of the density of normal fluid, ρ_n. *See also* helium-4, superfluid; superconductivity, two-fluid model.

angle of incidence   When a ray of light is incident on a surface, the angle between the incident ray and the surface normal is called the angle of incidence.

angle of minimum deviation   If a light ray is incident on a prism's front surface, the emergent ray will be deviated. The amount of deviation will vary with angle of incidence. For a specific value of the angle of incidence, the deviation will be a minimum. When minimum deviation occurs, the ray of light will pass symmetrically through the prism. Measurements of minimum deviation angle are used to calculate the refractive index, n of the prism:

$$n = \frac{\sin[(A + \delta)/2]}{\sin(A/2)} \ ,$$

where A is the prism angle, and δ is the minimum deviation angle. For small prisms:

$$\delta = A(n - 1) \ .$$

angle of polarization   For a light wave going from an optically rarer dielectric medium to an optically denser medium, that value of the angle of incidence for which the angle of transmission

$= 90° -$ angle of incidence. According to the Fresnel's laws, at this angle of incidence, the coefficient of reflection of the electromagnetic wave with the electric field vector lying in the plane of incidence (containing the incident ray and the normal to the surface) vanishes. According to Brewster's law, the tangent of the angle of polarization equals the relative refractive index n of the two media.

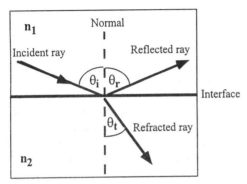

Angles of reflection, refraction.

angle of reflection   When a ray of light is incident at an interface dividing two uniform media, part of the incident ray will be reflected back. The angle of reflection is the angle between the surface normal and the reflected ray. The angle of reflection equals the angle of incidence and this is known as the Law of Reflection.

angle of refraction   When a ray of light is incident at the surface between two uniform media, the transmitted ray (also known as the refracted ray) remains within the plane of incidence and the angle between the refracted ray and the surface normal is called the angle of refraction.

angstrom unit   Unit of length equal to 10^{-8} cm or 10^{-10} m; symbol: Å.

angular dispersion   The angular separation of the wavelengths of a diffracted beam. For a grating with the spacing a and order of diffraction m, the angular dispersion is given by $\frac{d\theta}{d\lambda} = \frac{m}{a\cos\theta_m}$.

angular frequency   For a harmonic wave with a frequency of oscillation f (unit: Hz) and period T (unit: s), the angular frequency ω is given by $\omega = 2\pi f = \frac{2\pi}{T}$. (unit: radian per second).

angular magnification of eyepiece   Ratio of the angle subtended by the eye (with the aid of the eyepiece) with the virtual image of the object to the angle subtended by the object without the eyepiece. For an eyepiece with a focal length f, the angular magnification is $\frac{N}{f}$ for the eye relaxed and $\frac{N}{f} + 1$ for the image viewed at the near point (nearest position of accommodation) N.

angular magnification of microscope   If f_o and f_e are the focal lengths of the objective and the eyepiece separated by a distance d of a microscope, the angular magnification is $\left(\frac{N}{f_e}\right)\left(\frac{f_e+f_o-d}{f_o}\right)$ where N is the near point distance (approximately 25 cm for young adults).

angular magnification of telescope   If f_o and f_e are the focal lengths of the objective and the eyepiece of a telescope, the angular magnification is $\left(-\frac{f_o}{f_e}\right)$.

aniosotropic, materials   These materials exhibit different physical properties in different directions within the material.

aniosotropy (energy)   Energy stored in a ferromagnetic crystal by virtue of the work done in rotating the magnetization of a domain away from the direction of easy magnetization.

anisotropic media   Media in which certain properties are different along different directions (as opposed to isotropic media in which all directions are equivalent). For example, certain crystals have different values of elastic constants or refractive indices along different orthogonal directions leading to differences in propagation of sound or light velocities; the crystal structure or the periodic arrangement of atoms determines this property. Cubic crystals (e.g., diamond) are optically isotropic, uniaxial crystals (e.g., quartz) have two different refractive indices and bi-axial crystals (e.g., mica, topaz) have three.

anode   The electrode from which positive current enters a device. This may also be seen as the point where electrons leave a medium or device.

anomaloscope   An optical instrument that allows an investigator to display one solid color in a half field of view and a mixture of two complementary colors in the second half field of view. The patient's perception or lack of the similarities of the two fields of view allows a measure of the degree of color blindness of the patient.

anomalous dispersion   The refractive index, n, of a dielectric in the wavelength regions of transparency decreases slowly with the wavelength of light. This leads to "normal dispersion" with the red being refracted less than the blue. However, in the vicinity of absorption bands, n increases rapidly with increasing wavelength leading to the so called "anomalous dispersion" (see figure). Longer wavelengths will be refracted more than the shorter ones. This effect can be demonstrated in the visible wavelengths by observing refraction through some dyes placed in an empty glass prism. If the absorption is not too high to black out the light, a reversed spectrum of the visible light becomes observable.

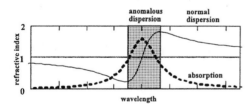

Anomalous dispersion.

antenna   A device constructed to radiate or intercept electromagnetic energy. Sometimes called an *aerial*.

antenna, aperiodic   An antenna having a roughly uniform input impedance and antenna pattern over a wide band of frequencies. Examples include traveling-wave antennas such as the Rhombic antenna. Also called a "non-resonant" antenna. *See* antenna, rhombic.

antenna array   A group of individual antennas called elements acting in unison to provide a desired antenna pattern through constructive and destructive interference. Arrays can be used to provide a much higher gain than available from a single element, to shape an antenna main beam or sidelobes, and to provide a means of steering the antenna pattern without physically moving the antenna.

Antenna arrays are classified by a variety of characteristics. They can be classified according to configuration as linear arrays (all elements aligned along a single line), planar arrays, circular arrays, or three-dimensional arrays. Linear arrays are often classified as end-fire or broadside, depending on whether the main beam is centered along, or perpendicular to, the axis of the array. Parasitic arrays (such as the Yagi–Uda) make use of the interaction among elements to provide individual element excitation, rather than using individual feeding structures. Phased arrays use precise control over the amplitude and phase of each element to provide electronic steering of the main beam and placement of pattern nulls. Adaptive (or "smart") arrays use feedback information to automatically steer their main beam toward a desired signal, while nulling out undesired signals.

If all of the array elements are identical, and if the elements are assumed to have no interaction between them, the antenna pattern of an array is the product of the antenna pattern of a single element multiplied by an "array factor" that takes into account the geometry of the array configuration. Pattern synthesis is the process of selecting the element placement to provide a prechosen pattern.

Many different types of antenna elements may be used to form arrays. Yagi–Uda, log-periodic, and curtain arrays are usually composed of dipoles. Satellite antennas often use arrays of helices, and phased array radars commonly use slots or patches. Even large reflecting antennas may be used in arrays, such as the very-large-array interferometer radio telescope that employs 27 25-m dishes arranged in a Y-configuration of 20-km long legs. *See* antenna, Yagi; antenna, steerable.

antenna backlobe   The radiation side lobe of an antenna pattern located spatially opposite the antenna main beam. *See* antenna beam; antenna sidelobe.

antenna beam   Also called a *radiation lobe,* a portion of the antenna pattern containing a localized maximum, bounded on all sides by significantly lower power density. The "main beam" is the beam with the largest local maximum, while all other beams are called pattern *sidelobes.* The sidelobe level is the ratio of the maximum sidelobe power density to the main beam maximum, and is usually expressed in dB.

Beams are often classified by their shape and their spatial extent (*see* antenna beamwidth) and include "pencil beams" that are highly concentrated along both angular directions and "fan beams" that are narrow along one angular direction and wide along the other.

antenna beamwidth   A measure of the angular extent of an antenna beam in a chosen cross-sectional plane, usually given in degrees. Several methods exist for describing beamwidth. Most often it is taken as the angle between the adjacent 3-dB (half-power) points on the power pattern, but it is also described in terms of the 10-dB or 20-dB points. For patterns with many sidelobes, the beamwidth is sometimes taken as the angular width between nulls (zeroes) adjacent to the beam.

In general, the narrower the antenna main beam, the higher the gain and the directivity. *See* antenna pattern; antenna beam; antenna gain; antenna directivity.

antenna, dipole   An antenna consisting of two segments called "legs", generally made from straight wires and of equal length l, and fed from the center by a two-wire transmission line. A dipole may be used either alone or in arrays such as log-periodic and Yagi-Uda, and as such is the most commonly used antenna element.

The pattern of a dipole depends on the length of the dipole legs as a fraction of the operating wavelength, l/λ, and the input impedance depends on both l/λ and the radius of the wires, a/λ. Dipoles are usually operated in the resonant mode, with l/λ slightly less than 1/4 for $a/\lambda \ll 1$. In this case the input impedance is real and approximately 72 Ω and the pattern is

very nearly a sine of the angle measured from the antenna axis. *See* antenna, half-wave.

Coaxial cable can also be used to feed a dipole antenna if an appropriate "balun" is used. *See* antenna feed system.

A monopole antenna operating above a ground plane (or approximately so, in the case of a monopole above the earth) acts as a dipole that radiates one half the power and has one half the input impedance of a dipole in empty space.

antenna directivity   Symbol: D. A dimensionless parameter, greater than or equal to unity, describing the effectiveness of a transmitting antenna in concentrating its radiation intensity along a certain direction. Expressed as

$$D(\theta, \phi) = \frac{U(\theta, \phi)}{\frac{W}{4\pi}}$$

where $U(\theta, \phi)$ is the radiation intensity (power radiated per unit solid angle) and W is the total radiated power. Thus directivity relates the radiation intensity along a particular direction to the radiated power averaged over the total solid angle 4π (the radiation intensity of an isotropic radiator). Often the word directivity is used to describe the maximum value of $D(\theta, \phi)$ over all angles.

Directivity is related to antenna gain and antenna effective area. *See* antenna effective area.

antenna effective area   Also called *effective aperture,* a parameter with dimensions of m^2 describing the ability of a receiving antenna to capture the power carried by the wave impinging on it. Specifically

$$A_e = \frac{P_L}{W_i}$$

where P_L is the power in W delivered to the load attached to the antenna, W_i is the power density of the impinging wave in W/m^2 and A_e is the effective area. Thus, a larger A_e implies a greater ability to intercept power.

Effective area depends on several variables, including the orientation and efficiency of the antenna, the polarization of the impinging wave, and the value of the load. The effective area is maximized when the orientation of the antenna is polarization matched to the impinging wave, the load is conjugate matched to the antenna input impedance, and the antenna is lossless. The maximum effective area of an antenna acting as a receiver, A_{em}, is related to the maximum directivity of the same antenna acting as a transmitter, D_m, by

$$A_{em} = \frac{\lambda^2}{4\pi} D_m \ .$$

Although the effective area of aperture-type antennas (such as horns and reflector antennas) is often of the same order as the physical antenna aperture area, there is little correlation between the effective area of wire-type antennas (such as dipoles and loops) and the physical area they present to the impinging wave.

antenna efficiency   A dimensionless parameter, less than or equal to unity, describing the ability of a transmitting antenna to convert input power into radiated power. Defined through

$$e = \frac{P_r}{P_{in}}$$

where P_r is the radiated power and P_{in} is the input power to the antenna. The radiated power will be less than the input power for antennas exhibiting conductor or dielectric loss, so that $P_{in} = P_r + P_l$ where P_l is the power loss.

Often the impedance mismatch between the feeding line and the antenna is included as a loss mechanism and the mismatch efficiency is expressed as

$$e_m = 1 - |\Gamma|^2$$

so that the total antenna efficiency is given by $e_t = e \cdot e_m$.

antenna feed system   The physical connection between an antenna and the transmission line or waveguide supplying or drawing power. Often the feed is assumed to include all or part of the transmission line.

Antenna feed structures are as varied as the antennas they are designed to feed. A feed structure is usually constructed to match the input impedance of the antenna with the characteristic impedance of the feeding transmission line. It may also be constructed to provide a transition between a balanced antenna (such as a dipole)

and an unbalanced transmission line (such as coaxial cable). In this case it is called a *balun*.

For a reflector antenna, the antenna feed refers to the primary radiator and its own associated feeding system.

antenna gain   Symbol: G. Antenna directivity corrected for efficiency. A dimensionless quantity, usually expressed in dB, corresponding to the ratio of the radiation intensity in a certain direction to the input power averaged over the total solid angle 4π. Expressed as

$$G(\theta, \phi) = e_t D(\theta, \phi) \ .$$

The word gain is sometimes used to describe the maximum value of $G(\theta, \phi)$ over all angles.

antenna, half-wave   A dipole antenna of total length $2l$ equal to one half wavelength. A very thin half-wave dipole has a sinusoidal current distribution, a maximum directivity of 1.64, and an input impedance of $73 + j42.5\,\Omega$. A very thin dipole antenna can be made resonant by reducing its length slightly, to $l/\lambda = 0.24$, producing an input impedance of approximately $70 + j0\,\Omega$. For a dipole of larger diameter, the length must be reduced more, and the resulting resonant input impedance is also reduced. *See* antenna, dipole.

antenna, horn   A class of high-gain aperture antennas used extensively in the microwave and millimeter-wave bands. The simplest horn antennas are constructed by flaring out the mouth of a circular or rectangular waveguide. The flare geometry controls the pattern shape and thus the antenna gain. Rectangular horns are classified as sectoral E-plane, sectoral H-plane, and pyramidal, depending on whether the flare is along the wide side, narrow side, or both sides of the guide, respectively. A typical standard-gain pyramidal horn has a gain of about 20 dB.

Several modifications can be made to the simple horn geometry to improve gain, sidelobe level, bandwidth, and polarization characteristics, resulting in such types as the ridged horn, the corrugated horn, the aperture-matched horn, and the TEM horn.

Horn antennas are commonly used as single elements or in arrays, and as feeds for reflector antennas such as microwave satellite dishes.

Other important applications include use as a standard for calibrating other high-gain antennas and as a radio-astronomy telescope.

antenna, isotropic   A hypothetical antenna that radiates uniformly in all directions. An isotropic antenna cannot exist physically, but is useful for defining various antenna properties such as directivity and gain. The directivity of an isotropic radiator is unity and its gain is 0 dB. *See* antenna directivity; antenna gain.

antenna, lens   An antenna combining a primary radiating element or array with a converging lens to produce increased gain. The lens acts much the same as the reflector of a dish-type antenna, collimating the radiation from the primary radiator and narrowing its main beam.

The primary element of a lens antenna is often a low-gain antenna such as a dipole, slot, or small horn. The lens may be constructed from dielectric material, as with an optical lens, or from artificial dielectrics composed of small conducting objects imbedded in foam. For lower frequencies the weight of the lens becomes a significant factor and alternative lens designs using stacked metal plates or wire meshes may be more appropriate.

Lens antennas are classified both by the type of material used to construct the lens, and by the geometry of the lens. Important lenses categorized by shape include the Luneberg lens, which is spherical with a radial grading of the dielectric constant, and the Schmidt lens, which is used to correct aberrations in spherical reflector antennas.

antenna, loop   An antenna consisting of one or more turns of wire, often contained in a plane, formed into typically a circular, square, or rectangular shape. Small loop antennas are those whose perimeters (number of turns times circumference) are generally less than a tenth of a wavelength. These have sinusoidal antenna patterns regardless of the shape of the loop, with a sharp null on the loop axis useful for direction finding, station nulling, and radiowave navigation. Small loop antennas have low antenna efficiency due to large resistive losses, strong mismatches due to a highly inductive input impedance, and low input resistance. The input

resistance can be enhanced by increasing the number of turns or by winding the antenna on a ferrite core, forming the "loop-stick" antennas commonly found in AM radio receivers. Small loops are also used for probing near-zone fields and currents.

Loops can also be made of resonant size, with perimeters approximately one wavelength in extent. A resonant square loop has an input impedance of about 100 Ω, and is often used by radio amateurs in Yagi–Uda arrays called *cubical quads*.

antenna, paraboloid   A reflector antenna in which the main reflector is paraboloidal (parabolic surface of revolution). These antennas are often used for radio astronomy because of their narrow main beams, and are used as the "dish" antennas in many home satellite television systems. *See* antenna, reflector.

antenna pattern   The angular variation of the radiation-zone electromagnetic field of an antenna, generally expressed in terms of the spherical-coordinate variables θ and ϕ. Usually plotted in polar coordinates and normalized to the maximum value using either natural units or dB. Either the field magnitude (field pattern) or the power density (power pattern) may be plotted, with power density proportional to the square of the field magnitude.

Occasionally, the spatial variation of the near-zone field is considered. Describing this using a near-zone antenna pattern is complicated by the need to include distance as an additional parameter.

antenna, reflector   Antenna combining a primary radiating element or array with a large conducting surface to produce increased gain. The reflector acts much as the mirror in an optical telescope, collimating the radiation from the primary source and decreasing the width of its main beam. It may be constructed from solid metal plating, or from perforated plating or wire mesh to reduce weight. The primary source is usually a low-gain antenna such as a dipole, slot, or small horn placed at the focal point of the reflector. More complicated primaries such as circular corrugated horns and dielectric conical antennas provide more efficient illumination of the reflector.

Reflector antennas are often categorized according to the reflector shape, and include planar reflectors, corner reflectors, spherical reflectors, parabolic cylinder reflectors, and paraboloidal reflectors.

Shaping of the reflecting surface can improve the performance of a reflecting antenna by increasing its gain or decreasing its sidelobes. A dual-reflector Cassegrain antenna allows for further improvements by providing shaping on two surfaces, but only at the cost of increased "aperture blockage" — the tendency of the feed to block a portion of the impinging wave.

antenna, rhombic   A traveling-wave wire antenna used mostly at medium and short-wave frequencies, consisting of four straight wire legs placed parallel to the ground in the form of a rhombus. One narrow apex of the rhombus is fed by a two-wire transmission line while the opposite is terminated by a resistance chosen in value (typically $600 - 800 \Omega$) to eliminate reflections of the traveling waves generated at the feed. When the legs are several wavelengths long a highly directional pattern is produced, with the main beam aligned along the rhombus, and elevated at an angle to the ground due to the ground effect. The elevation angle allows the reception of sky waves reflected by the ionosphere. Rhombic antennas typically have two large sidelobes adjacent to the main beam, and may have a backlobe produced by traveling waves reflected from an imperfect apex termination.

antenna sidelobe   Any antenna beam that is not the main beam of the antenna pattern. *See* antenna beam.

antenna, steerable   A highly directive antenna with a main beam direction that can be changed either mechanically or electronically. A reflector antenna is often mounted on a gimble, allowing it to be mechanically rotated through azimuth and elevation. A reflector antenna may also be steered by moving its primary feed. The 305-m reflector antenna in Arecibo, Puerto Rico is mounted in a natural valley, and steered by moving a feed suspended from three towers. In contrast, a phased array antenna is steered by precisely controlling the phase of the signal supplied to a fixed-position array elements such as patches. This type of array is commonly used in radar applications, where rapid scanning of the main beam is required.

antenna subreflector   The smaller reflecting surface in a dual reflector antenna system, such as the Cassegrain. *See* antenna, reflector.

antenna, Yagi   An antenna array consisting of one driven element and several parasitic elements, arranged in a linear configuration to produce a main beam aligned with the antenna axis. Also called a *Yagi–Uda* array after the two inventors, Hidetsugu Yagi and Shintaro Uda, who developed the array in the 1920s. Usually consisting of dipole elements, Yagi–Uda arrays may also be constructed from loops or other simple elements. These easily constructed arrays are used heavily in the short-wave bands up through the low microwave bands.

Dipole Yagi–Uda arrays consist of several parallel elements, including a single resonant dipole (called the "driver") fed by a two-wire transmission line. A single parasitic (short-circuited, undriven) element, called a "reflector", is located behind the driver and has a length slightly greater than the driver. Several parasitic elements called "directors" are located in front of the driven element.

The Yagi–Uda array works on the principle of current induction. Through very careful choice of length and placement of the parasitic elements, the driver induces the proper current to create constructive interference of the individual element patterns along the antenna axis. The gain of the array increases with the number of directors used, but little additional gain is realized beyond the 11 dB of a five-director Yagi. Beamwidth and sidelobe levels are also important design considerations for Yagi–Uda arrays.

See antenna array; antenna, dipole.

antiferromagnetism   This is a weak magnetism similar to paramagnetism insofar as it is characterized by a small positive susceptibility.

anti-jamming   Anti-jamming refers either to the capacity of a device (typically a radar, navigation guidance system, or communication system) to resist jamming without critical deterioration in its effectiveness, or to measures undertaken by the device to mitigate the effects of such jamming. One common anti-jamming measure is the broadening of the information-bearing signal's spectrum using various spread spectrum modulation techniques.

antinode   That portion of a standing wave where its amplitude is maximal. Antinode is also called a loop. *See also* node.

anti-reflecting films   Thin layers of dielectric films deposited on a material so that light of desired wavelengths is not reflected due to destructive interference. One application is to minimize reflected glare from window panes and glass covering paintings and photographs. A perfectly anti-reflecting film for light of vacuum wavelength λ should have a thickness of $(\lambda/4n_1)$ and satisfy $n_1 = \sqrt{n_0 n_s}$ where n_1, n_0 and n_s are the refractive indices of the film, the incident medium (air, in most cases) and the substrate material, respectively. Multiple layers of materials of high and low refractive indices in a quarter-wavelength-thick stack can also achieve anti-reflecting properties over a broad range of wavelengths. For a three-layer stack with materials of refractive indices n_1, n_2 and n_3 on a substrate n_s, the condition for anti-reflectance is $\frac{n_1 n_3}{n_2} = \sqrt{n_0 n_s}$.

antiresonance   A regime of an acoustic system consisting of two or more parts with interaction between each other when the effective *impedance* of the system is very high (in limit, infinite). An example of such a system is a membrane vibrating in the water.

anti-Stokes lines   The inelastic scattering of light by matter leads to frequencies that are higher than those of the incident photons called anti-Stokes lines (in contrast to the lower-frequency Stokes lines and elastically scattered Raleigh lines at the same frequency). If f_M are internal excitation frequencies of a material and f_I the incident photon frequency, then the anti-Stokes frequencies are $f_{\text{anti-Stokes}} = f_I + f_M$. If f_M correspond to sound waves or acoustic frequencies, the scattering is named *Brillouin*

scattering; for vibrational, rotational and electronic excitations in molecules or crystals the phenomenon is called *Raman scattering. See* scattering, Brillouin; scattering, Raman.

aperiodic vibrations   Vibrations without repetitive pattern. This term is used as opposite to that of periodic vibrations. An example of aperiodic vibrations is random vibrations, when amplitude, phase and/or frequency of vibrations are changed randomly.

aperture   The diameter (usually measured in inches) of the opening or objective of the telescope, camera, etc. that determines the amount of light ultimately reaching the image. Sometimes, the aperture is quantified as the angle between the lines from the opposing ends of a diameter of the objective to the principal focus.

aperture acoustic   (**1**) Surface that effectively radiates sound. An example of acoustic aperture is the mouth of a horn. An area occupied by an *array of acoustic sources* is often called an acoustic aperture too.

(**2**) An opening in a screen through which sound waves can propagate from a source located behind the screen.

aperture ratio   Determines the light passing power of a lens for a non-parallel beam from a nearby object. Its value is given by $2n \sin(\theta)$ where θ is the angle between the lens axis and that emergent ray in the image space, which starts from an axial object point and passes through the rim of the lens.

apertures, complementary   *See* complementary apertures.

aplanat   A lens or an optical system that is free of both spherical aberration and coma. Such a system satisfies the *Abbe's sine condition*. A lens made of a dielectric of refractive index n with the radii of curvature r_1 and r_2 can have an image free of spherical aberration and coma if the object is located at a distance r_1. The image that is virtual will be at a distance nr_2 should the radii of curvature satisfy the condition $r_1 = \left(\frac{n+1}{n}\right) r_2$. Such a lens is called aplanatic and the object and image points are called *apla-*

natic points. Aplanatic lens systems made of immersion oil and index-matched thick plano-convex lens are used next to the objective lens of microscopes.

aplanatic achromatic doublets   Two lenses cemented together to correct for spherical aberration, coma and chromatic aberration. *See* aplanat; aberration, chromatic.

aplanatic points   The two conjugate points of an *aplanat. See* aplanat.

aplanatic refraction   Refraction at a spherical interface in which *the Abbe's sine condition* is satisfied for incident and refracted rays.

apochromatic correction   *See* aberration, chromatic.

apodization   Literally, *to remove the feet.* Any process in which the aperture function is altered to produce a redistribution of the energy in the diffraction pattern. It is usually employed to reduce the secondary diffraction maxima. This procedure is accomplished by altering the aperture with suitable masks so that the resultant diffraction pattern has reduced secondary maxima resulting in cleaner images as shown in the figure below. The diffraction pattern due to a grating without a mask (a) and with a standard mask (b) is shown in (c) in solid and dotted curves, respectively. The intensity scale is drawn in the log scale to visualize the secondary maxima that are a few percent of the principal maximum without the mask. It is clear that they are substantially reduced in the apodized spectrum. The principal maximum suffers some broadening.

Apodization can also be done after the observation by a mathematical operation. The measured diffraction pattern of the aperture is convoluted with the Fourier transform of a suitable mask. In *Fourier transform spectroscopy, apodization* is performed by multiplying the *interferogram* with the mask function and obtaining the spectra by a Fourier transform.

apparent size   The size of the retinal image of an object. It is proportional to the angle that

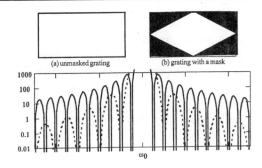

(a) unmasked grating   (b) grating with a mask

(a) unmasked grating, (b) grating with a mask, and (c) the secondary maxima without mask (solid curve) and with mask (dotted curve).

the object subtends at the first principal point of the optical system.

applique   An auxiliary circuit equipment appended onto a standing communication system to offer substitute or supplementary utility.

arithmetic operations   Operations that treat the inputs or operands as numbers. In digital circuits, binary numbers can be added and subtracted. It is also possible to multiply the binary-represented numbers, in addition to division, etc. These operations are arithmetic. *See also* logic operations.

array of acoustic sources   A number of acoustic sources coupled together. Radiation pattern of an array of acoustic sources can be quite different from that of an individual source, e.g., the former can be much narrower than the latter. This property of an array of acoustic sources is often used in forming a narrow radiation beam.

Arrhenius equation $k = Ae^{-E_{act}/RT}$ where k is the rate constant, A is the Arrhenius factor (a constant for a given system), e is the exponential function, R is the universal gas constant (8.315 J/mole-K), T is the absolute temperature of the system, and E_{act} is the activation energy of the reaction. *See* activation energy.

Arrhenius plot   A generalized plot of any physical parameter that is a function of temperature. Following the form of the Arrhenius equation (*see* Arrhenius equation), one makes a plot of the physical parameter, say for example, correlation time τ_c, vs. 1/T (the reciprocal of the absolute temperature). One often observes the data follow a straight line on such a plot, the slope of which is the activation energy for that process. Another utility of such a plot is that where phase transitions occur in the system, the phase transition is observed as an abrupt change in slope at the temperature of the phase transition.

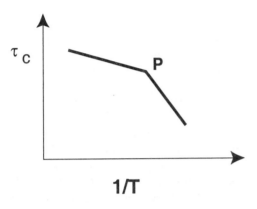

Arrhenius plot.

arteriography   The use in the bloodstream of a dye opaque to x-ray radiation (for example, iodine) which allows a screen display (fluorescent or computer) of artery systems exposed to x-ray radiation. Sometimes called x-ray imaging or angiography.

artificial inductor   A circuit designed to synthesize the effects of an inductor without having a physical inductor coil. This may be an active device or a passive network device. If the effective admittance approximates $Y \simeq \frac{\sqrt{-1}}{\omega L_{\text{effective}}}$ for any simulated inductance $L_{\text{effective}}$ near the angular frequency of operation ω, the device can be used in place of an inductor. Often useful in integrated circuits, in which capacitor and resistor elements are easy to construct in small areas using planar processing techniques, while inductors are not. *See* admittance.

artificial reactor   A device constructed to mimic the complex admittance desired in a circuit. For instance, examine the possibility of an artificial capacitor without having a physical

capacitor in the circuit. The admittance Y will be nearly completely imaginary, and scale linearly with operating frequency. The effective capacitance $C_{\text{effective}}$ can be found from the relationship $Y = \sqrt{-1}\omega C_{\text{effective}}$. The artificial capacitor may be constructed of an active circuit, or may be a passive network design, and can be used in many circuits in place of a capacitor. Of course, capacitors are among the easiest of elements to construct in the normal ways, but the example shows the possibility of tuning the amplitude and phase response as a function of frequency for any desired impedance. *See also* artificial inductor; admittance.

aspect ratio   In images, the aspect ratio is the ratio of one dimension of the total image (say, horizontal) with the other (in this case, the vertical). For video image reconstruction or any CRT process, the aspect ratio should be tuned to the design value so the images appear realistic. It is well known that current television design has a different aspect ratio for the viewed image than that used in motion picture applications. Now the aspect ratio for high-definition television (HDTV) is different than that for current video as well.

aspherical mirrors   Paraboloids, ellipsoids and hyperboloids which produce perfect images between a pair of conjugate points corresponding to their two foci.

association constant   When macromolecules such as proteins self-assemble by thermodynamic means, i.e., due to a minimum condition in free energy provided by hydrogen bonds between the macromolecular assemblies, then the self-assembly reaction constant is referred to as the *association constant.*

association kinetics   When macromolecules self-assemble in response to the kinetic pathway of the reactants, the final association is not dependent solely on the minimum in free energy due to nearest neighbor interactions; rather the final association is dependent on the reaction kinetics of the individual processes making up the self-assembly. The study of these reaction processes is referred to as *association kinetics.*

astigmatic surface   The effect of astigmatism is that the rays of a narrow oblique bundle, instead of being brought to a focus at a single point, pass through two small focal lines at right angles to the path of the chief ray in the image space. If the chief rays proceeding from the various object points lying in the meridional plane of a symmetrical optical instrument are constructed, and if along each of these rays, the positions of the image points of the pencils of the meridian and the sagittal rays are determined, the loci of these points will be two curved lines, both symmetric with respect to the axis, which touch each other at a common vertex on the axis. These curved lines are the traces in the meridional plane of the two astigmatic image surfaces generated by revolving the traces around the axis of symmetry. The focal lines of a narrow pencil of meridional rays lie on one surface, and the sagittal rays are focused on the other surface.

astigmatism   When an object point is away from the optical axis of a lens or a mirror by a considerable distance, the cone of incident rays are asymmetric with respect to the optical system leading to the aberration called *astigmatism.* The image of a point object results in two mutually perpendicular line images displaced from one another. Rays in the vertical (or meridional) plane and in the horizontal (or sagittal) plane lead to these two images. The image is disc shaped at some intermediate point called the *circle of least confusion.* Projection systems and photographic enlargers suffer from this defect due to the closeness of the lens from objects over a large area. Two or three lens systems that correct for this defect also flatten the curvature of the field. *See* curvature of field.

astigmatism of the eye   In the case of vision, astigmatism occurs in the eye because the cornea is not a perfect sphere, and hence there is a contribution due to additional cylindrical curvature. This could occur due to oblique incidence of light on the cornea or lens. If there is an astigmatic refractive error, rays of light from a single point object are focused as 2 line images at different distances from the system, at right angles to each other. This is due to different refraction of the incident light by the dioptric system in different meridians. Astigmatism of the eye is in general classified as:

1. *against-the-rule:* Astigmatism in which the meridian of greatest refractive power of the eye is in or within 30° of the horizontal.

2. *with-the-rule:* Astigmatism in which the meridian of greatest refractive power is in or within 30° of the vertical.

3. *irregular:* Astigmatism in which the two principal meridians of the eye are not at right angles to each other.

astigmatism, radial   This is a monochromatic aberration of a spherical lens. For an optical system imaging an off-axis point, the chief ray (or principal ray) will go from object point through the center of aperture of the system. The plane perpendicular to the plane containing the chief ray (the tangential plane) is called the sagittal or radius plane. When evaluating the image at the tangential conjugate, there will be a line in the sagittal direction. A line in the tangential direction will be formed at the sagittal conjugate. In between these conjugates, the image will be either elliptical or circular. The separation of these conjugates is called radial astigmatism.

atmospheric acoustics   The discipline that deals with sound radiation, propagation, and scattering in the atmosphere, and use of sound waves for the remote sensing of the atmosphere. There are several factors that can simultaneously affect a sound wave in the atmosphere: absorption of sound in air, interaction of a sound wave with the ground, temperature and wind velocity stratification resulting in refraction of a sound wave, scattering of sound by atmospheric turbulence, terrain and different obstacles such as barriers, houses, etc. Among modern concerns of atmospheric acoustics are studies of noise propagation from highways, factories, airports, and supersonic aircrafts; source detection, ranging and recognition by means of acoustical systems; acoustic remote sensing of the atmosphere, etc.

atmospheric duct   A layer of the earth's atmosphere that traps and guides electromagnetic waves by reflection and refraction. Electromagnetic waves propagating in the earth's troposphere normally bend concave down toward the earth due to the negative gradient in the refractive index of the atmosphere with increasing height. When the refractive index of the atmosphere changes rapidly, or in a discontinuous fashion, the waves are in essence reflected from that region of the atmosphere. A "groundbase" duct occurs when the reflecting layer is close to the earth so that the reflected waves bounce repeatedly between the layer and the earth. An "elevated" duct occurs when the reflecting layer is high above the earth, and the reflected waves are bent back upward before reaching the earth's surface. In a fashion similar to metallic waveguides, cut-off frequencies may be computed for various ducted modes, giving the lowest possible frequency of a ducted wave. For a typical duct a few hundred meters thick, waves with frequencies in the VHF range and higher are allowed, while waves with lower frequencies are cut off.

Atmospheric ducting is often associated with temperature inversions. Bounding layers may extend for over 1800 km along stationary weather fronts. Ducting over water has provided VHF communication distances of over 4500 km.

attenuation, acoustic   A decrease in amplitude and intensity of a sound wave in its reflection from a surface or in its propagation through gaseous, fluid or solid media. This decrease is caused by both *absorption* of a sound wave and its scattering due to inhomogeneities on the surface or in a medium. For propagation of a sound wave in a medium, its attenuation is described by the formula $A = A_0 e^{-\delta x}$, which is similar but not identical to that for absorption of sound. Here, A and A_0 are the amplitudes of sound pressure at two fixed points, x is a length of a sound path between these points, and δ is the extinction coefficient which is a sum of the absorption coefficient and the scattering coefficient.

attenuation coefficient   When a plane wave travels through a medium, the intensity of the wave will drop exponentially as a function of distance traveled, x, $e^{-\alpha x}$, due to the absorption of the medium. The coefficient α is called the *attenuation coefficient.*

attenuation constant   A constant that describes the exponential decrease of the intensity

of an EM wave traveling through a medium due to absorption.

attenuator   A device designed to attenuate the input signal without distorting the waveform. It can be an electric circuit to attenuate the electric signal, or it can be an absorbing material to attenuate the optical input.

audibility, limits of   Frequency and intensity ranges in which a sound can be heard by ear. A human ear is able to detect sound in the range 15 Hz to 20,000 Hz. At a given frequency in this range, sound can be heard if its intensity is above the threshold of audibility and below the threshold of feeling. Both thresholds depend on frequency and other factors, for example, an age of a person. A sound with intensities above the threshold of feeling cause pain and may cause trauma.

audio frequency   Sound frequency that can be heard by ear. Audio frequency is in the range 15 Hz to 20,000 Hz.

audiogram   A plot of hearing loss as a function of frequency for the audible range. Audiograms are widely used in *audiometry.*

audiometry   A procedure of investigating impaired hearing. Both pure tones and speech are used in audiometry to measure the threshold of audibility of a person. As a result, the hearing loss is determined as a ratio (in decibels) of the measured threshold to that of the normal ear. If pure tones are used in audiometry, it is customary to plot the hearing loss vs. frequency for the audible range. Such a plot is called *audiogram.*

audiometry, bone conduction   Bone conduction is the process of conducting acoustic signals to the inner ear through the cranial bones rather than through the ear canal. Audiometry probed with an oscillator placed on the forehead or head to produce a sound response in the auditory nerve is referred to as bone conduction audiometry. This technique is considered to be problematical for lateralization effect (does not isolate one ear) and masking. Oscillator placement strongly affects the results.

audiometry, brain-stem electric response
Audiometry detected not by the patient relaying whether or not they heard the response; rather, by detection of the response via electrodes implanted in the auditory nerve or in the brainstem.

audition limits   The same as *audibility, limits of.*

auditorium acoustics   *See* acoustics of rooms.

aurora borealis   Also known as *northern lights;* This phenomenon of shimmering lights in the northern skies is due to the charged particles of the solar wind that stream through space and ionize air molecules in the upper atmosphere which in turn emit light. The magnetic field lines of the earth trap the charged particles. The electrons and protons swirl around the field lines and proceed towards the poles due to the Lorentz force, which is normal to both their velocity and the magnetic field direction. The colors in the auroras are due to the emission of excited oxygen atoms in the red (630 nm) and green (558 nm) spectrums. Excited nitrogen atoms emit a number of lines between 391 and 470 nm, and 650 and 680 nm. A number of atmospheric factors on earth and the solar wind lead to a variety of color displays.

autocollimator   Any optical system with the property that the incident parallel light emerges as parallel light but is travelling in the opposite direction. This is accomplished, for example, in the Kellner-Schmidt optical system by placing a small-aperture convex mirror at the focus of a large concave mirror, both mirrors having the same center of curvature. Parallel light enters the system through a correcting lens (to correct spherical aberrations) placed at the center of curvature and, after double reflection, emerges in the opposite direction. For an autocollimating eyepiece, *see* eyepiece, Gaussian.

automatic bias   The bias voltage of an amplifier element produced by voltage differences due to the device current. This may be accomplished by placing resistors in the circuit, so when the current flows, the voltage drop across the resistor is sufficient to bias the device elements.

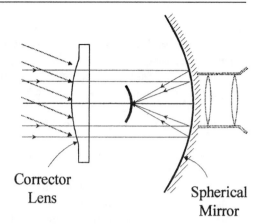

Corrector Lens   Spherical Mirror

Kellner-Schmidt optical system.

For example, in a common emitter amplifier, instead of providing two voltages for the correct bias of both junctions, a resistor from the base to common will carry most of the collector current, providing the voltage difference (the automatic bias) needed for the base-emitter junction. This circuit can be made to work with only one applied voltage.

automatic frequency control   The technique of automatically adjusting the local oscillator or intermediate frequency to compensate for frequency shifts in the input signal. Often used in radio receivers. A by-product (in FM reception) is that the control voltage for the adjustment can be the amplitude of the demodulated signal.

automatic gain control (AGC)   The technique of automatically adjusting the amplifier circuit gain to give a constant output level. Thus when signal strength diminishes (as may be the case in radio reception), the amplitude of the output (or the volume) will not change dramatically. In actual AGC circuits, the output is allowed to change slightly with signal variation to facilitate tuning.

automatic volume control (AVC)   Automatic adjusting of audio amplifier gain to give a constant volume level. *See* automatic gain control.

autoradiography   Visualization of the spatial distribution and concentration of tissue radioactivity. Usually detected by placing the tissue (human body) in close proximity to photographic film and upon developing the film the distribution of tissue radioactivity is recorded.

auxiliary channel   An auxiliary channel refers to a channel adjunct to the main transmission channel but with a transmission direction independent of that of the main channel.

Avrami equation   Used as a model for crystal growth rate as a function of the density of the crystalline and the melt phase (m) with respect to time. Generally expressed in two equivalent forms:

$m = m_o e^{-kt^n}$ or,

$y = 1 - m/m_o = 1 - e^{-kt^n}$.

Where m is melt mass of the crystal, y is the fraction of crystallized crystal, k is the rate constant, t is the time, and n is an integer.

axial modes, of laser cavity   Also known as *longitudinal modes*. These are the resonant modes at which a laser can oscillate. If L is the length of the laser cavity resonator, the axial mode frequencies are $f_m = m \left(\frac{c}{2L} \right)$ where m is an integer (usually very large) and c is the speed of light. The separation between successive modes is the free spectral range of the resonator and a typical laser transition line is broad enough to accommodate several modes. A single axial mode can be sustained at the expense of others by inserting an etalon of appropriate length in the cavity.

axicon   A refracting element that can image a point as an axial line. A common form of an axicon is a refractor in the shape of a plano-convex shallow cone. Axicons are used in auto collimators and alignment telescopes to detect misalignment of illuminated point objects.

B

Babinet's compensator   Also known as Babinet-Soleil compensator. A device that can produce a continuous change in phase retardation between two orthogonally polarized beams of light. In this device, two quartz wedges — the slow direction in one being perpendicular to the slow direction in other — can be gradually slid against one another. The light passing through such a device suffers a retardation that is proportional to the distances in each wedge.

Babinet's principle   A principle that relates the fields diffracted by a screen with an *aperture* and by the complementary screen. Babinet's principle states that the sum of these fields is equal to a field that would be in the absence of the screen. The complementary screen is formed from the original one by replacement of all its transparent parts by opaque parts, and by replacement of all opaque parts by transparent ones.

back emf   The electromotive force (emf) generated in an AC electric motor that is out of phase with the initial applied voltage of the motor. Back emf is due to Lens law; namely, a change in the magnetic flux inside a closed loop will induce an emf to oppose the change in magnetic flux.

back focal length   This is the distance from the back (or secondary) vertex to the secondary (or rear) focal point.

backlash   The hysteresis inherent in a device (such as a tuning element) will cause slightly different dial readings for the same operation depending on the direction of travel. The magnitude of the difference is termed the *backlash*.

backscatter   Electromagnetic waves propagating in a direction directly opposite their direction of origination. In radar, the backscatter signal is the reflected wave returned to the sending antenna.

backward channel   A backward channel transmits supervisory control signals, error-control signals or acknowledgment signals in a direction reverse of the instantaneous direction of the information signal in the forward channel.

backwave   Wave propagating in the backward direction. *See* backscatter.

baffle   Surface that is an extension of the diaphragm of a loudspeaker or an *aperture* of a source. Baffles are used to increase the power output of an acoustic source and to change its radiation pattern, especially at low frequencies. For example, a vibrating circular disk radiates primarily as a dipole if its radius is less than the wavelength of the emitted sound. The same disk baffled (surrounded) by a large surface radiates primarily as a monopole and with much greater output.

balance, amplitude/phase   In splitter or coupler circuits (especially in microwave devices), the balance is a measure of the symmetry of the signal or power division process. The amplitude balance (usually specified in decibels) approaches 0 dB if the signal strength is identical in the two output paths. The phase balance (often stated in radians or degrees) is zero if there is no phase lag between the two outputs.

balance, bridge   A bridge is a circuit designed to make some measurement by setting up a series/parallel connection. Imagine a square circuit, with each side of the square containing components, one of unknown admittance, the others known, some of which can be tuned. A driving circuit is connected across one of the diagonals, and the current (or voltage) is measured across the remaining diagonal. The sides are tuned until there is no current flow (or voltage difference) measured, allowing calculation of the unknown admittance.

   The process of nulling the measured quantity across the bridge is called balancing.

Balmer lines/series; band   These are spectral lines observed in the emission or absorption spectra of hydrogen in the visible and near ultraviolet. They correspond to transitions between $n = 2$ and $n = 3, 4, 5 \ldots \infty$ energy levels. The first three lines of the Balmer series are H_α, H_β, and H_γ lines at wavelengths of 6562.79 Å, 4861.33 Å and 4340.47 Å, respectively. Transitions to higher levels get closer with increasing n and the continuum (for $n = \infty$) appears as a band. *See also* hydrogen spectra.

band analyzer   A device that allows measurement of amplitudes or intensities of a complex sound in many contiguous frequency bands. Many acoustic signals that are dealt with in practice and technique have complicated spectra. Band analyzers are widely used to measure these spectra.

band gap   In semiconductors, the electronic energy states consist of completely filled bands of energies (valence band) followed by unoccupied bands (conduction band). The separation in energy is called the *band gap*. *See* absorption edge.

band pass filter   A device that can transmit a narrow band of frequencies of light while reflecting or absorbing the rest. Usually it is constructed by depositing thin layers of dielectrics on a transparent substrate. The desired wavelength of transmission by constructive interference is achieved by a proper choice of film thicknesses. Two dielectric mirrors (each consisting of a stack of high and low refractive index materials) separated by a spacer of another dielectric film can produce a very narrow band pass filter. *See* interference.

bandwidth   The frequency band useful for the proper operation of a circuit will have a width, measured as the difference between the maximum usable frequency to the minimum usable frequency. High bandwidth devices are either tunable over a large range of frequencies, or can pass an entire range at once. Low bandwidth devices find use in filters and clock or oscillator circuits.

A more general use of the term bandwidth signifies the rate of information transfer.

band width   The range of frequencies contained in a wave packet. The band width, Δv, is inversely proportional to the average duration of the pulses. *See* coherence time.

bandwidth constrained channel   A bandwidth-constrained channel (or narrowband channel) passes only selected spectral components of the transmitted signal within the channel's frequency passband. Other spectral components of the signal lying outside the channel passband will be significantly attenuated. The bandwidth-constrained channel effectively functions as a bandpass filter. If the transmitted signal possesses significant spectral power outside the channel's passband, then the transmitted signal will undergo serious spectral distortion, resulting in inter-symbol interference (ISI) in the time domain.

bandwidth unconstrained channel   A bandwidth-constrained channel (or wideband or broadband channel) passes all spectral components of the transmitted signal with little or no spectral distortion. No actual channel is completely unconstrained in bandwidth, but it may be effectively considered as so for particular classes of transmitted signals.

Barkhausen criterion   The condition for oscillation in a feedback amplifier circuit.

Consider an amplifier with gain g_1. A voltage signal input (V_{in} of magnitude V_0) is amplified giving the voltage signal output V_{out} with magnitude $g_1 V_0$. Next, a feedback path through a second amplifier (with gain g_2) is added. The product $g_1 g_2$ is called the loop gain, since it is the total gain due to going once around the feedback circuit loop. The effective gain of the circuit is

$$g' = \frac{g_1}{1 - g_1 g_2},$$

where $V_{out} = g' V_{in}$, which is to say that for a given desired output voltage the input voltage required is

$$V_{in} = \frac{(1 - g_1 g_2)}{g_1} V_{out}.$$

Note that when the loop gain is identically one ($g_1 g_2 = 1$, the Barkhausen criterion), the output voltage is achieved with no input voltage level.

This is exactly what happens when an audio system with the microphone too close to the speaker screeches with no other input to the microphone. To stop the oscillation, the gain of the circuit must be decreased, either by moving the microphone away from the speaker or by reducing the volume.

In the design of real oscillator circuits, the loop gain is greater than one, so that the oscillation amplitude increases until the amplifiers no longer respond linearly. At saturation of this circuit, the loop gain can only reach unity; thus a nearly constant output level is maintained. In general, real gains are complex, so the two conditions for oscillation must include phase information:

Barkhausen criteria: 1) The feedback loop gain must be greater than one at the frequencies of interest. 2) The sum of phase shifts around the feedback loop must be an integer multiple of 360°.

A sine wave oscillator can be constructed with the appropriate filters at the output of each amplifier. However, smooth oscillations are not the only waveforms that can be generated in this manner. For instance, a stable multivibrator is a two stage amplifier feedback circuit that oscillates between quasi-stable states if the Barkhausen criteria are met, yielding nearly square-wave pulse trains.

Oscillator design is not the only field where these conditions become important. In many applications it is important to ensure there is not oscillation. The criteria were named after physicist G. Heinrich von Barkhausen (1881–1956).

Barkhausen effect   This effect occurs when the grid of an amplifier tube is sufficiently capacitively coupled to the plate that oscillations at very-high or ultra-high frequencies may take place.

bar magnet   A rectangular block of magnetic material producing a static magnetic field.

barrel distortion   An image distortion resulting in decreasing magnification for the rays away from the axis. It results from the limitations of some ray bundles by aperture and stops.

barrier capacitance   The capacitance of the depletion region in a diode junction. Under reverse bias, the depletion region (also called the barrier region or space-charge region) does not allow current to flow (except for the dark current of the device). The depletion depth depends on the magnitude of the reverse bias voltage and the dopant profile of the junction. The capacitance is inversely proportional to the depletion depth.

barrier, insulating   Insulating material placed between signal lines or electrodes of a device, electrically isolating them by increasing the interelectrode impedance.

barrier potential   The intrinsic voltage produced at the junction of two materials with different energy bands (say, p and n-type silicon, or a metal-semiconductor junction). Before being joined the materials would have a different chemical potential (or Fermi level). Upon reaching thermal equilibrium, however, there will be only one chemical potential. Diffusion of charge carriers in the semiconductor will allow equilibrium to be reached, but in the process will produce a space-charge region near the junction, depleted of carriers (the depletion region or barrier region). There will be a non-zero electric field in this region, thus a difference in electrostatic potential. The magnitude of the barrier potential depends upon the resistivities (or dopant profiles) in the materials, but for many photodetectors is measured in tenths of volts.

bars, vibration in   Studies of different kinds of vibrations that can occur in bars. A bar is a solid elastic object the length of which is much greater that its characteristic transverse size. After excitation, longitudinal and transverse vibrations can occur in a free bar. Transverse vibrations can further be subdivided into those due to twisting and bending of a bar. Any vibration in a bar can be represented as a sum of longitudinal and transverse vibrations. Studies of vibrating bars are important in practice and technique. For example, vibrations in construction beams are modeled as vibrations in bars. Furthermore, vibrating bars are used as parts of many musical instruments.

base   In a bipolar transistor, the base is the center electrode and/or bulk, which is separated by junctions from the emitter on one side, and the collector on the other. In an *npn* transistor, the base is the *p*-type semiconductor material, while for *pnp* transistors, the base is the *n*-type. Often, the lead connecting to the base bulk is also termed the base. *See* transistor, bipolar.

base-emitter breakdown   In a bipolar transistor, when the base-emitter junction is reverse biased past the peak voltage, avalanche conduction occurs. This is called *breakdown*, and in most transistors will destroy the device.

basilar membrane   A soft partition that divides the cochlea located in the inner ear lengthwise. The cochlea is a cavity in a form of a snail shell, which is connected to the middle ear by two membranes called the oval window and the round window. The cochlea is filled with a fluid (the cochlea fluid). The basilar membrane has about thirty thousand nerve endings. The cochlea fluid and the basilar membrane are set into vibrations by movements of the oval window caused by sound traveling from the outer ear. The nerve endings "feel" these vibrations and pass on information about them to the brain. Vibrations of the basilar membrane have maximal amplitude at a certain point along the membrane, the position of which depends on a frequency of sound. This is a mechanism that allows distinction between different frequencies.

bats, sound from   Ultrasound emitted by bats to orientate and to find prey. Bats not only emit ultrasound but also hear echoes from objects and flying insects that enables them to fly around these objects and find prey in darkness. In other words, bats use a principle of echolocation. Bats radiate ultrasound through the mouth or the nostrils. The radiated ultrasound is usually in the band 20 − 100 kHz. The level of the radiated ultrasound can reach the value of 120 dB at a distance of 10 cm from a bat.

battery   Two or more cells connected together to form one unit that can convert chemical energy directly into electric energy. There are many different types of battery, two major types of which are (1) dry battery and (2) wet battery.

baud   The baud represents the minimum time interval between successive signaling symbols. It derives from the name of Emile Baudot, a Frenchman considered by many as the father of automatic telegraphy. The baud embodies the shortest unit of modulation rate in a particular signaling scheme. The baud rate equals the number of discrete signaling events in unit time and, as such, determines the signal bandwidth. The baud rate always exceeds or equals the bit rate in bi-level signaling schemes.

beacon   A coded signal transmitted for use in identification or for navigational use in the determination of position, direction, or distance. The signal may be optical or in the form of radio or radar waves.

Optical beacons, in the form of lighthouses and channel buoys, are used for guiding ships. Aircraft guidance is aided by rotating optical airport and airway beacons. Airport beacons are color coded to identify the airport as civilian or military and land or sea.

Radio beacons are passive stations radiating a coded signal for use in bearing determination or the analysis of radiowave propagation conditions. *See* beacon, radio.

Radar beacons may be either active or passive. A passive radar beacon consists of reflectors installed at a lighthouse or buoy to enhance the reflection of radar signals transmitted by a ship. The known locations of the beacons are used to find the ship's bearing and position.

Active radar beacons are used for identifying and locating aircraft. The ATCRBS (air traffic control radar beacon system), also called SSR (secondary surveillance radar), consists of a ground-based radar system and an airborne transponder unit. The ground-based system is used to query the transponder, which then responds by transmitting a selected code, and possibly other data such as the aircraft altitude. Because the transponder is actively transmitting, the signal received by the ground station is generally larger and more reliable than that returned by a radar echo.

beacon, marker   A radio beacon used for identifying a specific region or position. Aircraft marker beacons are used to designate critical positions on precision instrument approaches to airport runways. These beacons radiate a 75 MHz low-power signal of 2 W in an upward-pointing fan-shaped pattern. The signal is modulated to allow audio identification.

beacon, radio   A beacon consisting of a signal transmitted at radio frequencies. These beacons are generally used in air or sea navigation to identify position or bearing. A system of coastal beacons operating in the 285–325 kHz band exists for nautical direction finding. These stations radiate signals of 100 W to 10 kW, providing ground-wave coverage up to 1800 km over the sea.

Aviation radio beacons transmit azimuthally uniform signals in the band 200–1600 kHz using from 10 W to 2 kW of power. Aircraft automatic direction finding systems (ADFs) can use the signals from these non-directional beacons (NDBs) to determine bearing at distances up to 320 km.

Radio beacons are also used to analyze propagation conditions. By monitoring beacons in a range of frequencies and from a variety of locations, an optimum radiowave propagation channel can be determined.

beam   The totality of all ray pencils emanating from a source or aimed toward an image. In the case of a point source, there is only one pencil and the beam is made up of a single pencil. With extended sources, the beam consists of all pencils emanating from every point on the source.

beam divergence   The radius of the spot size w at a distance z from the beam waist in a *Gaussian beam* is $\frac{\lambda z}{\pi w_0}$ where λ is the wavelength of light and w_0 is the *beam waist*.

beam expander   This is an optical system with two lenses of focal lengths f_1 and f_2 arranged so that the separation is equal to $f_1 + f_2$. Both lenses could be positive or one of them could be negative (see figure). The magnification is given by the ratio of focal lengths. The device can be used as a beam reducer if the beam traverses in the opposite direction.

beam, Gaussian   The output of a laser, for example, is the TEM_{00} or fundamental mode,

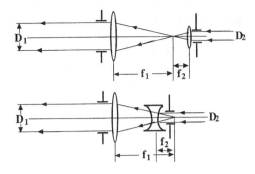

has a perfect plane wavefront and a Gaussian transverse irradiance profile. When considering a Gaussian beam propagating along the Z-direction, the intensity distribution at the $Z = 0$ (flat wavefront) plane is given by:

$$I(x, y, o) = I_o \exp \left[\frac{-2 \left(x^2 + y^2 \right)}{\omega^2} \right]$$

$$= I_o e^{-2r^2/w_0^2} \,,$$

where I_o is the intensity at the beam center and ω_o is the measure of beam width, known as spot size of the beam; it represents the distance at which intensity falls off to I_o/e^2 (13.5%).

As the beam propagates along the z axis, diffraction occurs and the transverse intensity distribution after a propagation distance z is

$$I(x, y, z) = I_o \frac{\omega_o^2}{\omega^2(z)} \exp \left[\frac{-z \left(x^2 + y^2 \right)}{\omega^2(z)} \right]$$

$$= I_o \frac{\omega_o^2}{\omega^2(z)} e^{-2r^2/\omega^2(z)}$$

where $\omega(z)$ is the z-dependent spot size of the beam given by:

$$\omega(z) = \omega_o \left(1 + \frac{\lambda^2 z^2}{\pi^2 \omega_o^4} \right)^{1/2} .$$

beam, radio   Beam of an antenna radiating radiofrequency electromagnetic waves. *See* antenna beam.

beam splitter   Any device (the simplest being a partially silvered mirror) providing transmitted and reflected beams of desired relative intensity. One may use frustrated total reflection by fixing the distance between two accurately parallel hypotenuses of two simple prisms (for example, by coating the hypotenuse of one prism with a solid film of low refractive index material of desired thickness and placing the other prism in contact with it). A typical application is for microscopes where one could adjust the fractions of light from the source going to the eyepiece, photographic film and to the light meter.

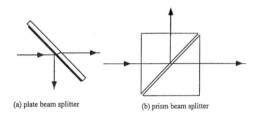

(a) plate beam splitter   (b) prism beam splitter

Beam splitter.

beam, waist   The smallest radius (i.e., at the focal plane) of a Gaussian beam. *See* beam, Gaussian.

beam waist   The minimum transverse size of the beam after reflections from the concave mirrors inside a laser cavity.

bean critical state model   In type II superconductors, magnetic flux quanta, vortices form when the magnetic field is above a critical field, H_{c1}. These vortices move from the edges of the superconductor into the interior until they are pinned by defects or impurities. The bean critical state model assumes that the pinning is as strong as possible, such that the vortices are unable to move from the pinning sites. As a result of this assumption, the flux density always produces the critical current density for the superconductor. That is,

$$|\nabla \times \mathbf{B(r)}| = \mu_0 J_c$$

where $\mathbf{B}$ is the flux density averaged over many vortices, μ_0 is the permeability of free space, and J_c is the critical current density. Physically, this means that as the magnetic field is increased, more vortices form, "pushing" the other vortices around until they produce a maximal screening current over a minimal area of the superconductor.

beat frequency   Superposition of two waves of closely spaced frequencies results in a wave whose amplitude is modulated by the beat frequency which is the frequency difference of the two waves. One can determine an unknown frequency by beating it with a reference frequency and detecting the beat frequency. One popular application of this idea is the tuning of acoustical instruments. Advent of lasers has made it possible to detect a beat frequency of a few Hertz out of 10^{14} Hz. The ring laser gyroscope takes advantage of such precision.

beat reception   When two periodic signals are close in frequency, the sum of the signals exhibits an interference that is sometimes constructive, and sometimes destructive. This results in an oscillation of the amplitude envelope of the sum that has a frequency of the difference of the two initial signals. This low frequency oscillation is termed the beat frequency. There is also another higher beat frequency that has a frequency of the sum of the two initial frequencies.

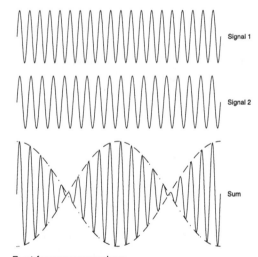

Beat frequency envelope.

One method of receiving an amplitude modulated signal is to utilize the beat frequency to convert the signal back to audio frequencies. This can be done without a transistor-based mixer, by adding an appropriate, locally generated frequency. However, a mixer is still often used in beat reception, so that the local oscillator is at the intermediate frequency of the receiver. In either case, this reception method allows demodulation of a single sideband by using the local oscillator to provide the carrier synthesis (or carrier insertion).

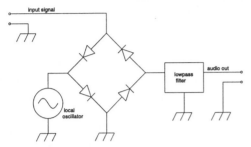

Circuit diagram for a simple beat frequency receiver.

beats   Periodic variations in the amplitude of a sum of two harmonic oscillations whose frequencies are close to each other. In the simplest case, these harmonic oscillations have the same amplitudes B and different angular frequencies ω_1 and ω_2 so that they are given by: $B\cos(\omega_1 t)$ and $B\cos(\omega_2 t)$, where t is time. The sum of the oscillations is given by:

$$B\cos(\omega_1 t) + B\cos(\omega_2 t) = A(t)\cos(\omega t).$$

Here, $\omega = (\omega_1 + \omega_2)/2$ is the angular frequency of the resulting oscillation, and $A(t) = 2B\cos((\omega_1 - \omega_2)t/2)$ is its amplitude. If $|\omega_1 - \omega_2| \ll \omega_1 + \omega_2$, the amplitude $A(t)$ periodically and slowly varies in time in comparison with fast variations with the frequency ω. This slow periodic dependence of A on t is called a beat, and the frequency of this dependence $(\omega_1 - \omega_2)/2$ is called the beat frequency. Beats are an example of amplitude-modulated oscillations.

Beer-Lambert relation   A combination of two separate laws relating the amount of light passing through an absorbing medium to the properties of the medium. Lambert's law states that equal paths in the same absorbing medium absorb equal fractions of the light passing along those paths. Beer's law states that the absorption coefficient of a medium is directly proportional to the concentration of the absorber. Put together the Beer-Lambert Relation is

$$I = I_o e^{-Kx}$$

where I_o is the light intensity incident on the medium, I is the light intensity exiting the medium, K is the absorption coefficient (related to the extinction coefficient), and x is the path length in the medium.

Beer's law   *See* absorption, Beer's law of.

bel   A logarithmic unit of the ratio of two quantities having dimensions of energy, intensity, power, etc. If I_2 and I_1 are such quantities, then their ratio in bels is given by $N = \lg(I_2/I_1)$. Bel is named after A. Bell, American scientist and inventor, and is abbreviated by B.

bell   A widening object of tapered shape, closed at the narrow end and open at the wide one. Bells are made of metals (copper, tin, etc.) and are usually of nonuniform thickness that increases to their open ends. Bells are used as sound sources and musical instruments. They are set into vibrations by hitting them close to the open end. A vibrating bell radiates as a quadrupole.

Bernstein model   A mathematical model of the heredity of human blood factors based on the triple allelic theory.

betatron   A device for accelerating electrons to speeds approaching the speed of light. Electrons are injected into strong magnetic fields maintained in a toroidal-evacuated chamber. The device is used to produce X-rays by having the electron beam impact on a metal target.

Bethe-Slater curve   This is the relationship between the exchange energy for the transition elements vs. the ratio of the interatomic distance to the radius of the $3d$ shell.

B-field   In magnetic induction, the number of lines of magnetic flux per unit area of a surface perpendicular to the field.

bias circuit   The circuit that provides the needed DC bias for device operation, often separate from the input and output active circuit paths.

bias current   The current through a device that is due solely to the bias voltage. Since the bias voltage is necessary for operation but does not provide information, the bias current is often a main source of inefficiency in a device. In photosensitive applications, the bias current is sometimes referred to as the "dark current", since it flows with no light signal present.

bias, forward   A real diode, any diode junction, or any rectifying portion of a device (such as the emitter to base path in a transistor) has an asymmetry in the current response depending upon the polarity of the applied voltage. When operated far from breakdown, one polarity of applied voltage will allow more current flow (affording a lower effective resistance) while the other polarity will yield a lower current (a higher effective resistance). An applied voltage with polarity yielding the lower resistance (more current flowing) is termed a *forward bias voltage,* or just *forward bias*. *See also* bias, reverse.

bias, reverse   The voltage applied to a circuit element or portion of a device that is of a polarity yielding higher effective resistance is termed the *reverse bias voltage,* or merely *reverse bias*. *See* bias, forward.

bias voltage   The voltage needed to operate a device, provided by the bias circuit. For many applications, the polarity of this voltage is important. *See* bias, reverse; bias, forward.

binary circuits   Logic circuits with only two logic states, which may be labelled 0 and 1. *See* circuit, logic.

binary coded decimal (BCD)   A way of representing decimal numbers in binary format. Instead of being a true binary (base 2) number, each decimal (base 10) digit is separately represented in a binary format.

binary symmetric channel   A binary channel represents a communication channel over which signals are transmitted only as a sequence of binary-valued symbols. A binary symmetric channel is a memoryless channel (i.e., each unit of channel output depends only on the corresponding unit of channel input but not on other channel input units) with equal conditional error probabilities (i.e., the probability that a transmitted '1' symbol becomes received as a '0' equals the probability of a '0' received as a '1'). The channel may thus be fully characterized by a single error probability parameter, typically denoted as p. Such a simple channel model often suffices for many practical applications.

binaural   Pertaining to sound, process or system that deals with listening with two ears by humans and animals. Binaural listening allows the listener to determine the direction of a sound source up to 3° in the horizontal plane. This phenomenon is based on the binaural effects, i.e., the ability of humans and animals to distinguish the intensity and time arrivals of sound from a source at both ears. The ear that is closer to a source "hears" more intense sound and earlier than the other. Binaural effects are a background for performance of stereophonic audio systems.

binoculars   Instruments (e.g., binocular telescopes) offering comfortable telescopic-enhanced viewing of distant objects while allowing both eyes to remain active. The final image is made erect with the help of Porro or other types of prisms. Thus, the distance between the objective lenses can be made larger than the interpupillary distance. The designation, e.g., 6×30, means that the angular magnification is $6\times$ and the diameter of the objective lens is 30 mm.

bioelectricity   Electrical energy (current/voltage) produced within a biological organism, as in muscle tissue (*see also* action potential), nerve synapses, photosynthetic pathway, etc. Bioelectronics focuses on external electronic control of physiological response in plants and animals.

biofeedback   A learned response whereby a physiological output such as heart rate, blood pressure, metabolism, anxiety is controlled by conscious monitoring of the output (feedback) leading to control of the physiological process.

biological control theory   The theory of organism population control through the use of naturally occurring enemies, pests, pesticides, predators, etc.

biological effects, electric fields (1) *Static:* Static electric fields inside an organism may cause muscle reaction, nerve stimulation, death, or accommodation. External electric fields below the dielectric breakdown value (about 3 million volts/meter in air) appear to little damage larger organisms but do cause a stimulus to appear in the sensory nerves.

(2) *Time dependent:* Time dependent electric fields can have varying degrees of effects on biological form and function, leading to deformity and death. These effects are not strong if the electric field strengths are weak or the frequency of oscillation is low.

biological effects, electromagnetic fields Low frequency electromagnetic fields (less than 1000 MHz) appear to have no distinguishable effect on biological form and function. At higher and higher frequencies, the biological effects go from burning (infrared), sensory perception (visible), DNA and cellular damage (UV, X-ray), to death (gamma ray).

biological effects, gravity For all forms of life on earth, the gravitational field is always present and the organisms adapt to the presence of the gravitational force (mg). This force is apparently responsible for plants knowing which way to grow "up", for the bone size and distribution of walking mammals, for the limit to the size of animals, a limit to the maximum height of animals, and a limit to the longevity of animals by the work needed from the heart muscle to pump blood through the organism against the gravitational force.

biological effects, ionizing radiation High frequency electromagnetic waves in the UV, X-ray, and gamma ray end of the spectrum possess enough energy to break atomic bonds and to liberate electrons from atomic orbitals. This electron liberation is called ionization. Ionization leads to damage to DNA, RNA, protein structure and function and cell death by breaking these atomic and molecular bonds allowing the constituent atoms to recombine in non-functioning configurations. Some of this damage may be re-paired via RNA function; however, if the damage is extensive or the DNA/RNA templates are destroyed, cell death will be the end product of ionizing radiation. If the repair is not correct, cell deformity, including cancer, may result.

biological effects, magnetic fields Static magnetic fields, more so than electric fields, appear to strongly affect some biological organisms. It is believed that some birds and mammals know North and South by sensing the earth's magnetic field. Stronger magnetic fields appear not to be dangerous unless the organism is in possession of a ferromagnetic component, in which case the organism will be accelerated toward and held at one of the poles. The movement of an organism in a magnetic field will cause induced electric currents (Faraday Effect). If the magnetic field strength is small and the organism's speed is low, then the induced currents are very small and cause no damage or disruption. However, care must be exercised around large magnetic fields, such as with magnetic resonance magnets, for even moderate speeds near these magnets will induce appreciable currents.

biological effects, microgravity Recent experiments in the space shuttle program and the Mir program have provided a wealth of data on the effect of small gravitational fields (microgravity) on the form, development, and function of biological organisms. These space ships have very small gravity because they are essentially in free-fall orbit about the center of the earth. There is a residual gravitational field due to the moon, sun, and the nonspherical shape and the nonuniform density of the earth. Significant effects of microgravity on humans, for example, include muscle atrophy, loss of bone mass, disorientation on returning to the earth's gravitational field, etc.

biological effects, noise (1) Signal processing noise is a normal component of signal transmission within an organism. Optic nerve transmission is an example. If the noise level becomes too large, then the receptor is confused about the signal and loses its proper response. This can lead to mental confusion for processed signals; or to disablement and death if the nerve signals to the heart, for example, have too much noise.

(2) Sometimes referred to as unwanted genetic mutations (not recommended).

(3) Noise has psychological and physiological effects depending on amplitude, threshold, or timbre. Physiological response may include sleeplessness, anxiety, elevated blood pressure or heart rate. Psychological responses may include stress, anger, depression. Severe acoustical noise may lead to deafness, dysfunction, or death.

biological effects, non-ionizing radiation
(1) *Photosynthesis.* Visible light supplies the energy for photosynthesis in green plants.

(2) *Infrared.* Infrared radiation leads to local heating and may assist or hinder plant/animal growth and sustenance.

(3) *Radiofrequencies.* Appears to have little to no effect on biological systems. Studies are still in progress.

biological effects, statics   Statics, the science of equilibrium structures relating to their forces and moments of interaction, attempts to understand the structure of biological organisms with respect to their structure. For example: the variance of blood pressure with height, the maximum size of a mammal for bone structure integrity, or bone joint differences resulting in different walk patterns for various species.

biological effects, ultrasound   Ultrasound is acoustic frequencies in the frequency range above human hearing, approximately 20 kHz. Ultrasound is used for two-dimensional imaging of internal body structure because of the apparent nonharmful character of low intensity ultrasound. Sometimes referred to as echography or sonography.

biological effects, ultraviolet radiation   Ultraviolet radiation is electromagnetic radiation with frequencies beyond the visible spectrum. At these frequencies radiation is usually identified by wavelength — from approximately 185 nm to 390 nm. Ultraviolet radiation is usually harmful since it can energetically break DNA bonds leading to cell death or harmful muta-

tions. Some ultraviolet radiation is necessary for normal growth and calcium metabolism.

biological effects, X-rays   Similar to the harmful effects of ultraviolet radiation; however, there are no beneficial effects of X-ray radiation since X-rays are so much more energetic than ultraviolet radiation that chemical bond breakage leading to cellular death and harmful mutation is guaranteed.

biological half-life   The time required for one half of an injected radioactive substance to be excreted by a biological organism.

biological kinetics   (1) A study of the processes and rates of change of biological organisms.

(2) A study of the motion of biological systems.

biological rhythm   A regularly occurring process in the maintenance or growth of a biological organism. Heartbeat for example is a cardiac rhythm.

biomaterials   Materials derived from, or at least compatible with, biological organisms. A specific class of biomaterials are those developed for synthetic prostheses.

biomechanics   Sometimes referred to as biophysics. However, more properly biomechanics is the science devoted to elucidating the underlying forces of interaction responsible for the growth, maintenance, function, and form of biological organisms.

biorthogonal code   A set of $2K$ biorthogonal codewords may be formed from a set of K orthogonal codewords by including the negative of the K orthogonal codewords to the K original codewords. The correlation coefficient between any two members in a biorthogonal code set equals either 0 or -1.

biosphere   (1) That part of the earth compatible with living organisms.

(2) A closed thermodynamic model of a functioning ecological system.

biosphere, energy stored in   The amount of free energy available for support of life functions. Without any free energy, life would not exist. The energy is composed of mechanical, chemical, thermal, and nuclear parts.

biostimulation   Any process, heat, light, touch, chemical contact, etc. that induces a reaction in a biological organism.

biotelemetry   Remote monitoring of the conditions within a biological organism without any direct connection to the organism.

Biot-Savart's law   This law gives the differential contribution to the magnetic field dB produced at a distance r from a differential line element dl that carries a current I. In SI units

$$dB = \mu_o \frac{I\,dl \times r_1}{4\pi\,r^3}$$

and r_1 is a unit vector in the direction of r.

Biot's law   The rotation of polarization of plane-polarized light in an optically active medium is nearly proportional to the inverse square of its wavelength. The specific rotation ρ (degrees/mm) is given by Biot's law as $\rho = A + \frac{B}{\lambda^2}$ where A and B are constants specific to the material.

bipolar code   A bipolar code, also called an *alternating binary code* or an *alternate mark inversion* (AMI) code, represents a tertiary code wherein a "low" bit is signified by a 0 and a "high" bit is signified by a 1 or -1, such that successive "high" bits would have opposite signs. Bipolar coding possesses limited innate error self-detection capability because error must have occurred if the aforementioned alternating sign rule is violated. Bipolar coding is characterized by a spectral null at DC.

biprism   Biprisms consist of 2 prisms placed base to base.

biprism, Fresnel   Fresnel used a biprism to show interference phenomenon. It consists of two acute angled prisms placed side by side, and is constructed as a single prism of obtuse angle, with the acute angles on both sides about 30°.

bird acoustics   The discipline that studies songs and other sounds produced by birds, mechanisms of their production and their functions. Acoustic source of bird's songs and sounds is the syrinx located near the trachea and bronchi. Songs and sounds produced by birds are very complicated acoustic signals with rapid modulation in both amplitude and frequency. Frequency range of songs and sounds is from 100 Hz to 10,000 Hz. Functions of songs are believed to be territorial maintaining, individual recognition, mate attraction, and stimulating reproduction.

birefringence   When a beam of light passes through a uniaxial or biaxial crystal, it undergoes *double refraction* with an ordinary and extraordinary ray polarized in orthogonal directions. Birefringence is a measure of the difference between the refractive indices of the optical indicatrix of a crystal. If the refractive index parallel to the optic axis is larger than that at right angles to it, the crystal is said to be *positive uniaxial birefringent* (e.g., ice, quartz). The opposite is said to be *negative uniaxial birefringent* (e.g., calcite). Optically active substances, such as quartz, possess different refractive indices for left and right circularly polarized light. This phenomenon is called *circular birefringence* which leads to a rotation of polarization of linearly polarized light as it passes through the material.

Bitter patterns   These reveal the domain structure of ferromagnetic materials. In the technique used by Bitter, a drop of colloidal suspension containing fine ferromagnetic colloidal particles is applied to the ferromagnetic crystal surface. The pattern of magnetic structures revealed by the colloidal particles are called *Bitter* or *powder patterns*.

blackbody radiation   A blackbody is a perfect absorber or emitter of radiation of all frequencies incident or emitted at all angles. The spectral intensity distribution M per unit area in a unit wavelength interval (units: W/m^2 - μm) of a blackbody as a function of wavelength λ (unit: μm), at a temperature T (unit: degree

Kelvin), is given by the Planck's formula

$$M(\lambda, T) = \frac{(3.7415)10^8}{\lambda^5} \left(\frac{1}{e^{\frac{14388}{\lambda T}} - 1} \right)$$
$$\left(W/m^2 - \mu m \right) .$$

The wavelength at which the intensity peaks is inversely dependent on temperature and is given by the Wien's law

$$\lambda_{max} T = (2.8978)10^3 \, (\mu m - K) .$$

The total radiation (unit: W/m^2) emitted by a blackbody at the temperature T is given by the Stefan-Boltzmann law

$$\int_0^\infty M(\lambda)d\lambda = (5.6697)10^{-8} \, T^4 .$$

The blackbody radiation spectrum at 300, 1500 and 3000 degrees Kelvin is shown below. The peak wavelengths λ_{max} are at 9.7, 1.93 and 0.97 μm as given by the Wien's law.

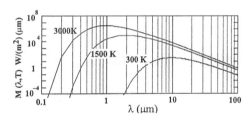

Blackbody radiation.

blackout, radio Also called *fade-out* or *short-wave fadeout,* a loss of short-wave communication along a specific propagation path due to prolonged fading of the radio signal. Blackout conditions occur whenever the lowest usable frequency (LUF) exceeds the maximum usable frequency (MUF) along a particular ionospheric propagation path.

Radio blackouts are often caused by ionospheric storms triggered by magnetic disturbances, resulting in a dramatic increase in ionospheric D-layer absorption. These storms may build slowly and last up to a week. Sudden ionospheric disturbances (SIDs) cause shorter blackouts, erupting very quickly and usually lasting less than an hour. They are precipitated by solar flare activity causing a reduction of F2-layer MUF and increase in D-layer absorption. Polar-cap absorption is caused by high-energy solar protons in high-latitude regions, resulting in blackouts lasting from a few hours to several days.

blanking In radar, turning off the receiver or transmitter to reduce interference from a particular direction or during a particular time. Target information can be lost if it arrives during the blanking period, or originates from the blanking direction.

In communications, silencing a receiver during a short period to reduce impulse noise. The period is chosen so that the loss of information does not significantly degrade the received signal.

In television, the use of a pulse waveform to render the return trace of the raster scan invisible.

blazing, of grating The technique of shaping the grooves of a ruled grating so that the maximum of the diffraction envelope due to the width of each groove coincides with the desired order of diffraction of the grating. As an example, a grating with the reflecting grooves making the blaze angle θ_B with the grating surface would have the grating equation

$$a \left(\sin \theta_i + \sin \theta_m \right) = m \, \lambda \, ; \theta_m = 2 \, \theta_B - \theta_i .$$

Here θ_i, θ_m and θ_B are the angles of incidence with respect to the normal to the grating surface, angle of diffraction in the mth order, and the blaze angle, respectively. Typically spectrographs are designed so that either $\theta_B = \theta_i$ (Littrow mount) or $\theta_i = 0$ (Normal mount).

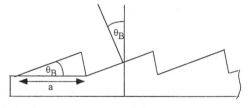

Blazing.

blindness, color The inability of the eye to distinguish different colors. Due to genetic factors or disease, about 3% of the human population does not have one or more types of cones, which have three different absorption spectra due to the three pigments present in them. If only one type of cone is present, then the person is a *monochromat* and can see only black and white. This occurs in 0.003% of the population. If two types of cones are present (about 5% males and 0.4% females) the person is a *Dichromat*. The most common among these are those who lack green-sensitive cones. They cannot distinguish red, yellow, and yellow-green. They are called *Deuteranopes*.

blind spot A small region of the retina, typically oval in shape, approximately 7.5° along its vertical axis and 5.5° along the horizontal axis, with its center located approximately 15.5° to the temporal side of the visual field. This corresponds to the point of exit of the optic nerve and is insensitive to light stimulation because it is devoid of photoreceptors (rods and cones). It is also called the *physiological blind spot* or *Mariotte's spot*.

blinking Alarm function in loran (long-range radio navigation) indicating the loss of signal integrity. Users are warned within one minute of signal loss. For aviation applications, an alert is given when the signal-to-noise ratio drops below −6 dB.

Bloch-Gruneisen formula The Bloch-Gruneisen formula is an approximate formula for the resistivity of a metal due to electron-phonon scattering. The equation is

$$\rho_{\text{phonons}} = A \frac{T^5}{\Theta_D}$$

$$\int_0^{\Theta_D/T} dx \frac{x^5}{(e^x - 1)(1 - e^{-x})}$$

for temperatures below the Debye temperature, where Θ_D is the Debye temperature and A is a material-dependent constant. This approximation is valid for a free-electron model of a Debye solid if Umklapp processes are negligible.

Bloch's equations Bloch's equations are the macroscopic equations of motion for a spin system in a magnetic field. For homogeneous, isotropic systems, Bloch's equations take the form

$$dM/dt = \gamma M \times B + \text{relaxation terms}$$

where M is the magnetization of the sample, γ is the gyromagnetic ratio, B is the applied magnetic field, and the relaxation terms will be discussed below. If the external field is constant and in the $\hat{z}$ direction, in equilibrium, the Bloch equations become $M_x = 0, M_y = 0, M_z = \chi_0 B_0$ where χ_0 is the magnetic susceptibility.

The relaxation terms take into account the spin-spin and spin-lattice interactions. If a system of spins, initially unmagnetized, is placed in a magnetic field, the magnetization approaches a new equilibrium value, M_0. Since this magnetization is in the same direction as the external field, it is the *longitudinal magnetization.* This relaxation to equilibrium takes place as energy flows from the spin system to the lattice system and is therefore known as *spin-lattice relaxation.* The spin-lattice relaxation time, T_1, must be included in the Bloch equations in order to describe these non-equilibrium processes. Another relaxation mechanism is present for transverse components of the magnetization. The spin-spin relaxation time, T_2, is a measure of the phase coherence of the spins. If, for example, a given spin, spin α, is in a local magnetic field B_α and another spin, spin β, is in local magnetic field B_β, these spins will precess at different frequencies. After some time, the magnetic moments of spins α and β will have different orientations which cause their magnetic moments to cancel each other out. At this point, these spins no longer add to the total magnetization.

Including relaxation effects in Bloch's equations, we obtain

$$
\begin{aligned}
dM_x/dt &= \gamma (M \times B)_x - M_x/T_2 \\
dM_y/dt &= \gamma (M \times B)_y - M_y/T_2 \\
dM_z/dt &= \gamma (M \times B)_z + (M_0 - M_z)/T_1 .
\end{aligned}
$$

Note that Bloch's equations are only approximate equations. No effort is made to determine the exact interactions in the system, which are only included via T_1 and T_2. As a result, the Bloch equations must be modified to describe accurately magnetic resonance in solids.

In nuclear magnetic resonance, an rf field is applied perpendicular to the static field. If the rf field is at a frequency $\omega = \omega_0$ where $\omega_0 = \gamma B_0$, the spins absorb energy from the rf field resonantly, thereby producing *nuclear magnetic resonance*. Nuclear magnetic resonance has become a powerful tool in chemistry, biology, and medicine.

Bloch's law   A relation between the fractional change in magnetization of a material and absolute temperature based on spin wave interaction and scattering in the material. Felix Bloch was able to show that the fraction change in magnetization should be proportional to $T^{3/2}$ where T is the absolute temperature.

Bloch walls   In 1931 Bloch showed theoretically that the boundary between magnetic domains is not sharp on an atomic scale but is spread over certain thickness wherein the direction of spins changes gradually from one domain to the next. This layer is usually called a domain wall or Bloch wall.

blood cell analysis, electrical impedance method   A sample of blood is placed between two apposing conducting plates across which an AC voltage is applied. The impedance is calculated from the response as a function of frequency and phase angle. From this information structure and content function of the blood may be inferred. Also referred to as *impedance spectroscopy.*

blood cell analysis, hydrodynamic method
Blood flow through a restriction creates a pressure differential across the flow and a velocity differential along the flow. From this information blood density, viscosity, and volume may be inferred.

blood cell analysis, photoelectric method
In this technique, a diluted blood specimen is passed through a laser beam. Each blood cell scatters the light; the amount of scattering and the intensity of the scattered light reaching the photodetector yields volume, optical density, and number of blood cells.

blood flow   The movement of blood past a given point. The motion of blood in the cardiovascular system.

blood flow measurement, Doppler method
The Doppler effect is the shift in frequency of sound upon reflection by a moving object. If the sound is reflected by an object moving toward you, the frequency increases in proportion to the object's speed. If the sound is reflected by an object moving away from you, the frequency decreases in proportion to the object's speed. In the Doppler method, a pulse of sound is passed into an artery and the frequency of the reflected sound is measured, yielding a plot of blood flow speed as a function of time and location. For example, this method is currently used to detect constrictions in the carotid arteries.

blood flow measurement, electromagnetic induction method   Blood is allowed to flow in an artery through a magnetic field. Because the blood plasma is ionic, the Faraday effect causes a transverse electric potential to appear across the artery, the strength of which is proportional to the flow rate of the blood.

blood flow measurement, fiber-optic method
A fiber optic probe is inserted into the artery allowing direct measurement of blood flow by count of the scattered light reflected back into the fiber-optic probe.

blood flow measurement, ultrasonic method
Short pulses of ultrasound (8 MHz) are applied to the skin via a transducer. The echo signals are picked up acoustically and the Doppler shift yields the blood flow rate.

blood pressure   The pressure, relative to atmospheric, in the arteries of the circulatory system. The larger pressure, systolic, occurs while the heart is pumping and the lower pressure, diastolic, occurs while the heart is at rest. The typical range for humans at heart level is about 130 mm Hg to 80 mm Hg. However, the blood pressure decreases with height, and is much larger at the feet than at the head of a standing person. Blood pressure also depends on mood, exercise, chemical intake, and conditioning.

blood viscosity   The resistance to flow of blood. A function of blood content, artery wall lining, and the presence of clotting factor which results in a marked increase in blood viscosity.

blue color of sky   When sun light traverses in a region of clear sky, the elastic or Rayleigh scattered light from air molecules appears blue when viewed in the lateral direction. This effect is due to the spectral dependence of scattered intensity (which is proportional to the fourth power of the frequency of light) and the sensitivity of the human eye. More violet and blue photons are scattered than red but the eye is less sensitive to violet. The scattered light is also partially polarized in the vertical direction, the degree of polarization being maximum when the direction of the sun light in the region of the sky and the viewing direction are perpendicular.

blue light, biological action   Blue light, being closest to ultraviolet radiation, appears to more strongly affect biological action than the other visible wavelengths. Blue light appears to set the circadian clock in mammals, stimulates cell proliferation, and appears to activate some of the transferase in cells. The mechanism appears to be proton transfer kinetics due to the absorption energy of the blue light.

Bode plot   Complex quantities that may vary with frequency can be visualized by simultaneously graphing a measure of the amplitude of the quantity (either directly or in decibels from the minimum or maximum amplitude) vs. frequency and the phase (measured in radians or degrees). Such a plot is termed a *Bode plot*. For a complex value $Z = |Z| \exp i\phi$, a Bode plot shows $|Z|$ and ϕ on separate traces as a function of frequency.

bolometer   A thermal detector of infrared radiation. The detection mechanism is the change in temperature produced by the absorption of incident radiation. It consists of a thin blackened slab whose impedance is temperature dependent. It can be used either in a DC (for steady signals) or AC (for periodic signals such as those from a pulsed laser or a chopped light beam) mode. The range of wavelengths and operating temperatures also vary. For some commonly

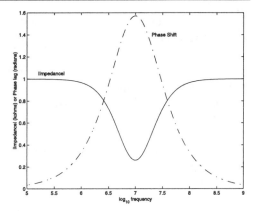

A Bode plot.

known bolometers. *See* bolometer, thermistor; bolometer, free electron; bolometer, low temperature.

bolometer, free electron   The mobility of free carriers in a semiconductor (e.g., indium antimonide) increases via absorption of the incident radiation leading to the bolometer effect. These detectors operate at liquid helium temperatures and have two to three orders of magnitude superior D star value compared to thermistor bolometers. The response time is in microseconds and the spectral range is in the mid- and far infrared.

bolometer, low temperature   Doped silicon or germanium bolometers operate over a very broad spectral range (1.7 to 1000 μm) with D star values of 10^{13} cm. $Hz^{1/2}$/W at 10 Hz. Operating temperature is 2 K. The time constant is inversely dependent on the thermal conductance which is typically 1 μW/K.

bolometer, thermistor   The bolometer element is made of thin flakes (10 μm thick) of polycrystalline oxides of Mn, Ni, and Co whose resistance can change by several percent per change of temperature of 1 K. The spectral response is in the mid-infrared and depends on coating. The time constant is in the millisecond range with the spectral D star values of 10^9 cm. $Hz^{1/2}$/W. The detector operates at room temperature.

boolean algebra   A set of rules for the formal representation of set or logic relationships. For instance, the union of two sets A and B can be represented as $A \cup B$, while the intersection is $A \cap B$. The empty set is $A \cap \bar{A}$ where $\bar{A}$ is the complement of A. Algebraic relationships and rules such as commutation are defined. When applied to logic, the empty set is analogous to false and the rules for formal logic can be stated in symbolic form, e.g., by replacing AND with the $\cap$ operator.

Common use of the term boolean algebra includes logical formalisms and some logic circuit design rules.

Bose-Einstein condensation (BEC)   As the temperature of a system of N bosons is lowered, more and more of the particles occupy a single quantum mechanical state, the ground state. Bose-Einstein condensation, the presence of macroscopic numbers of particles in the ground state, is a result of Bose-Einstein statistics only and will occur even in the absence of interactions. If the phase space density of a system of non-interacting bosons reaches

$$n_c = 2.612/\Lambda^3$$

where Λ is the de Broglie wavelength, $(2\pi\hbar^2/mk_BT)^{1/2}$, Bose-Einstein condensation takes place. If this equation is solved for the critical temperature, it is found to be

$$T_{BEC} = 115/\left(V_M^{2/3}M\right)$$

where V_M is the molar volume (cm^3/mol) and M is the molecular weight. The fraction of particles in the ground state is $(1 - -(T/T_{BEC})^{3/2})$. Superfluid ^{4}He and excitonic superfluids are both physical realizations of Bose-Einstein condensation, but the high densities involved result in highly interacting systems not easily described by weakly interacting BEC models.

BEC was observed in dilute gases of alkali atoms in 1995, and great progress has been made since then in a variety of studies. Currently, BECs consisting of tens of millions of atoms can be made routinely.

The techniques used to create a BEC vary, but most have several common features. The source of atoms is first heated in an oven, boiling atoms off the surface. These liberated atoms are then slowed (and thereby cooled) during loading into an optical or magneto-optical trap. Such a trap uses a set of lasers and/or a non-uniform magnetic field to confine the atoms. The magnetic fields are (usually) arranged such that only atoms in a particular hyperfine state are trapped, all others are rejected from the trap. These atoms are cooled further by "evaporation" of the high energy atoms. This is accomplished by inducing Zeeman transitions in the high energy atoms by using an rf field tuned to put the most energetic atoms into states that are not trapped. Eventually, the remaining atoms cross the critical phase space density, resulting in a BEC.

For the alkali gases, the transition temperatures are of the order of $0.1 - 1~\mu$K (but recall that the transition temperature is connected to the density of atoms, so a denser gas results in a higher T_{BEC}). Present and future studies of dilute gas BECs will probe for effects similar to those found in superfluid helium (e.g., the Josephson effect) and other new possibilities (e.g., an atom laser). *See also* helium-4, superfluid.

Bose-Stoner hypothesis   Stoner theory is a theory of metallic ferromagnetism. Bose excitations are spin waves in these systems.

Bouger's law   Describes the behavior of light transmission through an optical medium as

$$\log_{10}(\tau)/d = \text{constant}$$

where τ is the internal transmittance, d is the thickness of the medium, and the constant depends on the optical properties of the material.

boundary conditions   Conditions that are imposed on the pressure and the fluid velocity at an interface of two media in fluid-dynamics. The pressure and the component of the fluid velocity normal to an interface must be continuous across the interface. No conditions are imposed on the temperature. Continuity of pressure and the normal component of the fluid velocity result in relationships for acoustic pressure and fluid velocity at both sides of an interface, which are also called *boundary conditions*.

boundary conditions (magnetic field)
When an electromagnetic wave is incident on the boundary between two dielectric media such that the electric field is perpendicular to the

plane containing the media (the transverse electric mode), then the boundary conditions that have to be applied to the magnetic field component require that the net magnetic fields on both sides of the boundary be equal.

boundary resistance   Any time two materials are joined, a resistance (thermal and/or electrical) will occur at the joint. This resistance is due to imperfections in the joint and the differing properties of the two materials. In joints between two metals, electrons are scattered off the interface between the two materials; the same can be said for phonons in the case of two dielectric media. Thermal boundary resistances are a major concern in low temperature experiments as they can easily limit the sample temperature.

Whenever such a resistance exists, there will be a step in the temperature across the interface according to $\Delta T = \dot{Q} R_{Th}$ where $\dot{Q}$ is the heat incident on the surface and R_{Th} is the thermal boundary resistance. In metals, thermal boundary resistance can be minimized by minimizing the electronic boundary resistance. This can be done by maximizing the actual area of contact and ensuring clean surfaces free of oxide layers. When two metals are pressed together lightly, as little as one part in a million of the area will actually be in contact. Therefore, pressure must be exerted to improve contact.

Another technique is to weld the two pieces together, but only if the metals do not produce an alloy with large thermal resistance, of course. To guarantee clean, oxide-free surfaces for press joints, the metals are often gold-plated. It is also common practice to use the thermal contraction of materials to good advantage when designing the parts to be joined. When these techniques are combined, it is possible to get 10–$100\,n\Omega$ of boundary resistance between two metals.

For non-metals, heat is conducted primarily by phonons, and so transmission of phonons across the boundary is of utmost importance. In this case, however, there is not as much for the experimentalist to do except choose materials wisely and maximize contact area. At the interface between dielectrics, phonons scatter off the surface according to acoustic mismatch theory. In complete analogy with optics, a phonon is much more likely to be scattered when crossing the boundary between two very dissimilar materials than if the sound speeds are nearly equal. In general, it is better to plan not to rely on heat conduction through dielectrics unless absolutely necessary. *See also* Kapitza boundary resistance.

boundary waves   Also known as surface waves. When a ray of light is incident at the interface of two media at an angle larger than the critical angle, it is totally internally reflected. However, there is a tangential component of the electric field at the boundary of the interface. The amplitude of the boundary waves decays exponentially with distance. This field can couple with another nearby medium of higher refractive index leading to *frustrated total internal reflection*. *See* critical angle.

bound charge   Bound charges are charges due to the polarization of the material. There are two types of bound charges:

1. surface bound charge, σ_b, which is the vector product of polarization and surface normal unit vector; and

2. volume bound charge, ρ_b, which equals the negative of the divergence of polarization.

bow wave   A wave occurring in front of a ship in motion.

boy's method   A method of measuring the refractive index n of the material of a lens. The radii of curvature r and s of the two surfaces of the lens, and the focal length f are measured by determining the distances at which an object is coincident with its image produced after reflection from the respective curved surface of the lens, and after putting a plane mirror behind the lens. The relation $1/f = (n-1)(1/r + 1/s)$ then gives the desired refractive index.

Bragg's law   The diffraction of a beam of X-rays by the atomic planes of a crystal results in bright spots obeying the Bragg's law $2d\sin\theta = m\lambda$ where d is the spacing of atomic planes, θ the angle of the incident beam from the planes, and λ is the wavelength of radiation. The integer m refers to the order of diffraction. Different sets of atomic planes in a crystal diffract X-rays at different angles, as shown in the figure below, leading to a pattern of bright spots. Analysis of such data can yield information of the crystal structure. Other applications of Bragg's law include the scattering of light by a periodic refractive index grating generated by acoustic waves in crystals. A special case of such an acousto-optic effect is Brillouin scattering.

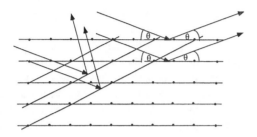

Bragg's law.

Bragg-Williams approximation The Bragg-Williams (B-W) approximation is a zero order mean field approximation often used to study the Ising model, binary alloys, and other similar systems. The primary assumption in the B-W theory is that spatial correlations between nearby spins (or lattice sites) are unimportant. The Bragg-Williams approximation is equivalent to the Weiss molecular field model. It most often produces qualitatively correct behavior, but fails when fluctuations and long-range correlations become important — in the critical regime near phase transitions, for example.

Braun tube Cathode-ray tube (CRT). It was invented by Braun (1850–1918) and was originally called *Braun tube.*

Bremsstrahlung A continuous spectrum of X-rays (photon energies in the 100 to 100,000 eV). A typical X-ray tube consists of a hot cathode and a rotating anode made of metals such as copper, molybdenum or tungsten. Thermally emitted electrons from the cathode are accelerated toward the anode by an applied voltage (of the order of kilovolts) between them. As the electrons colliding with the target nuclei are decelerated, X-rays are emitted. The minimum wavelength is inversely proportional to the applied voltage. At higher voltages, the electrons colliding with the anode will have sufficient energy to eject the electrons of the target atom from their inner shells. These ejected electrons relax to the available empty shells giving rise to X-rays of discrete energies. The sharp line spectrum of the so-called *characteristic* X-rays will be superposed on the continuous spectrum called the Bremsstrahlung.

Brewster angle The angle of incidence (measured with respect to the normal) of a ray of light traveling from a medium of refractive index n_1 to that of n_2, so that the TM (transverse magnetic) or p-polarized light, with the electric vector parallel to the plane of incidence, has zero reflection. This angle is also called *polarizing angle* because the reflected light is completely s-polarized or TE (transverse electric) with the electric vector perpendicular to the plane of incidence. The transmitted light will be partially TM polarized. In terms of the refractive indices, the Brewster angle, θ_B, is given by $\theta_B = \tan^{-1}(\frac{n_2}{n_1})$, where n_1 and n_2 correspond to incident and transmitted media, respectively. For the angle of incidence at θ_B, the reflected and transmitted beams will be at right angles.

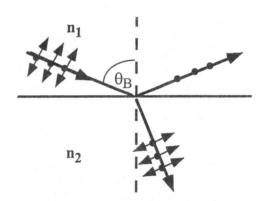

Brewster angle.

The figure below shows the reflectance of light for TE and TM polarizations as a function of the angle of incidence θ_i. The figure (a) is drawn for light traveling from air ($n_1 = 1$) to glass ($n_2 = 1.52$). The direction is reversed in figure (b).

Brewster's fringes Two plane-parallel Fabry-Perot plates with exactly the same thickness, or the thickness of one being an exact mul-

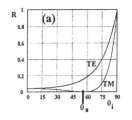 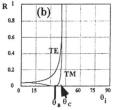

Brewster angle.

tiple of the other, are inclined at an angle of 1 to 2° and the interference due to white light is observed. Straight line fringes called *Brewster's fringes* appear if the ratio of interferometer spacings is an exact integer. These measurements are useful for accurate calibration of length such as the determination of the standard meter.

Brewster's law   Refers to the state of polarization of reflected and transmitted light from one optical medium to other. *See* Brewster angle.

Brewster window   Suppose a beam of p or TM polarized light beam is incident on a material, in the form of a plate with parallel faces: at *the Brewster Angle* it will be transmitted entirely with no reflection losses. Such perfect windows are used extensively in lasers. *See* Brewster angle.

brightness   Also called *luminosity* of an object; brightness is the power per unit solid angle per unit projected area emitted or scattered. The units used in radiometry are watt per steradian per meter square.

broadcast   Broadcast refers to a one-way transmission of information to the general public tuned to the particular transmission channel within a given geographical area or to the general public in a given user domain, with the receivers providing no acknowledgment of receipt to the transmitter. The broadcast channel thus consists of one transmitter but many receivers. Typical uses of broadcast includes commercial television and radio, and traffic station communication to ships or aircrafts. Radio broadcasting uses four different frequency bands: pan-European broadcasting in the longwave frequencies between 150 and 290 KHz, AM (amplitude modulation), broadcasting in

the medium-frequency range of 525 to 1700 KHz, long-distance international broadcasting in several short-wave sub-bands lying between 5950 KHz to 26.1 MHz, and FM (frequency modulation) broadcasting in the very-high frequencies from 88 to 108 MHz. FM broadcasting offers superior fidelity and robust reception over AM broadcasting.

broad-side on (magnets)   The broad-side on position is a point that lies on a line through the center of a magnet perpendicular to the magnetic axis.

bubbles, suppression of (low temperature)
At temperatures above the superfluid transition in liquid ^{4}He, the thermal conductivity is sufficiently poor as to allow local heating and thereby allow bubble formation. The thermal conductivity of superfluid helium is very high, comparable to that of a metal. As a result, the local heating necessary for bubble formation cannot occur, and superfluid helium is "quiet" with no bubbles formed in the interior of the fluid. This suppression allows visual identification of the superfluid transition in quite dramatic fashion. *See also* helium-4, superfluid.

bug   A bug may refer to an error in a computer program or a defect in an apparatus. A bug may alternately refer to a miniature concealed or secret electronic gadget for eavesdropping.

bus   A bus refers to a single conductor or a group of conductors for power or signal transmission. The signal bus transmits data, address and control commands to facilitate data exchange among various components in a communication or a computer system. In a computer system, a bus embodies a standardized circuit interface.

byte   A byte refers to a batch of binary digits (or bits) processed as one unit. Eight-bit bytes are commonly used to represent an alphanumeric character or a control signal in the American National Standard Code for Information Interchange (ASCII). Two or more bytes make up one word.

C

cable   A cable is a thick wire or a bundle of wires inside an insulated covering used for transmitting electric or electronic signals.

cable model, cell membrane   Intracell communication across intracellular membranes may be modeled after the theory of transmission lines in electronic engineering (cable model). Studies suggest that the transmission of voltage signals across membranes occurs with an effective cable length on order of 100 micrometers.

cadmium cell   The cadmium cell was developed by W. Jungner in the late 1890s and early 1900s. It is a rechargeable cell employing the following chemical reaction:

$$AgO + Cd + H_2O \longrightarrow Cd\,(OH)_2 + Ag\,.$$

The cadmium cell is still in common use today as a voltage reference, at 20°C its voltage is 1.0186 volts. Cadmium cells are noted for their high energy and power density, but the high cost has limited their applications.

cadmium red   Compounds of color pigments ranging from orange through red made of solid solutions of the semiconductors cadmium sulphide and cadmium selenide.

cadmium wavelength standard   An internationally agreed standard that in dry air at 15 degrees Celsius and a pressure of 760 mm of mercury the red line of cadmium has a wavelength of 6438.4696 Å.

calorimeter, adiabatic   (**1**) A calorimeter thermally isolated from all heat sources and/or sinks not actively involved in the experiment. If the temperature of a sample inside an adiabatic calorimeter were measured while no heat was being added to the sample, the temperature would remain constant indefinitely. Experimentally it is possible to create such calorimeters with heat leaks of below 1 nW.

(**2**) This is a bomb calorimeter for which the insulating jacket temperature is kept equal to that of the bucket. With the jacket and the bucket temperatures equal, there is no heat transfer out of the bucket and the necessary corrections for isothermal systems are removed.

calorimeter, Nernst vacuum   A Nernst calorimeter consists of a sample inside a vacuum can that is immersed in a liquid cryogen. A thermometer and heater are attached to the sample — in practice, they may be the same item. The sample is cooled through the thermal conduction of a gas admitted into the vacuum can. After reaching thermal equilibrium, the sample is thermally isolated by evacuating the vacuum can. Then, the temperature is recorded while known amounts of heat are added to the sample.

calorimetry, low temperature   Low temperature calorimetry is fraught with complexities beyond those found at higher temperatures. At low temperatures, most materials under study have very small heat capacities, making small heat leaks and the heat capacity of the surrounding materials very important. Much of modern low temperature calorimetry is designed to measure the specific heat of very small (meso-scopic – microscopic) samples attached to a substrate. In many of these experiments, the substrate and surrounding apparatus have a heat capacity which is much larger than that of the actual sample. This addendum heat capacity must be measured very precisely for such experiments to have any meaning whatsoever.

At low temperatures, the boundary resistance between the sample and its surroundings becomes quite large, requiring care to thermally anchor the sample properly. As at higher temperatures, in order to measure the specific heat, it is necessary to also understand all possible sources of heat input into the system. The low temperature experiment is sensitive to incredibly minute heat leaks. It is common in ultra-low-temperature calorimetry to be sensitive to stray heat at the tens of pW ($1pW = 10^{-12}$ W) level!

At low temperatures, it is also necessary to consider a wide variety of sources for heat input, including heat leaks from residual atoms in a vacuum chamber and the black body radiation from surfaces at higher temperatures. As in all low temperature experiments, it is necessary to measure the temperature accurately and with high precision. In measurements to determine the specific heat, it is also important that the thermometers do not deposit a substantial amount of heat into the sample. As mentioned earlier, sensitivity to very small heat sources makes this problem additionally important. *See also* heat switches; Kapitza boundary resistance.

camera   Any instrument used to form an image. The simplest camera is a pinhole camera (*see* camera, pinhole). A typical camera consists of a positive lens on one side of a box for forming a real image on a photographic plate or a screen situated on the other side of the box. This image is later developed or printed to get the final picture. For fast moving objects or for a handheld camera, shorter exposure time requires large aperture for the lens, thereby necessitating corrections for various aberrations. The materials of the lens and the plate depend on the wavelength of the radiation involved.

camera, aperture mechanism of   The *aperture mechanism,* which controls the throughput of light from the object entering the lens, can be set to a number of settings called f-stops or f-number. The f-number is the ratio of the focal length of the lens system to the diameter of the aperture. The sequence of f-stops denoted as f/1.4, f/2, f/2.8, f/4, f/5.6, f/8, f/11, f/16, f/22 and f/32 imply aperture diameters beginning with the largest at f/1.4 and decreasing so that successive stops have half the throughput of the former one. To keep the amount of exposure fixed, if the f-stop is increased by one step the shutter time should be doubled. *See* camera.

camera, depth of field of   The *depth of field* is the range of object distances in focus without blur. The depth of field is inversely related to the aperture diameter. In order to increase the depth of field and maintain the same throughput one should increase both the f-stop number and the exposure time. *See* depth of focus; depth of field; camera.

camera, field of view of   The *field of view* is the angular width of the object that can fit on the film. It depends on the film size and the focal length of the lens. For example, a typical 35 mm camera (i.e., the width of film is 35 mm) using a lens of focal length 50 mm has a field of view of 45 degrees. A telephoto lens of 200 mm focal length has 10 degrees. In contrast, a 28 mm wide angle lens will have a field of view of 75 degrees. *See* camera.

camera, lens of   The *lens system* usually consists of multiple lenses corrected for various kinds of aberrations. *See* aberration; camera.

camera obscura   Also known as a pin-hole camera. The earliest form of a camera with a light-tight box and a small hole. Light from an object enters through the hole and an inverted image is produced on a screen.

camera, pinhole   The simplest optical device consisting of a light-tight box with a very small hole (pinhole) on one side which is placed toward the object and a photographic plate or a screen on the opposite side. There is an optimal aperture size for a given distance of the aperture from the plate. With long time exposure, excellent inverted images of the extended objects can thus be formed without distortion. This camera is very useful in capturing the architectural details of buildings.

camera, shutter mechanism of   The *shutter mechanism* controls the duration of time the film is exposed, which can range from a few seconds down to a thousandth of a second in steps of two. A *leaf shutter* contains metal blades that swing open momentarily. Expensive cameras contain *focal plane shutters* that slide past the film. The exposure time is a critical parameter in fast action photography. *See* camera.

camera, television   Instruments for converting audio-visual information to electric signals that are ultimately used for display on a television screen. The basic process is either collection and acceleration of electrons when light falls on the surface of photosensitive cathodes (image-orthicon tubes) or recording changes in the electrical conductivity of a photoconductive layer deposited on one side of a transparent conducting film (Vidicon tubes).

camera, types of   Cameras can be classified into three types according to how the image is viewed by the photographer.

1. The oldest kind is the *View Camera,* built like an accordion, in which a large film can be used to record the details of a large landscape or a group of people. The photographer views the image on a glass plate and adjusts the parameters such as the focus and composition of the image. The glass plate is replaced by the film and the picture is exposed. The image is inverted on the viewing glass plate and can be disconcerting to a novice photographer. The advantage is that one can tilt the lens relative to the plane of the film moving the bellows thus correcting for distortions due to foreshortening and depth of focus.

2. The second type is a *reflex camera,* which permits the film in place all the time. A mirror between the lens and the film reflects the image on a viewing screen. The mirror is positioned such that the image position is identical on the film without the mirror as it is on the viewing screen with it in place. A single lens reflex (or SLR) also contains a five-sided prism that corrects the inversion of the image from left to right caused by the mirror reflection. When an exposure is made, the mirror swings out of the way momentarily to admit light on the film.

3. The third type is a *viewfinder camera* with a separate optical system to view the image and another to focus on the film. Due to their simplicity such cameras are inexpensive but suffer from parallax. The two images may not coincide exactly. A range finder that views the image from two different angles can reduce this defect. *See* camera.

candela   Unit for luminous intensity equal to one lumen per steradian. It is defined as the luminous intensity of one sixtieth of one square centimeter of the projected area of a black body radiator operating at the freezing point of platinum (2042 K). Abbreviation: cd.

candle, international   *See* candela.

candle power   Luminous intensity equal to one candela.

capacitance   Property of an electric conductor, or set of conductors, that is measured by the amount of electric charge that can be stored on it per unit change in electrical potential.

capacitance, cell membrane   A cell membrane, due to impermeability or diffusion-limited ion flow, has unlike charges distributed on either side. This is the basic structure of a capacitor; a physical separation of unlike charges and hence a capacitance may be associated with a cell membrane, the capacitance being the ratio of charge to potential difference across the membrane.

capacitance, distributed   The effective capacitance of a circuit network that is physically distributed through the circuit or device. For instance, microstrip antennas have some capacitance per unit length, as well as resistance and inductance per unit length. The proper calculation of all infinitesimal impedances will yield an effective reactance (possibly mostly capacitance) that equals the measured capacitance in the whole device, but which is not localized to a specific portion of the circuit.

capacitance, interelectrode   The (often unwanted) capacitance between the electrodes of a device.

capacitance, junction   In a diode, the capacitance across the diode junction is termed the junction capacitance. For abrupt junctions, this capacitance C_j is inversely proportional to the depletion depth, and therefore

$$C_j \propto (V_d)^{\frac{-1}{2}}$$

(for a reverse bias voltage V_d).

capacitive discharge   When two terminals of a charged capacitor are connected through a conductor, the charge on the capacitor will be annihilated by a current through the conductor. This phenomenon is called *capacitive discharge.* The time constant of the discharge depends on the capacitance of the capacitor and the resistance of the conductor. In the process, the energy stored in the capacitor is dissipated through the Joule heating of the conductor.

capacitive reactance   Capacitive reactance is defined as the inverse of the product of angular frequency and capacitance, or $X_c = 1/\omega C = 1/(2\pi f C)$.

capacitor, blocking   A capacitor placed in a circuit for the purpose of blocking a DC offset voltage from another portion of the circuit.

capacitor, by-pass   A capacitor placed in a circuit to allow the high frequency components to bypass a part of the circuit. This may be to eliminate unwanted signals to a power supply, direct the AC signal elsewhere, or to protect the biasing voltage of a device from dropping below the useful level when many devices become active at once. This eliminates power spikes and "brownouts" from the current drain of many devices clocking at once in a synchronous circuit, and is a crucial design feature in digital circuits.

capacitors   Capacitors are electric components that are made of two conductors embedded in a dielectric medium. Capacitors are indispensable electric components in an electric circuit. They can be used to store electric charges, to block direct current flow, and to pass alternate current. Capacitors can be combined with resistors to form an RC circuit.

capacitors in parallel   When two or more capacitors are connected in such a manner that one terminal of each capacitor is connected to the same common point A, while the other terminal of each capacitor is connected to another common point B, as shown in the first diagram, this is called *capacitors in parallel*. When capacitors are connected in parallel, the voltage difference between the two terminals of each capacitor is the same. The equivalent capacitance of capacitors in parallel is equal to the sum of each individual capacitance:

$$C = C_1 + C_2 + C_3 + \dots .$$

capacitors in series   When two or more capacitors are connected in such a manner that one capacitor is connected to the next sequentially as shown in the next diagram, this is called *capacitors in series*. When capacitors are connected in

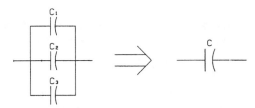

Capacitors in parallel.

series, the inverse of the equivalent capacitance is equal to the sum of the inverse of capacitance of each individual capacitor:

$$C^{-1} = C_1^{-1} + C_2^{-1} + C_3^{-1} + \dots .$$

Capacitors in series.

cardinal planes/points   A thick lens/lens system can be described in terms of six cardinal points allowing graphical determination of images for arbitrary objects. These six cardinal points on the axis of a thick lens consist of first and second focal points, first and second principal points, and first and second nodal points. The corresponding planes normal to the axis at this point are called *cardinal planes*.

cardinal points   The six cardinal points of an optical system are a) two focal points (F_1 and F_2), b) two principal points (H_1 and H_2), and c) two nodal points (N_1 and N_2). As shown in the diagram, ray 1 originating from the focal point F_1 emerges parallel to the axis on the other side. The intersection of these two rays defines the first principal plane, which cuts the axis at H_1. Ray 2, parallel to the axis, passes through F_2 on the output side. The intersection of these two rays defines the second principal plane, which cuts the axis at H_2. An off-axis ray 3, passing through one nodal point N_1 emerges parallel to it through another (N_2). Planes passing through the cardinal points normal to the axis are called *cardinal planes*. The sketch below identifies these points and planes in an optical system. V_1 and V_2 are the input and output planes. The refractive indices in input, optical system and out media are n_3, n_2 and n_1, respectively. In terms of the *ABCD matrix* elements, the distances to various cardinal points, as defined in the figure below, are

$$p = \frac{D}{C}, q = -\frac{A}{C}, r = \frac{D-\frac{n_3}{n_1}}{C}, s = \frac{1-A}{C}, v = \frac{D-1}{C}, w = \frac{\frac{n_3}{n_1}-A}{C}, f_1 = p - r \text{ and } f_2 = q - s.$$

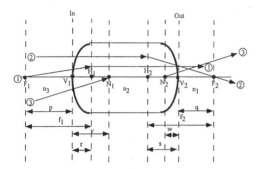

Cardinal points and cardinal planes.

Note that the distances obey the following sign convention: p, r and v are taken positive if they are to the right of V_1 and negative otherwise. q, s and w are positive if they are to the right of V_2. Similarly f_1 and f_2 are positive if they are to the right of their respective principal planes H_1 and H_2 and negative if to the left. As an example, H_1, H_2, N_1 and N_2 coincide at the center of a refracting sphere irrespective of its radius of curvature and refractive index. However, $p = -q = \frac{R(2-n)}{2(1-n)}$ for a sphere of radius R and refractive index n. *See* ABCD matrix.

carrier   A carrier is the basic continuous waveform or pulse train on which the information-bearing signal is to be modulated in order for many channels to share one transmission medium. The particulars about the carrier are defined by the particular communications system apart from the particular information-bearing signal to be transmitted. The carrier is often a sinusoidal wave at a specific frequency set orders-of-magnitude higher than the bandwidth of the information-bearing signal. The carrier may alternatively be a similarly high-frequency periodic pulse train or simply a constant direct-current (DC) voltage offset. A set of distinct carriers may be viewed as a strategy to partition the total available transmission

bandwidth into distinct transmission channels. In order for an information-bearing signal to be transmitted on any of these channels, the information-bearing signal modulates the carrier corresponding to the channel to be used. Unmodulated versions of the carrier may or may not be transmitted alongside the modulated carrier. If not, the carrier is said to have been suppressed. At the receiver, the information-bearing signal is decoupled from the carrier.

carrier binding   The attachment mechanism for the cotransporter binding to the transporter in carrier-mediated diffusion.

carrier frequency   The carrier frequency refers to the frequency of an unmodulated sinusoidal wave carrier or the pulse repetition rate of a periodic pulse-train carrier. In frequency modulation schemes, the carrier frequency is equal to the center frequency.

carrier level   The carrier level refers to the power level of the unmodulated carrier at a specific position within the communication systems; may be expressed in absolute terms in watts or in relative terms, in reference to some system ground level, in decibels (dB).

carrier, majority, of current   In doped semiconductors, one polarity of current carrier will provide the dominant charge transfer mechanism, usually by simply being the most plentiful. These types of carriers are the majority carrier.

For instance, in boron doped silicon (p-type), the boron is an electron acceptor. This allows holes to be the majority carrier of current. In n-type semiconductors, electrons are the majority carriers. *See also* acceptor; donor.

carrier mediated transport, macroscopic model   General description of the transport model using free energy, kinetic, and entropic modelling for the prediction and the description of the rate constants.

carrier mediated transport, microscopic model   Detailed molecular modelling of the stoichiometry and energy analysis of the carrier transport mechanism.

carrier mediated transport, steady state process   Transport of biomolecules through a membrane restriction not normally permeable to the biomolecules. The biomolecule reacts with a transport molecule A to form transportable molecule B which passes through the membrane. B is then decomposed to yield A and the original biomolecule. A diffuses back through the membrane to start the process again.

carrier, minority, of current   The minority type of charge carrier in a semiconductor. For example, in p-type boron doped silicon, most charge transfer is through hole migration. However, some electrons do diffuse in the crystal. These electrons are minority carriers. In n-type semiconductors, the holes are the minority carriers. *See* carrier, majority, of current.

carrier models (cell)   Modelling of the active transport of biomolecules across membranes upon reversible attachment to a transporter molecule.

carrier power   The average power of an unmodulated sinusoidal carrier over one sinusoidal period. The carrier power may deviate from the nominal value when flawed modulation produces unequal envelope amplitudes on the positive and negative sides. This phenomenon is called *carrier shift*.

carriers (cell)   A molecule to which another molecule may become reversibly attached for transport through a membrane, in effect, a molecule that can be reduced by attachment to the second molecule and after transport may be re-oxidized.

carrier-to-noise ratio   The carrier-to-noise ratio refers to the ratio of carrier power to noise power at the same position within the communication system. The carrier-to-noise ratio is typically expressed in decibels (dB).

carrier wave   A wave in which amplitude, frequency and/or phase are varied in time to produce modulated oscillations (signals). The frequency of the carrier wave is much greater than those of amplitude, frequency or phase modulations.

CARRY, half-adder   The flag in binary addition signifying the carry operation.

A half-adder is a binary logic circuit that gives two outputs for two binary inputs. One output S (for Sum without carry) is, in the single binary digit case, an exclusive OR operation of the inputs. The other output, C (for carry) is true if S is not the binary sum. In the simple binary digit case, this is the AND operation of the inputs. For this case, given single binary digit inputs A and B, the S and C outputs are

A	0	0	1	1
B	0	1	0	1
C	0	0	0	1
S	0	1	1	0

Since in general the carry is only a flag, two half-adders (plus some logic operations) are needed for true binary addition.

cartesian surface   Refracting or reflecting surfaces that form perfect images, named after René Descartes. Each object point requires its own surface. In case of reflection, such surfaces are invariably conic sections.

Cary-Foster bridge   Cary-Foster bridge is used to measure the small difference between two nearly equal resistances. The bridge is of the slide-wire type and the resistance of the slide wire per unit length, r, is accurately known. R_x and R_s are to be compared. A balance is first secured with the contact C at a distance a_1 from the D. Then R_x and R_s are interchanged and another balance obtained with C at a distance a_2 from D. It can be shown that.

$$R_x - R_s = (a_1 - a_2)\, r \ .$$

Cassegrain telescope   In this design of the reflecting telescope, the light from the primary mirror is incident on a secondary mirror shaped as a convex hyperboloid. The light reflected by the secondary mirror passes through an aperture in the primary mirror to secondary focus, f_s, for viewing. The primary f_p and secondary f_s focal points are adjusted to coincide with the foci of the hyperboloid which results in a distortion free image.

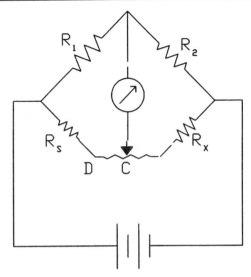

Cary-Foster bridge.

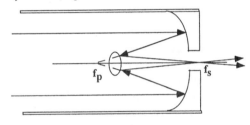

Cassegrain telescope.

cataracts A disorder that causes pockets of cloudy or opaque discoloration of the lens tissue of the eye; a surgery involves removal of the lens and replacing it with a plastic implant. Often the membranes that hold the implant become opaque and a corrective procedure known as *posterior capsulotomy* uses a high power laser to rupture the membrane and restore vision.

catheters, blood pressure measurement A catheter is any instrument designed for the free flow passage of fluid into or out of the body. When used for blood pressure measurement the catheter allows the blood to be in direct contact with a manometer thus allowing direct blood pressure measurement.

cathode The electrode that emits electrons into a space, medium or device. May also be viewed as the path for positive current to leave the medium.

cathode ray oscilloscope A device based on a CRT that allows visualization of input signal voltage waveforms. There are two main modes of operating the device.

1. One mode is a two-input mode where the x coordinate of the electron beam is proportional to the first input, while the y coordinate is proportional to the second.

2. The other mode uses a synthesized voltage for the x input signal that is proportional to time (modulo the repeat time). When the y position is proportional to the input voltage, a representation of the waveform (voltage vs. time) is traced out on the screen. If the signal waveform is periodic, triggering circuitry allows the phase matching of successive traces, producing a sustained trace of the waveform.

cathode ray tube An evacuated tube or bulb in which a voltage difference of separated conductors produces accelerated electrons (cathode rays). These electrons may be used in the production of X-rays, for physical measurements, or other purposes. However, the most prominent use of the cathode ray tube (CRT) is the focusing of the electron beam in a phosphorescent surface for the displaying of information in two dimensions.

Most television screens, computer monitor screens, and oscilloscope displays are the phosphorescent end of a CRT. The x and y position the beam strikes is controlled by a magnetic (or sometimes electric) field perpendicular to the electron beam direction. The intensity may be varied by control circuitry in the electron gun producing the beam. This electron gun also has the cathode/heater circuit, focusing elements and the accelerator system.

catoptrics Optics of reflecting surfaces. Telescopes and microscopes built entirely of reflecting optics are called *catoptric systems*. In contrast, systems with a combination of lenses are called *dioptric systems*. Combinations of lenses and mirrors are *catadioptric systems*.

Cauchy dispersion formula The refractive index of transparent material is dependent on the wavelengths. When the refractive index is plotted as a function of the square of the wavelengths, the resulting curve is known as the dispersion curve which appears to be asymptote in the *UV* region and is somewhat linear in the near infra red (up to $1\mu m$). This is called *normal dispersion*. When the refractive medium has characteristic excitation that absorbs light of wavelengths within the range of the dispersion curve, the curve in general will be monotonically decreasing, but will have a positive slope in the wavelength region of the absorption. This is called *anomalous dispersion*.

An empirical relation that gives n as a function of λ for normal dispersion is called the *Cauchy dispersion formula*:

$$n = a + b\lambda^2 + \frac{c}{\lambda^2 - 0.028} + \frac{d}{\left(\lambda^2 - 0.028\right)^2}$$

where a, b, c and d are empirical constants. A more general expression is

$$n(\lambda) = a + \frac{b}{\lambda^2} + \frac{c}{\lambda^4} + \dots .$$

Often, the first two terms in the above expression are sufficient to provide a reasonable fit and if we have experimental knowledge of n at 2 distinct wavelengths, then the constants can be determined. The dispersion is defined as $\frac{dn}{d\lambda}$, and is approximately equal to $\frac{-2b}{\lambda^3}$.

caustic curve   The geometrical envelope of the meridian section of a bundle of refracted or reflected rays. The points of intersections of pairs of consecutive rays lying in the plane of a meridian section of a refracting or reflecting spherical surface form a curve lying symmetrically above and below the optical axis, if the incident bundle is symmetrical with respect to the optical axis. This plane curve is called the *caustic curve* of the meridian rays. The two branches on the opposite side unite in a double point or cusp at the point on the axis where the paraxial rays intersect, so that the axis is tangent to both branches at this point. Each refracted or reflected ray in the meridian plane touches the caustic curve.

cavitation   Formation and collapse of cavities and bubbles in liquids, filled with gas and vapor. Cavities and bubbles may be formed by several mechanisms: due to working pumps or rotating turbines and ship propellers (hydrodynamic cavitation); due to intense sound radiated into a liquid (acoustic cavitation); and due to laser beams

and elementary particles propagation through a liquid. In the cases of hydrodynamic and acoustic cavitation, formation of cavities and bubbles occurs at points where the local pressure is below a threshold that allow so-called cavitation nuclei (tiny bubbles filled with gas or vapor) to grow. Collapse of cavities and bubbles produce intense noise. This collapse can also destroy materials of different kinds, for example, ship propellers.

cavity dumper   Energy can be built up in a laser cavity for a length of time and deflected outside the cavity to get a sudden burst of energy. Such a device is called a cavity dumper. An acousto-optic deflector placed inside a laser cavity can spoil the alignment temporarily when turned on and dump the energy. *See* acousto-optic deflector.

cavity modes of a laser   A laser resonator can sustain two types of resonant modes of oscillation depending on the separation and curvature of the mirrors. These are (a) *axial or longitudinal modes* and (b) *transverse modes*.

If L is the length of the laser cavity resonator, the axial mode frequencies are $f_m = m\left(\frac{c}{2L}\right)$ where m is an integer (usually very large) and c is the speed of light. The separation between successive modes is the free spectral range of the resonator, and a typical laser transition line is broad enough to accommodate several modes. A single axial mode can be sustained at the expense of others by inserting an etalon of appropriate length in the cavity.

The transverse modes, denoted by TEM_{mn} (transverse electric magnetic; m and n are integers), have characteristic intensity patterns in the plane normal to the beam direction. The lowest order TEM_{00} mode has a Gaussian profile of intensity over the cross section. Some of the other less desirable patterns, sketched below, have multiple spots. The electric field directions are shown by arrows.

cavity resonance   Resonant vibrations in a cavity. Cavity resonance occurs when the frequency of a sound wave incident on a cavity is equal to one of its natural frequencies. In this case, the ratio of the acoustic pressure amplitude inside the cavity to that of the incident wave

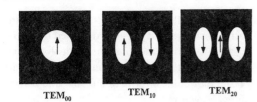

TEM$_{00}$ TEM$_{10}$ TEM$_{20}$

Cavity modes of a laser.

reaches its maximum. Cavity resonance is used in acoustic devices to amplify specific frequencies of a complex sound. A resonant cavity is called a *resonator.*

CCD Charge coupled device, a planar device for holding and moving charge in two dimensions. The charge is managed by potential wells controlled by pixel electrodes over (or under) the area of the well. These charge bins can be multiplexed out to an analog to digital converter or any other application.

CCDs are generally used in optics (like video cameras, telescope readout electronics, or infrared imagers). However, other uses are also possible, such as charged particle radiation tracking detectors.

cell An electric device that converts chemical energy into electric energy. There are many different types of cell depending on the materials or type of chemical reaction employed. A cell consists of a positive and a negative electrode immersed in a chemical solution (electrolyte). Cells are classified into two major types: primary (non-rechargeable) and secondary (rechargeable). In both types of cells, the electric energy released is derived from the chemical reaction that takes place between the electrodes and in the electrolyte.

centered optical system An optical system where all surfaces are rotationally symmetric about a common axis.

centrifugation Separation of molecular or particle species by placement in a rapidly rotating environment. The particles with the greater centripetal force flow to different depths in the gel medium. The centripetal force being equal to $m\omega^2 r$ where ω is the angular speed of the centrifuge, r is the radius from the center of rotation, and m is the mass of the particle.

centrifugation, isopycnic Centrifugation of a mixture when the separation is based on density, not mass difference.

centrifuges Instruments consisting of tube holders distributed in a circular array about a common axis of rotation and capable of achieving very high frequencies of revolution. This gives rise to a strong centripetal force (sometimes referred to by the non-inertial term centrifugal force) yielding separation of mixtures in solution.

ceramic magnets These are ferrimagnets composed of the hard magnetic material $BaO.6Fe_2O_3$.

Cerenkov radiation Radiation emitted by charged particles traveling in a medium at a speed faster than the speed of light in that medium. The wavefront of such a shock wave will be in the shape of a cone in three dimensions with the apex at the source. The half angle α is given by $\alpha = \sin^{-1}\left(\frac{V}{V_s}\right)$ where V and V_s are the speeds of light and that of the charged particle, respectively, in the medium in question. One can observe a blue shimmer of Cerenkov radiation in nuclear reactors with the core immersed in a pool of water. The speed of charged nuclear fragments can easily exceed the speed of light in water which is about two thirds the speed of light in vacuum.

channel A channel represents what separates the transmitter from the receiver in a communication channel. The channel embodies a communication connection for the transfer of information signals from the data source to the data sink. The channel may be unidirectional or bidirectional in how the information may flow. A channel may correspond to one particular carrier frequency in a frequency-division multiplex communication system, a particular time-slot in a time-division multiplex system, or a particular spreading code in a code-division multiplex system. The channel may distort the transmitted signal if the transmitted signal is altered by other than a real-valued constant multiplicative factor and/or a constant time delay, thereby modifying the signal shape.

channel (cell)   A corridor through the lipid bilayer of the plasma membrane often created by macromolecular proteins. This channel allows an imbalance of electrical potential across the membrane resulting in the assisted flow of ions through the plasma membrane.

channel bank   A channel bank embodies that part of a carrier-multiplex terminal that multiplexes a set of channels onto a higher frequency band or demultiplexes a higher frequency band into distinct channels.

channel capacity   The channel capacity of a channel refers to that channel's theoretical maximum information transfer rate, typically denoted as C. If the information rate of the information signal (typically symbolized by R) remains below the channel capacity, then arbitrarily small error probability may be attained by suitable signal coding. Otherwise, transmission error is unavoidable, regardless of any coding. For a bandlimited channel (with a bandwidth of B Hz) affected by additive white Gaussian noise with signal-to-noise ratio at S/N, the channel capacity (C) equals $C = B\log_2(1 + S/N)$ bits per second. Channel capacity rises with increasing bandwidth or improved signal-to-noise ratio. The bandwidth and the signal-to-noise ratio thus become two design variables to be traded off each other for any particular communication system and application. For example, limited power resources onboard a space satellite means wider bandwidths may be used for a lower signal-to-noise requirement. It is false that channel capacity would become unlimited as channel bandwidth grows towards infinity. This is because the wider the bandwidth, the more channel noise there would be while signal power remains constant, thereby decreasing the signal-to-power ratio. If noise is completely absent, S/N equals infinity, and C becomes unlimited regardless of channel bandwidth. If the additive channel noise is other than white Gaussian, then channel capacity may exceed that given by the above expression.

channel capacity and distortion   Channel distortion occurs whenever the transmitted signal is altered other than by a real-valued constant multiplicative factor and/or a constant time delay, thereby modifying the signal shape. The information signal may also be distorted by signal transients during modulation; such distortion, called characteristic distortion, depends in part on the particular information signal and carrier signal concerned.

channel capacity and interference   For a Gaussian interference channel with power constraint, the channel's capacity region in the presence of very strong interference is surprisingly equivalent to the case when there exists no interference.

channel capacity and rate distortion   If the source information rate R exceeds the channel capacity C, then distortion must necessarily occur, regardless of the type of source coding and channel coding used to process the source data stream. The distortion incurred in representing a source alphabet $\mathcal{X}$ by a reproduction alphabet $\hat{\mathcal{X}}$ is measured by the distortion function D of that particular source alphabet-representation alphabet set. The rate distortion region of a source is defined as the closure of the set of achievable rate distortion pairs (R, D). The rate distortion function, $R(D)$, is defined as the infimum of rates R such that (R, D) lies in the rate distortion region of the source for a given distortion D.

channel capacity and self-information   The self-information $I(X = x_n)$ of the event $X = x_n$ (that is, symbol X is valued at x_n) is defined as:

$$I(x_n) = \log_2 \frac{1}{P(x_n)}.$$

The self-information, when averaged over all I gives the entropy $H(X)$ of a source:

$$H(X) = \sum_{n=1}^{N} P(x_i) I(x_n).$$

Both $I(x_n)$ and $H(x_n)$, as defined above, have units of bits per symbol. If the source's symbol rate equals R symbols per second, then $H(X)R \leq C$ or transmission error must occur despite any encoding.

channel capacity, energy per bit   To successfully transmit information from an information source with information rate R through a channel of capacity of C bits per second, bandwidth B Hz and additive Gaussian noise level N_o, the minimum signal power level in energy per bit equals:

$$E_b = \frac{N_o B}{R} \left(2^{\frac{C}{B}} - 1\right) \frac{\int_0^T \{s(t)\}^2 dt}{K} .$$

channel kinetics (cell)   The study of the motion of ions and particles in membrane channels.

channel noise level   The channel noise level equals the noise power in the channel, typically expressed in decibels (dB). The channel noise level may also be defined relative to the noise power at some reference point. Channel noise may arise from thermal noise, intermodulation products, adjacent channel crosstalk, or other unwanted interference. The channel noise power density measures the noise level per hertz of channel bandwidth.

Chapman-Kolmogorov relation   A predictive mathematical relation between system parameters subject to random, or chaotic, noise generation.

character   A character embodies a uniquely defined cluster of consecutive bits representing an alphanumeric mark (such as A, 9), a non-printable control token (such as the *escape* character or the *carriage-return* character). A finite ordered set of distinct characters constitute a character set. Eight-bit bytes are commonly used to represent a character in the American National Standard Code for Information Interchange (ASCII) character set. The character mean entropy of a particular information source, in units of shannon per character, measures the information-content of that source.

character generator   The character generator embodies a device used to control a display writer to display any character graphically on the display device's display surface (such as the display screen of a cathode-ray tube).

characteristic curve   A plot showing the relationship of two variables, which may show useful information about a device. For instance, current vs. voltage for a diode, collector current vs. base voltage for a transistor with a specified emitter voltage, or the capacitance vs. reverse bias voltage for a diode junction.

characteristic function   The characteristic function $C(f)$ of a random variable x with probability distribution $p(x)$ is defined as:

$$C(f) = \int_{-\infty}^{\infty} p(x) e^{j2\pi f x} dx .$$

In other words, the characteristic function $C(f)$ embodies the Fourier transform of $p(x)$. The characteristic function facilitates the determination of the moments (i.e., expected values of powers) of the random variable as follows:

$$E\{x^n\} = (j2\pi)^{-n} \frac{d^n C(f)}{df^n} |_{f=0} .$$

Moreover, the characteristic function $C_{x+y}(f)$ of a sum of two random variables (x and y) equals the product of the two random variables' respective characteristic functions ($C_x(f)$ and $C_y(f)$):

$$C_{x+y}(f) = C_x(f) C_y(f) .$$

charge coupled device   Consists of a two-dimensional array of photo diodes on a substrate. The incident radiation produces electric charges that are stored in a potential well created by the gate voltage. The stored charge in each pixel is proportional to the irradiance which is later scanned and read out along rows of such devices referred to as CCD. The charge is cleared and the device is reset for the next measurement. By a sequential reading of the stored charge a two-dimensional image is constructed. Such CCD arrays are used in detectors in telescopes and spectrographs, television and video cameras. *See also* CCD.

charge density   Charge per unit volume is called charge density. Charge density is an important physical quantity that appears in many physical laws. For example, Gauss law states that the divergence of the electric field is equal to 4π times charge density.

charge, diffusion   When p-type and n-type semiconductors join together to form a diode, some electrons from n-type semiconductors will migrate into p-type semiconductors, while some holes from p-type semiconductors will migrate into n-type semiconductors. These minority charge carriers are called diffusion charges.

charge, electric   A fundamental characteristic of matter. All fundamental particles can be either positively charged, negatively charged, or neutral. Charge is quantized, which means that charges come in multiples of 1.602×10^{-19} Coulomb, the amount equal to the magnitude of the charge of one proton, or one electron.

charge injection device   This is similar to a charge coupled device (CCD) with the exception of the method of processing of the photo-induced charge. The charge is injected to the underlying substrate semiconductor. *See* charge coupled device.

charge transfer device   Any device capable of storing signal charge, and transferring that charge from one capacitor to the next at some clock rate. A CCD is a specific type of charge transfer device, which uses field-produced energy wells without discrete capacitors.

charging by friction   When two different substances rub against each other, one may give up its electrons while the other may gain electrons. Once they are separated, one carries positive charges while the other carries negative charges. This process is called *charging by friction*.

charging/discharging capacitors   When a battery is connected to a capacitor, the positive charge flows into one terminal of the capacitor while the negative charge flows into the other terminal until the potential difference between the terminals reaches the emf of the battery. This is called the *charging of a capacitor*. In this process, energy is converted from chemical energy stored in the battery to the electric energy stored inside the capacitor, $\varepsilon = 1/2QV = 1/2CV^2$. Discharging a capacitor is the reverse process of charging a capacitor. When a resistor is connected to the two terminals of a charged capacitor, a current will flow through the resistor. This process is called the *discharging of a ca-*

pacitor. The energy stored in the capacitor will be dissipated through Joule heating.

chemiluminescence   The luminescence, or emission of light, due to chemical reactions taking place. If the chemical reactions are of an organic nature it is also referred to as *bioluminescence*.

Child-Langmuir Law   The relationship between current in a tube (thermionic diode) and the applied voltage and distance. For the cathode current I limited by space charge

$$I = GV^{\frac{3}{2}}$$

where V is the applied anode voltage and G is termed the perveance. In general the perveance depends on the geometry, and, for instance, may be inversely proportional to the distance between cathode and anode.

chip   A chip may refer to an integrated circuit or to one basic time unit between signaling transitions for one digit such as in a pseudo-random (PN) sequence used in code-division multiple access (CDMA) communications.

chirality   A property of left/right, or mirror, asymmetry of a molecule. The molecule is distinguishable in such a way that it cannot be superimposed on its mirror image. Chirality is a cause of asymmetric optical scattering.

chirp   A short, high pitched sound.

choke   An inductor in some specific application is called a choke. It consists of many turns of wire wound on a support.

cholesteric crystals   These are organic compounds that can flow like liquids yet maintain their molecular orientations. Such liquid crystals have a helical structure and exhibit very large optical rotatory powers. The polarization axis of an incident beam can rotate by as much as 40,000 degrees per mm of the liquid crystal. The pitch of the screw-like molecular structure is much smaller than that in crystals such as quartz.

chopper   A simple switching circuit that opens and closes the primary circuit at some clocking rate. The time open need not be any set fraction of the period.

chromatic resolving power R The term used to denote the limit of resolution between two neighboring wavelengths as observed in a diffraction grating with N slits. Application of the Rayleigh criterion yields

$$R = mN \, ,$$

for the mth order of the spectrum.

chromatography   A percolation procedure where particle, chemical, or pigment separation is achieved by passing the substance through a two phase (usually liquid/solid) system in the earth's gravitational field.

chromatography, exclusion   Chromatography in which the gel phase is a stationary phase of controlled pore size. Hence, molecules are separated as a function of their size and their shape (exclusion).

cineradiography   A motion picture radiograph, or in today's terms a video radiograph of a moving biological organ such as the heart, the lungs, or blood flow through a constriction. Another related process is cinefluorography.

cipher   A cipher represents a cryptographic system transposing or displacing alphanumeric characters in a plain text with other alphanumeric characters in accordance with predetermined code. A cipher may also refer to such a cryptosystem or a message written or transmitted in such a cryptosystem.

ciphony   Ciphony (or cyphony) refers to the use of ciphering for telecommunication signals, typically for confidentiality.

circle of least confusion   When a circular object is imaged by a sphero-cylindrical or cylindrical lens, the circle of least confusion is the circular section found in between the two images forming the Conoid of Sturm. It is dioptrically halfway between the two images and has the smallest cross-section of the images.

circuit   A circuit consists of various electric/electronic components connected together to perform a specific task. It could be as simple as a flash light that consists of batteries, resistor, and a switch, or it could be as complex as a computer motherboard. When a circuit is built monolithically on a single semiconductor wafer, it is called an integrated circuit, IC.

circuit, active   A circuit that contains one or more active elements or devices. These devices do not merely store or dissipate energy, but may introduce energy into the primary circuit. *See* active device.

circuit, astable   A circuit that oscillates between unstable states of the device. Devices that alternate between the states at a regular frequency can be used as a clock for synchronous switching circuitry.

circuit, bipolar   A circuit including a bipolar element or device. *See* transistor, junction; bipolar code.

circuit, bootstrap   An amplifier circuit using resistors to lower the effective input impedance.

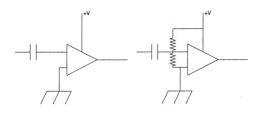

An AC amplifier and a bootstrap version of the same circuit.

circuit breaker   A switch that automatically disconnects the circuit when the current flowthrough becomes larger than a preset value.

circuit, clipping   A circuit that limits the maximum or minimum voltage level of a signal that traverses it.

circuit, logic   A switching circuit in which the quantized states represent logical states. This allows the definition of circuit elements analogous to logical operations (AND, OR, NOT, etc.). These can also be seen as logical ones and zeros for use in any type of binary numbering system.

All computer circuits are logical circuits. An example of an actual (in this case binary) logic circuit type is the standard TTL, for bipolar transistor-transistor logic, where an output about the high voltage represents true or logical one, and no voltage (or below the low cutoff) represents false or logical zero. Additionally, other values may be chosen to represent the logical states, including current and optical intensity.

circuit, master-slave   A segmented circuit in which one portion (the master) provides the time base or other specification for the other segments (slaves). For instance, a master clock may provide timing signals, to which the slave clocks must synchronize, providing the master time base to the slave circuit instead of an independent clock pulse.

circuit, parallel tee   A bridge circuit containing at least five known impedances, with a sixth unknown impedance element from one diagonal to common. Balancing the known impedances allows for the calculation of the unknown value.

So named because the oscillating signal sent through the other diagonal can be seen as travelling through two T-circuits in parallel.

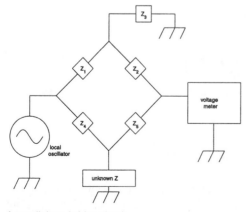

A parallel tee bridge circuit.

circuit, passive   A circuit that contains only passive elements, storing and releasing energy applied to the circuit. *See* passive device, active device.

circuit, switching   Early switching circuits were devices that selected which additional circuits to complete. The limited number of (quantized) selections possible, along with the history of the development of digital circuitry, has allowed an extension of the term to include any circuit that handles quantized (or digital) information or outcomes.

Relay circuits and phone dialing systems are switching circuits, but so are digital computers. Since the success of digital computing, most switching circuits are purely binary.

There are two main types of switching circuits. Combinational, in which the inputs uniquely define the switch/output state, and state-dependent (or sequential), where the internal state(s) or memory of a device may be set, changing the response to inputs.

circuit, synchronous   A logic switching circuit in which all the state changes occur on system clock pulses (at the same time or synchronously).

circular intensity differential scattering Optical spectroscopy using the scattering of circularly polarized light to elucidate structure and function of optically active materials. A nondestructive investigation.

circular polarizer   A device that can produce or analyze circularly polarized light. When unpolarized light passes through a linear polarizer followed by a quarter wave plate (90 degree phase retarder) with their principal axes at 45 degrees from one another, the emerging light is circularly polarized. The handedness (left or right) depends on whether the transmission axis of the polarizer makes +45 or −45 degrees to the fast axis of the retarder.

cisternography   A specific investigative tool using radioactive contrast imaging (roentgenography) for visualization of the subarachnoid spaces containing cerebralspinal fluid.

citizen's band (CB)   The citizen's band (or CB) represents a frequency band in the electromagnetic spectrum allocated by governments in the U.S., Canada, Germany and other countries for short-range voice communications among private individuals. In the U.S., CB occupies from 26.965 to 27.225 MHz and from 460 to 470 MHz. Due to government regulations on CB radio transmitter power and receiver antenna heights, CB radio in the U.S. may typically reach up to 15 miles when used at a mobile or up to 30 miles at a fixed location.

cladding, fiber optic   Fiber optic materials are usually encased in a material of lower refractive index than that of the fiber. This cladding material prevents degradation of light in the fiber by protecting it from surface scratches, dust, moisture, etc. Another important function of the cladding material is to prevent the frustrated total internal reflection from occurring. Under the condition of total internal reflection, a longitudinal electric field exists at the surface of the fiber and the electromagnetic energy, whose amplitude exponentially drops, penetrates across the boundary. It can couple with the external medium leading to leakage and cross-talk of the signal. The maximum acceptance angle, θ_m, of light that can transmit through an optical fiber of index n_1 enclosed in a cladding material of index n_2 is $\theta_m = \sin^{-1}\frac{\sqrt{(n_1^2 - n_2^2)}}{n_0}$. The refractive index of the surrounding medium is n_0.

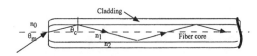

Cladding, fiber optics.

clamping diode   The diode used in a clamp circuit, which is capable of adding a DC offset to an input signal.

coagulation   The transformation of a liquid to a gel via chemical reaction rather than by evaporation. For example the coagulation of blood upon exposure to air.

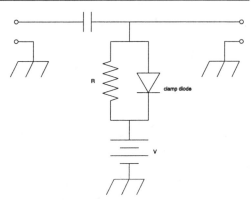

Clamping diode.

coaxial cable   A circular cross-section transmission line made up of a central conductor surrounded by a dielectric layer which in turn is covered by a metallic shield. An insulated coating or outer layer is often used to protect the shield conductor.

cochlea   A section of the inner ear looking like a spiral canal consisting of two and a half turns around the modiolus and containing the organ of Corti.

code   A code represents an unambiguous set of explicit rules to represent information. A code set means the set of all possible code values allowed by the code. A code book embodies a systematic ordering of the set of all codes in a particular coding system and the respective characters they encode. Encoding refers to the conversion of information via such a set of rules into a specific code value. The source encoder maps the original data from one alphabet set to another alphabet set, which may or may not be identical as the first set of alphabets. Decoding refers to the recovery of information given a particular code value.

code division multiple access (CDMA)
Code division multiple access (CDMA) represents a multi-user transmission coding scheme wherein each user is assigned its unique pseudo-random signature spreading code sequence at the transmitter to modulate (i.e., to spread) its information-bearing message bits and at the receiver to demodulate (i.e., to de-spread) the received digits back into the information-bearing message bits. Each user's information-bearing

signal is recovered at the receiver by correlating the received signal with that user's unique signature sequence. All system users simultaneously share the same frequency band but are mutually distinguishable by their respective unique signature sequences. Different users in a CDMA system are typically assigned orthogonal or near-orthogonal signature spreading code sequence, such that the transmitted digits from different users have no or little interference against each other at the receiver after de-spreading. CDMA represents a spread spectrum transmission technique because the bandwidth of the transmitted signal is spread to become much wider than the bandwidth of the original message signal. The spreading may involve amplitude modulation (as in direct sequence CDMA) or frequency modulation (as in frequency hopping CDMA). The aforementioned spreading process disperses signal energy over a very wide frequency spectrum, rendering it very difficult for hostile jammers to detect or to jam the transmitted CDMA signal.

coercive force   (*See* hysteresis.) Coercive force is the value of the magnetic field to be applied to magnetic materials that exhibit hysteresis in order to reduce the intensity of its magnetization to zero.

coercivity   This is a measure of the ease with which materials may be magnetized. Materials with a high coercivity are difficult to magnetize.

coherence   The phase correlation between two distinct parts in time or space of the radiation field of electromagnetic radiation. If the correlation is due to the same field, it is described as *self coherence;* the correlation between two different fields is called *mutual coherence.* This is the property that enables interference between two fields resulting in a pattern of bright and dark fringes. The fringe visibility i.e., contrast between successive bright and dark fringes, is a measure of the *degree of coherence.* The phase correlation between the wave functions of subatomic particles (e.g., neutrons, electrons) can also exhibit coherence and interference.

coherence, acoustic   Correlated in space and/or time behavior of two or more acous-

tic waves (oscillations); correlated behavior of parts of a single wave, taken at different spatial points or at different time moments. Acoustic coherence is similar to optic coherence. Acoustic waves (oscillations) are coherent only if the difference between their phases is changed deterministically, not randomly. Random fluctuations in phases and amplitudes of acoustic waves occur, for example, when they propagate in a turbulent or random medium (in a turbulent atmosphere, ocean, etc.). One of the quantitative measures of coherence in a turbulent or random medium is the coherence function of a sound pressure field p, determined by $\Gamma(\mathbf{R}, \mathbf{R}', t, t') = \langle p(\mathbf{R}, t)\, p^*(\mathbf{R}', t') \rangle$. Here, $\mathbf{R}$ is the Cartesian coordinates, t is time, $\langle\rangle$ denotes ensemble average, and the asterisk denotes complex conjugation. Γ is 1 if $\mathbf{R} = \mathbf{R}'$ and $t = t'$, and decreases if the difference $|\mathbf{R} - \mathbf{R}'|$ (or $|t - t'|$) is increased.

coherence, degree of   The normalized correlation function, γ_{12}, between two radiation fields $E_1(t)$ and $E_2(t + \Delta t)$ as a function of Δt depends on the time variation of the phase difference between the two fields. For two fields with equal amplitudes the interference fringe visibility, V, is

$$V = \frac{I_{\max} - I_{\min}}{I_{\max} + I_{\min}} = |\gamma_{12}| \; .$$

The visibility, V, which is a measure of the degree of coherence, is zero for complete incoherence and one for complete coherence. In general it is less than one for partially coherent radiation fields and decreases as Δt increases. *See* correlation function.

coherence length   The temporal or transverse coherence length is the average length of wave-trains separated by abrupt changes in phase in a quasi monochromatic source. It is related to the *coherence time, τ,* by $\ell = c\tau = \frac{c}{\Delta f}$. c is the speed of light and Δf is the bandwidth of the radiation field. The coherence lengths can range from a few centimeters in gas-discharge lamps to several kilometers in lasers. The spatial or longitudinal coherence length in the lateral plane is a measure of the phase correlation between different parts of an extended source of radiation. Extended sources can be rendered spatially coherent by limiting their wavefront with apertures and masks. The spatial coherence length is (λ/θ) for slit-like sources emitting radiation of wavelength λ and subtending an angle θ. For sources with circular geometry spatial coherence length is $(1.22\,\lambda/\theta)$.

coherence, partial   A radiation field is partially coherent if the normalized correlation function at different points in time or space is between zero and one. The two extremes correspond to complete incoherence and complete coherence, respectively. *See* coherence, degree of and coherence.

coherence time   Quasi monochromatic radiation fields contain harmonic wavetrains of finite duration separated by others by abrupt changes in phase. The average lifetime of the wavetrains is called the coherence time. By Fourier analysis it can be shown that the coherence time is the reciprocal of the frequency spread of the field. *See* coherence; coherence length.

coherent bundles   Fiber optic assemblies in which the fibers are assembled in the same relationship at both ends. The signals entering at one end emerge at the other end with their mutual coherence in tact.

coherent sources   Sources of radiation field with temporal and spatial phase correlation. The temporal coherence is obtained by highly monochromatic fields such as those from lasers and masers. The temporal or transverse coherence length is $\frac{c}{\Delta f}$ where c is the speed of light and Δf is the bandwidth of the quasi monochromatic radiation. The spatial coherence in the lateral plane is obtained by a point source. Extended sources can be rendered spatially coherent by limiting their wavefront with apertures and masks. The spatial coherence length is $\left(\frac{\lambda}{\theta}\right)$ for slit-like sources emitting radiation of wavelength λ and subtending an angle θ. For sources with circular geometry spatial coherence length is $\left(1.22\,\frac{\lambda}{\theta}\right)$.

coil foil   A coil foil is a group of wires, laminated side by side, forming a close-packed flat array of wires. A coil foil is made by winding a wire around a cylinder, diameter d, several times with each turn of wire wrapped tightly against the previous one – making a solenoid with one layer of wires. After the needed number of turns (N) are wound, the coil is glued together using a varnish or epoxy and then cut lengthwise. This produces N wires, πd long, varnished together side by side. Such bundles of wire are often used in any cryogenic apparatus where the wires need to be wrapped around a post for heat sinking purposes.

coincidence circuit   An electric circuit that gives an output only if two input-signals appear simultaneously or within a specific time interval of each other.

coincidence method   The method for extracting information about different wavelengths of light emitted by a source using a Fabry-Perot interferometer. Suppose, in some given direction, maxima of two wavelengths coincide for a given separation of the plates. The plate separation is slowly increased till the two maxima coincide again in the same direction. One counts the number of fringes that pass a given mark in the field of view of the instrument, thereby determining the ratio of the two wavelengths. If one of the wavelengths is known, then the difference from the second wavelength can be calculated.

cold inactivation   A hardening and function ceasing of a biological organism subject to prolonged cold.

cold plate   A cold plate is a metal plate inside a cryostat that is anchored at a well-known temperature. In many cases, the cold plate temperature can be varied in a controlled manner.

cold resistance   A resistance to hardening and/or system shutdown of a biological organism subject to prolonged cold.

cold shock   Shock induced in a biological organism due to prolonged cold; includes constriction of blood vessels, contraction of involuntary muscles, reduced blood flow, and marked pallor.

cold stress   Negative effects on the functioning and efficacy of a biological organism due to experiencing prolonged cold.

collector In a bipolar transistor, the collector is one of the bulk regions on the outside of the bi-junction laminate of the device. In most amplifier applications, the collector is the path for the output signal. *See* transistor, junction; bipolar code; amplifier, bipolar.

collimator Any device that produces parallel rays for an optical system. The emergent beam is then called *collimated*. The simplest collimator is a device with a small aperture or pinhole at the principal focus of a well-corrected convex lens. For a beam of charged particles, a collimator is a heavy metal tube used to restrict the solid angle of the emergent beam. Also, a term used for a small telescope attached to a large telescope to set the line of sight.

collinear transformation The relationship between the object and image within the paraxial region, since for any plane object perpendicular to the optic axis, a plane image perpendicular to the axis is formed. The image can act as an object for the next system. Thus, one speaks of a collinear transformation between the object and the final image formed by an optical system.

color In general, electromagnetic radiation of different wavelengths can be said to have different colors. However, this term has a special meaning for those wavelengths in the 400-700 nm range where human visual perception can distinguish the amount of radiation of different wavelengths reflected by objects. The table shows approximate ranges of wavelengths of electromagnetic radiation and the corresponding names of colors.

Wavelength (nm)	Color
380–430	Violet
430–500	Blue
500–520	Cyan
520–560	Green
560–590	Yellow
590–630	Orange
630–740	Red

The perception of color is often described in terms of three attributes: *hue, saturation* and *brightness*. These three form the coordinates for each color in a three-dimensional *color space*. The attributes of human visual perception that give rise to color names, such as blue, green,

yellow, purple, etc., is called the *hue*. *Saturation* refers to how much a color differs from white. *Brightness* is the perceived intensity of light. Black through different grays to bright are called achromatic colors. They lack hue and are characterized by brightness alone. As an example, the bright sun seen at noon has a yellow hue which is unsaturated whereas the dim red hue at sunset is strongly saturated. *See* colors, primary.

color code A method of representing circuit element parameters by bands of color. Usually, the first two color bands represent the first two significant digits of the value of interest. The third band is an exponent for the tens multiplier, and the fourth is the tolerance. The digits 0 through 9 are represented by the colors:

digit	color
0	black
1	brown
2	red
3	orange
4	yellow
5	green
6	blue
7	violet
8	gray
9	white

In general, the value represented is $(10d_1 + d_2) \times 10^{d_3}$ where d_k is the kth digit. Special colors for the tolerances are gold ($\pm 5\%$) and silver ($\pm 10\%$), and if no fourth band exists $\pm 20\%$. If the third band is gold use (-1) for the digit in the multiplier, and for silver use (-2). The units are natural units for real devices, i.e., ohms for resistors, picofarads for capacitors.

color, complementary Any two colors that produce white when added together are called complementary. The complement of a primary color is called secondary. Red (R) and cyan (C) are complementary; so are green (G) and magenta (M). Blue (B) and yellow (Y) are another pair. *See* colors, primary.

color, desaturated Color saturation refers to the hue referenced to white light. For example, a desaturated hue in white background makes it indistinguishable. Desaturation of colors in rocks and dry wood surfaces comes from the partial specular reflection by tiny facets in contrast to the diffuse reflection from the texture of the surface. For example, moistening a surface darkens it and provides more saturated colors. Glossy surfaces are more desaturated when observed at angles at which they reflect light specularly; otherwise they are more highly saturated than matte surfaces. *See* color.

colorimetry The Commission of Internationale d'Eclairage (CIE) has determined an objective system of colorimetry in which a colored object illuminated by a standard source will have a spectral power distribution (SPD) from which the three color coordinates — luminance, which determines *brightness,* and chromaticity, which determines *hue* and *saturation* — are derived. The field of color specifications in terms of these color matching functions valid for observers with normal color vision is termed colorimetry. *See* color match.

color match A specific color in consideration can be obtained by an additive mixture of three primary colors in appropriate values known as *tristimulus values.* The color-matching functions are the tristimulus values as a function of wavelength. The normalized (i.e., $x+y+z = 1$) tristimulus values are called the trichromatic coefficients or chromaticity coordinates. The color match for a standard observer were determined by CIE, an international organization of colorimetry in 1931 and a supplement was published in 1964. It can be found in handbooks such as the *CRC Handbook of Chemistry and Physics.* The graph below shows the trichromatic coefficients as a function of wavelength. *See* color, colorimetry.

color printing The process involves several steps that can be described as follows: The picture is photographed through a fine mesh screen using a blue filter. The screen is rotated by approximately 30° and is photographed through a red filter. The screen is rotated again and the process repeated with a green filter. Three plates are made by exposing the film to ultraviolet light and etching away the exposed areas. Inks of exact complementary colors of the filters used for

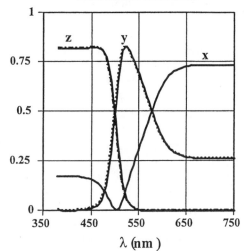

Trichromatic coefficients.

each plate (yellow, cyan and magenta, respectively) are laid down on the unexposed areas. To reconstruct the picture, the three plates are printed on the same sheet. In some cases, a fourth print using black ink is made to enhance the color contrast. The degree of overlap of the dots of the three colors produces different shades of colors. The quality of reproduction is determined by the number of inks used, the degree to which the inks and their corresponding filters are complementary and the number of dots per inch.

colors, primary The additive primaries are red (R), green (G) and blue (B). These colors have very little overlap and a suitable additive mixture of them can produce all other colors including white (W). The *subtractive primaries* are cyan (C), magenta (M) and yellow (Y) which essentially remove red, green and blue colors from white light. One can obtain red, for example, by using yellow and magenta filters in front of a projector producing white light. If the brightness of each color is the same, the following relations hold good: $G + R = Y$; $B + R = M$; $B + G = C$; $B + G + R = W$.

color stimulus Radiant flux of a given spectral composition that produces a sensation of color. *See* color match.

color, surface   It is the common primary color of the light falling on the surface and the color reflected by the surface under illumination of white light. For example, if a surface appears yellow (Y) under white light, and is illuminated by cyan (C), the color of the surface is green (G). Since $Y = G + R$ and $C = B + G$, the common color is green (G).

color temperature   The color temperature of a specimen is the temperature of the blackbody whose spectrum is closest to that of the specimen. For example, the spectral distribution of the sun can be best fit by the blackbody spectrum at 5500K. Hence one can specify the color temperature of the sun as 5500K. *See* blackbody radiation.

coma   A distortion of the image due to oblique and non-paraxial rays incident on a lens or mirror. It derives its name from the comet-like image of a point object located off-axis. Rays striking different parts of the lens or mirror lead to image points with differing magnification. If the magnification for the outer rays is greater than that compared to the central rays, the coma is said to be positive. If the reverse is true the coma is termed negative. For positive coma the tail of the comet spreads away from the axis as shown in the figure below. This defect is corrected by satisfying the *Abbe's sine condition.*

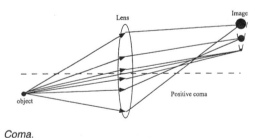

Coma.

combinational logic   Operations that produce the same result or output from a specific array of inputs. Thus the output only depends on the input vector now, not at any time previously, nor on an internal state history of a device. For instance, logic operations like AND and OR are combinational. Flip-flop circuits, where an input changes the state and output of the device, are not combinational devices, neither are computer CPUs or memory arrays.

common   A connection to a common voltage of the circuit. Often called common ground or local ground, the common is not necessarily related to earth ground at all. For instance, the common may be connected to the metal chassis of an instrument, allowing all voltages within the circuit to be defined in reference to the chassis. However, if the chassis itself is not connected to earth ground, this voltage level may take any voltage. Since the difference between circuit elements and common remain unchanged, the voltage to common remains a good local reference for many devices.

communication by balloon   A balloon may be used as a low-altitude satellite to relay telecommunication signals over a wide geographic region. One of the most famous primitive satellites, Echo I, launched on August 12, 1960, was a plastic balloon thinly coated with aluminum. Echo I was a passive experimental satellite for voice and data telecommunications. Echo II, launched in 1964, embodied another metalized Mylar balloon functioning as a passive communication satellite. High-altitude balloons, more easily launched than traditional satellites and may stay aloft for years during their lifetime, can function like very-low-altitude satellites to offer wireless cellular service in a stratospheric telecommunications networks. For example, a communication balloon platform at 70,000 feet can cover about 625 miles in diameter on the ground. There also exist balloon-borne meteorological telemetry instruments to relay data collected onboard the balloon to ground-based communications base stations.

communication processing unit (CPU)   The message control and processing unit of a communication network switching center. CPU also stands for the *central processing unit* of a digital computer, which houses arithmetic and logic processing electronic circuitry that carries out the execution of instructions.

communication satellite   A communication satellite, which may range from a few kilograms to several metric tons, embodies a space-based re-transmission and routing station orbiting the earth to link disparate geographic points on earth. A communication satellite receives a radio communication signal from one earth-based station, amplifies and processes this received signal, and then re-transmits it to another earth-based station. The function played by a communications satellite, thus, resembles that of a ground-based microwave repeater and routing station but covers a much wider geographical expanse interconnecting many more mobile as well as fixed nodes in the communication network.

Communications satellites often orbit the earth near the equator, thereby covering the most densely populated regions. A low earth orbiting satellite, with an orbit about 100 to 300 miles in altitude and about 17,500 miles per hour in speed, can circle the earth in about 90 minutes. A geostationary or synchronous satellite, in contrast, possesses an orbit 22,280 miles (or 35,860 km) high and synchronized with the earth's own rotation, thus allowing the satellite to remain roughly fixed above a particular geographic point. Communication satellites need to continually adjust their orbit and pointing directions to compensate for gravitational and other influences from the sun, the moon, the earth, and other planets. The very high orbital speed of low earth orbit satellite is to avoid having the satellite pulled out of orbit by gravity, but geostationary satellites can nonetheless maintain their orbital position more readily than low earth orbit satellites. Several satellites on complementary orbits may form a group to provide continuous communication service over a wide geographic expanse on earth. The higher altitude of geostationary satellites permit them to cover about a third of the earth at any one time. As few as three synchronous satellites are sufficient to cover the entire earth continually. Thus, many fewer geostationary satellites than low-earth orbit satellites are needed to cover a given geographic region on earth. Geostationary satellites suffer no Doppler effects and are less affected by radiation, eclipse-induced thermal stresses and battery drain, or perturbation by the earth's magnetic fields. However, the geostationary satellite's higher altitude requires greater transmission power and longer transmis-

sion delays — significant disadvantages for real-time two-way voice and data communications. Both low-orbit and geostationary satellites are vulnerable to catastrophic failure, because there essentially exists no economic way to repair them after launch.

Other essential characteristics of a communication satellite, besides the choice of its orbit, are its antenna system, transponders, mechanism for position and orientational control, link budget, and the mechanism that places the satellite into its orbit.

communication satellite, passive   A passive communication satellite has no active transmission power source onboard and thus only reflects back, and does not re-process and re-transmit the signals it receives. Hence, only a very modest proportion of the signal energy transmitted by the ground station or other satellites to the passive communication satellite is forwarded, thereby restricting the communications capacity of the satellite network. While most early communication satellites were passive, essentially all current communication satellites are active satellites.

commutative law   An algebraic statement that the order of two operations does not matter. If $AB = BA$, then the operators A and B are said to commute. Commutation for a single operation means that the order of the operands does not matter. Addition commutes, and the AND and OR operators are commutative. Translations do not commute with rotations, a fact that makes parallel parking possible.

commutator   A device employed in a generator or motor to provide (1) electric connection between the rotating armature winding and the stationary terminal, (2) a mean to reverse the current.

comparator   Comparators, like operational amplifiers, are high gain difference amplifiers. The schematic symbol is shown below. It is a device that ideally provides one of two fixed output levels depending on the relative values of its input levels, V_+ and V_-. Specifically, if $V_+ > V_-$ then the output V_{out} will be some fixed DC value, say V_1. On the other hand, if

$V_+ < V_-$ then $V_{out} = V_2$. Usually, V_1 and V_2 are the positive and negative saturation voltages operating the circuit.

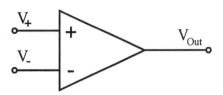

Schematic symbol for a comparator.

When V_- passes through V_+, V_{out} will therefore make an abrupt change. These output voltage characteristics are summarized in the presented transfer function.

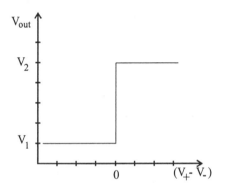

Comparator output V_{out} as a function of the input voltages V_+ and V_-.

Comparators often form the basis of analogue to digital converters. They provide an input voltage of one of two levels to a logic gate; i.e., one of two logic levels. Thus the logic level transmitted depends on if $V_+ < V_-$ or if $V_+ > V_-$.

compass   A device consisting of suspended magnetic material that rotates under the influence of the earth's magnetic field to point due magnetic north or south.

compensation   The shaping of an op-amp frequency response in order to obtain stable operation.

Operational amplifiers (*see* operational amplifier) generally have high frequency limitations. This is due to either frequency limitations of the constituent active components or due to

stray capacitance inherent in the construction of the circuit. Typically, when the input signal frequency is increased, the magnitude of the open loop gain will decrease. Also, the output signal will suffer a phase shift as a function of frequency. If the open loop gain is greater than one when the loop phase shift is 2π radians, unwanted oscillation of the amplifier can occur. This will also depend on the feedback network in the amplifier circuit. By definition, an inverting amplifier already has a loop phase shift of π.

It may not be necessary to have an input signal with high frequencies to cause an uncompensated op–amp to oscillate. If the gain is much greater than 1 where the phase shift is 2π, the oscillation can grow out of noise in the amplifier circuit.

Therefore, it is desirable to make the gain less than 1 when the phase shift is 2π. Then it is guaranteed that the amplifier will not oscillate and it is said to be *unconditionally stable*. This is accomplished by adding non-essential frequency sensitive circuits to provide gain. Some op-amps are internally compensated; they are built in. Others provide connections for external compensation circuits, such as a capacitor. Frequency compensation, however, does reduce the high-frequency gain of the amplifier.

compensator   A plate, usually of variable thickness, to provide additional optical path length for a ray and for production and analysis of elliptically polarized light. In a Babinet compensator, for example, one uses two narrow angle quartz wedges, one fixed and the other movable, with parallel refracting edges and hypotenuses facing each other. The optical axes of the wedges are respectively parallel and perpendicular to the refracting edges, so as to have opposing effects on the E (extraordinary) and O (ordinary) rays passing through them. At any one point, proper sliding of the lower wedge can thus change the phase between the emergent rays. In a Soleil compensator, with the addition of a parallel plate, the effective thickness of the plate is not dependent on the point of incidence.

complementary apertures   Two apertures are said to be complementary if the openings and opaque sections in one are exactly reversed in the other. The figure below shows an example. *See* Babinet's principle.

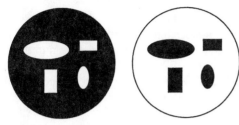

Complementary apertures.

complementary emitter follower   Or a class–B push pull amplifier stage. Uses two complementary transistors, e.g., *npn* and *pnp*. Each transistor, considered separately, is used in an emitter follower scheme. An emitter follower stage has high input resistance and low output resistance, hence it will increase the power level of a signal. In the complementary emitter follower, one transistor conducts for positive values of the input voltage, while the other complementary transistor conducts for negative input voltages. This type of configuration is more efficient than other amplifier schemes and has near zero power loss at zero input signal. However, its disadvantage is that distortion may be greater, particularly crossover distortion.

This transistor connection scheme is also useful in improving the transient response in transistor switches (e.g., logic gates). In single transistor switches, the transient response can be limited by stray capacitance. Generally, the stray capacitance is easily charged through the low impedance of a turned–on transistor, but discharges more slowly when the transistor is off. By adding a second complementary transistor to be turned on when the other is off (and vice versa), any stray capacitance will be discharged more quickly.

complementary error function   The error function of an independent variable z is defined as

$$\text{erf}(z) = \frac{2}{\sqrt{\pi}} \int_0^\infty e^{-z^2}\, dz$$

and represents the integrated probability of a Gaussian probability function with unit standard deviation. The *complementary error function* is defined as

$$\text{cerf}(z) = 1 - \text{erf}(z)$$

and is plotted in the following graph. The complementary error function, under certain boundary conditions, is the solution of the differential equation describing diffusion. Thus it represents the dopant density as a function of distance from the surface in a semiconductor doped by diffusion process.

complex radius of curvature   The radius of curvature of the wavefront of a Gaussian beam is often expressed as a complex quantity in which the real part is related to the radius of curvature R and the imaginary part is related to the beam waist (i.e., beam radius at the narrowest part) w as shown: $\frac{1}{q} = \frac{1}{R} + \sqrt{-1}\frac{\lambda}{\pi w^2}$. Here q is the complex radius of curvature and λ is the wavelength.

compole   This is an auxiliary pole arranged between the main poles of a commutator in order to produce an auxiliary flux to assist in commutation.

compressibility   The property of a substance to change its volume due to the application of pressure. Compressibility is quantitatively defined as $-\frac{1}{V}\frac{dV}{dP}$, where dV is the change in the volume V of a substance due to change dP in the pressure. Compressibility depends on conditions under which the pressure is changed, and may be adiabatic, isothermal, etc.

compression, acoustic   An increase in the density of a medium at a point at which the pressure peak of a progressive sound wave arrives. A half period later, this pressure peak is followed by the peak underpressure so that the density of the medium is rarefacted at this point. Because of compression and rarefaction of a medium at a given point due to passage of sound, sound waves are called compressional waves.

Compton effect   First observed by A.H. Compton in 1922, this effect consists of the scattering of monoenergetic X-ray photons by weakly bound electrons in a metal target. The scattered photon has a reduced energy that depends on the scattering angle. This process can be treated as a collision of a photon with a free electron consistent with the conservation of energy and momentum. The difference in wavelength, $\Delta\lambda$, between the scattered photon and the incident photon is called the Compton shift and is given by $\Delta\lambda = \frac{h}{m_0 c}(1 - \cos\theta)$. Here h is the Planck's constant $(6.626 \times 10^{-34}$ J.s), c, the speed of light $(3 \times 10^8$ m/s), m_0, the rest mass of the electron and θ, the angle between the direction of the scattered photon to that of the incident photon.

computer   A device consisting of hardware for the input, processing, and output of data or information. Input of raw data can be from a variety of sources, by human input via a keyboard, for example, or by remote kinematic sensing as in a robotic application. The processing of data is usually accomplished digitally with a scheme of gates, memory, processing units, etc. The computer also has a list of instructions, the software, which details how the computer is to gather and process the information. Thus, by simply changing the software (instructions), the computer can perform a variety of tasks.

computer, analog   A system that uses analog components such as operational amplifiers to perform specific computational tasks. Simple operations such as addition, subtraction, multiplication, differentiation, and integration are easily accomplished with operational amplifier circuits. These can be combined into a larger, more sophisticated system to solve more difficult problems. As an example, the differential equation

$$\frac{d^2V}{dt^2} + k_1\frac{dV}{dt} + k_2V = V_{app}$$

or

$$\frac{d^2V}{dt^2} = -k_1\frac{dV}{dt} - k_2V + V_{app}$$

can be solved with the use of some basic operational amplifier circuits. The analysis of a block diagram of a circuit to perform this task starts by assuming the signal $\frac{d^2V}{dt^2}$ is available. It is integrated twice with integrating circuits, each with a time constant $RC = 1$, to obtain $\frac{dV}{dt}$ and V as a function of time. These are scaled and added with an applied signal as per the above equation. Since this sum is equal to $\frac{d^2V}{dt^2}$, it is sent back to the first integrator, as initially assumed.

Initial values of $\frac{dV}{dt}$ and V at $t = 0$ can be programmed into the circuit by charging the capacitors of the respective integrators to the appropriate value. The solution of the above differential equation is obtained by recording the output V as a function of time.

computers, use in communication   Computers, defined as programmable electronic devices that perform high-speed mathematical, logical operations and that process and store information, are essential and ubiquitous components of a communication system. Computers are used to acquire, analyze, organize, store, disseminate, and present information.

On the physical level, computers are used as source encoders to condense raw information into an compact source code. They then serve as channel encoders to add in coding redundancy to render the message more robust against channel distortions and noises. Computers also act as modulators to transform these digital channel codes into continuous waveforms for continuous channels. They then further multiplex various modulated signals to share one multiple-access communication system. The reverse actions — channel demultiplexing, signal demodulating, channel decoding, and source decoding are also effected by computers at the receiver. On the network level, computers serve as routing switches to process each transmitted message packet's routing control signals to direct such message packets from network node to network node toward their final network destination. These computer switches may temporally store the message packet if the next network link is momentarily congested.

On the other hand, the various traditional parts of a computer — a keyboard/screen for human/machine interface, computation process and memory devices may be geographically dispersed but linked via a local-area communication network. For example, instead of providing each user on the network one fully implemented computer, individually equipped software library, printers, modem, and storage disks, each user may instead have access to only a minimally equipped work station linked to network servers that have access to centrally administered software libraries, printers, modems, storage disks, or high-power corporate main-frame computers. Different users can communicate with each other and access each other's data and files via this network.

concave grating   A grating on a concave surface (rather than plane). The advantage of such a grating is that the diffracted beams come to a focus without the need for additional focusing optics. It is useful in spectrographs in which the entrance and exit slits are located at the radius of curvature of the grating. The disadvantage is the expense of fabrication.

concentrator   A concentrator embodies a functional unit in a communication system allowing a common path to handle more transmission sources than the number of currently accessible channels within that path. A concentrator can often handle numerous low-speed asynchronous channels with diverse transmission speeds, coding and protocols using contention schemes with buffering, or one or more high-speed synchronous channels.

condensation, counterion   The binding of counterions to the phosphate groups of nucleic acids in solution.

condensation in longitudinal wave   Quantitative measure of change in the density of a medium due to progressive sound wave. The condensation s is mathematically defined as $s = (\varrho - \varrho_0)/\varrho_0$, where ϱ_0 is the *ambient* density of a medium, and ϱ is the instantaneous value of the density that is different from ϱ_0 due to propagation of a sound wave.

condenser (Abbe)   A condenser obeying the *Abbe sine condition.*

conductance   The real part of admittance of a material or a circuit. It is a measure of the ability to conduct electricity of a material or a circuit. In the DC case, it is the reciprocal of resistance.

conductance, acoustic   The real part of the *acoustic admittance;* i.e., if $Y = G + iB$ is the admittance, G is the conductance. The imaginary part of the admittance, B, is called the susceptance.

conductance, membrane   The ability of a membrane to pass flowing charge (electrical current) be it electrons or ions. Membrane conductance is a function of frequency, with DC conductance on the order of 10^{-2} to 10^{-5} siemens. Membrane conductance is often measured by impedance spectroscopy.

conductance, state attribute   Treating the conductivity of a subsystem (biological) as a state variable. That is, conductivity suffices to unambiguously identify the state of the biosystem.

conduction current   Movement of electric charge carriers constitutes a conduction current. The charge carriers can be electrons, protons, ions, or holes. Conduction current in a metallic wire is the result of electrons movement in the conductor. The direction of the current is conventionally defined as the direction of positive charge carriers. The MKS unit of current is ampere.

conduction in metals   In metals, valence electrons of atoms are free to move and are not associated with any particular atom; they are called free electrons. The free electron in the metal can move freely inside the metal, which makes metal a good conductor.

conduction, nerve   Usually by ion flow where the modeling is a series of resistors (impedances) due to the conductivity jumps at the ion channels. This is a very complicated process. An introduction may be found in K.S. Cole, *Membranes, Ions, and Impulses,* Univ. Calif. Press (1972), an older but useful introductory text.

conduction, saltatory   Discontinuous conduction of nerve impulses in myelinated nerve fibers.

conductor, cytoplasm   Cytoplasm is richer in potassium ion concentration and poorer in calcium and sodium ion concentration as compared with the extracellular fluid. Hence the cytoplasm has an "ionic pump" for transporting ions across it. This pump mechanism increases the conductivity of the cytoplasm relative to the basic membrane conductivity. Hence, active transport is important in making cytoplasm a conductor.

cones   Cones are one of two kinds of photoreceptor cells found in the human retina. It derives its name from its microscopic appearance. Cones are responsible for photopic vision. There are three distinct types of cones, each having absorption spectra within peaks around 440 nm, 530 nm and 560 nm. They are called S-cones, M-cones and L-cones, respectively. S-cones ("blue" cones) refer to short wavelength sensitive cones, M-cones ("green" cones) refer to mid-wavelength sensitive cones, and L-cones ("red" cones) refer to long-wavelength sensitive cones. In the human eye, there are approximately 8 million cones.

confocal microscopy   A special kind of microscopy that can image one plane at a time in a thick transparent medium. This is particularly useful in imaging biological specimens with features in different layers requiring contrast in the axial direction as well as in the object plane. Confocal microscopy takes advantage of the property of conjugate points in an optical system. A point source S, the object O, and a pin hole in the image plane I are all made conjugate or confocal to one another (*see figure*) so that only the object is illuminated by the source and its image passes through the pin hole. Another object in an adjacent plane is neither illuminated nor is its image passed through the pin hole. A scanning confocal microscope images one object plane at a time when the specimen is translated on the axis.

confocal resonator   An optical resonator with two identical concave spherical mirrors separated by a distance equal to their radius of curvature. The focal points will be coincident on the axis midway between the mirrors. The surfaces of constant phase match the curvature of the mirrors in the vicinity of mirrors while they are planar at the focal point. The beam waists at

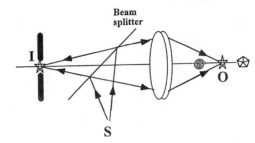

Confocal microscopy.

the position of mirrors and at the center of the resonator are $\sqrt{\frac{\lambda d}{\pi}}$ and $\sqrt{\frac{\lambda d}{2\pi}}$, respectively. Here λ is the wavelength and d is the mirror spacing.

conjugate distances/conjugate planes
When imaging by a lens (or lens system) for every object point, there is an image point. By the principal of reversibility, one can consider the image point to be imaged by the lens to the object point. These two points, object and image, are the conjugate points for the lens. The distance from the lens to these points are called conjugate distances. Planes passing through these points are called conjugate planes. In an ideal optical system, every ray from the object intercepted by the lens also passes through the image.

conjugate points   The point object and the corresponding point image for an ideal optical system. The rays from an object point take equal time to converge to the image point. According to Fermat's principle, the reverse is also true.

conjugate rays   The incident ray in the object space and the corresponding emergent ray in the image space.

conjunction (logic)   A fundamental boolean logic operation over two or more logical variables. Sometimes referred to as **AND**, it is the logical product

$$A \otimes B \equiv A \textbf{ AND } B ,$$

where the variables A and B are elements of the set $\{1, 0\}$ (or, equivalently, the set { **true, false** }). In other words, the conjunction of these two variables is true if *and only if* A and B are true.

The postulated rules of the logical product are

$$0 \otimes A = 0$$

and

$$1 \otimes A = A .$$

Also, for every variable A there exists the product inverse, $NOT\ A \equiv \bar{A}$, such that

$$A \otimes \bar{A} = 0 .$$

The logical product is commutative and associative:

$$A \otimes B = B \otimes A$$

$$A \otimes (B \otimes C) = (A \otimes B) \otimes C .$$

Given these rules, all possible values of the conjunction $Y = A \otimes B$ can be tabulated in a 'truth table' and are presented in the accompanying table.

Conjunction (logical product) of all possible values of the variables A and B

A	B	$Y = A \otimes B$
0	0	0
0	1	0
1	0	0
1	1	1

In implementing this operation in digital electronics, voltage signals are used to represent the logical variables A and B following a predefined logic convention. A circuit, frequently referred to as a *gate,* is used to determine the conjunction, and the result is provided on an output, Y. The symbol representing such an electronic circuit performing an **AND** operation is shown below. *See also* disjunction (logic).

*Symbol representing conjunction (**AND**) in digital electronics.*

conservation of charge   The law of conservation of charge states that in any type of interaction, electric charges cannot be created or destroyed. This is one of the most fundamental laws of physics.

consonance   When two or more notes played simultaneously produce a harmonious and pleasant sound. Consonance occurs if the frequency of tones are in simple ratios, i.e., 2 : 1, 3 : 2, 4 : 3. These frequency ratios are called the octave, the fifth, and the fourth. When two or more tones do not produce a harmonious and pleasant sound, they are said to be in *dissonance.*

constant current source   An electronic component that allows or forces a fixed amount of current to pass.   Ideally, this current source maintains the prescribed current flow regardless of the voltage dropped across it. This can be contrasted with a constant *voltage* source that maintains the voltage dropped independent of the current flowing through the device.   A schematic symbol for a *constant current source* is shown below as well as a realization based on a transistor. This realization assumes the collector-emitter current of a transistor is equal to the base–emitter current times the current gain, independent of the collector-emitter voltage.

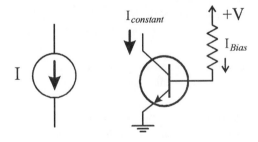

Schematic symbol for a constant current source and a realization using a transistor.

constant deviation   Arrangements, usually with the help of three prisms (e.g., Pellin-Broca prism, Abbe prism), to render the emergent ray at a constant angle of deviation from the incident ray regardless of wavelength. Thus, in contrast to the ordinary arrangement for examining the spectra where the telescope is moved to view the different parts of the spectrum, in the constant deviation configuration, the collimator and telescope are fixed and the prism arrangement is rotated.

constant deviation prisms   Prisms that have a constant angle of minimum deviation (which corresponds to maximum dispersion) for a range of wavelengths of light. They are used in spectrographs in which the incident light direction is at a fixed angle to a photographic plate or a viewing telescope. Different wavelengths are selected by simply rotating the prism about an axis normal to the plane of the beam. The angle of rotation can be calibrated to identify the wavelength. Most common constant deviation prisms are the Pellin-Broca and the Abbe prisms for which the angle of deviation is 90 and 60 degrees, respectively.

constringence   *See* aberration, chromatic.

contact electrification   The transfer of charge from one object to another (one molecule to another) due to physical contact between the two objects (molecules). The natural tendency of like charges is to repel; hence the charge will redistribute itself across two objects (molecules) if in contact with both simultaneously.

contact interaction   Chemical, biological, or dynamic interaction due to physical contact between the interacting species, as opposed to action at a distance (induced interactions).

contin.   Standard abbreviation for the Latin *cintinue'tur* meaning to let be continued.

continuity conditions, acoustic   Conditions governing change in acoustic pressure, displacement, fluid velocity and density across an interface between two media. An example of such media are two homogeneous layers with different values of the sound speed, density and medium velocity, divided by a plane interface. Acoustic pressure, displacement and density are continuous across an interface. The component of acoustic fluid velocity normal to a boundary is continuous across this interface in motionless media, but is in a general case discontinuous if at least one of the two media is moving.

contrast   For a system of fringes with the maxima and minima in irradiance denoted by I_{max} and I_{min}, respectively, the contrast (also

known as *visibility* or *modulation*) is defined as

$$\frac{I_{max} - I_{min}}{I_{max} + I_{min}} \, .$$

convection current   Convection current is produced by the motion of unneutralized change in plasma.

convention (logic)   An assignment of voltage levels used to represent the two logical states **true** and **false** in the implementation of Boolean algebraic computation with electronic signals. In practice, there is usually a range of allowed voltage levels for each state with an unambiguous voltage range separating the two. There are two distinct conventions:

1. Positive logic convention: Voltage levels that represent logical **false** are frequently denoted by L (for low) and are in a range near zero volts,

$$0 \leq V_L \leq V_{L_{max}} \, .$$

Similarly, voltage levels said to represent logical **true** are denoted by H (for high) and are in a range greater than that for V_L,

$$V_H \geq V_{H_{max}} \text{ with } V_{H_{min}} > V_{L_{max}} \, .$$

For example, the most frequently used logic convention in TTL (transistor-transistor logic) circuits, has $0 \leq V_L < 0.8$ volts and $V_H \geq 2$ volts and is sometimes referred to as a *5-volt positive logic system*. These voltage ranges are illustrated in the accompanying line graph.

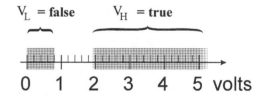

Voltage ranges corresponding to logical **true** and **false** in the positive logic convention employed by most TTL circuits.

2. Negative logic convention: Voltage levels representing **false** are denoted by H and are in the range

$$V_{H\,min} \leq V_H \leq 0 \, .$$

Voltages representing logical **true** are denoted by L. These voltages are in the range

$$V_L \leq V_{L\,\text{max}} \text{ with } V_{L\,\text{max}} < V_{H\,\text{min}} \,.$$

Logic circuits based on ECL (emitter-coupled logic), especially those of early discrete semiconductor circuits, use a negative logic convention. Modern ECL integrated circuits nominally assume $-0.9 \leq V_H \leq 0$ volts and $V_L \leq -1.7$ volts as illustrated in the following line graph.

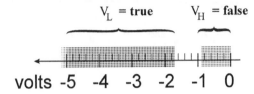

*Voltage ranges assumed for logical **true** and **false** in typical ECL circuits; an example of a negative logic system.*

Logic gates based on one logic convention can be made to operate using the opposite logic convention by redefining 0 volts. However, this inverts the definition of **true** and **false** relative to V_H and V_L and, therefore, the logical operation of the gate must be converted by applying the principle of duality. *See also* duality principle.

convention, sign of current and voltage   The relative sign of voltage difference between any two points in a circuit is determined by the relative electrostatic potential energy felt by a *positive* test charge, e.g., a proton. An applied voltage difference in a circuit creates an electric field. The electric field direction is determined by the direction the positive test charge moves under the influence of this field. As the test charge moves in the direction of the applied electric field, it will lose potential energy. Thus, the positive test charge will tend to move away from the point with greatest electrostatic potential, the "+", and move towards the point with lesser electrostatic potential, labeled "−".

This can be compared with a ball rolling down a hill under the influence of a gravitational field; it moves towards the point with lesser gravitational potential energy. Of course, the direction of movement and relative potential energy difference for a *negative* test charge

will be opposite, but the convention assumes a positive test charge as above.

The sign or direction of current flow in a circuit is determined by the direction that positive test charges would flow under the influence of the applied electric field. In metals, it is well known that negatively charged electrons are the mobile charge carriers and are hence responsible for current flow. However, the defined direction of current flow is still the direction that positive charges will flow.

converging wave   A wave in which amplitude and energy are increased with the distance of propagation. Most sound sources emit *diverging* waves in which amplitude and energy are decreased with the distance of propagation. Sound waves can be converging only under certain circumstances or conditions. For example, converging waves can be produced by using converging lenses and mirrors, can be radiated by specially designed transducers, and can appear in an inhomogeneous medium due to random focusing.

converter   A machine or device for changing alternating current to direct current, or the converse. If a conversion is made from DC to AC, the machine is called an invented converter, with an alternator DC generator combined in one machine having a single-field circuit. Converter losses consist of friction, resistance heat and core losses. The converter is frequently called a *synchronous converter* or *rotary converter.*

converter, analog/digital   Analog/digital converters are used to interface digital equipment such as computers with their binary operating scheme to real-world analog signals as from a sensor. Applications include digital voice communication where analog voice signals are converted to digital information and reconstructed after transmission.

(**1**) A *digital–to–analog converter* (DAC) generally converts a given digital number into a corresponding analog voltage level. This correspondence depends on the binary encoding scheme used (e.g., sign-magnitude, 2's complement, etc), the resolution, and the voltage range. The resolution, or equivalently the smallest effectual output change, of the DAC is determined by the number of bits in the digital number. For example, if the converter can accept 8 bits of digital input data, the smallest change available is $\frac{1}{256}$ of the output voltage range. The voltage range is the minimum and maximum output voltage levels.

For an ideal DAC, the output voltage is linearly dependent on the input digital information. Each increment in digital number δn will yield an equal analog increment δV. The slope of this linear dependence is given by $\Delta = \frac{\delta n}{\delta V}$. In the non–ideal DAC, however, each n may have an error ϵ from the ideal linear dependence. These points are illustrated in the graph.

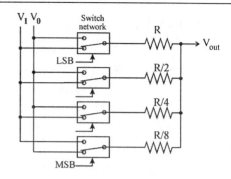

An example of a 4–bit weighted resistor digital–to–analog converter. The digital information chooses between one of two voltages, V_0 and V_1, corresponding to logical **false** and **true**. Details of the switching arrangement are not given.

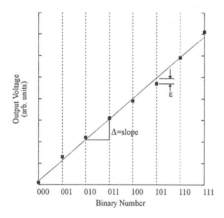

Example output voltages (solid squares) for a 3–bit DAC. The line represents a least squares fit.

Two common types of digital–to–analogue converters are briefly described below. In the description, the digital voltage levels are not applied directly in the conversion scheme; rather, they are used to select between one of two set voltage levels. This is accomplished with, for example, a transistor switching network, the details of which are not important in describing the converter.

The *weighted–resistor* DAC applies the selected voltage levels through resistors whose value is inversely proportional to the numerical significance of the corresponding digit. An example of a converter based on this scheme is shown below.

The resistance effected by the least significant bit (LSB), bit 0, is given by R and the re-

sistance associated with the nth bit is $R/2^n$. As may be noted, this scheme requires a wide range of resistor values.

The *R–2R Ladder* only requires two different resistor values, R and $2R$, but requires twice as many resistors as the previous scheme. The digital information is converted by providing inverse-proportionally weighted current division for each of the corresponding bits. The voltage selected by the LSB has the most significant attenuation while the one controlled by the most significant bit (MSB) has the least.

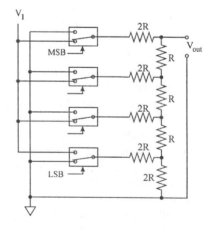

An example of a 4–bit R–2R ladder digital–to–analog converter. Again, digital information chooses between one of two voltages: V_1 and analog ground, in this example.

(2) An *analog–to–digital* converter (ADC) performs the opposite conversion of a DAC, i.e., it converts an analog signal into a digital number. An ADC must be able to sample and hold the input analog voltage long enough to determine its value. It must then quantize the value and represent it using some binary coding format. Quantization errors are introduced because of the limited amount of information that can be conveyed in the binary coding. Hence, a digital value provided by an ADC actually represents a range of analog voltages as determined by the quantization error. The resolution of an ADC is similar to that of a DAC: more digital bits imply more, and smaller, voltage ranges that can be discerned.

The most elementary converter, and the fastest, is the *comparator* ADC. It comprises $2^n - 1$ analog comparators, where n is the number of digital bits available to encode. There is a resistor chain that determines the reference voltage of each comparator. The reference voltages are usually set up such that the quantization error is minimized for any input voltage. The outputs of the comparators drive a system of digital gates that encode the information into a usable binary format.

As an example, the figure below illustrates the concept behind a 2–bit comparator ADC. With only 2 bits of digital information, there are four voltage ranges to be considered by the converter. To minimize the quantization error, the first and last cover a range of $V_0/6$ while the middle two ranges cover $V_0/3$. The relationship between input voltage range and output digital information is given in the table. Extension of this scheme to output more digital bits, and hence better resolution, is straightforward.

The *successive approximation* converter is popular as it is relatively fast and can provide

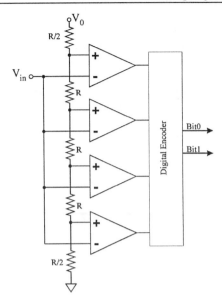

An example 2–bit comparator type ADC.

good resolution with less hardware than the comparator type. It works by successively adding and comparing known voltages, V_i, with the one to be measured. The values of the known V_i's are determined by their corresponding bit significance. For an n–bit converter, the known voltages are $V_0/2$ (associated with the MSB, $V_0/4$, $V_0/8$, ..., and $V_0/2^n$ (associated with the LSB). To convert the input voltage, it is first compared with $V_0/2$. If $V_{in} > V_0/2$ then the MSB is set true, and the next smaller voltage $V_0/4$ is added to the first. Otherwise, the MSB is set to false. The next most significant digit is determined by evaluating $V_{in} > (V_0/2 + V_0/4)$ or $V_{in} > V_0/4$, depending on the result of the MSB evaluation. This process is continued until the LSB is reached. To minimize the quantization error, the input voltage is shifted smaller by one–half the voltage associated with the LSB.

Input voltage ranges for the indicated 2–bit comparator-type ADC and its output information

V_{in}	Binary Encoding	Interpreted Value
$5V_0/6 < V_{in} < V_0$	11	V_0
$3V_0/6 < V_{in} < 5V_0/6$	10	$2V_0/3$
$V_0/6 < V_{in} < 3V_0/6$	01	$V_0/3$
$0 < V_{in} < V_0/6$	00	0

Input voltage ranges for an example 2–bit successive approximation-type ADC and its output information

V_{in}	Binary Encoding	Interpreted Value
$5V_0/8 < V_{in} < 7V_0/8$	11	$3V_0/4$
$3V_0/8 < V_{in} < 5V_0/8$	10	$V_0/2$
$V_0/8 < V_{in} < 3V_0/8$	01	$V_0/4$
$0 < V_{in} < V_0/8$	00	0

Other common types of analog–to–digital converters are the *counting* and *dual slope* converters.

convolution   Also known as *folding* or *superposition*. Mathematically, convolution of a function $f(x)$ with the function $g(x)$ results in another function $h(X)$ such that

$$h(X) = \int f(x)\, g(X - x)\, dx \ .$$

The result can be extended to higher dimensions and several variables.

convolutional code   For every frame of k consecutive information symbols, a *convolutional code* generates a resulting frame of n consecutive symbols (where $n > k$) such that these n encoded symbols' values depend also on the previous $m - 1$ frames of k information symbols. The convolutional encoder thus possesses memory and stores the m most recent frames of information symbols; when the encoder outputs a new frame of n encoded symbols, the oldest frame of k information symbols is discarded and a new frame of k information symbols is input to the encoder. The rate of the convolutional code is defined as k/n and typically ranges from 1/4 to 7/8. The $n - k$ parity bits in each encoded frame provide a convolutional code error detection as well as error correction capability. The Viterbi algorithm and the Fano algorithm represent well-known convolutional decoding algorithms.

convolution theorem   If two functions $f(x)$ and $g(x)$ have their respective Fourier transforms $F(k)$ and $G(k)$, the *convolution theorem* states that the convolution of $f(x)$ with $g(x)$ is equal to the Fourier transform of the product of $F(k)$ and $G(k)$. In some situations it is more convenient to carry out Fourier transforms than the convolution integral. The *convolution theorem* allows circumvention of this problem. One important application of this theorem is in signal smoothing. A spectrum can be smoothed by convoluting it with a smoothing function. Instead the Fourier transform of the spectrum is obtained and multiplied with the Fourier transform of the smoothing function which is band limited. The noise at high frequencies is thus eliminated in the spectrum which is smooth when a reverse transform is performed.

cooling, magnetic   A system of noninteracting magnetic moments, μ, can be used as a coolant in low temperature experiments. Such moments in thermal equilibrium at temperature T and magnetic field B will have an entropy which is a function of $\mu B / k_B T$. The cooling process begins with increasing the magnetic field while keeping the temperature constant. The sample is then thermally isolated, and the magnetic field is decreased isentropically. As a result, the sample temperature will be

$$T_f = \frac{B_f}{B_i}\, T_i \ .$$

In reality, a magnetic moment is effected by the magnetic field produced by the surrounding moments, changing all of the magnetic fields listed above to $\sqrt{B^2 + B_0^2}$ where B_0 is the field produced by the magnetic moments. At temperatures of a few Kelvins, paramagnetic salts can be used as the coolant, but most of these order at temperatures above several milliKelvins. In order to reach even lower temperatures, adiabatic nuclear demagnetization using *nuclear* paramagnets is necessary. *See also* adiabatic nuclear demagnetization.

cooling, nuclear   *See* adiabatic nuclear demagnetization.

cooperativity   In general, subsystems working together for a common effect. In the biological realm, organisms working together for mutual survival. In the chemical/physical realm, molecular movement in tandem rather than in random motion.

Cooper pairs   In a system of fermions with long-range interactions, it is possible for the interactions to become attractive at sufficiently low temperatures. This causes two fermions to form a composite boson, a Cooper pair. The temperature at which this occurs is the superfluid or superconducting transition temperature, T_c. In low temperature superconductors (e.g., Pb, Sn, Al), the pairs form as an s-wave pair, with one up- and one down-spin of equal and opposite momenta. In the high temperature superconductors, the pairing is d-wave, while superfluid ^{3}He is a p-wave superfluid. Some heavy-fermion superconductors show evidence of also being in a p-wave state. In all cases, the Cooper pairs behave collectively and coherently, leading to the superfluid and superconducting properties associated with such materials. In reality, a given electron will not remain paired with another electron for very long, rather the first electron will be paired with first one, then another, and then yet another electron. An analogy often made is with a dancer who, while continuing to dance, changes partners quite often but never misses a beat. *See also* superconductivity; helium-3, superfluid.

CO$_2$ laser   Emits infrared radiation of wavelength 10.6 μm in the continuous mode. The output power can be very large. The laser is used in medical surgery due to the high absorption of 10.6 μm radiation in water to make precise incisions and to vaporize malignant tissue. Industrial applications such as welding and cutting are also common.

cords, vocal   Triangular shaped folds of tissue located in the larynx. Vocal cords consist of two pairs: the tone cords and the false cords. There is no gap between them at the front, and there is a varying gap at the back, which is called the glottis. The voice is produced by air stream from the lungs through larynx and mouth, which is affected by vibrating vocal cords.

core   Generally some ferrite or powdered-iron material placed inside a coil or transformer to increase its inductance.

core, fiber optic   The inner portion of an optical fiber. *See* cladding, fiber optics.

core loss   Loss of energy due to induced circulating currents and hysteresis effects produced in the core.

cornea   A transparent tissue devoid of blood vessels but full of nerve cells through which light enters the eye. It is 12 mm in diameter and 0.6 mm thick at the center with a refractive index of 1.376.

corner cube reflector   Three mutually perpendicular reflectors meeting at the corner of a cube. Such a device reflects incoming rays back along their original direction.

Cornus spiral   This is a useful way of graphically evaluating the Fresnel integrals that appear in diffraction theory. In the integral

$$\int_0^s \exp\left(\frac{i\pi w^2}{2}\right) dw = C(s) + i S(s),$$

a graph of $C(s)$ vs. $S(s)$ is called the *Cornu spiral*. A straight line segment drawn from two points corresponding to s_1 and s_2 gives the value of the integral between those two limits. The real and imaginary parts are given by the projections on the $C(s)$ and $S(s)$ axes, respectively.

corona discharge   The discharge of electricity causing a faint glow adjacent to the surface of an electric conductor.

correlation function   The time correlation function between two radiation fields $E_1(t)$ and $E_2(t + \tau)$ is defined as the time average over a period $T (> \tau)$,

$$\Gamma_{12}(\tau) = \frac{1}{T} \int_0^T E_1(t) E_2^{\#}(t + \tau) \, dt .$$

The *normalized correlation* function, γ_{12}, is obtained by dividing the irradiances of the two fields from $\Gamma_{12}(\tau)$.

correspondence principle   A principle enunciated by Niels Bohr in 1923, which states that the predictions of quantum theory for the behavior of any physical system must agree with those of the classical physics in the limit the quantum numbers specifying the system become very large. This principle reconciles the apparent differences in the behavior of microscopic (described by quantum mechanics) vs. macroscopic (described by classical mechanics) phenomena. Another situation where this principle is applied is to reconcile the predictions of the relativistic (Einsteinian) mechanics which should agree with those of non-relativistic (Newtonian) mechanics in the limit the speed of light is very large compared to the speeds of particles in the system.

cosine law of emission   Suppose the radiant intensity (unit: W/sr) from a plane-diffuse radiator is viewed at a fixed distance from it, the intensity at an angle θ from the normal to the surface, $I(\theta)$, decreases with θ according to the Lambert's cosine law: $I(\theta) = I(0)\cos\theta$. The radiance or the intensity per unit solid angle per unit of projected area (unit: W/sr-m^2), however, is constant with θ. A surface with a constant radiance independent of the viewing angle is called a *Lambertian surface.*

cosine law of illumination   Suppose a surface of area A_2 is illuminated by a diffuse radiating surface of area A_1, and the line joining one element of the radiator to one of the receiver (at a distance r) makes angles θ_1 and θ_2 with the normals to the radiating and receiving elements, respectively: the total radiant power Φ (unit: Watts) received by A_2 is

$$\Phi = \int\limits_{A_1}\int\limits_{A_2} \frac{L\,dA_1 dA_2 \cos\theta_1 \cos\theta_2}{r^2}.$$

Here L is the radiance (unit: W/m^2-sr) of the source.

cotransporter   A substance that is actively transported across a plasma membrane and that brings another substance across the membrane with itself in the same direction.

Cotton-Mouton effect   (1) Magnetic field induced double refraction in optically isotropic substances. This is a magnetic analog of the Kerr electro-optic effect. It is observed in liquids and is proportional to the square of the applied magnetic field.

(2) This effect occurs when a dielectric becomes doubly-refracting when in a magnetic field H. The ordinary ray becomes retarded relative to the extraordinary ray by an amount δ given by $\delta = C_m \lambda l H^2$ where λ is the wavelength of the light, l is the length of the path and C_m is the Cotton-Mouton constant.

coulomb   Unit of electric charges in mks. One coulomb is defined as the amount of charges transported in one second through a wire carrying a current of one ampere.

coulomb field/force   The electric field (or electric force on a test charge) produced by a point charge q. Coulomb field is proportional to the inverse of distance square, $E = kq/r^2$.

Coulomb's law   This is an empirical law discovered by Charles Coulomb. It states that the force between two charges Q_1 and Q_2 separated by a distance d is proportional to the product of the two charges and inversely proportional to the distance square. In mks, the proportional constant in vacuum is given by $k_o = 8.987551 \times 10^9$ N m^2/C^2.

counter   A basic digital counter counts an input strobe or *clock* signal. It is assumed that the clock signal makes regular and periodic transitions from logical **low** to **high** and back. The counter counts clock transitions and provides an output, in some format, representing the count. Counters are used in diverse applications; computers, industrial applications such as counting nuts and bolts, and measuring speed are a few examples.

The fundamental components of a counter are set of J–K flip–flops (FF). There are two basic topologies for connecting the flip–flops, and hence ways of encoding the output.

1. A *ring counter* is a ring of J–K flip–flops connected in a manner similar to a shift register, except that the output of the last FF, Q_n, is returned to the input of the first, J_1.

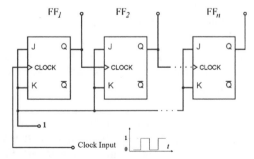

A simple ring counter using n flip–flops (FFs).

Initially, all outputs of the FFs are set to logical **0** except the first, which is loaded with a logical **1**. Upon each clock transition, the **1** is shifted to the next FF. Hence, the count is represented by determining which FF is set to **0**. The maximum count with n FFs in a ring counter is n. For example, consider the output of a ring counter with 4 flip–flops as a function of the number of pulses received:

Clock	J–K Output			
pulse	1	2	3	4
0	1	0	0	0
1	0	1	0	0
2	0	0	1	0
3	0	0	0	1
4	1	0	0	0

A *twisted ring, Moebious,* or *Johnson* counter is similar, except that the complement of the last FFs output, $\overline{Q}_n$, is returned to the first FF. Initially, all FFs are set to **0**. Upon the first clock transition, a **1** will be loaded into the first FF due to the "twisted" ring. The next n clock pulses continually load a **1** into the first FF while shifting the 1s across the ring until all FFs are set to **1**. The next clock will load a $\overline{1} = 0$ into the first FF and the next n pulses will successively load 0s into each FF. Thus, this counter can count to $2n$ with n flip-flops. Again, consider the output of a counter with 4 flip–flops as a function of the number of clock pulses:

Clock	J–K Output			
pulse	1	2	3	4
0	0	0	0	0
1	1	0	0	0
2	1	1	0	0
3	1	1	1	0
4	1	1	1	1
5	0	1	1	1
6	0	0	1	1
7	0	0	0	1
8	0	0	0	0
9	1	0	0	0

2. The second topology encodes the count modulo 2^n, since each FF can have two states. Hence, each FF represents a binary order of magnitude, and the counter can count to 2^n. An *asynchronous* counter uses the output of each FF, Q_i, for the clock input for the next higher significant bit. This type of counter suffers from

propagation delays, i.e., an input clock transition must "ripple" through each flip–flop. A *synchronous* counter, on the other hand, has the input clock signal going to each FF as in the ring counter/shift register. Proper counting is controlled by logic gates between each stage. Thus, each flip-flop changes states simultaneously and the propagation delay of the counter is minimized. *See* counter, asynchronous.

counter, asynchronous   A basic counter using a set of J–K flip–flops. Each J–K output is used to represent a binary order of magnitude. The clocking signal is sent to the first flip–flop representing the least significant bit (LSB). This output is used as a clocking signal for the next higher significant bit; *see figure.* Thus, the first flip–flop's output is toggling between **1** and **0** in response to the clock, causing the next flip–flop to toggle. Each flip–flop makes a toggle only when the preceding flip–flop has made a **1** to **0** transition. Each J and K input are tied to logical **1** and hence have no effect on the counting. This type of counter arrangement is relatively simple, however it can suffer from propagation delays causing decoding errors as the LSB will complete its change before the last flip–flop.

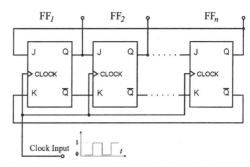

An asynchronous counter using n flip–flops (FFs).

counterions   Counterions are ions of opposite charge to that of the colloidal particles in suspension.

countertransporter   A substance that moves across a biological membrane in the opposite direction to active transport.

coupled circuit   A coupled circuit is a circuit that consists of several subunits that are connected through a capacitor or an inductor. Typically a coupled circuit is designed such that only the AC components of the signal are transferred.

coupling, intercellular   A pairing, or joining, of one or more cells in a biological system.

creep   *See* film, Rollin.

critical angle   When light passes from a medium of high index (n_1) to low index (n_2), it will bend away from the surface normal. The angle of refraction is given by Snell's Law. At the critical angle of incidence θ_c, the angle of refraction will be 90°. The critical angle is given by:

$$\theta_c = \sin^{-1}\left(\frac{n_2}{n_1}\right).$$

For angles of incidence greater than θ_c, the ray will experience total internal reflection, i.e., it will be reflected back into the medium.

critical damping   *See* damping factor.

critical magnetic field   This is the field below which a superconducting material is superconducting and above which the material is normal at a specified temperature and in the absence of a current.

cross talk   (**1**) Leakage of light from one optical fiber to another by frustrated total internal reflection. This can be caused by inadequate or defective cladding material. *See* cladding, fiber optic.

(**2**) Crosstalk in a multi-channel communication system refers to the undesirable spill-over of the transmitted signal in one communication channel onto another channel. Crosstalk may be caused by the nearness of the transmission media, such as electromagnetic mutual coupling between different pairs of twisted copper wires in a telephone cable. Crosstalk may arise from intermodulation distortion in frequency-division multiplexed systems. Crosstalk may also occur as inter-symbol interference for time-division multiplexed systems due to channel non-linearities such as that in frequency-selective mobile communications channels.

Crosstalk in these various cases may be reduced by, for example, better insulation among the twisted pairs, a wider frequency guard-band between adjacent frequency-division channels, and pulse-shaping, respectively. *Crosstalk resistance* refers to the capability of a multiplexed communication system to forestall crosstalk. Although various channels are separated from each other in frequency, time and/or space, transmission efficiency and systems economics are at times maximized at the expense of crosstalk resistance. In an optical transmission system, crosstalk refers to the leakage of optical power from one optical conductor to another optical conductor.

crown glass   Optical crown glass is a low index, commercial quality glass. It is designated as B270 and has a dispersion $n_F - n_C$, equal to 0.0089. The Abbe factor (constringence) is equal to 58.8.

cryobiology   The study of the effects of low temperature on biological systems.

cryogen, biological and medical uses of
Cryogens and low temperature technology have found their way into hospitals and doctors' offices in several areas. Specifically, liquid nitrogen is now used to remove small growths such as moles. Both liquid helium and liquid nitrogen are used in magnetic resonance imaging (MRI) machines. A relatively new field, magnetic encephalography (MEG, the magnetic analog to EEG) uses superconducting quantum interference devices (SQUIDS) to measure processes in the brain. These SQUIDS need liquid nitrogen to work. In most hospitals, the oxygen used for patients throughout the hospital is now shipped to the hospital as liquid oxygen where it is boiled off as necessary before use. In biological research laboratories, liquid nitrogen is often used to preserve cell cultures for future study and use.

cryogenics   The science of producing and measuring extremely low temperatures, usually temperatures below roughly 100 K. Low temperature physics utilizes cryogenic techniques to study the properties of matter at reduced temperatures.

cryogens   Liquids used in cryogenic applications, also known as liquefied gases. Typical cryogens include nitrogen, argon, hydrogen, helium. The cryogens used most frequently are liquid nitrogen, with a liquefaction point of 77 K, and liquid helium, with a liquefaction point of 4.2 K.

cryopumping   Cryopumps use physi-adsorption to decrease the pressure in a vessel. The physi-adsorption is accomplished through the use of a material with large specific surface area, e.g., activated charcoal, contained in a closed container. This closed container is often attached to a dipstick to allow easy immersion in a storage dewar of liquid helium. The container is connected to the region of interest via a tube, and the cryopump is then immersed in liquid helium. When the cryopump becomes very cold, gases adsorb onto the surface of the charcoal, thereby decreasing the pressure in the system. One gram of charcoal can adsorb half a liter of helium gas (STP) when the charcoal is at 4.2 K, producing a final pressure of 10^{-4} to 10^{-5} mbar.

cryostat   A cryostat is a device that allows a region to be maintained at low temperatures for extended periods. In its simplest form, a cryostat is often a dewar containing a cryogen in which an experimental sample is immersed. More complicated cryostats may include multiple cryogens separated by vacuum spaces (accurately described as "dewars-within-dewars") to allow operation at very low temperatures for extended periods of time. *See also* cryogens, dewar.

cryotron   A three-terminal electronic device in which the control element is controlled through a magnetic field. The control element is a superconductor, and the magnetic field limits the current through the control element.

crystal   A particular form of solid, characterized by periodicity of building blocks called *cells* in all three directions. Many crystals show fascinating optical properties. Partial crystals can exist in one or two dimensions.

crystal, biaxial   Triclinic, monoclinic and orthorhombic crystalline systems possess two different optic axes and are called biaxial crystals. These crystals, such as crypsum, feldspar, mica, or topaz have three distinct indices of refraction.

crystallography   The study of the architecture of atoms in a crystal constitutes *crystallography*. Scattering and diffraction studies using X-rays or neutrons is a technique to determine the structure and dynamics of crystals. *For details of this extensive field, see* "Crystal Structures" by R.W.G. Wyckoff, Vol. 1-5, New York, John Wiley & Sons. *See also* Bragg's law.

crystal, optically negative   In crystals showing double refraction or birefringence, an incident unpolarized ray of light causes the emergence of two refracted beams in addition to a reflected beam. One of the refracted rays does not follow Snell's law of refraction and hence is called an E (for extraordinary) ray in contrast to the other which is called the O (for ordinary) ray. In an optically negative crystal, e.g., calcite, the index of refraction of the E ray is less than that for the O ray. In biaxial optically negative crystals (e.g., mica, aragonite), β is nearer to γ than to α. (For a biaxial crystal, α, β and γ are the principal refractive indices such that $\alpha < \beta < \gamma$.)

The axes in a birefringent crystal in which the ordinary and extraordinary rays of light propagate with the same velocity. In other directions, the light passing through the crystal is divided into two polarized rays that pass with different velocities. The difference of the speed increases from zero along the optic axis to a maximum for propagation normal to the optical axis. An uniaxial crystal and a biaxial crystal have one and two optical axis, respectively. *See also* optical axial plane.

crystal, optically positive   If the index of refraction for the E ray is greater than that for the O ray, the uniaxial crystal (e.g., quartz) is called an optically positive crystal. In biaxial optically positive crystals (e.g., topaz, turquoise), β is nearer to α than to γ. *See* crystal, optically negative.

crystals, quartz piezoelectric   Quartz crystals that exhibit the piezoelectric effect. At the end of the last century, quartz crystals were used for detailed study of piezoelectric effect. Piezoelectric properties also have many other natural and artificially grown crystals and piezoceramik materials (polycrystal solid solutions exposed to polarization in electric field). Piezoelectric crystals and materials are widely used in piezoelectric transducers such as loudspeakers, microphones, etc. *See* piezoelectric effect.

crystal symmetry An ideal crystal contains an infinite regular repetition of identical structural units. A periodic lattice with a group of atoms (called the basis) situated at each lattice point constitutes a crystal. The crystal structure remains unchanged under certain symmetry operations of translation, rotation and reflection (or a combination of these) due to the property of crystal symmetry. For example, a crystal with a cubic symmetry remains indistinguishable if it is translated by one or integer multiples of the lattice spacing. *See* crystallography.

Curie law For paramagnetic materials the magnetization M produced by a magnetizing field H is proportional to H and inversely proportional to the temperature of the material. The constant of proportionality is the *Curie constant.*

Curie temperature This is the temperature above which the arrangement of electron spins in a ferromagnetic material becomes randomized by thermal agitation. At that point, the spontaneous magnetization of the material vanishes.

Curie Weiss law This gives the relationship between the paramagnetic susceptibility χ of a material well above the Curie temperature T_c at which a ferromagnetic material becomes paramagnetic, i.e., $\chi = C/(T - T_c)$, where T is the absolute temperature and C is the Curie constant of the material.

current, acoustic Mean flow in a fluid or gas caused by absorption of intense sound. Propagation of a sound wave in a medium always results in acoustic displacements of medium particles. These displacements can be converted into a mean flow in the medium if the *absorption* of sound in the medium and the *intensity* of sound are sufficiently high. The acoustic cur-

rents can occur near walls and away from them, and always have a form of eddies.

current balance This is a type of balance in which the force required to prevent the movement of one current-carrying coil in the magnetic field of a second coil carrying the same current is measured by means of a balancing mass.

current density A physical quantity that measures the amount of charge that passes through a unit area per unit time, $J = Q/(At)$.

current generator A type of generator that generates a constant current through a load and is independent of the resistance of the load.

current, membrane A flow of ions and polar molecules across the membrane. Movement of ions and polar molecules may be facilitated by active transport or restricted by proteins embedded in the membrane.

current source *See* constant current source.

curvature of a surface The reciprocal of the radius of a circle that most nearly approximates the section of a surface. It is one of the geometric factors determining the reflective and refractive properties of a surface. Only for a spherical surface, the curvature (measured in diopters) is the same in all meridians and is equal to the inverse of the radius of curvature (measured in meters), the sign depending on the sign convention used.

curvature of field An aberration caused by off-axis rays leading the image plane to be curved rather than flat. This defect is undesirable in cameras, enlargers and projectors. If the image is obtained on a flat surface, the central region will be in sharp focus and blurred at the edges. In a two-lens system, correction is obtained by meeting the Petzval condition $n_1 f_1 + n_2 f_2 = 0$, where n_1, n_2 are the indices of refraction and f_1 and f_2 are the focal lengths of the lenses.

cutin voltage For diodes this is the approximate voltage for which it begins to conduct. Current will always flow through an ideal diode for any applied voltage. However, given the exponential current dependence on the voltage and practicalities in dealing with real, physical devices, currents of the order of 1mA or 1% of maximum rated current may be considered as a turn-on current. The corresponding voltage for a silicon diode is $\approx 0.6V$ and $\approx 0.2V$ for a germanium diode.

cut-off region   Region in frequency where an amplifier's gain has fallen by 3 dB. *See also* half-power frequency.

c-w wave   Also known as continuous wave. Lasers that produce a steady output with time as opposed to a pulse or series of pulses of short duration.

cybernetics   Cybernetics represents the theoretical study of control and communication within large-scale complex engineering systems, human organizations, or human society as a whole. Cybernetics also study the control and communication between engineering devices and humans, and further studies the similarity between human-made machines and biological organisms. Man-made engineering systems studied in cybernetics may be electronic or mechanical, and human systems investigated in cybernetics may relate to corporate organization and management, education, public health care, urban development, socio-economics, national policies, the environment, or human ecology. Cybernetics also investigates the behavior of a complex system involving as its pivotal component a human being monitoring and responding to the dynamic behavior of the rest of the complex system. Cybernetics models, simulates, optimizes, tests, and evaluates various electronic or mechanical assemblages or human organizational systems using techniques in traditional numerical analysis, automatic control theory, artificial and computational intelligence, pattern recognition, adaptive and learning systems, cognitive sciences, game theory, data fusion, neural networks and fuzzy logic.

cycle, acoustic   One complete oscillation (vibration). Suppose that a one-dimensional oscillation is mathematically described as $\xi = \xi_0 \sin(2\pi t/T - 2\pi x/\lambda)$. Here, ξ is an oscillating quantity, ξ_0 is its *amplitude, T* and λ are a period and a wavelength of oscillations, t is time, and x is a spatial coordinate. Then, at a given point $x = x_0$, one time cycle of oscillations is a dependence of ξ on t from any time moment t_0 to the time moment $t_0 + T$. Similarly, for a fixed time moment $t = t_0$, one spatial cycle of oscillations is a dependence of ξ on x from any point x_0 to the point $x_0 + \lambda$.

cyclotron   A particle accelerator in which positively charged particles are accelerated in D shaped magnets (dees); the energy is supplied by a high frequency voltage applied across the dees. When the radius of the paths of the particles reaches that of the dees, they are electrically deflected out of the dees to form a high energy beam for use in nuclear experiments.

Czerny-Turner mount   An arrangement of diffraction grating and other optical elements in a spectrometer. In this system, the light from the entrance slit is collimated by a concave mirror, dispersed by the grating and focused on to the exit slit by another concave mirror.

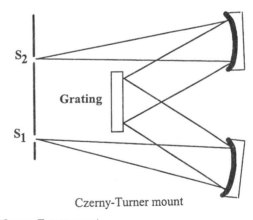

Czerny-Turner mount

Czerny-Turner mount.

D

damped oscillations Oscillations in which the amplitude decreases over the course of time t. The decrease is caused by a loss of energy in the oscillations due to friction and/or other mechanisms. The simplest model of damped oscillations is one-dimensional damped harmonic oscillations, which are described by the following second order differential equation:

$$M\frac{dx^2}{dt^2} + R\frac{dx}{dt} + Sx = 0 .$$

Here, x is a coordinate of a physical quantity under oscillation, M is the mass of the quantity, R is the *damping coefficient,* and S is the stiffness coefficient. For $R < 2\sqrt{SM}$, this equation has the following solution

$$x = Ae^{-\frac{Rt}{2M}} \sin\left(\sqrt{S/M - (R/2M)^2}t + \phi\right) ,$$

where A and ϕ are arbitrary constants. This formula describes damped harmonic oscillations in which the amplitude exponentially decreases over the course of time t.

damping coefficient A coefficient R of velocity term in the equation for *damped* harmonic *oscillations.* The damping coefficient determines a rate at which the amplitude of oscillations is decreased over the course of time.

damping factor A dimensionless index used in second order systems. This index describes the system's tendency to oscillate. Examples of mechanical second order systems are driven vibrating strings and pendulum with a means of mechanical energy loss, i.e., friction. The damping factor would influence the response of the system to the driving frequency, in contrast to the natural, undamped frequency.

This parameter is used in the analysis of amplifier and filter circuits with a double-pole transfer function. It thus governs the overall frequency response of the circuit. Considering a general filter with an input (driven) voltage, the output voltage (response) as a function of frequency for a low pass circuit is given by

$$\frac{V_{\text{out}}}{V_{\text{in}}}(dB) = -20\log\sqrt{(1 - \omega^2)^2 + 4\delta^2\omega^2} ,$$

where ω is the ratio of the input frequency (ω_{driven}) to the natural, undamped frequency (ω_n) and δ is the damping factor. As illustrated in the graph, δ controls the overall shape of the system's frequency response.

For $\delta = 1/\sqrt{2}$, the response is maximally flat. For values of $\delta < \sqrt{2}$, the response of the system is peaked near the natural frequency. The smaller δ, the closer the peak response will be to the natural frequency. In filter design, practical values are between $\delta = 0$ and 2. A filter–amplifier with zero damping will thus tend to oscillate at its natural frequency.

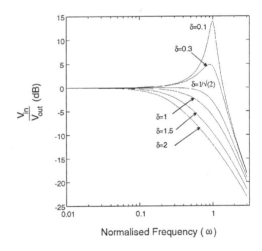

Low pass frequency response of a second–order filter for various values of the damping factor, δ.

The damping factor also determines the system's transient response to an instantaneous step input. In solving the transfer function of a second order system and thus determining the output due to a step input, there arise three special cases for the value of δ.

1. *Underdamped* $\delta < 1$: The output overshoots the step response and tends to oscillate with an exponentially decaying sinusoidal wave, i.e., ringing. The output response is proportional

to

$$\frac{V_{out}}{V_{in}} \propto 1 - \left[\frac{\delta}{\sqrt{1-\delta^2}} \sin\left(\sqrt{1-\delta^2}\omega_n t\right) + \cos\left(\sqrt{1-\delta^2}\omega_n t\right) \right] e^{-\delta\omega_n t}.$$

2. **Overdamped** $\delta > 1$: The output slowly approaches the final output value. The response function is given by

$$\frac{V_{out}}{V_{in}} \propto \frac{1}{2\sqrt{\delta^2-1}} \left[\frac{1}{d_1}e^{-d_1\omega_n t} - \frac{1}{d_2}e^{-d_2\omega_n t} \right]$$

with $d_1 = \delta - \sqrt{\delta^2-1}$ and $d_2 = \delta + \sqrt{\delta^2-1}$.

3. **Critically damped** $\delta = 1$: The output also approaches the final output asymptotically. This, however, is the fastest response without oscillation.

Usually, a damping factor of $\approx \frac{1}{\sqrt{2}}$ is acceptable as a best compromise between ringing and slow response.

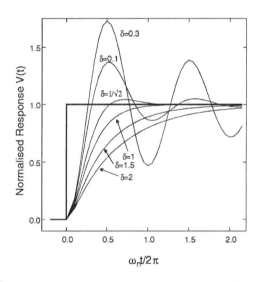

Temporal response of a second–order system to an instantaneous step input for various values of the damping factor, δ.

Daniell cell   A type of cell that was originally invented by J.F. Daniell in 1836. It consists of a zinc anode and a copper cathode immersed in sulfate electrolyte containing zinc ions and copper ions separated by porous wall. The chemical reactions during discharging include the release of electrons at the Zn electrode and the recombination of electrons and copper ions at the Cu electrode.

Darcy's law   In percolation theory, this law yields a model of gravitational flow of a liquid through a permeable membrane. Let J be the rate of flow of the liquid through the membrane in cubic meters per second, G be the amount of hydraulic pressure lost per flow distance (hydraulic gradient), and A be the cross sectional area of membrane through which the liquid passes. Then Darcy's law states: $J = kGA$ where k is the proportionality constant.

Darlington connection   A useful connection scheme for a direct coupled transistor amplifier. Shown in the figure are two transistors connected in the common–emitter (CE) form of the Darlington compound connection. The main advantages of this connection scheme are increased input impedance and improved power handling capacity as compared to a single transistor. The overall current gain is approximately the product of the CE current gain, h_{fe} (*see* h-parameters), of each individual transistor. Transistor T_1 is assumed operated as a common collector (CC) stage, thus it has a relatively high input resistance. The base current of T_2 is supplied via T_1 emitter current. So T_2 carries most of the current and is usually a higher power transistor.

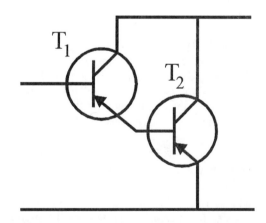

Two transistors connected in the common–emitter Darlington connection.

DC bias   A constant voltage or current that is superimposed with an AC signal. Vacuum tubes generally require a bias voltage added to the signal applied to the control grid. Transistors, on the other hand, generally require a current bias applied to the base in addition to the signal to be amplified. This is done, in both cases, to ensure that the device operates in the intended area of its characteristic curve.

DC feedback   The constant, or average, direct current component of the return signal in a feedback control system. *See also* feedback.

DC source   Provides voltage to power electronic circuits, e.g., V_{CC} or V_{EE} voltages for operational amplifiers. The voltage source maintains a prescribed DC voltage level, independent of current loading. The *compliance* is the current range for which the source can maintain the desired voltage level.

DC voltage regulator   A device that conditions a poorly specified input voltage source and provides a stable, well-specified DC voltage. The regulator's output voltage should be independent of such things as load current, temperature changes, or any temporal behavior of the input voltage source. Sometimes, protection circuitry is incorporated in the regulator to prevent overload conditions.

deafness   Partly or completely impaired hearing. There are three main types of deafness: cortial, nerve, and conductive deafness. Cortial deafness is the inability of a human brain to appropriately "understand" nerve signals from the inner *ear* even if these signals are unimpaired. Cortial deafness is typical mainly of elderly people. Nerve deafness is due to impairment of the cochlea located in the inner ear. Conductive deafness is caused by defects of sound transmission from the outer ear to the cochlea. Deafness is studied by *audiometry*. Hearing aids are commonly used devices to compensate deafness.

Debeye-Huckel constant   Peter Debeye and E. Huckel described the behavior of strong electrolytes in dilute solutions yielding the distribution of and the interaction forces between ions and counterions [Debeye, P. and Huckel, E.Z.,

Phys. vol. 24 (1923)]. For low potentials, the decay of the potential away from a membrane surface follows an exponential law in which the decay length (*Debeye-Huckel constant*) is proportional to the charge density of the counterions and inversely proportional to the square root of the temperature.

De Broglie wavelength   The wavelength, λ of a particle of momentum p given by the formula $\lambda = \frac{h}{p}$ where h is the Planck's constant $(6.626 \times 10^{-34} \text{ J.s})$. For sub-atomic particles, this wave aspect is significant, resulting in the observation of wave-like phenomena such as interference and diffraction. This wavelength also describes the uncertainty in the position of a particle with the uncertainty in momentum of the order of p.

debugging   The process whereby faults in software are corrected. If the software is faulty, the problem, which may be in just a small module, is corrected and the software generally released as a newer version. If there is a significant amount of reworking, it is then released as a higher version.

Debye-Scherrer rings   Diffraction of X-rays by a powder sample (usually contained in a fine glass tube and rotated about a vertical axis) produces a set of reflections for different atomic planes satisfying *Bragg's law*. As shown in the drawing, a monochromatic beam of X-rays is scattered by a sample in the middle of the circle. The film is a circular strip, and diffracted beams are cones coaxial with the incident beam and intersect the film in arcs. The exposed film will contain rings, named *Debye-Scherrer rings*, centered around the incident beam direction. By measuring the diameter (s) of a given ring and the radius (R) of the film cylinder, one can deduce the Bragg angle, which is $\theta = \frac{s}{4R}$, and hence the atomic spacing of the corresponding planes that cause diffraction. This technique is used for identification of samples by comparing the ring pattern to that of known substances.

decay of sound   A decrease of acoustic energy density E in a room over the course of time t after the shut-down all sound sources. In most cases, sound decays exponentially $E =$

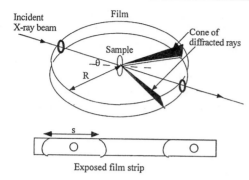

Exposed film strip

Exposed film strip; Debye-Scherrer ring.

$E_0 e^{-t/\tau}$. Here, E_0 is acoustic energy density at time $t = 0$, and τ is the characteristic decay time.

DECCA   A radio navigation system operating in the range of 70 to 130 kHz that is used by the British. Positions in air or at sea can be determined by comparing the phase differences received from two or more fixed synchronized radio stations. The operating range is about 400 km.

decibel   One ten*th* of *a bel.* 1dB = 0.1B, where dB and B are abbreviations for *decibel* and *bel.* If I_2 and I_1 are quantities having dimensions of energy, intensity, power, etc., their ratio in decibels is given by $N = 10 \lg(I_2/I_1)$. In acoustics it is customary to measure and report sound pressure levels in decibels.

declination, magnetic   The angle between the magnetic meridian and the geographical meridian is known as the *magnetic declination.*

decoding   In communication, a message in source symbols at the transmitter can be recovered at its destination from the string of values or symbols from its coded representation. This is done by using an algorithm for decompressing data at the destination. It is of great importance that the decoded symbols be error free.

decoding, decision   Transmission can cause a signal to degenerate and interfere with the task of the receiver to establish the transmitted sequence. The accuracy with which the decision threshold must be placed depends on the severity of the distortion suffered during transmission. A single decision threshold is therefore required for binary sequence.

decoding, feedback   A type of decoding in which each decoding decision on transmitted information is fed back to affect future decisions. This method can cause the undesirable property of error propagation caused by an incorrect decision in a previous step.

decoding, hard decision   A type of decision from the demodulator when regenerating the information sequence. The Hamming distance between the received symbols and the estimated transmitted symbols in the trellis are used as a measure of confidence, known as the metric. This decoding method is optimum since it minimizes the probability that the entire sequence is in error. The demodulated signal at the demodulator output is sampled and hard limited to regenerate the binary signal for channel decoding.

decoding, likelihood ratio   Noise degenerates the decision-making process at the output-stage of transmitted signals. It is therefore necessary that the decision-making is statistically quantified by indicating the probability of making an erroneous decision. The error probability is generally treated as that due to Gaussian noise. The *likelihood ratio* is the ratio of two Gaussian point distribution functions of receiving a signal x under the hypothesis that y and z are transmitted. Small changes in signal to noise greatly affect the error probability. The path with the smallest distance, from all the paths in the trellis, is selected, in Viterbi algorithm. This results in a minimum bit error rate. Equalizers are employed in circuits to reduce such problems. *See* decoding, Viterbi.

decoding, metric   The test statistic that is used in maximum likelihood decoding for each complete path in a tree, and determines the path for which certain optimized values will be largest. This is useful for minimizing errors associated with decoding. *See* decoding, trellis diagram.

decoding, soft decision   In this approach, the signal variations at the output of the demodulator are sampled and quantized. The demodulator passes a sequence of quantized levels to the decoder instead of a sequence of data bit. Sometimes incorrect decoding that occurs with hard decision decoding can be rectified by using soft decision decoding. *See* decoding, hard decision.

decoding, state diagram   A convolutional code tree with k inputs can be represented by this type of diagram. In general, each state is associated with a previous $k - 1$ input bits leading to it and the transition between states by the output sequence produced by the input bit causing the transition. *See* decoding, trellis diagram.

decoding, trellis diagram   This method is used for more effective utilization of available bandwidth and power, where convolutional coding and modulation are treated as a single entity. There is dependency between successive signal points such that only certain patterns or sequences of signal points are permitted. This produces the trellis structure. Maximum likelihood of decoding trellis codes consists of finding that path through the trellis with minimum squared Euclidean distance to the received sequence. The coding of points is done to maximize the chance of detecting errors. *See* decoding, Viterbi.

decoding, Viterbi   An algorithm procedure that involves considering paths through a trellis diagram, in comparing the received sequences of codes with all the possible sequences that can be obtained with the encoder. This procedure involves considering the retained paths in the trellis diagram so that a continuous path is formed through the trellis with a minimum aggregate Hamming distance. The decoding algorithm makes use of the repetition property of the convolutional code tree to reduce the number of comparisons.

de-coupling   Removing the inter-relationship between two entities. For example, in a multistage amplifier, it is necessary to de-couple the power supply of the input stage from the remainder of the amplifier. The reason for this is that the supply voltage changes with current because of the effective internal impedance of the power supply. Any small change in the power supply voltage alters the bias on the first stage, and is amplified in the same way as an input signal.

dedicated line   Generally refers to phone lines in which the path is set up from source to destination such that it is assigned exclusively to a particular connection or call.

defibrillator   An instrument that provides an electric shock in such a way as to restore a normal heartbeat by arresting fibrillation of the ventricular muscle.

deflector coil   The coil used in a cathode ray tube to deflect the direction of electron beam is called the *deflector coil.*

degeneracy acoustic   Existence of different normal modes in a vibrating system that have the same frequency. For example in a vibrating square membrane, there can be two different normal modes corresponding to the allowed frequencies. Various linear combinations of these normal modes give an infinite number of possible vibrations in a square membrane for the given frequency.

degree of coherence   *See* coherence, degree of.

de Haas van Alphen effect   If the magnetic susceptibility of metals is measured at low temperatures in a magnetic field, χ is found to oscillate as a function of magnetic field. Careful analysis shows that the oscillations are actually periodic in $1/H$, not in H. It can be shown that the length of the period in $1/H$ is inversely proportional to an extremal cross-sectional area of the Fermi surface normal to the magnetic field. For an ideal Fermi sphere, the extremal area is simply a circle of radius k_F, not providing much information. In real metals, however, the shape of the Fermi surface can be fairly convoluted. It is possible, by varying the direction of the magnetic field, to reconstruct the Fermi surface using this effect, the *de Haas van Alphen effect.* This can be a complicated task when the Fermi surface has several extremal cross-sections in a given direction, so it is often easier to develop a theoretical Fermi surface, and then match a theory to the data. Similar oscillations are found in the electrical conductivity (where they are known as *Shubnikov-de Haas oscillations*), thermal conductivity, magnetoresistance, sound attenuation, and all other physical properties of metals that depend on the electronic density of states.

dehardening   The thawing of a biological system that had been subjected to prolonged cold.

delay   Generally refers to the transient time involved in switching networks or digital gates in response to a stimulus. When considering the switching properties of a device (e.g., a transistor), the delay time is the time required for the output to rise to 10% of the full asymptotic output in response to a step input. Other parameters used to describe the response are the rise, storage, and fall times. These are illustrated below.

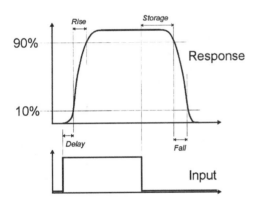

Definition of delay, rise, storage, and fall times for a device responding to a step input.

delay distortion   A type of transmission impairment caused by different Fourier components travelling at different speeds. For digital data, fast components from one bit may catch up and overtake slow components from the bit ahead, thereby increasing the probability of incorrect reception.

delay equalizer   This is used as a solution to the problem of a receiver getting several signals from a transmitter, each of which has traveled a different path between transmitter and receiver.

In this method, delayed and attenuated images of the direct signal are subtracted from the actual received signal.

delay line   A communications or electronic circuit that has a built-in delay. Acoustic delay lines were used to create the earliest computer memories by using tubes of liquid mercury that would slow down the digital pulses long enough, e.g., a fraction of a second, to serve as storage. An optical fiber of a precise length can also be used to introduce a delay in a light wave pulse which is equal to the time required for the pulse to propagate from beginning to end.

delay, signal   Associated with multipath dispersion, where multiple signals originating from the same transmitter follow different paths to the receiver. This causes signals relating to a previous bit/symbol to interfere with the signals relating to the next bit/symbol. *See* delay equalizer.

Dellinger effect   A sudden fade-out of radio signals that can occur in the band from about 1 MHz to about 30 MHz. This is caused by partial or complete absorption in the ionosphere often as a result of abnormal solar radiation affecting the transmission path.

delta connection   In a three-phase, three-wire motor, an arrangement such that the phases between any two wires are $\pm 120°$ apart. There are several advantages including:

1. compactness of the device,
2. greater initial torque,
3. ease of starting, and
4. minimization of the Joule loss on the lead wires.

A schematic diagram of a three-phase circuit is shown in the figure below. The name *delta connection* comes from the geometry of the circuit.

delta network   One of two common connection schemes in the generation and loading of three-phase electrical power, the other being *Y–network*. These are a system of three sinusoidal voltage sources, each with the same magnitude and frequency, but 120° out of phase with each other. The delta connection scheme is shown below for both source and load. If the currents

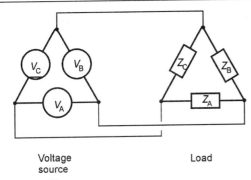

Voltage source   Load

Delta network.

in each "side" of the delta are equal, then the system is said to be balanced. In this case, the equivalent Y–network can be easily computed.

demagnetization The process of rendering the orientation of magnetic domains randomly to reduce the magnetization to zero.

demagnetization, adiabatic One technique for demagnetizing in which the material is heated to its Curie point and then cooled in the absence of an external magnetic field.

demodulation The process of extracting meaningful information from a composite waveform by performing the inverse process of modulation. Knowledge of the modulation scheme (e.g., amplitude, frequency, or pulsewidth modulation) is required to decode the waveform. *See* modulation.

de Morgan's laws A statement of the relationship between the elementary operations conjunction and disjunction in boolean algebra. De Morgan's laws are a direct implication of the duality principle and are useful in simplifying complicated boolean expressions. Consider an arbitrary number of boolean variables $A, B, C, \ldots$ each of whose value is either 1 or 0 (**true** or **false**) and can represent an input variable, constant, or functional result. Then, the two forms of de Morgan's law are

1. the inverse of a product of variables is equal to the sum of inverses of the individual variables,

$$\overline{A \otimes B \otimes C \otimes \ldots} = \bar{A} \oplus \bar{B} \oplus \bar{C} \oplus \ldots .$$

2. and the inverse of a sum of variables is equal to the product of inverses of the individual variables.

$$\overline{A \oplus B \oplus C \oplus \ldots} = \bar{A} \otimes \bar{B} \otimes \bar{C} \otimes \ldots$$

In the above equations, the bar $\overline{(\ldots)}$ represents the inverse of that variable or expression (**NOT**), $\otimes$ represents the conjunction (**AND**) operation, and $\oplus$ is the disjunction (**OR**) operation: the three elementary boolean operations and their implementations in digital electronics.

These rules serve as a mathematical basis for constructing arbitrary logic functions based on a subset of the elementary gates. For example, in TTL (transistor-transistor logic) circuitry, it is usually more economical to construct **NOR≡NOT OR** circuits. Thus, the **AND** gate can be formed by applying de Morgan's law:

$$A \otimes B = \overline{\bar{A} \oplus \bar{B}} ;$$

that is to say,

$$A \text{ **AND** } B = (\textbf{NOT } A) \textbf{ NOR } (\textbf{ NOT } B) .$$

This functional equivalence of the electronic gates is illustrated below.

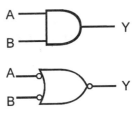

Equivalent digital electronic circuits as shown by de Morgan's laws.

depletion layer The electrostatic dipole layer formed at a p-n semiconductor junction. It is formed by electrons in the n-type region near the junction diffusing to the p-type side, leaving behind positively charged donor ions. Additionally, holes will flow from the p-region to the adjacent n-type side leaving behind negatively charged acceptor ions. Therefore, there are two adjacent layers of fixed equal-but-opposite charges at the p-n junction. *See also* diode junction.

The effective width of the depletion layer can be calculated by solving Poisson's equation:

$$\frac{d^2 V}{dx^2} = -\rho ,$$

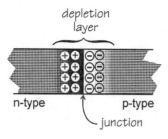

Pictorial representation of the depletion layer formed at a p-n semiconductor junction.

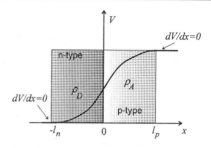

Calculation of the width of the depletion layer (see text).

where ρ is the charge density and V, the electrostatic potential. The junction is assumed to be positioned at $x = 0$ and the depletion layer extends into the p- and n-type sides by $-l_n$ and $+l_p$ respectively (see accompanying figure). Integrating Poisson's equation in the n-type region yields

$$\frac{dV}{dx} = -\rho_D x + \text{constant},$$

where ρ_D is the donor ion density. Given the boundary condition $\frac{dV}{dx} \equiv 0$ at $x = -l_n$, then

$$\frac{dV}{dx}\bigg|_{x=0} = -\rho_D l_n.$$

By a similar argument on the p-side of the junction,

$$\frac{dV}{dx}\bigg|_{x=0} = \rho_A l_p,$$

thus implying $-\rho_D l_n = \rho_A l_p$, or, owing to the opposite charges of ρ_A and ρ_D,

$$N_A l_d = N_D l_p,$$

where N_A and N_D are the number densities of the acceptor and donor ions, respectively. The depletion layer $(l_d + l_n)$ therefore depends on the relative degree of doping (see also doping) of each side of the junction.

depolarization   The effect that leads to the loss of the polarization state of a beam of light as it interacts with a medium. When light is reflected from mirrors or transmitted through interfaces of dielectrics at angles other than normal incidence, the ratio of TE (s-type) to TM (p-type) polarized light changes according to Fresnel's equations. Polarization-dependent absorption and scattering can also lead to depolarization.

depth gauges   See liquid refrigerant level; surface detection.

depth of field   The range of distances of an object from an optical system to produce an image considered to be in focus (see depth of focus). The depth of field is greater for smaller apertures (larger f-stop number) and longer object distances. See camera, depth of field of.

depth of focus   The greatest distance through which an image screen (and hence image of an object) can be moved with a tolerable blur (or noticeable lack of sharpness of an image). This is similar to depth of field, which is the greatest distance through which an object can be moved with tolerable blur. Depth of field and depth of focus depend on the aperture of the system.

depth sounding   Finding water depth by using an echo sounder.

de Sauty bridge   A type of AC variation of Wheatstone bridge. It is used to measure unknown capacitance. The balance condition for the de Sauty bridge is given by the following equation;

$$C_X = C_S \left(\frac{R_2}{R_1}\right).$$

desorption   Desorption, the reverse of adsorption, is when an adsorbed atom or molecule leaves the surface of the substrate and moves into the gas phase. Desorption and re-adsorption of gases at low temperatures can cause problems for the experimentalist due to heat transport. The desorbed molecules can increase their thermal energy by coming into contact with warmer surfaces, then deposit that energy on a cold surface, providing another source of heat into the

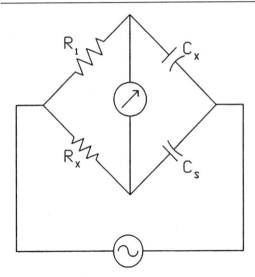

de Sauty bridge.

cal energy. The operating principle is the piezo-electric effect; the unbalancing of the positive and negative charges in the crystal due to mechanical stress.

detector, phase sensitive   *Phase sensitive* (synchronous) *detection* is a useful technique for measuring small signals that are obscured by larger and/or noisy background signals. Phase sensitive detection is the basic operating principle of *lock–in amplifiers.*

This detection scheme requires an excitation, or reference, modulation signal with frequency ω. In an experimental measurement, this signal is used to modulate a parameter of the experiment and hence indirectly modulate the measurement signal, as exemplified by the figure.

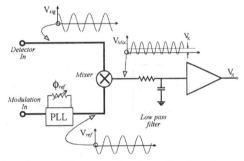

Representative measurement illustrating phase sensitive detection.

experiment. Desorbed gases can also be used to control the temperature of a sample located in the path of gas flow by controlling the rate of desorption.

detectivity   It is the reciprocal of the minimum detectable power, called the *noise equivalent power* (NEP), of a detector. The NEP (unit: $W/Hz^{1/2}$) of a detector is the rms value of sinusoidally modulated monochromatic radiation in a 1 Hz bandwidth, which gives rise to a signal voltage equal to the noise voltage of the detector. The detectivity, $D(\lambda)$, is limited by the inherent noise mechanisms such as radiation noise (that result from statistical fluctuations of photons) and Johnson noise (caused by thermal fluctuation of charge carriers). Spectral D star, $D^*(\lambda, f)$, is obtained by normalizing the effects due to the detector area and bandwidth (unit: $cm.Hz^{1/2}/W$).

detector   (**1**) In communication, a device that recovers information from a transmitted signal. Also referred to as a *demodulator.*

(**2**) A unit or device used to measure the presence of a given entity, such as the emission of energy, flux of particles, or static electric or magnetic fields. Usually, the output of the detector is an analog voltage proportional to the strength or amount of that which is detected.

detector, crystal   A crystal used as a transducer to convert mechanical energy into electri-

The principle of phase sensitive detection is based on mixing (multiplying) the detected signal with a sine wave in phase with the modulation reference, as illustrated in the following block diagram:

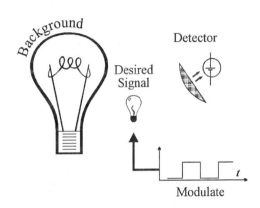

Block diagram illustrating phase sensitive detection.

It is assumed that the detected signal is $A = V_{sig} \cos(\omega t + \phi_{sig})$ where ϕ_{sig} is the phase difference measured relative to the modulation signal. A reference sine wave is generated from the modulation signal, usually with a phase-lock loop (PLL), and is given by $B = V_{ref} \cos(\omega_{ref} t + \phi_{ref})$. The product of the two signals is determined by the trigonometric identity:

$$cos(A) \times \cos(B)$$
$$\equiv \frac{1}{2} \left[\cos(A - B) + \cos(A + B) \right] ;$$

thus, the mixed output signal is

$$V_{mixed} = \frac{1}{2} V_{sig} V_{ref}$$
$$\left[\cos(\omega_{ref} t + \phi_{ref} - \omega t - \phi_{sig}) \right.$$
$$\left. + \cos(\omega_{ref} t + \phi_{ref} + \omega t + \phi_{sig}) \right] ,$$

and there will be two beat frequencies in the output, one at the difference and one at the sum frequency. Since $\omega_{ref} = \omega$, then

$$V_{mixed} = \frac{1}{2} V_{sig} V_{ref} \left[\cos(\phi_{ref} - \phi_{sig}) \right.$$
$$\left. + \cos(2\omega t + \phi_{ref} + \phi_{sig}) \right] .$$

A low pass filter is used to block the double frequency component, therefore leaving a DC signal proportional to the desired signal. Noise signals far from the reference frequency are attenuated by the low pass filter. Signals with frequencies close to ω_{ref} will yield a very low frequency AC component in the output. The attenuation depends on the bandwidth of the low pass filter, and the bandwidth of detection for the whole device is determined by the filter. Selecting a longer time constant for the filter will improve signal-to-noise ratios; however it will also reduce response times.

With the final output ideally given by

$$V_X \propto V_{sig} V_{ref} \cos \left(\phi_{ref} - \phi_{sig} \right) ,$$

the phase of the reference-generated signal is adjusted such that $\phi_{ref} = \phi_{sig}$ for a maximal V_X. Alternatively, the detected signal can be independently mixed with another reference $V_{ref} \cos(\omega_{ref} t + \phi_{ref} + \frac{\pi}{2})$, yielding

$$V_Y \propto V_{sig} V_{ref} \sin(\phi_{ref} - \phi_{sig}) ,$$

after proper filtering. Then the final, phase-independent, output can be determined from $V_{out} = \sqrt{V_X^2 + V_Y^2}$. This alternate scheme is not shown in the figure.

detector, square law   (1) In communication, noncoherent AM demodulation can be accomplished by rectifying the input signal (AM modulated carrier). If the rectifier has a characteristic

$$V_{out} = \text{ constant } \times V_{in}^2 ,$$

then the detector is said to follow a square law. However, the detector also generates second harmonic frequencies and thus leads to some distortion of the demodulated carrier.

(2) *Inverse square law:* The signal detected as a measuring device is moved away from the source it is viewing, e.g., an optical power meter detecting the light emitted from an incandescent light source. Assuming the source emission is isotropic in space, the square law indicates that the detected signal will decrease as a function of $1/r^2$ where r is the distance from the source. This is because the emitted power, or flux, is being distributed over larger and larger (imaginary) spheres centered on the source. So a detector that views the source with a fixed area of detection will be intercepting less and less radiation as the detector is moved away because the solid angle of detection is decreasing.

deuterium arc lamp   Deuterium gas (also called heavy hydrogen) under high pressure in a high voltage discharge tube produces intense continuous ultraviolet (UV) radiation from 180 nm to 400 nm. The emission in the 400 to 700 nm range contains broadened line spectra. The lamp is used as a source of radiation in UV spectroscopy.

deviation ratio   Ratio of the maximum allowable deviation in the frequency to the maximum allowable modulating frequency for FM transmission.

dewar   A container that holds a cryogenic coolant such as liquefied nitrogen or helium. *Also called a cryostat.*

dextrorotatory   When linearly polarized light propagates through some optically active substances, the polarization direction rotates with distance. Viewing the beam head-on, if the rotation is clockwise, the substance is called dextrorotatory (as in a type of sugar called dextrose). In contrast, levorotatory materials produce a counter-clockwise rotation.

D-field   The electric displacement field given by: $D = \varepsilon_o E + P$ where ε_o is the permittivity of free space, E is an external electric field and P is the polarization produced.

dial   A device used to generate the pulse signals needed for the automation of telephone exchanges. The rotary dial used contacts within the dial to make and break an electrical circuit from a battery in exchange-through loop made by the line to the customers' premises and through the phone itself.

dialysis   The separation of salts from solution by placement of the solution on one side of a permeable membrane with water on the opposite side. The ions diffuse across the membrane due to osmotic pressure; however, larger molecules are held back in solution.

diamagnetic materials   Materials that exhibit diamagnetism and that consequently have a negative susceptibility.

diamagnetism   A weak magnetism in which a material exhibits a magnetization that is opposite in direction to the applied field.

diaphragm   A flexible membrane or a thin plate that is used in transducers to radiate or receive sound. In order to radiate a sound wave by loudspeaker or other source, a diaphragm is set into vibrations. In receiving transducers such as a microphone, a diaphragm is set into motion by an incident sound; this motion is then transformed into an electrical signal. The theory of a vibrating diaphragm is that of a vibrating membrane or a thin plate.

dichroic mirror   A mirror that can reflect a specific color of light. Such mirrors are used in color television cameras. The principle of operation depends on the property that the color of some dyes is concentration-dependent. *See* dichromatism.

dichroism   A class of anisotropic media that polarize light by selective absorption of only one of the two rectangular components of vibration of the electric field vector. The wavelength of the absorption edge of a crystal depends upon the linear polarization of the light. A common dichroic crystal is the mineral tourmaline; some organic compounds such as Polaroid also exhibit this effect.

dichroism, circular   Unequal absorption of left- and right-handed circularly polarized light. First observed in solutions by Cotton in 1895; the anomalous rotatory dispersion observed is called the *Cotton effect.*

dichroism, circular, fluorescence detected (FDCD)   Anomalous dispersion of circularly polarized light near an absorption edge in a chemical substance and leading to fluorescence light with preferred circular polarization.

dichroism, linear   Dichroism is polarizartion of light (electromagnetic radiation) by selective absorption of the radiation along one preferred axis of two referred to as the O and E axes. Dichroism results from asymmetry in the molecular structure of the substance.

dichromate cell   A primary cell in which poles of carbon amalgamated zinc are immersed in a solution of potassium dichromate ($K_2Cr_2O_7$) in dilute sulfuric acid. The emf of a dichromate cell is 2.03 volts.

dichromatism   The presence of two absorption bands in an optical material with different absorption coefficients. This effect is seen in, for example, the material used for green sunglasses that look red when doubled over so that observation is through twice the normal thickness.

dielectric   A material that does not conduct electric charge is called a *dielectric*. It is also known as an *insulator.* There are two types of dielectric: polar and non-polar. A polar dielectric consists of polar molecules that have permanent dipole moment. When an external electric field is applied to a polar dielectric, the molecules can be aligned, while in the non-polar dielectric, the applied field will induce an electric dipole moment in the atom or molecule and align the dipole moment.

dielectric breakdown   When the electric field applied to a dielectric material exceeds the dielectric strength of the material, the electric charges will force themselves through the dielectric material. This is called a *dielectric breakdown.*

dielectric constant   The ratio of permittivity of the material to that of the free space, $\kappa = \varepsilon/\varepsilon_o$, is known as the *dielectric constant* of the material.

dielectric heating   Heating of a dielectric material through the use of radiation of high frequency electromagnetic wave.

dielectric hysteresis   The dependence of the polarization of ferroelectric materials on their previous history is called *dielectric hysteresis.* The dielectric hysteresis in dielectric materials is analogous to the magnetic hysteresis in ferromagnetic materials. It is also known as *ferroelectric hysteresis.*

dielectric strength   The maximum electric field a dielectric material can withstand without breakdown is called *dielectric strength* of the material.

difference frequency   One of the signal components obtained by mixing two signals with different frequencies. Ignoring phase and amplitude differences, the two signals can be described as

$$V_1 = \cos(\omega_1 t)$$

$$V_2 = \cos(\omega_2 t)$$

The mixed signal will yield two components:

$$V_1 \times V_2 \propto \cos(\omega_1 - \omega_2) + \cos(\omega_1 + \omega_2) ,$$

a component at the difference frequency and one at the sum frequency.

differential conductance   The inverse of the differential resistance. *See* differential resistance.

differential input   The voltage difference between the two input terminals of an amplifier, particularly an emitter-coupled amplifier stage as shown below. Referring to the diagram, the differential input voltage is $V_{\Delta\ in} = V_{in\ 1} - V_{in\ 2}$.

Most operational amplifiers use an emitter-coupled amplifier as their first amplification stage.

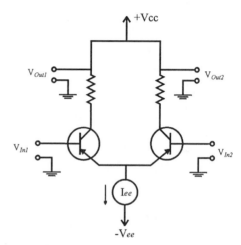

A basic emitter–coupled transistor amplifier.

differential output   The difference between the two output voltages of an emitter–coupled pair. Referring to the above figure, the differential output is $V_{\Delta\ out} = V_{out\ 1} - V_{out\ 2}$. *See also* differential input.

differential resistance   The effective resistance between the two input terminals of an amplifier, particularly an emitter-coupled amplifier stage. Contrast this with the common–mode input resistance, which is the resistance from either input to analog ground. An ideal operational-amplifier will have these resistances infinite. *See also* differential input.

differential voltage gain   The change in differential output voltage per unit change of differential input, expressed as:

$$g_\Delta = \frac{V_{\Delta\ \text{out}}}{V_{\Delta\ \text{in}}}\ .$$

See differential input, differential output.

differentiator   A basic circuit, commonly based on an operational amplifier, that differentiates an input voltage with respect to time. Shown below is a simple differentiator circuit. The input capacitor does not allow any direct current to flow, only the displacement current which depends on the time rate of change of the voltage across the capacitor. Since the voltage at the − terminal of the op-amp is an effective null, it can be shown that the output voltage of the presented ideal differentiator is given by

$$V_{\text{out}} = -RC\frac{dV_{\text{in}}}{dt}\ .$$

Thus, a differentiator circuit is useful for measuring the rate of change of an input voltage.

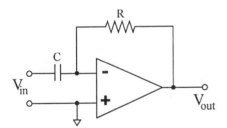

A simple differentiator.

diffraction   The propagation of light waves in any manner that departs from rectilinear propagation predicted by the laws of geometrical optics. The term originates from the observation that light bends around opaque obstacles, resulting in shadows that have slightly blurred boundaries. Patterns are produced near the edges of the shadow that depend on the size and shape of the obstacle. This breaking up of the light as it passes the object is *diffraction* and the observed patterns are *diffraction patterns.*

Since the dimensions of obstacles encountered by light are not large compared with the wavelength, the effects are subtle but ubiquitous in our common experience. The luminous border outlining a mountain profile in the first

seconds prior to the sun rising behind it and the streaks of light seen with half shut eyes viewing a strong light source are but a few of the many common examples of the diffraction phenomena. It was first commented on by Grimaldi in his book published in 1665.

Diffraction is now known to be a direct consequence of the wave nature of light. Much of the phenomena can be quantitatively described by a mathematical form of Huygen's principle formulated by Kirchoff that is an approximation to the wave equation, making it unnecessary to rigorously solve the wave equation to understand diffraction. There are two convenient categorizations of diffraction phenomena, depending on whether a parallel beam of light passes the diffracting object. When either the light source or the observing detector, or both, are a finite distance from the diffracting obstacle or aperture, the diffraction is classified as *Fresnel diffraction.* When either the source or detector, or both, are at infinite distance, effectively making the beam of light passing the obstacle parallel, the diffraction is termed *Fraunhofer diffraction,* for historical reasons. Fresnel diffraction is easiest to observe of the two as no lenses are needed and it was the first to be investigated. However the mathematical theory is much more difficult than the plane waves of Fraunhofer diffraction.

diffraction, crystal   Destructive and constructive interference of waves scattered by the periodic placement of electrons, nuclei or force fields in the lattice of a crystal, resulting in a pattern of discrete spectra.

diffraction, Fraunhofer   The diffraction phenomena observed when both the source and observation point are at very large distances from the diffracting object. In this limit, the mathematics of diffraction is much simpler than Fresnel diffraction. A point source at the focal point of a converging lens or collimator is frequently substituted for an infinitely far light source. The diffracted light may be collected by lenses and fringes observed in the focal plane of the lens, rather than from an infinite distance. For diffraction from a slit, it is readily found that the diffraction pattern has an absolute maximum at the center of the line appearing on the focal plane of a lens, and a diffraction pattern that is symmetrical about this center. The width of the principal maximum is twice the width of the secondary maxima, and both the principal and secondary maxima are inversely proportional to the width of the slit. For a point source, the Fraunhofer diffraction pattern is a line perpendicular to the slit, whereas with Fresnel diffraction it would be a band. More elaborate mathematics is required for Fraunhofer diffraction from a circular aperture than from a single slit. In this case, the diffraction pattern consists of a bright circular disk surrounded by a series of dark and bright fringes that rapidly decrease in intensity. The detailed results for the circular aperture are of practical use for the properties of optical instruments. Typically a lens is limited by the circular rim, so a converging lens does not produce an exact point image of far-away point sources, despite careful correction for aberration. Concave spherical mirrors such as those used for telescopes also exhibit this spreading. These diffraction disk images limited by the resolving power of the optical instrument, and for this reason objective lenses and mirrors of telescopes are made with large diameters.

diffraction, Fresnel   Diffraction phenomena observed when either the source and observing screen or both are at a finite distance from the diffracting object. In the case of a circular aperture, one or more Fresnel zones are uncovered, and the amplitude of the optical disturbance is estimated from the area of the zone, and neighboring zones have opposite signs. The light intensity goes through a series rings of maxima and minima due to the appearance and disappearance of successive positive and negative Fresnel zones. Near the axis of an aperture with dimensions comparable to the distance from the observation point, the illumination is nearly identical to that produced by the unobstructed wave. If a circular obstacle is used rather than an aperture, then given the contributions of various Fresnel zones, the pattern consists also of concentric bright and dark rings, but at the center the intensity is always a maxima.

diffraction, Kirchoff's formula   A mathematical form of Huygen's principle formulated by Kirchoff that is an approximation to the wave equation, making it unnecessary to rigorously solve the wave equation to understand diffraction phenomena. Using Green's theorem, the light wave at any point in space is expressed as integral over a closed surface surrounding the point. The differential contributions from surface elements provide the Huygen's secondary wavelets. In principle, if the part of the closed surface coincides with the diffracting screen, then the solution of any diffraction may be obtained by evaluating this integral with suitable boundary conditions.

diffraction of waves   Propagation of acoustic and electromagnetic waves that follows laws of geometrical acoustics and optics. Many phenomena in wave propagation are the result of diffraction, for example: penetration of a wave into a region of geometrical shadow; penetration of a wave through a small opening in a screen; and propagation of waves along a surface (surface and greeping waves). Usually diffraction effects become important when $\lambda > d$, where λ is a wavelength, and d is a characteristic geometrical scale of a problem such as the diameter of an *aperture,* the size of an inhomogeneity in a medium, etc. Diffraction pattern also depends significantly on the distance x of sound propagation. If $x \sim d^2/\lambda$, rays from the opposite side of the aperture, inhomogeneity, etc. have different phase increments at the point of observation. This case is called *Fresnel diffraction* (diffraction in converging rays). If $x \gg d^2/\lambda$, these phase increments are almost the same. This is known as *Fraunhofer diffraction* (diffraction in parallel rays).

diffractometer   An apparatus used in conjunction with optical diffraction methods to solve problems of X-ray structure analysis. A source of light is imaged by a lens onto a pinhole and the emerging light from the pinhole is made parallel by a lens after the pinhole. The interfering beams are made to focus with an additional lens identical to the first lens after the pinhole. A plane mirror is then used to reduce the length of the instrument, and either a diffracting screen or a microscope is used to view the resulting diffraction pattern.

diffuse spectra   There are three main sequences of lines in the spectra of neutral alkali atoms: the *principal, sharp* and *diffuse* series. The *principal* series are the strong lines, the *sharp* series are the very narrow lines, and the *diffuse* series spectra quite broad. This terminology applies as well to series arising from the same types of electron transitions in other atoms. The physical characteristics of the lines, however, may be very different from the simple observational character of the alkali atomic spectra.

diffusion, cell membrane (**1**) *Lateral diffusion:* Two-dimensional effective transport of molecules within the cell interior. Mechanisms consist of Brownian motion and percolation, in addition to active transport.

(**2**) *Translational diffusion:* One-dimensional transport of molecules across a membrane. Mechanisms include facilitated transport and active transport.

diffusion coefficient, translational A measure of the rate of flow across a permeable membrane due to diffusion, having units of square meters per second and relating the ratio of flux to concentration gradient. That is diffusion coefficient, $D = -J/(dC/dx)$, where J is the flux and dC/dx is the concentration gradient.

diffusion, cytoplasm The specific diffusion of potassium, calcium, and sodium atoms across the cytoplasm giving rise to an action potential. The kinetics of this diffusion.

digital arithmetic Digital arithmetic is performing mathematical operations on numbers using digital electronic circuits. The operations are performed in a binary number system because they are easiest to implement with logic circuits. Addition and subtraction are the two basic operations on which all others are based.

Since only the two states **true** and **false** are available to represent a digit in digital electronics, decimal numbers are represented in their base 2 (binary) equivalent. Instead of the characters 0, 1, 2 . . . 9 to represent digits, there are only 0 and 1. For example, "13" in base 10 represents $1 \times 10^1 + 3 \times 10^0$. Using the *natural* representation scheme, this is written in binary as 1101 implying $1 \times 2^3 + 1 \times 2^2 + 0 \times 2^1 + 1 \times 2^0$.

However, to represent signed numbers, i.e., negative numbers, a convention must be used. The three most common schemes to represent signed numbers are *signed magnitude, 2's complement,* and *1's complement.* Positive numbers in all three schemes are identical to the natural representation. In *signed magnitude,* one of the number digits is reserved to indicate the sign. It is usually the first digit with 1 indicating a negative number. So, 1011 represents −3 in this scheme. In *1's complement,* a positive number is represented with 1 in the first digit and the remaining digits inverted. Therefore, 1011 implies negative ($\overline{011}$) which is −4. Negative numbers are represented in *2's complement* by using 1 in the first digit position and $1 - X$ for the numerical part.

Example representation of decimal numbers in different, 3–digit binary representations

Scheme	000	001	010	011
natural	0	1	2	3
sign–mag	0	1	2	3
1's comp.	0	1	2	3
2's comp.	0	1	2	3

Scheme	100	101	110	111
natural	4	5	6	7
sign–mag	−0	−1	−2	−3
1's comp.	−3	−2	−1	−0
2's comp.	−4	−3	−2	−1

Digital arithmetic operations are based on the addition and/or subtraction of single digits, starting with the least significant digit position and propagating a carry or borrow digit to the next higher position. First consider addition of two single-digit binary numbers, an addend and augend. The addition is defined for all possible values as: Thus, aside from the sum, a carry is

$$0 + 0 = 00$$
$$0 + 1 = 01$$
$$1 + 0 = 01$$
$$1 + 1 = 10$$

addend ⌐ | | ⌐ sum
augend ——— └ carry

generated as well when adding the single digits.

In adding multiple digit numbers, however, consideration must be given to a carry from the next least significant digit. Let x_i be a digit of the addend and y_i a digit of the augend. We also want to consider a carry from the $i-1$ sum, c_{i-1}. Then, the sum s_i and carry c_i from adding these two digits are defined in the accompanying table. A subtraction truth table can be defined as well.

Digital addition truth table

i	y_i	c_{i-1}	c_i	s_i
0	0	0	0	0
0	1	0	0	1
1	0	0	0	1
1	1	0	1	0
0	0	1	0	1
0	1	1	1	0
1	0	1	1	0
1	1	1	1	1

Digital subraction truth table

i	y_i	c_{i-1}	b_i	d_i
0	0	0	0	0
0	1	0	1	1
1	0	0	0	0
1	1	0	0	1
0	0	1	1	1
0	1	1	1	0
1	0	1	1	1
1	1	1	0	0

As an illustration, consider adding the binary numbers 0110 and 0111 (which represent 6 and 7 respectively in the natural scheme). Using the tabulated rules, which is consistent with $7 + 6 =$

```
      1↖  1↖  0↖      carry
    0 | 1 | 1 | 1     addend
  + 0 | 1 | 1 | 0     augend
    1  1  0  1         sum
```

13. The rules for subtraction vary somewhat depending on which number-coding scheme is being used.

The implementation of the addition and subtraction tables with electronic circuits is done with a full–adder. A full–adder comprises two half–adders (*see* half–adder), and will accept x_i, y_i, and c_{i-1} inputs. It will generate s_i and c_i as per the truth table. Multiple digit addition can be carried out serially or in parallel. In serial addition, the digits to be added are delimited by a clock pulse. The addition is carried out by only one full–adder with the carry being held in a D flip–flop for the next digit. In parallel addition, there are separate full adders for each digit.

digital circuit   *See* digital electronics.

digital combining   In digital transmission, some processing tasks are involved from the data source via a communication channel to a distant data terminal. The two main categories are source coding and channel coding. Greatly improved efficiency can be obtained by combining some tasks. *See* multiplex.

digital communication   Digital communication refers to a mode of communication wherein the transmitted information signal is defined only at a discrete set of time instances, when the signal may take on any one value from a discrete set of values.

A common block-diagram model of a digital communication system involves the following blocks connected in series: a source, source encoder, discrete channel encoder, digital modulator, communication channel, digital demodulator, discrete channel demodulator, source decoder, and a user. The source outputs a stream of information-bearing signal, which may be analog (for example, speech, music, a photographic image) or digital (for example, digitized data or computer files). The source encoder takes the source output and reconstructs it into a stream of binary bits with a lower data rate while imposing minimum distortion of the information contained therein. The source encoder achieves this data compression task by removing the redundancy in the information-bearing signal output by the source. The channel encoder accepts and transmutes the binary bit stream from the source encoder into another stream of discrete-time and discrete-valued symbols with a higher data rate. This increased data rate aims to incorporate certain intelligent redundancy in the data stream to endow it with the capabilities of transmission error self-detection and transmission error self-correction against possible distortions caused by the modulator-channel-demodulator unit. The modulator inputs and transforms the digital data stream from the channel encoder into an analog waveform for the multi-user multiplexed analog channel. Note that all physical communications medium, including optical fibers carrying laser light pulses, must necessarily be analog. This transmitter unit (consisting of the source encoder, channel encoder and modulator, connected in series) sends onto the channel the modulator's analog output, which is received by the receiver. Each component in the receiver — demodulator, channel decoder, and source decoder — performs the reverse function of its corresponding counterpart in the transmitter.

Digital communication contrasts with discrete-time analog communication, wherein the discrete-time signal may take on any one value from an uncountably infinite set of values. Digital communication also differs from fully analog communication wherein the information signal is defined over a continuous time duration and may take on any one value from an uncountably infinite set of values. An analog communication system would need to perform any source compression, channel coding and modulation all in analog — a difficult task, which often means the source output would be directly modulated without source compression or channel error self-detection/self-correction coding. The major advantages of digital communications are easy error detection and correction, easy signal manipulation, and increased dynamic range. These advantages arise partly because the source/channel encoder/decoder transmutes the wasteful redundancy in the source output into an intelligent redundancy that facilitates transmission error self-detection and self-correction by the received signal itself. The main disadvantages of digital communications are the requirement for wider bandwidth than analog communication and the need for signal synchronization.

digital electronics   Electronics applied to the processing of binary variables or numbers. The processing circuits which employ components such as diodes, transistors, resistors, etc. to construct gates, which perform logical operations. These *logic circuits* are generally operated in a non-linear, switching mode to accomplish their intended design. They are the building blocks of larger and more complicated functions like arithmetic operations and memories, also encompassed in digital electronics. *See also* gate.

digital signature   Used for message authentication as a trailer at the end of a message. The encrypted trailer is analogous to a signature at the end of a letter since it verifies the person who actually sent the message.

digital switch   A mechanical or electronic device that can be used to direct the flow of electrical or optical signals from one side to the other. Switches with more than two ports, such as a LAN switch or PBX, can be used to route traffic. The semiconductor switch, known as a transistor, performs the same function as the familiar on/off light wall switch. The switch is electronically closed by pulsing the semiconductor material, which makes it conduct.

digital television   Digital TV or DTV debuted in major cities in the United States in 1988. In order to receive DTV, a new digital TV set or a set-top box for an existing analog TV is needed since transmission in the radio frequency is done digitally. Digital TV sets will support analog TV transmission, which is expected to be broadcast until at least 2006. Digital TV offers 18 formats from SDTV (Standard Definition TV), without snow or ghost. More SDTV channels can be transmitted within the same bandwidth. Therefore, it will be up to the broadcasters, cable providers and satellite companies to determine the amount of content versus quality. Digital TV can provide 14 progressive scan and 4 interlaced formats.

digital voltmeter   Device that can sample an analog signal and quantize the voltage level. It converts the analog signal to a usable digital number with an analog–to–digital converter. The digital number is then presented to a user indicating the measured voltage in decimal format. *See* converter, analog/digital.

diode An electronic circuit element. A diode is the simplest integrated semiconductor structure. It consists of a junction between an n-type and p-type semiconductor. Ideally, current is only allowed to flow in one direction. Shown in the figure are the diode's electronic schematic symbol and its semiconductor construction. Also shown is an illustration of a typical discrete diode; the n-type side is marked with a band. *See* diode junction.

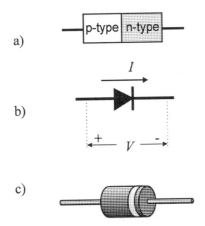

Diode, a) semiconductor junction, b) schematic symbol, and c) discrete diode with marking scheme. See text.

When the voltage of the p-type side is positive with respect to the n-type side, the diode is forward biased. Then, current flow is defined as positive from p– to n–, the magnitude of which depends on the magnitude of the applied voltage. When V is such that the p-type side is more negative than the n-type side, only a very small amount of current flows through the diode. The current flow through a diode as a function of applied V (for either direction) is given by

$$I = I_s \left(\exp\left(\frac{eV}{kT}\right) - 1 \right)$$

with I_s, the reverse saturation current, $e = 1.602 \times 10^{-19}, C$ the electronic charge, $k = 1.38 \times 10^{-23} J/K$ Boltzmann constant, and T in Kelvin. A typical diode $I - V$ curve is plotted below.

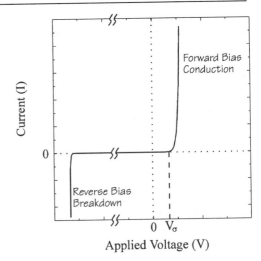

Quintessential diode I-V curve.

From a practical standpoint, a current flow of 1% of rated current can be considered as a nominal "turn–on" current for most applications. For typical silicon diodes, this corresponds to an applied forward bias voltage of $V_\sigma \approx 750mV$; the *cutin voltage*. In reverse bias, only a small amount of current is allowed to flow, I_s independent of reverse bias. However, if the reverse bias is large enough, then reverse breakdown will occur (*see* diode, avalanche) and current will flow in the reverse direction. Provided that power dissipation in the diode does not exceed design limits, reverse bias breakdown will not harm the diode.

If a diode is being used in a high speed application where the reverse bias voltage can change rapidly, then consideration must be given to the capacitance of the reverse biased diode junction. If this *barrier* or *transition capacitance* is large enough, then the current intended to be stopped in reverse bias will still flow due to the displacement current of the capacitance. The reverse bias capacitance varies as $\frac{1}{\sqrt{V}}$ and this behavior can be exploited in frequency locking or modulating circuits or in parametric amplifiers.

There are different special types of diodes engineered to have special properties. Examples include the *Zener*, the *tunnel*, and the *light emitting diode*.

diode, avalanche A diode with sufficient power dissipation capabilities to be intentionally operated in a reverse bias breakdown condition. Such a diode is useful in a voltage regulation circuit. A simple voltage regulation circuit is shown below illustrating this. Provided that R_x, I_{Load}, and V_{Supply} are such that the design V_{Load} is in the breakdown region (V_B) of the diode's I–V curve, the diode will accommodate moderate changes in V_{Supply} or I_{Load} thus regulating V_{Load}. *See also* diode.

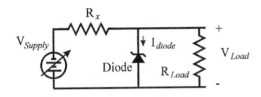

A basic voltage regulation circuit based on an avalanche diode.

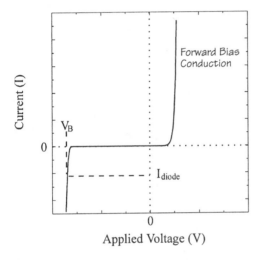

Characteristic I–V curve for an avalanche or Zener diode.

A similar device to the avalanche diode is the Zener diode. It has a qualitatively identical I–V curve as the avalanche diode and behaves similarly in circuit. However, the physics behind the Zener diode are quite different.

In a reverse bias situation, an extremely large electric field can be imposed in the region of the p–n junction. Considering a genuine avalanche situation, the charge carriers (electrons and holes), which constitute the relatively small reverse conduction current, can be accelerated to large kinetic energies. If the energy

is large enough and they spend sufficient time in the junction, these carriers can ionize atoms upon impact with the crystal lattice thus *exciting* free electron-hole pairs. These extra carriers are also subject to the large electric field and can continue the process until a very large (reverse) current is obtained hence the term *avalanche multiplication*. This effect strongly depends on the applied electric field, therefore accounting for the rapid increase of reverse current at the breakdown voltage. The degree of doping determines the physical distance of the depletion layer and how much time a charge carrier will spend under the influence of the electric field. This, in turn, determines what V_B will be.

On the other hand, by tailoring the doping of each semiconductor, it is possible to make the junction between p– and n– become physically very narrow. Then at some particular value of reverse bias, it is possible for the electrons to quantum-mechanically tunnel from the valence band of the p–side to the conduction band of the n–side. This is essentially the Zener effect and is shown in the figure below.

In silicon diodes, breakdown voltages in the range of 10 to 100V is due to avalanche multiplication. The Zener effect is responsible for diodes with $V_B \approx 1$ to 2V. Diodes with V_B in the range of 6 to 8V are due to both mechanisms simultaneously.

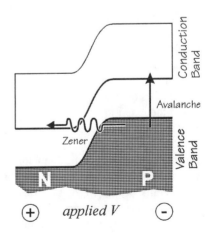

Energy band levels in a diode junction illustrating the mechanisms behind reverse bias avalanche and Zener operation.

diode, Gunn A diode with I-V characteristics similar to that of a forward biased tunnel diode. The electrical characteristics of a Gunn diode, however, arise from the peculiar band structure of *bulk* GaAs (gallium arsenide) rather than any characteristics due to a p-n junction.

The band structure of GaAs essentially permits the existence of two different types of electrons. One, which is normally responsible for conduction in the semiconductor, has an effective mass of $\approx 0.08m_e$, where m_e is the normal free electron mass. The other allowed electron has an effective mass of $\approx 1.2m_e$.

Under normal electric fields, the semiconductor behaves normally, the light electron being responsible for current flow. However, at higher applied electric fields ($\approx 10^4 V/cm$) some of the light electrons can be excited into the other energy band, thus becoming a *heavy* electron. This causes the conductivity to decrease since the effective mobility of the electrons has decreased with increasing effective electron mass. This, in turn, decreases the current through the device although the applied voltage is increasing. Hence, the device exhibits negative differential mobility.

Under these conditions, moving domains of high electric field strengths are created within the material. The thickness of the active material can be chosen so that the frequency of these "electron-waves" is of the order 10 GHz. Gunn diodes find uses in high frequency oscillator circuits (e.g., police radar) as well as high speed logic switching circuits.

diode junction A junction formed by the mutual contact of an n– and p–type semiconductor. This junction has unique and useful electrical properties (*see* diode) that result from the physics of the junction.

In the bulk of an n–type material, an abundance of conduction electrons can be found, neutralizing the space charge of the donor ions. Similarly, there will be an abundance of holes available in the bulk of the p–type material. When the n– and p–type materials are joined, the electrons and holes cannot remain separated unless there is an electric field at the interface.

When the two materials are in contact, charge transfer will occur until the Fermi levels are equalized. Initially, the excess electrons on the

n–side will diffuse into the empty electron states on the p–side. Simultaneously, excess holes on the p–side will diffuse and fill vacant hole states on the n–side. This initial flow of charge will leave behind the negatively charged donor ions in the n–side and positively charged acceptors on the p–side of the junction. This will form an electrostatic dipole layer at the junction. (*See* depletion layer.) Therefore, the resulting electric field (and associated potential energy difference) will generally oppose further diffusion.

In the p– and n–type materials, the Fermi energy (E_F) lies approximately at the acceptor and donor levels respectively. When the junction is formed, charge transfer equalizes the Fermi levels. Thus, the positions of the valence and conduction band edges must vary relative to E_F within the transition region of the junction.

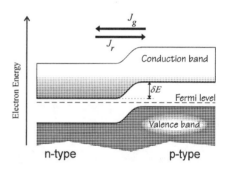

A p–n junction (diode junction) in thermal equilibrium and zero applied voltage. Shading suggests electron population in filling of energy bands. For clarity, potential energy, current flow, and band filling shown are for electrons only.

A very simple argument explaining the rectification nature of the diode junction is now given. In what follows, only electron currents and energies will be considered. There is also hole transport in the junction as well. The description for hole current parallels that for the electron case since the energy of a hole is measured opposite that of an electron and because its charge is opposite of the electron.

Even though the depletion layer produces a potential energy difference δE, which prevents electrons from flowing to the p–side, there will be a small number of electrons with non-zero thermodynamic probability to have an energy greater than δE. These electrons can flow to the p–side where they recombine with holes. This is the recombination electron current flow, J_r, and its magnitude will depend on the doping in the material. There will also be conduction electrons thermally generated on the p–side as well, but they can easily drift "down" the potential hill at the junction; this is the generation current J_g. If there is zero voltage across the diode (as in the figure), then these two currents must balance.

$$J_r(V = 0) \equiv -J_g(V = 0) \, .$$

Now, if a voltage is applied such that the n–side is more positive than the p–side then the potential barrier will be raised. This condition is *reverse* bias. Again, the electron current flow through the junction is determined by the thermodynamic probability of an electron having sufficient energy to overcome the barrier. The recombination current will therefore be reduced by the Boltzmann factor

$$J_r(V_{\text{reverse}}) = J_r(V = 0) \times \exp\left(\frac{-e|V|}{kT}\right) \, .$$

The generation current J_g is nominally unaffected by the reverse bias as this current is going down the potential hill anyway.

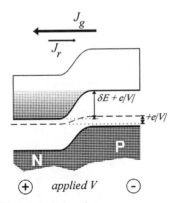

Reverse biased diode junction.

If the applied voltage is now *forward* biased, the potential barrier is reduced and more current can flow. The increase in flow is again determined by the thermodynamic probability factor

$$J_r(V_{\text{forward}}) = J_r(V = 0) \times \exp\left(\frac{+e|V|}{kT}\right) \, .$$

And again, the generation current is mostly unchanged.

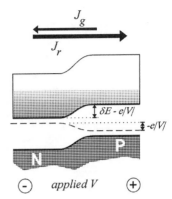

Forward biased diode junction.

There will also be a contribution to the *total* conduction current by the hole current as well. Hole current through the p–n junction will behave similarly and the currents due to electrons and holes will add. The total current through the diode as observed in a circuit is then given by

$$I = I_S\left(\exp\left(\frac{eV}{kT}\right) - 1\right)$$

where I_S is the sum of the two generation currents and is essentially the reverse bias current. The current I is defined using common circuit conventions as positive flowing from p– to n–side with positive V indicating the p–side potential with respect to the n–side. Thus the diode junction current flow depends on the relative bias of the applied voltage. *See also* diode.

diode-transistor logic (DTL) A realization of logical gates using discrete or integrated electronic components; particular *logic family;* detailed electronic circuit implementation of logic gates exhibiting a common theme or convention in the operation of the circuit.

A simple example circuit employing DTL is shown below. This discrete component circuit operates as a NAND gate assuming the positive logic convention. It has two inputs, A and B, as well as the output C. The basic electronic operation of the DTL gate is now described. If either input A or B is a logical **0** (i.e., either V_A or V_B is near 0 volts) then the current flowing through

R_x is shunted by the diodes DA or DB respectively. In this situation, the base–emitter current of the transistor, I_{BE}, is small and the transistor is "turned-off". Thus the output voltage V_C is near 5 volts, logical **1**.

On the other hand, if both A and B are logical **1** (i.e., V_A and V_B are near 5 volts) then the current flowing through R_x is no longer shunted and is divided between resistor R_z and the transistor. Then, I_{BE} is such that the transistor is saturated and the output V_C is near zero volts. The resistor R_z is used to remove charge from the transistor during the transition time from saturation to off.

When driving other gates, the transistor must be able to sink the current provided through the input diodes of the other gates. This consideration leads to the maximum number of gates which can be driven — the fan-out — by the circuit. The DTL logic family has a greater fan-out capability over RTL (resistor-transistor-logic) but is somewhat slower.

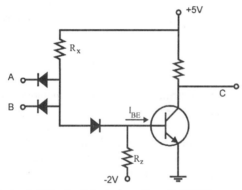

A simple DTL circuit implementation of a NAND gate.

diopter   The curvature of a wavefront at a given distance from the source is given in terms of the diopter and is expressed in reciprocal meters. Powers of lenses and other optical systems are usually expressed in terms of diopters.

dip   A perfectly freely suspended magnet would in general align itself with the direction of greatest field strength. The angle between this direction and the horizontal is called the *dip*.

dip circle   This is a mechanical device consisting of a thin steel magnet suspended so that it can rotate in the vertical plane and also be rotated about the vertical axis to determine azimuth. The instrument is used to measure dip.

diplex operation   The use of a single circuit, carrier or antenna for the simultaneous transmission or reception of two signals.

dipole acoustic   Two identical monopoles located at a distance d with the amplitudes equal in magnitude and opposite in sign. A dipole is called the point dipole if $kd \ll 1$, where k is a wavenumber of a sound field radiated by both monopoles. An acoustic pressure field p of a dipole is the superposition of those of the two monopoles. For a point dipole, p contains a factor $(1+i/kr)\cos\theta$ that is not present in the field of a monopole. Here, r is the distance to the observation point, and θ is the angle between the radius vector to this point and the dipole axis (the line connecting the two monopoles). Due to this factor, a directivity pattern of a dipole (the dependence of p on θ) and its near field (acoustic pressure field for $kr \ll 1$) is different from that of a monopole.

dipole, in dielectric   In dielectric material, each atom or molecule can be either a permanent dipole or an induced dipole. These dipoles can be aligned by an external electric field and contribute the properties of the dielectric material.

dipole magnetic   A permanent magnet, current loop or particle with angular momentum that experiences a torque when placed in a magnetic field. It acts as if it consists of two magnetic poles separated by a small distance.

dipole magnetic moment   A vector whose cross product with the magnetic induction of a magnetic field is equal to the torque exerted on the system by the field.

dipstick   A dipstick is a colloquial term used to describe a small experimental probe that is designed to be inserted directly into a storage dewar of a liquid cryogen, most often liquid helium. Such a probe consists of whatever electrical connections and wiring are necessary attached to a tube or rod with the experimental sample clamped at one end. The other end contains the electrical connections that interface the experiment with any peripheral equipment.

direction finder A device used for the determination of one's terrestrial location. It is usually a radio receiver and becomes an integral part of a larger system of radio transmitters of known location. Early direction finders employed a rotating loop antenna; the direction to an established transmitter could be determined from the known detection pattern of a dipole antenna. Modern location usually involves an array of transmitters; location is determined using a form of hyperbolic differential distance ranging (*see figure*).

The transmitters may be one of two basic types:

1. *Pulsed with a common carrier.* The differential distance is determined from differences in pulse arrival times, or

2. *Continuous wave using different, but related carrier frequencies.* Here, the difference is determined from phase differences.

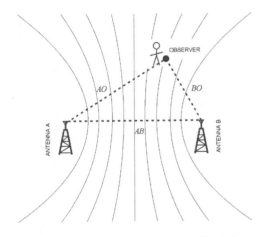

Hyperbolic ranging to determine position. The hyperbolic ranges are such that $AB + BO - AO$ is a constant.

directivity of sound Angular distribution of radiated or received sound power by an acoustic antenna, array, loudspeaker or microphone. Directivity of an antenna, etc. characterizes its ability to radiate (receive) sound from one direction better than from others. This property is mathematically characterized by the directivity function which is defined as the ratio of the radiated (received) power in a particular direction to the maximum possible power in the direction known as the acoustic axis.

discharge The release and eventual recombination of opposite electric charges of a battery or a capacitor when a load is connected to its terminals. Energy is released during the discharge process.

discharge tube Enclosure used in the production of a glow discharge. The inside is evacuated and partially filled with the intended gas to be used in the discharge. *See also* glow discharge.

discrete channel A discrete channel may refer to a communications channel that allows the transmission and reception of information represented as a discrete sequence of alphabets. A discrete channel may alternately refer to a communication channel whose effect on the transmitted signal may be represented as a discrete-time filter. A discrete-time channel refers to a communication channel whose response is defined only at a set of discrete time instances.

disjunction (logic) A fundamental boolean logic operation over two or more logical variables. Sometimes referred to as **OR**, it is the logical sum

$$A \oplus B \equiv A \ \mathbf{OR} \ B \,,$$

where the variables A and B are elements of the set $\{1, 0\}$ (or, equivalently, the set $\{\mathbf{true}, \mathbf{false}\}$). In other words, the disjunction of these two variables is true if *either A or B* is true. The postulated rules of the logical sum are

$$0 \oplus A = A$$

and

$$1 \oplus A = 1 \,.$$

Also, for every variable A there exists the sum inverse, $NOT \ A \equiv \bar{A}$, such that

$$A \oplus \bar{A} = 0 \,.$$

The logical sum is commutative as well as associative:

$$A \oplus B = B \oplus A$$

$$A \oplus (B \oplus C) = (A \oplus B) \oplus C \,.$$

Given these rules, all possible values of the conjunction $Y = A \oplus B$ can be tabulated in a *truth table*. This is illustrated in the accompanying table.

Disjunction (logical sum) for all possible values of the variables A and B

A	B	$Y = A$ **OR** B
0	0	0
0	1	1
1	0	1
1	1	1

In implementing this operation in digital electronics, voltage signals are used to represent the logical variables A and B following a predefined logic convention. A circuit, frequently referred to as a *gate*, is used to determine the disjunction and the result is provided on an output, Y. The symbol representing such an electronic circuit performing an **OR** operation is shown below. *See also* conjunction (logic).

*Symbol representing disjunction (**OR**) in digital electronics.*

disk of least confusion   *See* circle of least confusion.

dispersion   The variation of refractive index, n, with frequency of electromagnetic fields traversing a material medium. The dielectric constant is a function of the frequency of the fields and phase velocity is not the same for each frequency component. In a nondispersive medium, the index of refraction is independent of the frequency, and the phase and group velocities are both c/n. In a dispersive medium the velocity of energy flow differs greatly from the phase velocity, or may even lack a precise meaning. Dispersion is explained by taking into

account the actual motion of the charges in the optical medium traversed by the light. This motion is modeled using the damped forced oscillator of the charge bound to a fixed atom site. The solution gives a polarizability which is inversely proportional to the mass, so it is the electrons that determine the index of refraction in the transparent regions of the optical spectrum. The ionic polarizabilities only dominate in determination of the refractive index in resonant regions.

dispersion, anomalous   In the vicinity of an absorption band, the index of refraction cannot usually be measured because the substance will not transmit radiation of this wavelength. On the long wavelength side of the band, the index is quite large, but decreases very rapidly with increasing wavelength. On the short wavelength side the opposite behavior of the index is observed — it is very small but increases rapidly as wavelength is decreased. The index of refraction thus has a large discontinuity in the vicinity of an absorption band, which causes anomalous dispersion. The longer wavelengths have a higher value of the refractive index and are more refracted than the shorter wavelengths in this region. This situation is anomalous, unlike the rest of the dispersion curve where index of refraction increases as wavelength decreases.

dispersion, chromatic   The decomposition of a beam of white light into separate beams of color that spread out to produce spectra.

dispersion, normal   Away from an absorption band, the index of refraction exhibits normal dispersion. The index of refraction increases with decreasing wavelength, and the rate of its increase becomes larger at shorter wavelengths. For a variety of materials, the refractive index versus wavelength curve is steeper if the material has a larger refractive index, but the curve for one substance cannot be merely shifted in ordinate scale to obtain the curve for another. This latter property of normal dispersion means that the spectra from prisms of different substances never agree exactly in the relative spacing of spectral lines. Since there is no simple relation between the normal dispersion curves of different substances, their dispersion is termed irrational. In the visible region, all transparent substances that are not colored exhibit normal dispersion.

dispersion, partial   Difference in the index of refraction at two specified wavelengths. Used by optical designers to compare various pairs of glasses to determine which will give the least secondary spectrum in an achromat. The specified wavelengths are usually at the so-called Fraunhofer lines that are designated by the letters $A, B, C, \ldots$, starting at the extreme red. C denotes the red C line of hydrogen at 656.3 um, D denotes the average wavelength of the two yellow D lines of sodium, agreed on as 589.3 um, F denotes the blue F line of hydrogen at 486.1um. Glass catalogs denote the various partial dispersions at these wavelengths by $nD - nC, nC - nA, nF - nD$, the subscripts denoting the spectral lines at which the indices were determined.

dispersion, sound   Dependence of the phase velocity of a harmonic wave on the frequency of this wave. Sound dispersion can occur in ducts, waveguides, and *dispersive media*. In the latter case, sound dispersion is caused by inclusions in a medium (i.e., bubbles in water), by effects of thermoconductivity and viscosity on sound propagation, or by acoustic *relaxation*. Sound dispersion due to acoustic relaxation is well studied; in this case the dependence of the sound speed c on the frequency ω is given by $c^2(\omega) = c_0^2 + (c_\infty^2 - c_0^2)\frac{\omega^2}{\omega_r^2 + \omega^2}$. Here, c_0 and c_∞ are the values of c at $\omega = 0$ and $\omega = \infty$, respectively, and ω_r is the relaxation frequency.

dispersive medium   A medium in which the phase velocity of a harmonic sound wave depends on a frequency of this wave. Propagation of a sound wave through a dispersive medium results in *sound dispersion.*

dispersive power   A measure of the way the refractive index varies with wavelength in a sample of glass. Let F and C denote the blue F line and the red C line of hydrogen. The dispersive power is define by the equation $1/nu = nF - nC/nD - 1$ where the subscripts denote the spectral line at which the index of refrac-

tion is determined. The reciprocal of dispersive power, denoted by the Greek letter v, is between 30 and 60 for most optical glasses.

displacement, acoustic   Deviation of a particle in a medium from its equilibrium position because of the passage of a sound wave. This wave also causes deviations (fluctuations) in *pressure, fluid velocity* and *density* in a medium.

displacement current   Is the partial time derivative of the displacement, i.e., $\frac{\partial D}{\partial t}$. It was first introduced by James Clark Maxwell to complete Amperes Law.

dissipation, acoustic   Transformation of energy of acoustic oscillations into other forms of energy, such as heat.

dissonance   Nonharmoniuos sounding of two or more tones played together. The term dissonance is opposite to *consonance*. Dissonance occurs when unpleasant beats occur between partials of tones that form an interval.

distortion   When the output of an amplifier is not simply a scaled exact replica of the input signal; i.e., the transfer function of the amplifier is not perfectly linear for all inputs.

Different types of distortion can be identified based on how the amplifier's imperfections affect the output signal. *Linear distortion* describes *amplitude distortion*: different frequency components are amplified by different effective gains, and *phase distortion*: different frequency components have unequal phase shifts. *Non-linear distortion* produces new frequencies in the output of an amplifier that were not present in the original signal. *See* distortion, harmonic.

distortion (optical)   In the third order theory of aberrations, distortion is the fifth of the five Seidel sum representing deviations from the path prescribed by the Gaussian ray tracing formulas. A system is free of distortion when it has uniform lateral magnification over its entire field. Pinhole cameras show no distortion as all straight lines connecting each pair of conjugate points pass through the opening. Lenses show barrel distortion when the magnification decreases toward the edge of the field, and *pincushion distortion* when there is greater magnification towards the edge.

distortion acoustic   Any undesired change in an amplitude, frequency or phase of a signal in its transmission or reproduction in devices such as microphones, earphones, loudspeakers, etc. The main types of acoustic distortion are *amplitude distortion, frequency distortion,* and *non-linear distortion. Amplitude distortion* occurs when the ratio between the amplitudes of the output and input in a device is not constant for different values of the input amplitude for a fixed frequency. *Frequency distortion* occurs when the ratio between the amplitudes of the output and input is not constant for different frequencies of the harmonic input signal. *Non-linear distortion* occurs when there is no linear relationship between the input and the output. This distortion can result in production of harmonics in the output even if they are not presented in the input.

distortion, cross over   In push-pull amplifiers that employ two Class B transistor amplifiers, one transistor is used to amplify the positive going portion of an input signal while the other is used to handle the negative going portion. In principle, the amplifier should operate linearly for all portions of an input wave form. However, due to the initial curvature of the emitter-base diode characteristics of the transistor (i.e., increased input impedance for low signal levels), small signals are not accurately reproduced. Therefore, a sinusoidal base-voltage excitation will not yield a sinusoidal output current.

For example, shown in the figure is a typical dynamic characteristic transfer function for the two transistors of a push–pull amplifier. It is assumed that the bases of the transistors are given a pure sinusoidal input signal. The peaks of the sine wave are reproduced in the transistor output currents, but the signal near zero is distorted. This shows up as distortion in the output wave form as the output swings through zero volts: cross-over distortion.

distortion, harmonic   A convenient means by which the deviation from a sought wave

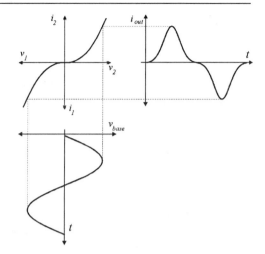

Dynamic transfer showing cross-over distortion of push–pull transistor amplifier.

form from an amplifier or signal generator can be measured. For example, it is a measure of how accurately an amplifier (or generator) can reproduce (generate) a sinusoidal wave. The deviation from a pure sinusoidal wave is expressed by the *total harmonic distortion* as

$$THD = 100\% \times \sqrt{A_2^2 + A_3^2 + A_4^2 + \cdots}$$

where A_k is the ratio of the amplitude of the kth harmonic to the amplitude of the fundamental frequency in the Fourier series representation of the given waveform.

diverging wave   A wave in which amplitude and energy are decreased with the distance of propagation. Point source, dipole, quadrupole and most other sources emit diverging waves. Two important particular kinds of diverging waves are a *spherical diverging wave,* the amplitude of which decreases with the distance r of propagation as $1/r$, and *cylindrically diverging wave,* the amplitude of which decreases as $1/\sqrt{r}$.

diversity system   This refers to the technique of providing more than one path for the establishment of a channel or circuit, thus providing a more reliable service.

divider   (1) Voltage or current divider. A system of resistors, or loads, intended to reduce a given voltage or current to a desired fraction of the original value. A simple voltage divider is shown below. The output voltage is

$$V_{out} = \frac{R_2}{R_2 + R_1} V_{in} \,,$$

assuming no current is drawn on the output.

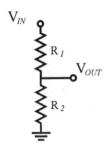

A voltage divider based on two resistors.

(2) Frequency divider. *See* frequency divider.

D-lines of sodium   The emission spectra of sodium vapor consists of two bright yellow spectral lines called the sodium D_1 and D_2 lines. Their wavelengths are 5889.95 Å and 5895.92 Å, respectively. They result from transitions between the spin-orbit split $3p$ excited states and the $3s$ ground state.

DNA structure   DNA (deoxyribonucleic acid) is a polymeric molecule in which the monomer is composed of a ribose sugar, a phosphate group and one of four bases (the purines *adenine* and *guanine,* and the pyrimidines *cytosine* and *thymine*). The deoxyribose and phosphate groups form the backbone of the polymer. Double-helical DNA is formed by hydrogen-bonding of the bases on two strands of DNA. The nature of the hydrogen bonding is such that adenine always binds with thymine and guanine always binds with cytosine. The stability of the double helices are affected both by the hydrogen-bonding within base pairs and by stacking interactions (van der Waals in nature) of the planar bases. Four different right-handed conformations (denoted A, B, C and D) have been observed and one left-handed conformation (called Z) has been reported. DNA in solution adopts the B conformation, making it clearly important for living cells. The A conformation might have biological relevance since RNA is always found in the A conformation. Z-DNA has been associated with gene regulation. The C and D conformations appear to exist only in the laboratory. The right-handed conformations differ by their helical pitch as well as the position and orientation of the base pairs within the double helix. The genetic code itself is contained in the sequence of the bases along one of the two strands (known as the *sense strand*).

dominant wavelength   One of the three objective parameters (dominant wavelength, colorimetric purity and luminosity) used in colorimetric representation of the psychological sensation of light. In the monochromatic method of colorimetry, half the photometric field is illuminated by the color to be matched, and the other half by a mixture of controllable amounts of monochromatic light of adjustable wavelengths and white light of definite spectral quality. The dominant wavelength is the wavelength of the monochromatic light. The luminosity is the sum of the monochromatic and white light luminosities. The colorimetric purity is the ratio of the luminosity of the monochromatic radiation to the total luminosity.

Donnan equilibrium   Gibbs-Donnan equilibrium exists when, on opposite sides of a semipermeable membrane, the product of diffusable anion and counterion concentrations are equal, and when the sum of the concentrations for diffusable and nondiffusable anions and the sum of the diffusable and nondiffusable counterions are equal. This creates the membrane potential.

donor   A donor is an impurity added to a semiconducting material, either during the crystal growing process, or later by diffusion. The donor atoms are substituted for the original semiconductor atoms within the crystaline structure. However, in forming the covalent bonds with neighboring atoms, the donor atom has one (or more) extra electron(s) that can act as extra charge carriers in the bulk of the crystal.

For example, pure silicon (Si) has four electrons in its outermost unfilled shell (valence electrons). In forming the crystal, Si forms four covalent bonds with four different neighbors; each bond is two electrons with opposite spins.

The geometrical arrangement of the four bonds is tetrahedral. Thus, the crystaline structure is *diamond-like*.

If a pentavalent atom (like P, As, Sb, or Bi) is substituted for one of the original Si atoms, four of the five valence electrons will participate normally in the diamond-like structure and there will be an extra electron. (*See the first accompanying figure.*) Normally this electron would be tightly bound to the donor ion. However, because the electron moves in the bulk of the semiconductor, the binding energy of this atom to the donor ion is greatly reduced. This is because the electron now moves in a material with a dielectric constant, ϵ, much higher than vacuum and thus feels a reduced Coulomb attraction, and also because the effective mass of the electron, m_e^*, is reduced within the solid.

the electron and $\hbar$ is Planck's constant divided by 2π.

In the solid state, the allowed energies of the electrons form *bands* and for the pure semiconductor, most of the electrons are in the valence band and cannot contribute to the bulk conductivity. However, the electrons associated with the donors are only $\approx 0.02 \, eV$ (*see above*) from the conduction band. Since thermal energies at room temperature are $\approx \frac{1}{40}$ eV, this electron can easily be ionized by thermal agitation and participate as a charge carrier. (*See the second acompanying figure.*) Thus, donor electrons also have the effect of raising the Fermi level (the energy of a state for which the probability for occupation is $\frac{1}{2}$) closer to the conduction band. *See also* acceptor; doping.

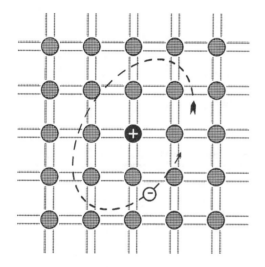

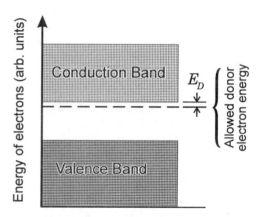

Qualitative illustration of allowed electron energies in a donor doped (n-type) semiconductor.

Two-dimensional projection of a tetrahedral crystaline structure. Also shown is a donor atom with its (extra) loosely bound electron.

An estimation of the binding energy of this electron to the donor can be calculated by approximating the donor ion and electron pair as a hydrogenic system. Then, from elementary quantum mechanics, the binding energy is given by

$$E_D = \frac{-e^4 m_e^*}{2\epsilon^2 \hbar^2} \,,$$

or $\approx 0.02 \, eV$, using numbers for a typical semiconductor. Here, e is the elementary charge of

doping   Doping is the deliberate addition of impurities to a semiconductor. If the impurity atom is approximately the same size as the indigenous semiconductor atoms and does not disrupt the normal crystalline structure, then adding impurities has the effect of replacing some of the original atoms with the impurity atoms. This is done to change the available number of charge carriers in the bulk semiconductor, thus enhancing the conductivity. When this is done, the conductivity changes from *intrinsic* to *extrinsic* conductivity and can have vastly different temperature dependences. *See also* intrinsic conductivity.

If an impurity contributes extra electrons to the semiconductor, it is called a donor and the crystal as a whole becomes an *n–type* semiconductor. If, on the other hand, the impurity contributes a hole, then it is called an acceptor and the semiconductor becomes *p–type*. This is summarized in the accompanying table.

Summary of basic semiconductor doping schemes

impurity class	type	contributed carrier
donor	n-type	electron
acceptor	p-type	hole

Electrical devices with desirable electrical characteristics can be constructed by judiciously forming junctions of different types and concentrations of impurities. For example a suitable junction between an n–type and a p–type semiconductor forms a *p–n junction* — a basic diode. *See also* donor; acceptor; diode.

Doppler effect (1) The change of the apparent frequency of the source of electromagnetic or acoustic radiation due to the relative motion of the source and observer. If the source emits light of frequency ν, wavelength λ, and its motion is towards the observer with a velocity u, the observer receives u/λ waves in addition to the number $\nu = c/\lambda$ that would reach the observer in the absence of relative motion. If the motion is away from the observer, then u/λ fewer waves are counted.

(2) Difference in frequencies of a sound or electromagnetic wave radiated by a source and received by an observer which are in relative motion. In acoustics, the Doppler effect in a homogeneous non-moving medium is given by the formula $\omega' = \frac{\omega}{1-\mathbf{n}\cdot\mathbf{u}/c}$. Here, ω and ω' are the angular frequencies of the radiated and received wave, $\mathbf{u}$ is the source velocity, c is the sound speed, $\mathbf{n}$ is the unit vector in the direction of wave propagation, and it is assumed, for simplicity, that the receiver is at rest. The difference $\Delta\omega = \omega' - \omega$ is called the *Doppler shift*. The Doppler shift $\Delta\omega$ is positive (negative) if the source is moving toward (away from) the receiver. In electrodynamics, the Doppler effect in a vacuum is given by $\omega' = \frac{\omega\sqrt{1-u^2/c^2}}{1-\mathbf{n}\cdot\mathbf{u}/c}$, where the notations are the same, except c, that is, light velocity. The difference between the formulae for the Doppler effect in acoustics and electrodynamics is the factor $\sqrt{1 - u^2/c^2}$, which is due to the Lorentz transformation in the theory of relativity.

Doppler shift The change in frequency seen by an observer of a source of sinusoidal waves when the observer and source are in relative motion. For electromagnetic waves, all inertial frames are physically equivalent, so the Doppler shift depends only on the relative velocity. For sound waves, the medium is the preferred reference frame, and the shift can depend on the velocity of the medium as well.

dosimetry A procedure for measuring absorbed radiation dose.

dosimetry, thermoluminiscent Measurement of radiation dose absorbed by lithium fluoride by quantification of the light output of the heated material.

double Kevin bridge Double Kevin bridge is a type of resistance bridge designed to minimize the effects of lead or contact resistance. It is used to measure low values of resistance precisely. A diagram of double Kevin bridge is shown below. Here, the unknown resistance, x, and the standard resistance, S, are connected in series with a battery, variable resistor, a Galvanometer, and low resistance wire, l. The resistances r_1, r_2, r_3, and r_4 are contact resistances. When the balance is achieved, the value of X is given by

$$X = S\frac{A}{B} + Bl(a + b + l)\left(\frac{A}{B} - \frac{a}{b}\right).$$

Balance is achieved by adjusting variable resistance on the other two arms until they are equal.

double refraction The presence of two refracted beams from an unpolarized incident beam on an anisotropic material in place of the usual single refracted beam observed for isotropic materials. By measuring the angles

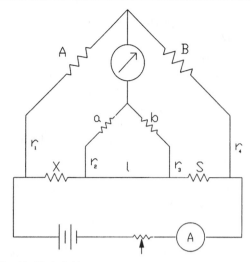

Double Kevin bridge.

of refraction, one of the double refracted beams is found to obey Snell's law, and is termed the *ordinary ray,* while the other does not and is termed the *extraordinary ray.* Double refraction in crystals of calcite and quartz allow the production of polarized light over a wider range of wavelengths than is possible using dichroic materials such as Polaroid. The frequency of the light appears to the observer to be increased by the ratio of the propagation speed to the propagation speed reduced by the relative speed. If the motion is away from the observer, the frequency appears to the observer to be decreased by the ratio of the propagation speed to the sum of the propagation speed and the relative speed.

doublet   A lens combination of opposite signs used for the elimination of spherical aberration. The amount of spherical aberration introduced by one lens of such a combination must be opposite to that introduced by the other. Neutralization is possible in a doublet because the spherical aberration varies as the cube of focal length and therefore changes sign with the focal length.

downlink   The transmission of data from a space vehicle, e.g., a missile or satellite, to the ground. It is usually modulated onto subcarriers and then onto RF carriers.

drain   One of the connections to the channel of an FET (field effect transistor). *See also* transistor, field effect.

drift   (**1**) Slow changes in operating parameters or conditions that affect the output of, for example, an electronic amplifier. This would be indistinguishable from a very low frequency input signal.

(**2**) Electron motion in a conductor or semiconductor under the influence of an applied electric field. The drift velocity is given by

$$v = \mu E$$

where μ is the (material-specific) mobility coefficient and E is the applied electric field.

drift velocity   Mean velocity of the current carrying particles upon the application of an electric field. It is given by $v = J/(ne)$, where J = current density, n = density of charge carriers and e = their charge. Typical values for metals are $10^{(-5)}$ m/s.

drum   An early high-speed, direct access storage device that used a magnetic-coated cylinder with tracks around its circumference. Each track had its own read/write head.

drum factor   The ratio of the length of the drum usable for scanning to its diameter, for a receiver or transmitter drum.

drum receiver   Facsimile apparatus in which the recording medium is attached to a rotating drum. This is scanned helically by a recording head.

drum speed   The speed with which the magnetic coated cylinder, used as a direct access device, rotates. *See* drum.

drum transmitter   Facsimile apparatus in which the document to be transmitted is attached to a rotating drum. This is scanned helically by a reading head.

dry cell   A cell in which the electrolyte is soaked by absorbing material to prevent the spilling or leakage of the electrolyte is called a *dry cell*. A common example of a dry cell is a carbon-zinc (Laclanche) cell.

duality principle   The duality principle states that the logical value of an boolean expression remains unchanged if

1. every conjunction is replaced with a disjunction,

2. every disjunction is replaced with a conjunction, and

3. every variable, term, and functional result is inverted.

An example of the duality principle is de Morgan's laws. To illustrate the principle, the expression

$$Y = A \otimes B$$

is logically equivalent to

$$\overline{Y} = \bar{A} \oplus \bar{B} ,$$

where the bar $\overline{(\ldots)}$ represents the inverse of that variable, $\otimes$ represents conjunction, and $\oplus$ is the disjunction operation.

One application of the duality principle is determining the logic function of a circuit if one changes the logic convention. If a circuit is defined in, say, positive logic, then the corresponding function implemented by the same circuit using negative logic is found by replacing each conjunction with a disjunction and vice versa. Note, however, that by "inverting" the logic convention, the truth value of each input signal (the input boolean variable) and output (functional result) is necessarily inverted. This can be illustrated by using the equations above as an example.

Consider a circuit with two inputs A and B which, when operated with a positive logic convention, acts as an **AND** gate. Shown in the first section of the acompanying table is the output of the gate, Y, for all possible states of the inputs using 1 and 0 to represent logical **true** and **false**. The second section is obtained by re-writing this information in terms of voltage levels H and L (*see* convention; logic gates) with the view that the gate is now just an arbitrary circuit. Now, if the circuit is operated using a negative logic convention (by redefining the relative potential of 0 volts and being mindful of proper power supply connections), the truth table shown in the third section is obtained. Thus, by inspection of the table, the circuit now operates as an **OR** gate in negative logic.

Operation of a circuit used in positive or negative logic

Positive logic			Arbitrary circuit			Negative logic		
A	B	Y	A	B	Y	A	B	Y
0	0	0	L	L	L	1	1	1
0	1	1	L	H	L	1	0	1
1	0	1	H	L	L	0	1	1
1	1	1	H	H	H	0	0	0

duct (waveguide), acoustic   (1) A pipe or tube along the interior of which sound is transmitted. These ducts can be of various shape and form (i.e., rectangular, cylindrical) and are widely used in practice.

(2) A medium where sound can propagate only between two layers or surfaces. Examples of these waveguides are an atmosphere between the ground and the height of temperature inversion, where sound waves can be trapped, and an ocean where sound is trapped between its surface and bottom.

duplexing   Term used to describe two systems that are functionally identical. They both may perform the same functions, or one may be standby in the event the other fails.

duty cycle   The capacity of a machine to work under normal conditions. For example, for a printer, it would indicate the number of pages that can be printed per month without a problem.

dynamic characteristics   The dynamic characteristic curve determines the output of a circuit or circuit element for a given input voltage. Usually, characteristic curves are a simple, one-to-one functional relationship between the input and output. With more complex characteristic curves, the output may be multi–valued and depend on the previous value of the input, e.g., hysteresis.

As an example, consider the diode–resistor circuit shown below. A voltage from a source is applied, V_s, across the series diode and resistor circuit, and the voltage developed across the resistor will be considered the output voltage. From elementary circuit analysis, the voltage dropped across the diode is

$$V_{\text{diode}} = V_s - I_{\text{diode}} R \ .$$

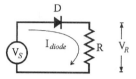

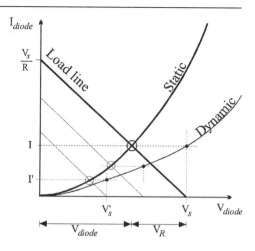

Example circuit for determining a dynamic characteristic curve.

To actually determine the voltage dropped across the diode, its *static* characteristic is required, relating the voltage across and current through the diode. The solution can be determined graphically by plotting, on the same graph, the *load line,*

$$I_{\text{diode}} = \frac{V_s}{R} - \frac{V_{\text{diode}}}{R}$$

where V_{diode} is treated as the independent variable and plotted on the x–axis and I_{diode} is plotted on the y-axis (see the second figure). Thus, the intersection of the static characteristic and load line determines the actual voltage and current for the given (and instantaneous) value of applied V_s.

To determine V_R, the output voltage, the *dynamic* characteristic I_{diode} as a function of V_s is to be determined. This is done by drawing a line from the solved value of I_{diode} and intersecting with a line extended from the applied V_s (this is shown in the figure by the dashed lines). Different values of applied voltage, V_S' for example, will yield a family of load lines and thus a family of currents corresponding to the applied voltage. In this manner, the dynamic characteristic of the circuit can be determined from the static curve of the diode. This curve is also shown.

Given the dynamic characteristic of the circuit, the resulting output for a given input can be determined. For time varying signals, the input is plotted as a function of time below the horizontal axis of the dynamic characteristic curve.

This is illustrated in the third figure, using the dynamic characteristic above as an example and assuming a sinusoidal input. The output is determined by reflecting the input waveform about the dynamic characteristic curve, also shown.

Graphical construction of dynamic curve from static characteristic and load line for the example circuit above.

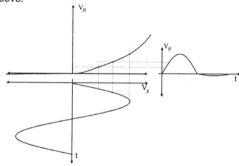

A dynamic characteristic curve used to determine the output voltage as a function of time for a given input waveform.

dynamometer A device, consisting of a set of parallel rollers, which allows the wheels of a vehicle to be driven while the vehicle remains stationary. Typically used for engine diagnostics.

dynamotor A device that contains a motor and one or more generator(s). The motor and the generator(s) have a common magnetic field and separate armature windings. One of the armature windings receives direct current and operates as a motor. The motor rotates and the other armature windings operate as generators. It is used for transformation of DC voltage.

E

ear   A component of a complex acoustical system of hearing. Three main parts of the human ear are the outer ear, the middle air, and the inner ear. The outer ear consists of the pinna and the auditory canal which is closed at the end by the eardrum. The middle ear has three small bones: the hammer, the anvil and the stirrup, which are connected to each other. The hummer and the stirrup are attached to the eardrum and to the oval window, respectively. The oval window is a membrane that divides the middle ear from the inner ear, which consists in part of the cochlea and *the basilar membrane.* A sound from the outer ear sets the eardrum into vibrations. These vibrations are further transmitted through the bones of the middle ear to the oval window and the basilar membrane, which has nerve endings connected to the brain. This process results in the hearing of sound. Human ears can localize a source of sound and distinguish its loudness, timbre and pitch. *See also* audibility, limits of; audio frequency; binaural.

ear, artificial   An artificial device to permit hearing for those whose ears do not function.

earthquake   Sudden motions in the earth's crust. The most common mechanism of earthquakes is displacement along a fault that generates seismic waves. Earthquakes can lead to devastating distractions on the earth's surface and tsunami in the ocean. Seismic waves generated by an earthquake can propagate through the crust, the mantle and the core of the earth, enabling retrieval of information about the internal structure of the earth.

echelette   Diffraction grating with large intervals and flat grooves inclined at an angle to reflect radiation in the direction of the order intended to be brightest. Term used by R.W. Wood for infrared grating of this type. Echelette gratings are now generally called *blazed grating.*

echelon   Diffraction grating for high resolution studies of a small portion of the spectrum, such as the hyperfine structure of lines or the Zeeman effect. Twenty to forty accurately plane-parallel plates are stacked together and staggered to form a series of steps with a constant offset of about 1 mm. The thickness of each plate is usually about 1 cm, so the grating space is very large and the concentration occurs in an extremely high ordering. The light is concentrated in a direction perpendicular to the fronts of the steps. At most, two orders of a given wavelength appear under the diffraction maxima. But the order is so large that the resolving power (order times the number of plates) is 100,000 to 1,000,000 even though the number of plates is relatively small. Echelon was pioneered by A.A. Michelson and was the first use of the principle of concentrating a light in particular order.

echo   A wave reflected or scattered by an object, surface or inhomogeneity in a medium. Echoes are a very common phenomenon in the propagation of sound. For example, a listener in a room hears not only direct sound from a source but also multiple echoes from walls (*see* acoustics of rooms). Bats use echoes to navigate and to find prey. Echoes are also used in acoustical instruments, such as *sonar* and *echo sounders,* to measure the distances to an object.

echocardiography   The use of ultrasound to study heart structure and motion. In this noninvasive technique, a transducer is held against the chest to send a beam of very high frequency sound waves to the heart. A certain fraction of these ultrasonic waves are reflected by the interface between two different types of tissue. The same transducer receives the reflected signal that is then displayed to reveal the structure of the heart and the motion of the various parts of the heart throughout the cardiac cycle.

echoes, flutter   A succession of echoes from a single impulse source that do not overlap and occur rapidly one after the other. Flutter echoes can appear in auditoriums and rooms with highly reflective walls. Flutter echoes significantly decrease the acoustic quality of auditoriums and rooms and, therefore, should be eliminated.

echo, harmonic An *echo* produced by scattering or reflection of an overtone rather than the fundamental frequency of a complex sound. Harmonic echoes can occur because scattering amplitude is proportional to a power of frequency and the frequency of overtone is always greater than the fundamental frequency. Therefore, the amplitude of the harmonic echo can be greater than that produced by scattering of the fundamental frequency.

echo, musical A specific kind of *flutter echo*. Musical echoes occur when scattering objects are located at approximately uniformly increased distances from a source, resulting in periodic multiple echoes.

echo sounder A navigational acoustic system that is used to measure water depth. An echo sounder consists of a transducer that emits an acoustic pulse, which propagates from the ocean surface to the bottom, reflects and comes back to the transducer. A measured characteristic is the time interval Δt between pulse emission and reception. The ocean depth h is calculated by using the formula $h = c\Delta t/2$, where c is the sound speed in water.

eddy currents Currents induced in a conductor due to the presence of an applied, changing magnetic field through the conductor. These currents always circulate in such a way that the produced magnetic field opposes the change in the applied field, in accordance with Lenz's law.

effusive beam A beam of molecules formed by leakage of gas through a fine orifice. Graham's law applies when the mean free path is small compared to the dimensions of the orifice, the flow being analogous to a fluid jet forced out by pressure. At low pressures, the mean free path is large compared to the dimensions of the orifice, the mechanism of escape of gas is different, and it is termed *molecular effusion*. In this case the volume diffusing per second into a vacuum is directly proportional to the area of orifice and inversely proportional to the square root of molecular weight.

Einstein function The Einstein model for the heat capacity of phonons assumes all of the phonons are at the same energy. This assumption produces a heat capacity C_V

$$C_V = 3Nk_B \left(\frac{\hbar\omega}{k_B T} \right) \frac{e^{\hbar\omega/k_B T}}{(e^{\hbar\omega/k_B T} - 1)^2}$$

where $\hbar\omega$ is the energy of an oscillator. At high temperatures, the heat capacity becomes $3Nk_B$, the classical Dulong and Petit value. At low temperatures, the model predicts the heat capacity will decrease as $\exp(-\hbar\omega/k_B T)$ in contrast to experiments which show the heat capacity following the Debye model (T^3 at low temperatures). The equal-energy approximation for the oscillator energies is valid for the optical branches of the phonon spectrum, and this is where the Einstein model is most often used.

electric attraction According to Coulomb's law, electric charges with opposite polarity will attract each other. This is known as *electric attraction*.

electric conductivity The ability of a material to conduct electric current. It is expressed in terms of the current per unit of applied voltage.

electric conductor An electric conductor is a material that conducts electric current. Typical electric conductors are metals with free electrons as carriers.

electric dipole When two charges, opposite in sign (q and $-q$), are placed a very short distance d apart, they form an electric dipole with electric dipole moment $p = qd$. An ideal electric dipole exists where q goes to infinity and d goes to zero while maintaining the product qd constant.

electric dipole moment A measure of the strength of an electric dipole. In the simple point charge model, the electric dipole moment is defined as $p = qd$, where q is the charge and d is the distance between the two point charges. In a more general form, $p = \int x\rho(x)\,dx$.

electric displacement The vector D, which is related to the sum of the electric field E and the electric polarization P, by $D = \epsilon_0 E + P$. By Gauss's law it can be shown that the flux of

D through a closed surface is equal to the free charge within the surface.

electric field   The force per unit charge on a test charge in a given electric field. For the case of a static charged point particle Q, its electric field is found via Coulomb's law $E(r) = F(r, q)/q = Q/(4\pi r^2)r$.

electric field, energy in   The energy stored in an electric field is given by $E = \epsilon_0/2 \int E.E \, dv$. Thus the energy per unit volume in the field is $\epsilon_0/2E^2$.

electric field gradient   The gradient of the electric field vector. The artificial electric field gradient is used for particle acceleration. The electric field gradient around an atom is affected by the anisotropy of the electric charge distribution of the atom. It can be observed by using the nuclear magnetic resonance (NMR) technique.

electric field, induced   The electric field produced by a time variance.

electric flux   By definition, this is $E.da$, where E is the electric field, and da is a vector normal to an infinitesimal surface of area da. Gauss's law states that the electric flux through a closed surface is related to the enclosed charge Q.

electric flux density   Also known as *electric displacement,* and normally denoted by a symbol D. The density of theoretical lines of force that extend in all directions from an electric charge or a charged body. It is measured in coulombs-per-meter squared. The permittivity ϵ of the medium is given by D/E. The electric flux density D can defined as:

$$
\begin{aligned}
D &= \epsilon_0 E + P \,, \\
&= \epsilon E \,,
\end{aligned}
$$

where ϵ_0, E, and P are the permittivity of free space, electric field, and polarization, respectively. The permittivity ϵ of the medium is given by D/E. The divergence of the electric flux density is from surface charge density ρ :

$$
\mathrm{div}\, D = \rho \,.
$$

The rotation of the electric flux density is zero; $\mathrm{rot}\, D = 0$. The energy stored in a capacitor U is calculated from E and D:

$$
U = \frac{1}{2} \int E \cdot D \, dV \,.
$$

There is a relationship between magnetic field strength H, magnetic flux density B, magnetism M, and D:

$$
\mathrm{rot}\, H = \mathrm{rot} \left(\frac{B}{\mu_0} - M \right) = \frac{\partial D}{\partial t} \,,
$$

where μ_0 is the permeability.

electric intensity   Normally electric intensity is denoted by a symbol E. Currently, it is called the *electric field strength* or *electric field intensity.* The electric intensity is the strength of an electric field at a given point in the field and is equal to the force exerted by the field on unit charge at the point.

It is measured in volts per meter. Electric field strength E is the total electric field strength due to the set of point charge q_i:

$$
E(r) = \frac{1}{4\pi \epsilon_0} \sum_i \frac{q_i (r - r_i)}{|r - r_i|^3} \,,
$$

where ϵ_0 is the permittivity of free space. A scalar function electrostatic potential $\Phi(r)$ is defined as:

$$
\Phi(r) = \frac{1}{4\pi \epsilon_0} \sum_i \frac{q_i}{|r - r_i|} \,.
$$

E is derived from *Phi* as:

$$
E = -\nabla \Phi \,.
$$

electric moment   The torque exerted on an electric dipole in an electric field. A distribution of charges can be regarded as a dipole and its electric moment can be calculated as a vector equal to the product of the magnitude of the charges and the distance between the charges of the virtual dipole.

electric polarization P, the electric dipole moment per unit volume, given by Np, where p is the dipole moment of the molecule/atom, and N is their number per unit volume. One can show that P is related to the volume density ρ_b and surface density σ_b of bound charges via the relationships $\rho_p = -\nabla.P$ and $\sigma_b = P.n$, where n is the surface normal.

electric screening   A screen of conductive material used for the reduction of electric fields entering a particular region. *See* electrostatic shielding.

electric shock   A physiological stimulation caused by electric current passing through tissue. It sometimes causes involuntary contractions of the muscles. A severe shock can damage heart and even cause death.

electroacoustics   The branch of acoustics and technique that deals with basic principles and design of electroacoustical transducers that are used to convert sound energy into electrical energy or vice versa. One of the main tasks of electroacoustics is to find a relationship between an acoustic (electrical) input and an electrical (acoustic) output of a transducer. To solve this task, *equivalent electrical circuits of acoustic systems* are often used. The other important task of electroacoustics is to achieve maximum efficiency of an electroacoustical transducer and minimal *acoustic distortion*.

electrocardiography   The study of electrical potential produced by the heart beat at various locations on the surface of the body. The rhythmic contractions of the heart are controlled by an electrical signal generated in a specialized region of the right atrium called the *sinoatrial node*, the heart's natural pacemaker. This electrical signal spreads throughout the heart, causing its contraction.

electrocorticography   The study of the electrical activity of the brain via electrodes placed directly on the exposed cerebral cortex.

electrode   A conductor for emitting, deflecting, or collecting electric charge carriers in a cell, an electron tube, a semiconductor device, and so on. A positive electrode is usually called an *anode* and a negative electrode is called a *cathode*.

electrode dissipation   The power dissipated in the form of heat by an electrode. It is caused by bombardment by electrons and/or ions.

electrode efficiency   The ratio of the actual yield of metal deposited in an electrolytic cell to the yield that could be deposited theoretically as a result of electric current passed through the cell.

electrode, implanted   A conductor that has been placed inside tissue in order to detect electrical activity or to supply exciting pulses. Implantation may be accomplished surgically.

electrodes, beveled   A conductor used to establish contact with a part of a circuit.

electrodes, intercerebral   A conductor attached to the cerebrum in order to monitor electrical activity of that part of the brain.

electrodes, pH sensitive   A conducting material that is sensitive to the concentration of positive hydrogen ions in solution.

electrode, surface   A conductor attached to the outside of a body in order to monitor the electrical activity inside.

electrodynamometer   Instrument for measuring small currents.

electroencephalography   The study of the electrical activity of the brain as measured by electrodes attached to the scalp. Potential differences up to 100s of microvolts are recorded with this technique. Spectral analysis of electroencephalograms (EEG) show wave-like phenomena in several different frequency ranges, particularly when the subject is asleep. The lowest frequency components (between 1 and 3 Hz) are termed *delta waves* and are associated with deep sleep. *theta waves* have frequencies between 4 and 7 Hz. Most brain waves are found in the frequency range between 8 and 13 Hz. These waves are called *alpha waves* and are associated with light sleep. The highest frequencies wave, called *beta waves,* are in the 13 to 30 Hz range.

Several diseases, such as cerebral tumors and epilepsy, can be diagnosed by causing unusual features in the EEG. Electroencephalography has also been used very extensively in research on sleep and its various stages.

electrogenic pump   The mechanism responsible for transferring electrical charge across a membrane, resulting in a potential difference across that membrane.

electrokinesis   The movement of charged particles through a continuous medium under the influence of an electrical field. Such physical phenomena are frequently associated with a charged surface in an electrolytic solution. The four principal electrokinetic phenomena are *electrophoresis, electroosmosis, streaming potential* and *sedimentation potential.*

electroluminescence   Process by which electrical energy is converted directly into light without thermal losses, e.g., electron recombination of a *pn* junction.

electrolysis   The production of a chemical reaction by an electric current passing through an electrolyte. The chemical action is a oxidation-reduction reaction depending on an electron transfer at the electrodes. Positive ions migrate to the cathode. Positive ions are reduced (gain electrons) to form a neutral species at cathode. Negative ions migrate to the anode to be oxidized at the anode. Negative ions are oxidized (lose electrons) to form a neutral species at cathode.

electrolyte   A liquid that contains positive or negative ions as electric carriers and conducts electricity. It contains substances that act on one or both of the electrodes and cause a chemical action to generate an electric current flow. Electrolytes are molten ionic compounds or solutions containing ions, i.e., solutions of ionic salts or compounds that ionize in a solution. Usually, a liquid metal is not regarded as an electrolyte.

electrolytic dissociation   The separation of a neutral ionic compound into positive and negative ions, usually caused by dissolution. In case of reversible dissociation, the equilibrium constant of the reaction is called *dissociation constant.*

electrolytic polarization   The phenomenon of the existence of the electrolytic polarization voltage, which is required for a steady current to flow through an electrolytic cell. The irreversible chemical reaction around the electrode causes the electrolytic polarization voltage. The delay of diffusion and transport of the substances to the reaction around the electrode also cause electrolytic polarization. Depolarizing agents are used to reduce the electrolytic polarization, also known as *electrochemical polarization.*

electrolytic tank   A device used to make a model for solving electrostatic problems analogously by measurements made on a model electrolyte in the tank.

electromagnetic pump   This type of pump operates on the principle that a force is exerted upon a conductor that carries current in a magnetic field. Liquid metals have high conductivity and can be conveniently used in such pumps.

electromagnetic units   Electromagnetic units (EMU) are a system of electrical units. For each system,

$$\epsilon_0 \mu_0 = \frac{1}{c^2} \, ,$$

where ϵ_0 is the electric constant (the permittivity of free space), ϵ_0 is the magnetic constant (the permeability of free space), and c is the speed of light in free space. It is used in the centimeter, gram, second (CSG) system. Usually, electromagnetic units have the prefix *ab-* attached to the names of conventional units. In the EMU system, the unit of electric current is defined by making the coefficient constant of the force between a pair of parallel electric currents equal to two:

$$F = 2 \frac{ii'}{r^2} \, ,$$

where F is the force between a pair of parallel currents i, i' with distance r. The ab-ampere is the EMU of a current 1 ab-coulomb per second. It also can be said that, in the EMU system, the unit of magnetic poles is defined to make the coefficient constant of the force between the magnetic poles equal to one:

$$F = \frac{gg'}{r^2} \, ,$$

where F is the force between a pair of magnetic poles g, g' with distance r. The dimension of the ab-coulomb is different from the *coulomb* of SI units. The quantity of electric charge measured in electrostatic units q_{EMU} is related to one measured in SI units q:

$$\begin{aligned} q_{EMU} &= \sqrt{\frac{\mu_0}{4\pi}} \cdot q_{SI} \, , \\ &= \frac{1}{c}\frac{1}{\sqrt{4\pi\epsilon_0}} \cdot q_{SI} \, . \end{aligned}$$

The unit of the electric charge is (s dyn$^{1/2}$) and its dimension is (length$^{1/2}$ mass$^{1/2}$).

electromagnets   Electromagnets are temporary magnets that make use of electric currents to generate the magnetic field.

electrometer   A voltmeter with a very high resistance (10^14 ohm), suitable for measuring voltages on small capacitors, etc.

electrometer, Hoffmann   A sensitive electrometer consisting of a half-vane (only one blade vane) in a pair of segmented metal boxes (binants).

It is also known as a *binant electrometer*.

electrometer, Lindemann   A highly sensitive electrometer. It has a light needle supported by a torsion fiber surrounded by metal plate quadrants on all sides. The opposite metal plate of the quadrants are connected. The voltage between the plates causes a force on the needle. A microscope measures the deflection of the needle tip.

electrometer, quadrant   An electrometer consisting of a set of quadrants and a light vane suspended by a quartz fiber between the quadrants. The quadrants are oppositely connected in pairs. A quadrant electrometer is used to measure voltages and charges. The voltage between the pairs of quadrants causes the deflection of the vane. The angle of deflection is proportional to the voltage. The Dolezalek quadrant electrometer is well known.

electromotive force   Old term for the induced electric potential given by Faraday's induction law.

electromotive force, motional   An emf that arises in a conductor in relative motion to an external magnetic field due to Faraday's induction.

electromotive force, self induced   An emf that arises in inductances when their self field changes.

electromyography   The study of the electrical signals associated with muscular activity, usually the skeletal muscles. The voltage signals are detected either with electrodes attached to the surface of the skin or with needle electrodes. This technique is useful in studying neuromuscular function, possible damage to nerves, and in kinesiology.

electron charge   The charge of one electron is equal to 1.60218×10^{-19} coulombs. This quantity is a fundamental constant of nature and a basic unit of charge.

electronic mail (email)   The ability to compose, send and receive mail via the computer using some type of email program. That is, terminals can transmit documents such as letters, reports, and telexes to other computers or terminals. Such services can be accessed using a public network through a host computer and can be retrieved at other terminals.

electronic musical instruments   Musical instruments that generate electromagnetic vibrations of desired form and spectrum that are then converted into sound by means of electro-acoustical transducers. Examples of electronic musical instruments are an electronic piano, guitar, organ, and carillon. Electronic musical instruments are used to imitate sound of "standard" musical instruments, to simplify their construction and minimize their size, and to develop new musical instruments. Sound produced by electronic musical instruments is called *synthetic sound*.

electron multiplier   A device primarily used for the detection of single, elementary atomic particles such as electrons, photons, and ions. An electron multiplier consists of a sequence of electrodes called *dynodes* and produces an output pulse of charge for each incident particle.

If the particle to be detected is incident on the first dynode and has sufficient energy, it will cause some electrons to be ejected from the dynode surface upon impact. These secondary electrons are accelerated toward the next dynode in the sequence (by proper arrangement of relative electrostatic potentials) and they eject more electrons upon impact. These tertiary electrons are accelerated to a fourth dynode, and so on. Assuming that, nominally, three electrons are ejected from a dynode surface per incident electron, a multiplier with n dynodes will have a gain of approximately 3^n. For example, an electron multiplier with 10 dynodes will have a shower of 10^6 electrons collected on the last dynode for a given *single* particle incident on the first dynode. Thus, there is now a sufficient amount of charge collected on the last dynode that can be processed with conventional electronics.

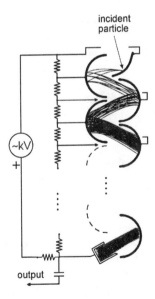

incident particle

The quintessential discrete dynode electron multiplier. Illustrated is electron multiplication of a single event (the incident particle) creating a shower of charge collected on the last dynode.

It is possible to have an electron multiplier with one continuous dynode instead of the discrete dynode chain described above. Here, the dynode is a film of moderately high resistance material with suitable secondary electron emission characteristics and is coated on the inside of a glass tube. A particle impinging on the beginning of the tube will eject secondary electrons and cause a shower of electrons to be accelerated and multiplied (in a manner similar to the discrete dynode multiplier) for collection at the end of the tube. A continuous dynode electron multiplier is usually physically smaller than a discrete multiplier with comparable gain.

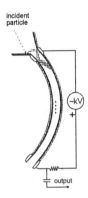

incident particle

A continuous dynode electron multiplier illustrating electron multiplication of a single event (the incident particle). Operation of the device is similar to the discrete dynode electron multipler.

electron nuclear double resonance (ENDOR) technique   A technique in which the magnetic resonance of a nucleus is detected by observing the resonance absorption of an associated unpaired electron. This effect is due to hyperfine coupling of the nuclear and electron magnetic moments. The statistical distribution of nuclear spin alignments within the sample causes each unpaired electron to experience a different local magnetic field. Because of the differences in the time scales for spin flipping, the electrons experience inhomogeneous broadening. To perform an ENDOR experiment, the sample chamber is bathed in both microwave and radio-frequency fields. The applied static magnetic field and the microwave radiation are tuned to an electron resonance. The intensity of the microwave field is sufficient to saturate the spin states for those matching electrons, a subset of all of the unpaired electrons. The frequency of the radio waves is slowly scanned and the microwave absorption of the cavity is monitored. When the frequency of the radio waves is correct to flip more nuclei into the orientation corresponding to the subset of unpaired electrons, the observed absorption of microwave radiation will increase. In this technique, the absorption of the microwave field is observed while the frequency of the radio field is scanned.

electron optics   Mathematical analogy between the passage of an electron beam through magnetic and electric fields and the passage of a beam of light through refracting media. A limited magnetic or electric field is considered to form a lens, which can be combined in ways analogous to optics to form various focusing instruments, such as an electron microscope.

electron-phonon interaction parameter
See superconductivity.

electron transport chain, photosynthesis
The chain by which an electron, excited by the absorption of light, moves through the pigment-protein complex of the light-harvesting structures to create a potential difference. All subsequent chemical reactions which convert CO_2 and water to carbohydrates are driven by this potential difference. In the purple bacterium *Rhodobacter sphaeroides*, optical excitation of a pair of chlorophyll molecules in the reaction center leads to the movement of an electron to a pheophytin molecule within a few picoseconds. This electron hops to a quinone molecule within about 200 psec. The electron hops to a second quinone molecule in about 100 μsec. After this process is repeated, a complex composed of two excited quinone molecules participates in chemical reactions which create the potential difference across the membrane in which the complex sits.

electron-volt   Energy equivalent to the kinetic energy gained by a particle with one electronic charge that is accelerated across one volt. It equals $1.6 \times 10^{(-19)} J$.

electrooptic effect   The changes in the propagation of light through matter due to the application of electric fields. Electrooptic effects include the *Kerr effect, electric double refraction, the Stark effect,* and the *inverse Stark effect.* In the *Kerr effect,* double refraction appears with the application of an electric field. Glass, some liquids (nitrobenzene), and gases exhibit this effect. The magnitude of this effect is proportional to the electric field strength squared. *Electric double refraction effect* refers to the appearance of double refraction at frequencies close to the absorption lines with the application of an electric field. In the *Stark effect,* spectral lines are split with the application of a strong electric field. The Stark effect with the lines appearing in absorption is called the *inverse Stark effect.*

electrophoresis   The movement of charged particles (usually macromolecules or colloidal particles) due to the presence of an electrical field. In most biochemical applications, a gel is used as the medium. Application of an electrical field to a collection of dissimilar charged particles in a gel will cause the particles to migrate in a direction determined by the sign of their charge and the polarity of the applied electrical field. The particles will move at different speeds depending on their net charge, size and shape. Since most species are in low-charged states (usually +1e and occasionally +2e), the geometric differences of the species are the dominant factor for determining the speed of the particle. The weight of the macromolecule itself is the usual cause for differences in size and shape, making electrophoresis an excellent method for separating macromolecules on the basis of weight.

electroretinography   The study of the electrical potential of the retina in response to stimulation via light.

electroscope   Instrument for detecting static electric charges. It consists of two thin sheets of conducting material that hang freely from a conducting pivot. When a statically charged object is brought near the pivot, the plates separate due to the mutual repulsion of the like charges that are on both plates.

electrostatic field   The electric field created by stationary charges. The force on test charges placed in such a field is given by Coulomb's law.

electrostatic focusing   Focusing of the electron beam by use of an electrostatic lens that varies the electric field. *See* electrostatic lens.

electrostatic hazards   General term applied to potentially dangerous situations that are caused by the build-up and eventual discharge of static electric charges. Such discharges can cause fires/explosions in environments with high concentrations of flamable or powered materials.

electrostatic induction   The generation of an electrical charge on a conductor by a electric field. A positive charge causes a negative charge on the uncharged object that is nearest to the original positive charge.

electrostatic lens   An device that generates an electrostatic field to cause electron beams to converge or diverge. It is used in a cathode ray tube (CRT) and a electron microscope. *See also* microscope, electron. In a CRT, the electrostatic lens consists of a focus anode, accelerating anode, and a control grid. The electrodes of an electrostatic lens have cylindrical form and concentric with the electron beam. The focus anode and accelerating anode are maintained at a positive potential with respect to the cathode. The potential of the accelerating anode is higher than that of the focus anode. The control grid controls the energy of the beam and consequently the intensity of the beam spot. The focal length of this lens depends on the potential of the focus anode. The potential of the acceleration anode also affects the focal length.

electrostatics   The branch of electricity study that studies electrical charges at rest, such as charge objects and stationary electric fields, and the electric fields associated with them.

electrostatic screening   An electrostatic shield that consists of a number of parallel conducting wires or rods connected at one end in order to obstruct electric flux, while permitting the passage of magnetic flux.

electrostatic shielding   A grounded conductive screen or enclosure placed around a device or between two devices to obstruct electric fields.

electrostatic units   Electrostatic units (ESU) are a system of electrical units in the centimeter, gram, second (CGS) system. For each system,

$$\epsilon_0 \mu_0 = \frac{1}{c^2} ,$$

where ϵ_0 is the electric constant (the permittivity of free space), ϵ_0 is the magnetic constant (the permeability of free space), and c is the speed of light in free space.

Electrostatic units have the prefix *stat-* attached to the name of conventional units. In an ESU system, the unit of electric charge is defined to make the coefficient constant of Coulomb's law equal to two:

$$F = \frac{qq'}{r^2} ,$$

where F is the coulomb force between a pair of charges q, q' with distance of r.

The ESU of electric charge is the *stat-coulomb*. ESU is based on this stat-coulomb, a unit of electric charge that exerts a force of 1 dyne on another unit charge at a distance of 1 cm in a free space. The dimension of the stat-coulomb is different from the *coulomb* of SI units. The quantity of electric charge measured in electrostatic units q_{ESU} is related to one measured in SI units q:

$$q_{ESU} = \frac{1}{\sqrt{4\pi \epsilon_0}} \cdot q_{SI} ,$$

$$= c\sqrt{\frac{\mu_0}{4\pi}} \cdot q_{SI} .$$

This unit system has been replaced for most purposes by SI units. Some of the relationships between a quantity in an ESU unit system and one in SI unit system are

$$E_{ESU} = \sqrt{4\pi \epsilon_0} \cdot E_{SI} ,$$

$$H_{ESU} = \frac{H_{SI}}{\sqrt{\epsilon_0/4\pi}} ,$$

where E and H are the electric field strength and the magnetic field strength, respectively,

Dimension of ESU units

	unit		dimension		
q	cm	$dyn^{1/2}$	$L^{3/2}$	$M^{1/2}$	T^{-1}
E	cm^{-1}	$dyn^{1/2}$	$L^{-1/2}$	$M^{1/2}$	T^{-1}
H	s^{-1}	$dyn^{1/2}$	$L^{1/2}$	$M^{1/2}$	T^{-2}
	L: length, M: mass, T: time.				

electrostatic voltmeter A voltmeter that measures the voltage applied between a fixed metal plate and a rotating metal plate placed close to each other. The applied voltage causes an electrostatic force that deflects the rotating plate against the torque of a spring. The arc of the plate rotated is proportional to the applied voltage.

electrosurgery Surgery performed using an active electrode as a cutting device or to promote coagulation. An alternating current in the MHz range is frequently used.

emitter One of three connections on a bipolar transistor. It is responsible for injecting the majority charge carriers into the transistor, i.e., electrons for *n-p-n* type transistors and holes for *p-n-p* type transistors. *See also* transistor, junction; bipolar code.

emmetropia A visual condition in which an infinitely distant object is imaged sharply on the retina without inducing any accommodation of the eye lens.

encoder (channel) A communication channel by which digital signals are transmitted. Channel coding is concerned with the characteristics of the transmission channel; the processed data must be compatible with the requirements of the channel. For example, a television camera produces a signal in two clear parts — *luminance* and *chrominance*. These two signals are combined at the transmitter by this process and the signals are decoded at the receiver.

encryption The process by which data is scrambled, or made unreadable, to prevent unauthorized access. It is the process of changing original data to ciphertext so that they cannot be understood until the ciphertext is decrypted into cleartext at the distant end of link.

end-of-pulsing signal *See* end-of-transmission character.

end-of-transmission character Also known as *end-of-pulsing signal*. A signal sent forward to indicate that the address signals have all been transmitted.

endoscopy The inspection of an internal cavity through a device made for that purpose. Endoscopy is important both in diagnosis of disease or injury as well as in surgery. The increased vision made possible via endoscopy permits surgeons to make much smaller incisions. This results in much less trauma to the tissues of the patient and speeds recovery time.

energy conservation, acoustic, law of The relationship between the acoustic energy density E and the acoustic energy flux $\mathbf{I}$: $\frac{\partial E}{\partial t} + \nabla \cdot \mathbf{I} = 0$. Here, t is time, and $\nabla = (\partial/\partial x, \partial/\partial y, \partial/\partial z)$. According to this relationship, acoustic energy is conserved in a process of sound propagation. The law of acoustic energy conservation holds in a non-moving medium, where E and $\mathbf{I}$ are determined by $E = p^2/(\varrho c^2)$ and $\mathbf{I} = Ec\mathbf{n}$. Here, p is the *acoustic pressure*, ϱ is the *ambient density*, c is the sound speed, and $\mathbf{n}$ is a unit vector in the direction of wave propagation. However, the law of acoustic energy conservation can be violated in a moving medium due to an exchange of energy between the mean flow and a sound wave. The total energy in a system of the medium and the sound wave remains, of course, the same.

energy, exchange *See* interaction, exchange.

energy, in magnetic field This is given by the volume integral of the energy density $1/2\mu_o H^2$ where μ_o is the permeability of free space and H is the magnetic field strength.

energy, zero point The energy present in a quantum mechanical system at absolute zero temperature. In the quantum harmonic oscillator, for example, the zero point energy is $1/2 \hbar\omega$ where ω is the angular frequency of the oscillator. In a molecule, the zero point energy can make up a substantial fraction of the binding energy. In a crystal, the zero point energy is found in the lattice vibrations.

entrance window The stop image on the object side that subtends the smallest angle at the center of the entrance pupil. Alternately, the image of the field stop formed by the part of the optical system that precedes it.

envelope A group of waves having slightly different frequencies travelling together characterized by the group velocity. In a public data network, a group of binary digits consisting of a byte together with additional digits required for the operation of the network, such as start and stop pulses. *See* envelop delay.

envelope delay Also known as *frequency delay*. Different frequencies arrive at the remote end at different times; frequencies around 1200 Hz are received first with the lower and higher frequencies arriving later. Frequencies in range 2900 Hz may come in more than 2 ms later; that is, bands of frequencies travel together in envelopes. It creates a degradation of the signal similar to what attenuation causes.

EOR An abbreviation for an *exclusive* **OR** logic function, where **OR** represents the disjunction. In other words, the output of this function is logical **true** if and only if one of the inputs are **true**. Compare this to the **OR** whose output is **true** if either (i.e., all) input variables are **true**.

The **EOR** is a basic (but not elementary) logic gate that frequently occurs as an independent unit. If A and B are input boolean variables, the output of the function **EOR** can be defined using the elementary **AND** and **OR** functions (*see* conjunction; disjunction) by the boolean expression

$$Y = A \textbf{ EOR } B \equiv (\bar{A} \otimes B) \oplus (A \otimes \bar{B}) \,,$$

where $\bar{A} \equiv$ **NOT** A is the logical inverse of A. Given this equation and the rules concerning **AND** and **OR**, all possible values of $Y = A$ **EOR** B can be tabulated in a truth table. Using 1 and 0 to represent **true** and **false** respectively, a truth table is presented below.

A circuit performing the **EOR** function in digital electronics is represented by a symbol similar to that for the **OR** — with an extra curved

Output values of EOR for all possible values of inputs A **and** B

A	B	$Y = A$ **EOR** B
0	0	0
0	1	1
1	0	1
1	1	0

line on the input side of the gate. This is illustrated in the following figure.

Symbol representing exclusive O R (E O R) in digital electronics.

episcope An optical system used to project a real, enlarged image of an opaque object (generally a flat picture). The object is illuminated as strongly as possible. Light reflected from the object is reflected from a mirror and then focused by a projection lens. The lens has to have the largest possible aperture to collect the scattered rays, resulting in rather poor image quality. (Also called *epidiascope*.)

equalization The process of compensating for frequency-dependent gain in the amplification, transmission, and/or reproduction of data, particularly voice data.

Transmission lines do not transmit all frequencies at the same velocity or suffer the same attenuation. So, to compensate for losses and distortion, equalization circuits can be inserted that compensate for the frequency-dependent transmission. This is accomplished by tailor designing the frequency response of the equalization amplifier.

Also, in sound recording and playback, different analog recording techniques will naturally have different frequency responses inherent to each process. Upon playback, equalization circuits can compensate the recorded signal. Recommended frequency responses to be used on playback with the various techniques (mag-

netic tape, magnet phonograph, crystal phonograph, etc.) have been provided by societies
such as the Record Industry Association of
America (RIAA).

equipotential surface   A surface in space
over which the electric potential is constant. For
a point charge, the equipotential surfaces are
concentric spheres centered on the charge. The
electric field E is always normal to the equipotential surface.

equivalent circuit   A circuit that can replace
a given circuit while maintaining the same func-
tionality with respect to measurable voltages and
currents. Usually such replacement is done in
the context of an approximation or simplifica-
tion for the purpose of analyzing a complex circuit. Circuits with two terminals can be replaced
with their Thevenin or Norton equivalents. Circuits with three terminals, particularly those in
electricity generation and distribution, can be
analyzed using Y or Δ equivalent circuits. An
example of the application can be found in the
analysis of transistors. *See* h-parameters.

**equivalent electrical circuit of an acoustic
system**   An electrical circuit that is described by the same differential equations as
those for an acoustic system under consideration. Many acoustic systems and electrical
circuits are described by analogous differential
equations. Therefore, an analysis of an acoustic system can usually be reduced to that of a
certain electrical circuit that is called the *equivalent electrical circuit*. Then, a consideration
of this equivalent electrical circuit is done by
well-developed methods of circuit analysis, and
the results obtained are used in analysis of the
acoustic system. Equivalent electrical circuits
are often used in theory and design of acoustic
devices and electroacoustical transducers.

ergodicity   Refers to behavior of random processes. Deals with correlations that may exist
between parts of the same signal or between
parts of one signal or another. If a process is
ergodic, the autocorrelation function obtained
from a member function, of sufficiently long duration, is the same as that obtained from the process as a whole, i.e., the autocorrelation function

may be expressed equivalently as an ensemble
average or as a time average.

erlang   Unit of measure of telephone traffic
engineering, which gives a measure of the total
traffic load on link, after Danish engineer A.K.
Erlang. For example, if the average number of
simultaneous calls in progress in a given period
over a particular group of trunks is N, then the
traffic intensity is N erlang. If there is one permanently engaged circuit, the traffic is 1 erlang.

etalon   Fabry-Perot interferometer that consists of two semi-silvered optically flat plates
that are fixed accurately parallel to each other,
with an air separation gap ranging from millimeters to centimeters. Etalons produce sharp
fringes and high resolving power, and are used
to accurately compare wavelengths and in the
study of hyperfine line structure.

ether   Medium filling all space that was believed necessary for propagation of electromagnetic waves. The medium had mechanical properties that were adjusted to provide a consistent
theory for electricity, magnetism, action at a distance, and the transmission of light and heat.
To account for the transmission of light, ether
was assumed to pervade all space and matter,
and have greater density in matter than in free
space, and be so elastic as to transmit transverse
waves with the speed of light. Michelson and
Morely attempted by optical means to measure
the motion of the earth through the ether, and
failed to detect any ether drift. Einstein's theory
of relativity has shown that these experimental
results and all theoretical ideas connected with
the concept of an ether can be systematized in a
self-consistent manner without reference to the
properties of the ether medium. From this point
of view, the ether is no longer required for an explanation of the empirical facts and has become
an unnecessary appendage of physical theory.

Ethernet   A method of connecting devices in
a local environment developed by Xerox corpo-
ration. It allows for transmission of data, using
network topology at up to 10 million bps for up
to half a mile. Workstations can exist on the
same cable but are only able to communicate
one at a time. To overcome these problems,

switched Ethernet and fast Ethernet were invented and combinations of them are also used.

evaporation (low temperature)   At low temperatures, evaporation, like so many other physical processes, takes on a quantum nature. Quantum evaporation occurs when an excitation propagates to a free surface of the material in question, annihilates, and emits one atom or molecule into free space. This can only happen when the energy of the excitation is greater than the binding energy of the material. In most solids, the atoms or molecules are bound too tightly to be liberated from the surface by a single excitation. Quantum evaporation is possible in superfluid ^{4}He, however. Both phonons and rotons can carry enough energy to cause quantum evaporation. Experiments studying quantum evaporation can provide direct measurements of phonons and rotons at high energies (E > 10 K) while at low temperatures (T < 0.2 K).

excimer   A molecular complex of two, usually identical, atoms or molecules that is stable only when one of them is in an excited state. Literally, a contraction of "excited dimer". An excimer laser is a rare-gas halide or rare-gas metal vapor laser emitting in the ultraviolet range (126 to 558 nm). It operates on electronic transitions of excimer molecules whose ground state is essentially repulsive. Lasing gases include the diatomic molecules ArCl, ArF, KrCl, KrF, XeCl and XeF. Excitation may be by electric discharge or electron beam.

exciplex   Strictly used, the term "excimer" refers to excited species made by the combination of two identical atoms or molecules. Excited complexes that do not fall into this category are more accurately called *exciplexes* or *heteroexcimers*.

excitation of vibrations   Setting an acoustical, electrical or mechanical system into vibrations. Excitation of vibrations always results in supplying energy to the system. Vibrations can be excited by direct action on a system, e.g., by a driving force that starts acting on a system (*see* forced oscillations). An example of such excitation is a pendulum pushed at some time moment. Vibrations can also be excited by changing parameters of a system, e.g., by changing a length of an oscillating pendulum.

excitons   A bound electron-hole pair found in nonmetallic solids. The two main types of excitons are *Frenkel* and *Wannier*. The *Frenkel exciton* usually exists in molecular solids and is highly localized. A Frenkel exciton may be confined to a single molecule, but can move through the solid by a hopping mechanism. The *Wannier exciton* usually exists in semiconductors and is very delocalized with the electron in the conduction band and a hole in the valence band. The coulombic interaction binding the electron and hole together is diminished by the dielectric screening of the material.

exercise testing   The process of performing medical examinations on a patient during physical exercise, such as walking or running on a treadmill. This allows for observation of bodily functions under a wide variety of physical stresses and helps with the diagnosis of disease conditions.

exit window   The image of the entrance window, as formed by the complete optical system. Alternately, the image of the field stop formed by the part of the optical system that follows it.

extinction coefficient   The imaginary part of the complex index of refraction. The extinction coefficient simplifies expressions of light intensity in the vicinity of the absorption band in dispersion theory, and refers to the extinction of electromagnetic waves during propagation through an absorbing medium. The term *attenuation index* is also used. The term *extinction ratio* refers to the ratio of the power of a plane-polarized beam that is transmitted through a polarizer with its polarizing axis parallel to the electric field vector of the beam, to the transmitted power when the polarizer's axis is perpendicular.

eye   It is the sense organ of vision.

eye diagram   Deterministic degradations such as errors in equalizing, offsets in decision timing, and gain errors in digital systems are assessed using this type of diagram. In the absence of noise, the width of the eye opening gives the time interval over which the received signal can be sampled without error. The rate of the closure of the eye gives the sensitivity of the system to decision timing errors. The height of the eye at a specified decision time determines the margin over the noise.

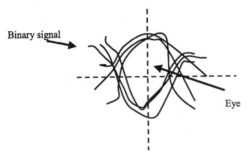

Eye diagram.

eye, far-sighted   Also known as *hyperopia*. In this condition, sharp focus in the relaxed (non-accommodating) eye occurs behind the retina, resulting in defective vision for near objects. This is a form of *ametropia*.

eye, near-sighted   Also known as *myopia*. In this condition, sharp focus, in the relaxed (non-accommodating) eye occurs in front of the retina, resulting in defective vision for far objects. This is a form of *ametropia*.

eyepiece   An eyepiece magnifies the image from a microscope objective and presents this image to the observer's eye.

eyepiece, compensating   Modifications of the Huygen's eyepiece in that both the lenses used are doublet lenses, thus avoiding the chromatic differences in the magnification.

eyepiece, erecting   Usually, a system consisting of two convex lenses to render the final image

erect. The use of lenses introduces aberrations and may make the optical system considerably longer. These disadvantages may be avoided by using a pair of erecting prisms.

eyepiece, Gaussian   A modification of the Ramsden eyepiece obtained by adding a thin parallel-sided plate of glass at an angle of 45° to the optical axis. When light enters an aperture on the side of the tube and, after partial reflection by the plate, travels through the axis of the eyepiece, it illuminates the crosswires at the foci of the objective and the eyepiece. When used in a telescope, such an eyepiece enables the telescope to be focused and to be placed perpendicular to a plane surface by requiring that the images of the crosswires formed from a plane mirror placed at the objective end be focused and coincide with the crosswires themselves.

eyepiece, Huygen   Also called the *negative type eyepiece*. It consists of two simple plano-convex lenses separated by half the sum of the focal lengths, and the focal plane falls close to the field lens. This eyepiece is usually used with low-powered standard objectives.

eyepiece, Kellner   A modification of the Ramsden eyepiece, in which the eye lens is replaced by an achromatic doublet.

eyepiece, Ramsden's   Also called the *positive type eyepiece*. It consists of two plano-convex lens facing each other and having the same focal lengths. The focal plane lies outside of the optics on the objective side and therefore a reticule can be placed in the tube of the eyepiece at the focal plane.

Eyring theory   A theory, based on statistical mechanics, that determines the rate of a given reaction in terms of the allowed energy levels and the temperature.

F

facsimile   A method of transmitting images or printed matter by electronic means. The image is optically scanned line by line at the transmitter, and the light and dark areas are converted into digital information. This is transmitted over the network using telephone lines or fax modems and reconstructed at the receiving station and duplicated in some form of film or paper. More sophisticated machines are able to skip over blank lines thereby reducing transmission times.

fading (signal)   Adverse transmission conditions can cause signals to be received simultaneously over more than one transmission path. In microwave transmission it can be caused by reflections from large objects such as buildings or aircraft.

farad   The unit of capacitance, equal to one coulomb per volt.

Faraday cage   A cage made of an electrically conducting material that is used to protect internal devices from outside electric fields. It works on the principle that there can be no electric field inside a conductor.

Faraday disk dynamo   A device consisting of a copper disk in which a radial EMF is induced when the disk is rotated between the poles of a magnet.

Faraday effect   Rotation of the plane of polarization of a beam of light passing through certain materials in the direction of applied magnetic lines of force. First discovered by Faraday in 1845 in heavy flint glass, which exhibits the effect markedly; it was one of the earliest indications of a connection between light and electromagnetism. The effect was also discovered in water and in quartz, and has since been observed for many solids, liquids, and gases. The effect is thousands of times stronger for thin transpar-ent films of ferromagnetic materials. The *Faraday effect* is not restricted to optical frequencies, and has been observed with microwave and radio frequencies. The rotation of polarization in the Faraday effect is independent of the sense in which the beam follows the field lines, which distinguishes it from the natural optical activity of the media — if the beam is reflected back along its path, the rotation is doubled only for the Faraday effect. The angle of rotation is proportional to the strength of the magnetic field and to the path length in the material. The proportionality constant is called *Verdet's constant,* and is the ratio of the angle turned to the distanced traversed evaluated for a unit magnetic field. The phenomenon can be understood in terms of a difference in the index of refraction for left- and right-hand circularly polarized light — the velocity of circularly polarized light depends on the direction of rotation. This variation of velocity comes about because the absorption frequencies and thus the dispersion of light depend on the polarization. The theory of the Faraday effect is closely related to the theory of the *Zeeman effect.* A well-known present day application of the Faraday effect is in protective devices used to prevent the destruction of high-power laser systems by back reflections from the target.

Faraday's law of induction   The law of electromagnetic induction states that the current induced in a conductor when it is subjected to c changing magnetic flux is proportional to the time rate of change of the magnetic flux.

far-field pattern   A region sufficiently far from an aperture or source (such as a light-emitting diode, injection laser diode, or the end of an optical waveguide) where the diffraction pattern is essentially the same as that at infinity. For all points within the far-field region, the diffraction pattern does not change significantly with distance.

fault current   The current that flows in a circuit under abnormal conditions. It typically occurs in a circuit in which there has been a loss of insulation between conductors or between a conductor and the ground.

Fechner fraction If the threshold in brightness (also known as *luminous flux* or *luminous sterance*) for the eye to just distinguish an object differs by an amount dB from a large background of brightness B, the contrast sensitivity is dB/B, a fraction of the total brightness, and is termed the *Fechner fraction*. It is the smallest difference of brightness that can be detected by the eye as a fraction of the total brightness when the two objects are side by side, as in photometry. *Fechner's Law* (1860) states that the sensation of brightness varies as the logarithm of the stimulus and can be deduced from the assumption that the Fechner fraction is a constant. For moderate degrees of brightness, it is only roughly constant. The term has been applied for sensations other than brightness.

feedback The principle of sending part of an output signal from a device to be re-evaluated with the original input signal.

A basic feedback system is shown in the figure. It consists of a summing network indicated by Σ, an amplifier with an open loop gain given by a, and a feedback network with transfer function b. The signals processed by such a system can be pneumatic or mechanical, but electrical voltage signals will be used in the following. There are two broad distinctions of feedback: In a *negative feedback* system, the feedback signal is subtracted from the input signal by the summing network and generates an error signal,

$$V_\epsilon = V_{\text{in}} - V_{\text{feedback}} .$$

On the other hand, if the summing network provided

$$V_\epsilon = V_{\text{in}} + V_{\text{feedback}} ,$$

then the system would be a *positive feedback* system. Negative feedback is normally used for general signal processing amplifiers, while positive feedback is used in non-linear systems such as comparators.

The amplifier in the system provides an amplified signal,

$$V_{\text{out}} = a V_\epsilon .$$

This signal is sampled by the feedback network and produces the feedback signal

$$V_{\text{feedback}} = b V_{\text{out}} .$$

Thus the amount of signal sent back is determined by b, and hence b is sometimes referred to as the feedback factor. The closed loop gain, G, of the entire system is found by eliminating V_{feedback} and V_ϵ from the above,

$$G = \frac{V_{\text{out}}}{V_{\text{in}}} = \frac{1}{b}\frac{1}{1+1/ab} .$$

If a is sufficiently large, as is usually the situation with operational amplifiers, then the gain reduces to

$$G = \frac{1}{b}$$

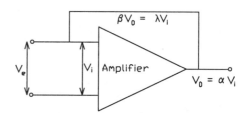

Basic feedback system.

feedback, acoustic Feedback in an acoustic system. Acoustic feedback can occur when a portion of the output sound in a system comes back to the input of the system.

feedback factor *See* feedback.

feedback network The circuit responsible for sampling the output of a feedback system and producing a feedback signal. The transfer function of the network, or sometimes the feedback factor, determines how much of the output signal is re-sampled by the system. It can also be reactive and thus make the entire system frequency dependent. *See also* feedback.

feedback voltage *See* feedback.

Fermat's principle Also known as the *principle of shortest optical path*. It states that the optical length:

$$\int_A^B n\,ds ,$$

of a ray between two points A to B in a medium of index n is shorter than the optical length of any other curve that joins these points. It is also known as the *principle of least time*.

ferrimagnetism A specific type of ordering in a system of magnetic moments. A material is ferrimagnetic when (a) all moments on a given sub-lattice point in a single direction and (b) the resultant moments of the sub-lattices lie parallel or anti-parallel to each other.

ferrite Ferrite, or ferrimagnetic oxide, is a ceramic material that is usually dark grey or black and very hard and brittle. The material has good magnetic properties and high bulk resistivities. It is therefore suitable for low loss transformer and inductor cores. The crystaline structure of a simple ferrite is cubic, mineral spinel. The magnetic properties arise from the metallic ion, usually a diavalent transition metal such as Mn, Fe, or Co, occupying a particular position relative to the oxygen atoms in the ceramic. Some ferrites have near-rectangular hysterisis curves, thus miniature toroidal ferrites have been used in early memory elements.

ferromagnetism Ferromagnetic materials exhibit a high magnetic permeability, e.g., ferrite and powdered iron.

fiber An optical fiber consists of a glass core that is completely surrounded by a glass cladding used in transmission of light pulses for telecommunications purposes. *See* fiber optic cable.

fiber absorption Transmission of light through fiber undergoes attenuation, which depends on the wavelength being used. The attenuation properties are of the order of about 5% loss per kilometer for bands centered on 1.3 and 1.55 microns. *Absorption* is one of two basic loss mechanisms in optical fibers. The other mechanism is *dispersion*. Small traces of metallic impurities, e.g., Fe, Cu, in the silica can increase the loss in the fiber considerably. A common impurity is water as hydroxyl ions.

fiber, inhomogeneous An optical fiber in which the refractive index changes throughout the fiber, as a function of the spatial coordinates.

fiber, low loss Optical fiber transmission systems have low transmission loss, which permits longer repeater sections than with a coaxial cable system, thereby reducing costs. Fibers are so free from impurities that very little energy is lost and there is usually an attenuation of less than 1 dB per kilometer; i.e., there is low energy loss per unit length of fiber. The energy loss is inversely proportional to the square root of the frequency being transmitted.

fiber, multimode Optic fibers with thicker cores, typically 50 or 62.5 microns, that allow many different modes of transmission since light travels in multiple paths such that it is reflected from the cladding back into the core as it travels down the core. Examples are *stepped index,* and *graded index multimode.*

Optic fibre

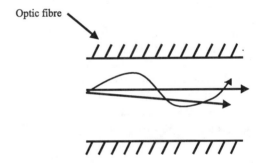

fiber optic cable Transmission systems employing the use of a pulsed light-wave as a carrier. This is sent in digital form using a binary code. The optical cable consists of a glass core that is completely surrounded by a glass cladding. The core transmits the lightwaves and the cladding minimizes surface losses and guides the lightwaves. It has low transmission loss, wide bandwidth, small cable size and weight, and immunity to electromagnetic interference. Its great advantage is that it can carry thousands of different frequencies without data loss.

fiber optics The branch of optical technology concerned with the transmission of radiant power through fibers made of transparent materials such as plastic, glass, or fused silica. A fiber is a thin filament of these materials that is drawn or extruded so as to have a central core and a cladding of a material of a lower refractive index so as to promote total internal reflection and travel along the length of the fiber without escaping reflection. It may be used as a single fiber to transmit pulsed optical signals (communications fiber) or in bundles of fibers to transmit light or images. Cables of optical fibers used for this purpose are smaller and lighter than conventional cables using copper wires or coaxial conductors, yet they can carry much more information, making them useful for transmitting large amounts of data between computers and for carrying data-intensive video or large numbers of simultaneous phone conversations. Optical fibers are immune to electromagnetic interference and to crosstalk from adjoining fibers. To keep a signal from deteriorating over long distances, optical fibers require fewer repeaters than does copper wire over a given distance. Most fibers for long distance purposes are made of quartz because of the low losses down to 0.1dB/km, while some short-distance fibers have less expensive and easier-to-handle plastics as the core material. In every case, the light guidance is provided by total reflection inside the fiber core, which has a slightly larger index of refraction than the rest of the fiber (the cladding). An additional outer plastic coating protects the fiber from mechanical or chemical damage. Standard telecommunication fibers have core diameters of 9 microns and an outer cladding diameter of 0.125mm (9/125 monomode fiber) or 50 microns (50/125 multimode fiber). For long distance telecommunications only monomode fibers are used, because of the absence of modal dispersion.

fiber optics communications   Fiber optics communications represent a form of optical communication through optic fibers made of silica glass. Because free unguided laser beams are highly vulnerable to obstruction by fog, rain or snow, laser beams need to be transmitted within protective pipes for earth-bound telecommunications. The efficacy of information transmission via optical fibers is a result of the orders-of-magnitude more superior transparency of silica glass to visible and infrareds, as compared to any other previously known solid medium for electromagnetic waves of any other frequency spectrum. Transmission is typically in baseband, with the information signal represented as a sequence of on-and-off light pulses. Transmitted signals may lose half their power only after having traveled along optical fibers for over 10 km or up to 50 km, depending on the optical fiber and optical source used. A signal repeater takes attenuated incoming on-off optical pulse trains, converts them into electronic pulses, then amplifies and re-times the electronic pulse trains and uses them to excite an optical transmitter to regenerate the received optical pulse trains to travel another length of the optical cable network. Semi-conductor photodiodes represent the most common optical communication receivers. Because the optical signal generally travels unidirectionally along the optical fiber, network nodes are typically arranged in a ring configuration (rather than, say, a star-shaped topology). Signal distortion, however, occurs in long-distance transmission as the light wave reflects off the boundary between the two glass layers in the coaxial cylindrical step-index fiber optical fiber, wherein the core layer is made of glass with a slightly higher refraction index than the coaxial outer layer. Because optical fibers would accept only light entering the fiber at a low angle from the fiber axis, laser sources (with their narrow spatial directivity) are typically used. Fiber optics communications using laser beams is characterized by lasers' very wide frequency spectrum, thereby permitting very high data transmission rates. Optical fibers can also be more easily insulated from outside interference, and are smaller and lighter than metallic cables. Fiber optics link the central telephone switching offices of major American and European cities.

fiber optics data link   An optical link capable of handling data in digital form. This deals with *layer 2* of the open systems interconnection (OSI) networking model, which concerns data packets and reliable data transfer. The data link layer detects and may correct errors in the physical layer (*layer 1*). The data link layer is responsible for several specific functions such as providing a well-defined service interface to the network layer as well as regulating the flow of frames.

Optic fibre

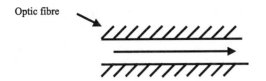

fiber scattering   Also known as *dispersion.* Light pulses sent down a fiber spread out in length as they propagate, the amount dependent on the wavelength being used. Rayleigh scattering of the light occurs within the molecules of glass material itself. This loss is independent of the light intensity but varies inversely with the fourth power of wavelength. It is therefore advantageous to operate at longer wavelengths. Favorable operating windows for fiber optics are in the region of 800 nm, 1300 nm, and 1500 nm. Dispersion effects can be canceled out by making pulses in a special shape known as solitons. These can be sent for thousands of kilometers without distortion.

fiber, single mode   Optic fibers with very small diameter core, typically 8 - 9 microns, that allow only a single mode of transmission of light pulses. An example is the *stepped index monomode.*

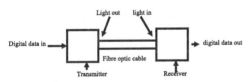

Single mode fiber.

fiber, W-type   A doubly cladded optical fiber with two layers of concentric cladding, in which the core usually has the larger refractive index and the inner cladding has the lower refractive index. There are several advantages of this type over conventional fibers, such as reduced bending losses.

Fick's first law   Entropy drives diffusion of a substance in the presence of a concentration gradient. *Fick's first law* connects the flux (J) of the substance to the concentration gradient ($\partial c/\partial x$ in one dimension) via the diffusion coefficient

D:

$$J = -D\frac{\partial c}{\partial x} .$$

Fick's second law   Connects the time rate of change of the concentration of a solute ($\partial c/\partial t$) in terms of the diffusion coefficient D and the spatial rate of change of the concentration gradient ($\partial^2 c/\partial x^2$):

$$\frac{\partial c}{\partial t} = D\frac{\partial^2 c}{\partial x^2} .$$

fidelity   The degree to which an electronic signal can be reproduced at the output compared to the input signal.

figure of merit   This indicates the performance of a device for a particular application.

1. For a *magnetic amplifier:* the ratio of power amplification to control time constant

2. For a *transistor:* the ratio of gain to the bandwidth

3. For a *galvanometer:* the ratio of the deflection to the current, also called *sensitivity*

4. The ratio of reactance to resistance.

film, Rollin   A Rollin film is the film of superfluid helium found on any surface in contact with bulk superfluid helium. This film flows, or creeps, along all surfaces that are below the lambda temperature, 2.1768 K. Due to the high thermal conductivity of the film, a superfluid film can increase the heat flow into a low temperature experiment. In such cases, heaters or knife edges are used to prevent such films. The heater boils off the film using less heat than would be carried by the film, and the knife edge is a sharp edge over which the superfluid film cannot flow. As the superfluid flows over the knife edge, it becomes thinner and thereby must flow faster to keep the mass flow the same. A sufficiently sharp knife edge will force the fluid to exceed the critical velocity, thereby turning the fluid normal. *See also* helium-4, superfluid.

filter circuits   Circuitry that selectively eliminates or passes.

filters (optical)   A homogeneous optical medium that is used for attenuating particular wavelengths or frequencies of light (including infrared and ultraviolet) while passing others with relatively no change. Filters are used to control or alter the relative energy distribution of a beam of light. In photography, filters may be placed over the light source or over some part of the optical path to the camera, frequently over the lens. Typically, transparent substances (colored glasses or films) are used. One type of colored filter is the *gelatin filter*, which can be dyed with a wide range of materials. Another type of colored filter is made from polyester and has some of the same characteristics. A filter based on a completely different principle is the *interference filter*, which uses destructive interference of waves transmitted directly through the filter and those reflected $2n$ times from the front and back faces of the filter to yield a narrower band of wavelength — 10 to 100 angstroms. A *linear polarizing filter* transmits light waves that vibrate in a single direction only. It eliminates various degrees of reflected light from glass, water, plastic, paper and similar surfaces. It can also eliminate light reflections from vapors or floating dust to emphasize a blue sky. *Polarizing filters* transmit light waves that vibrate in a single direction only; the effect can be seen in the viewfinder as the filter is rotated. A *circular polarizing filter* converts linear polarized light waves to circular polarized light waves. Required whenever polarizing is desired, as with use of autofocus cameras and cameras that have a semi-silvered reflex mirror. A *neutral density (ND) filter* reduces the amount of transmitted light without affecting color balance. Exposure change is rated by filter factors, such as $ND2$ or $ND4$. $ND2$ reduces light to 1/2 and $ND4$ to 1/4. A *color compensating (CC) filter* enables fine adjustments of color tone or color density in color photography. A *color (temperature) conversion filter* alters color temperature of light to make it suitable for the film in use, thereby enabling the photographer to use daylight film indoors or tungsten film outdoors. A *contrast-control filter* is used with black-and-white film to emphasize contrast in a picture. An *orthochromatic filter* is a green or yellow-green filter used with black-and-white films to compensate the difference between the color sensitivity of films and the relative luminous efficiency of the human eye. A *skylight filter* re-moves a portion of the blue light. Taking color pictures in the shade under a clear, midday sky results in an overall slightly bluish cast. This excess blue can be corrected with a skylight filter to produce a more natural effect. *UV filters* eliminate invisible ultraviolet light, rendering pictures with higher contrast, since ultraviolet rays of shorter wavelengths are easily reflected by vapors or floating dust, causing lower picture contrast. A *soft focus filter* diffuses light and imparts a slight flare to the image, providing a soft-focus effect, which is ideal for portrait photography.

filters, acoustic   Devices that significantly decrease amplitude of a complex sound in certain frequency bands while the amplitude in others remains almost unchanged. A low-pass filter cuts off all frequencies above a critical one. It can be designed as a number of cavities of the same form, arranged in a row and connected by narrow tubes. A high-pass filter cuts off all frequencies below a critical one. It can be designed as a tube with holes at regular distances along the tube. A band-pass filter cuts off all frequencies above and below a definite band of frequencies and can be designed as a combination of low-pass and high-pass filters. *Acoustic filters* are widely used in practice for noise reduction in air-conditioning systems, jets, and exhaust systems of internal combustion engines in motor vehicles.

fish, hearing organs of   The otolith organs, consisting of fine hairs. The *otolith organs* allow fish to have only a rough impression about a received sound without detailed analysis. The otolith organs can be considered as an early hearing mechanism in animals.

Fizeau fringes   The interference pattern that results from the interference of light transmitted by wedge-shaped thin film. These are the fringes obtained with the *Fizeau interferometer*, an optical arrangement in which light from a quasi-monochromatic source is collimated by a lens and falls on a planar-wedge-shaped thin film at nearly normal incidence. The light reflected by both surfaces of the film are collected by the same lens which converges it to an aperture in the focal plane. To an eye or film immediately behind this aperture, the Fizeau fringes are visible over the entire area illuminated by the lens when lines of equal optical thickness are followed. The fringes may also be obtained from an optically thick film, provided the source is sufficiently small. The method is used in optical workshops to test the optical thickness uniformity of transparent plane parallel plates.

Fizeau method (velocity of light)   The method used in the first successful measurement (1849) of the velocity of light not involving astronomical observations. A brief flash of light was sent out and the time to travel to a distant mirror and back was measured. A toothed wheel or cogwheel with 720 cogs was rotated at high speed so as to break up a light beam passing through its rim into a series of short flashes. The light was rendered parallel with a lens, and then focused with another lens onto a distant mirror (5.36 miles in Fizeau's original experiment). After reflection, the light retraced its path and was focused again onto the rim of the cogwheel by the first lens. If, during the time the light has traveled, the wheel had turned enough for a cog to be interposed, the flash would be blocked out. A further increase in speed would cause the flash to reappear as the returning light could then pass through an opening in the cogwheel. Fizeau was able to say that light moved at 313,300 kilometers per second which is close to what we now know to be 299,792.458. The largest uncertainty in this method is the determination of the condition of total eclipse by the cog. Young and Forbes overcame this difficulty by placing an additional lens and mirror at somewhat greater distance so that two images were observed simultaneously, and the speed of the cogwheel determined at the time the images appear of equal intensity, a more accurate experiment since the eye is very sensitive to the detection of slight differences in intensity of adjacent images. These experiments were soon improved by replacing the cogwheel with a rotating mirror by Fizeau and Focault independently in 1850.

flame, singing   A gas flame located at the lower end of a vertically positioned narrow tube opened at the upper end, that generates sound. The gas flame causes and upward current of air and vibrations of the air column in the tube with its natural frequency. This results in sound radiation from the open end of the tube. A singing flame is also called a *signing tube.* Signing tubes are often used in physics demonstration experiments.

flash   A type of converter with extremely rapid conversion times using comparators for coding voltages to give binary outputs.

Flemming's rule   A rule relating the direction of force, velocity and magnetic field felt by a moving charged particle. It can be summarized as $F = q(V \times B)$, where the direction of F is given by the cross product of V and B.

flip-flop   A sequential logical circuit whose output depends on the present values as well as the history of its inputs. It is a basic building block of larger and more complicated sequential circuits such as counters and shift registers.

The most important characteristic of a flip–flop is that it is capable of remembering binary signals. Thus, it is viewed as a memory storage element with control inputs. There are three different types of control inputs: *static*, where the flip–flops output will change according to the present inputs (and their history), *preparatory,* which set up the flip–flop but are not able to change the output directly, and *dynamic,* where a "clock" signal is required to instigate any changes of the output state as determined by the preparatory inputs.

There are three basic types of flip–flops, which are identified by the way their outputs respond to the inputs:

1. *Direct R–S.* This is the most elementary flip–flop. It has two static control inputs S and R labelled for *set* and *reset.* It also has an output Q as well as an ancillary output $\overline{Q}$. The normal state of the inputs is $R = S = 0$, and Q is allowed to be either 0 or 1. If S is changed to 1, then $Q = 1$ and will hold this value even if S returns to 0. If, on the other hand, R changes to 1, then $Q = 0$ and will remain so when R returns to 0. The output Q is not defined if both R and S are 1. The operation of the $R–S$ flip–flop can be summarized in the accompanying table where Q_{prior} refers to the prior state of Q before the R, S values changed. Thus, by examining the

output Q, one can determine the previous values of R and S: a rudimentary memory.

Truth Table of the R–S Flip–flop

Symbol	S	R	Q
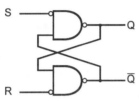	0	0	Q_{prior}
	1	0	1
	1	1	undefined

The operation above can be illustrated by considering the realization of an R–S flip–flop using **NAND** gates as shown in the figure.

Realization of an R–S flip–flop.

By using the truth table for the **AND** gate (*see* conjunction; logic gates) as a starting point, the output value of Q can be verified for any given valid input state. Observe that if both R and S change to 1 (from 0) simultaneously, Q is logically indeterminable as both 0 or 1 are equally valid outputs, hence this ambiguous input condition is avoided.

2. *D Flip-Flop.* This device also has two inputs, a preparatory input, D, and a dynamic input, *clock.* The output, Q, changes to the current state of D when the *clock* signal makes a particular state transition. Depending on the convention used by the particular circuit involved, the flip–flop will effect the change in Q when *clock* changes $0 \rightarrow 1$ or $1 \rightarrow 0$; the former is the most common. If we assume this convention for illustration purposes, then the truth table for the D flip–flop can be determined as shown in the accompanying table. Note that when *clock* is 0 (or even $1 \rightarrow 0$), the output does not depend on the input D, indicated in the table as X. Thus in operation, D is sampled at the clock transition instant and held indefinitely; hence D for *delay.*

3. *J–K Flip Flop.* This device has two preparatory inputs J and K, and a *clock* input

Truth Table of the D Flip–flop

Symbol	D	clock	Q
	0	$0 \rightarrow 1$	0
	1	$0 \rightarrow 1$	1
	X	0	Q_{prior}

to instigate a change in the output Q based on these inputs. The operation of this flip–flop is defined as follows: If $J = K = 0$, then the output will hold its previous value, even on a proper *clock* transition. If $K = 1$ while $J = 0$ during a *clock* transition, then Q is set to 0; similarly, if $K = 0$ while $J = 1$ then $Q = 1$. Thus, so far the J–K is operating in a manner similar to an R–S except that a *clock* is required. However, if $J = K = 1$, an ambiguous condition for the R–S flip–flop, then $Q = $ **NOT** Q_{prior} ("toggled") upon a *clock* transition. Assuming the same previous conventions, the operation of the J–K flip–flop is summarized in the accompanying table.

Truth Table of the J–K Flip–flop

Symbol	J	K	clock	Q
	0	0	$0 \rightarrow 1$	Q_{prior}
	0	1	$0 \rightarrow 1$	0
	1	0	$0 \rightarrow 1$	1
	1	0	$0 \rightarrow 1$	$\overline{Q}_{prior}$
	X	X	0	Q_{prior}

floating of circuit   Conduction wherein a circuit is not grounded or tied to an established voltage supply.

flow cytometry   A technique for counting, sorting or selecting individual cells as they flow through a tube of pipe.

flow impedance   Flow impedance is a device designed to regulate flow of a cryogenic gas in a low temperature apparatus. In a continuous ^{3}He refrigerator, the condensed liquid ^{3}He must have its pressure throttled from that at the condensation stage (of order 0.1 bar) to the low pressure found in the ^{3}He pot (a few mbar). This is usually accomplished by inserting a high impedance segment of the tube into the flow path. The most common material for such constrictions is cupro-nickel tubing with the proper conductance:

$$F = \frac{\pi a^4}{8\eta\ell} P_a$$

where a is the radius of the tube, ℓ is the length of the tube, η is the viscosity of the liquid or gas in question, and P_a is the average pressure in the tube. (The impedance is the reciprocal of the conductance.)

flow measurements   The determination of the rate of movement of a fluid.

flow measurements, continuous   The determination of the rate of movement of a fluid at all times.

flowmeters   Any device used to measure the rate and direction of movement of a fluid.

flow pattern, cerebral fluid   The particular manner in which the cerebral fluid moves.

fluidity, cell membrane   The degree to which the cell membrane is able to flow.

fluorescence   Radiation caused by a transition between two well-defined energy states of atoms of solids, liquids and gases due to absorption of incident light. According to *Stoke's law,* the wavelength of the emitted fluorescent light is always longer than that of the incident light.

fluorescent screen   Screen coated with fluorescent material used, for example, to detect the presence of ultraviolet light or X-rays. Fluorescent materials emit characteristic light when exposed to ultraviolet radiation.

fluorescent spectroscopy, nanosecond   The analysis of light given off by a sample during excitation by light of a higher energy. For pulsed excitation, light emitted within about 10 nanoseconds of the exciting pulse is considered to be fluorescent. The fluorescent light is emitted after the energy of the original excitation has been transformed by radiationless transitions to

different energy levels. Detailed information about the energy levels can be obtained by this technique.

fluorography   Photography of an image produced on a fluorescent screen by X-rays.

flutter   Rapid fluctuation of frequency of reproduced sounds due to fluctuations in speed during the processes of recording and reproduction. These fluctuations in speed can be caused by variations in speed of a turntable, for example.

flux ball   A test coil in which a series of coaxial cylindrical windings of different lengths are wound to form a sphere. It is used to measure *magnetic flux density.*

flux density, magnetic   A vector quantity that is used as a quantitative measure of a magnetic field; the force on a charged particle moving in the field is equal to the particle's charge times the cross-product of the particle's velocity with the magnetic induction.

flux gates   These are used to determine the magnitude and direction of an external magnetic field. They generate electrical signals whose magnitude and phase are proportional to the external field acting along its axis.

flux, luminous   In photometry, the time rate of flow of light per unit solid angle, weighted with respect to its efficiency to evoke the visual sensation of brightness. Sources of light emit radiant energy at a certain rate, which flows at the speed of light past any point, but only the part of this energy in the visible range of wavelengths will excite vision. *Flow* or *flux* can be expressed in mechanical units, ergs per second, but when visual stimulation is to be measured, the flow is expressed in luminous flux units (lumens). The unit lumen is defined in terms of the flux from a standard luminous source (which implicitly invokes luminous sensation, as for example the International candle, black-body radiation at 2046 K, etc.) into a unit solid angle. Equal luminous fluxes produce equal sensations of brightness, but different fluxes do not produce sensations of brightness in direct proportion. The luminous flux density gives the intensity of illumination in lumens per unit area.

flux, magnetic   Lines used to represent magnetic induction, B. Lines of flux are used to represent the field in magnitude and direction at any point. The number of lines of flux per unit area of a surface perpendicular to the field is equal to the magnitude of the magnetic induction. The total number of lines of induction through a surface is the *magnetic flux*.

flux, magnetic, changing   This produces an induced EMF in a conductor subject to the changing magnetic flux that results from Faraday induction.

fluxmeter   An instrument to measure magnetic flux. It consists of a moving-coil ballistic galvanometer with a long period. A search coil is connected to the galvanometer and the change in the flux that results from the motion of the search coil that is detected by the galvanometer.

fluxoid   The quantum of magnetic flux in a superconductor is one *fluxoid:*

$$\Phi_0 = 2\pi\hbar/2e = 2.06785 \times 10^{-15} \ T \ m^2 \ .$$

See also superconductivity.

flux quantization   The quantum mechanical expression for the current density in a superconductor located in a magnetic field is

$$\vec{j} = -\left[\frac{2e^2}{mc}\vec{A} + \frac{e\hbar}{m}\vec{\nabla}\Theta\right]|\Psi|^2$$

where $\vec{A}$ is the magnetic vector potential, Θ is the phase of the superconductor wavefunction, and Ψ is the amplitude of the superconductor wavefunction. If the superconductor is in the form of a closed ring, it is possible for the magnetic field threading the ring to be non-zero while the magnetic field inside the superconductor is zero. If we integrate the current density along a path inside such a ring, the integral simplifies to

$$\frac{2e^2}{mc}\Phi + \frac{2\pi ne\hbar}{m} = 0$$

where Φ is the magnetic flux, and n is an integer. This further simplifies to the flux quantization rule:

$$\Phi = \frac{nhc}{2e} = n\,\Phi_0$$

where Φ_0 is one flux quantum, or fluxoid. *See also* fluxoid; superconductivity.

flying spot   A moving spot of light that is controlled by external signals.

focal length   The distance along the optical axis between a focal point and its corresponding principal point, which, for a thin lens, is approximately the same as the distance of a focal point from the lens.

focal lines   For an off-axis cone of light, as a result of astigmatism, the fan of rays in the tangential (or meridional plane) and the sagittal planes come to focus on the so-called tangential focal lines and the sagittal focal lines, respectively. The *tangential* (or meridional or primary) *focal line* is perpendicular to the meridional plane, and has a smaller image distance than the secondary or sagittal focus. The *sagittal* (or secondary) *focal line* is in the meridional plane.

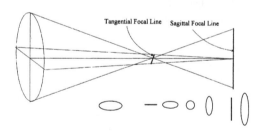

Tangential and sagittal focal lines.

focusing, acoustic   The convergence of sound waves. Acoustic focusing is similar to optic focusing. Due to diffraction, a sound wave cannot be focused to a point; it is rather focused to a spot. The point where the sound intensity is maximal is called *focus*. Since in most cases sound waves are diverging, acoustic focusing is achieved only under certain circumstances. It can be achieved by specially constructed transducers, by converging *acoustic lenses* and mirrors, and also as a result of random inhomogeneities in a medium. Acoustic focusing is widely used in technique and medicine for concentrating acoustic energy.

focusing, magnetic In some opto-electronic devices, magnetic fields are employed to produce an effective focusing of accelerated electrons to form electron images, for example.

focusing of electron beam Focusing of an electron beam by use of an electrostatic lens in a cathode-ray tube (CRT). In a CRT, electrostatic lens consists of a focus anode, accelerating anode, and so on. The focus anode and accelerating anode are maintained at a positive potential with respect to the cathode. The potential of the accelerating anode is higher than that of the focus anode. The focal length of this lens depends on the potential at the focus anode. The potential at the acceleration anode also affects the focal length. *See also* electrostatic lens.

focus, tangential *See* focal lines.

folded cascode A cascode constructed with two different types of transistors such as an *n*–*p*–*n* and *p*–*n*–*p* bipolar transistor pair, an *n*–channel and *p*–channel FET pair, or a FET and bipolar transistor combination. *See* amplifier, cascode.

footprint (communication) This concerns satellite communication, and refers to downward beam covering a substantial fraction of the earth's surface. It is therefore the area covered by the satellite and reached by its radio beams on earth. This area can be broad or narrow, covering up to hundreds of kilometers in diameter.

forced oscillations Oscillations that are due to a driving force. The simplest model of this is one-dimensional oscillations governed by the equation

$$M\frac{dx^2}{dt^2} + R\frac{dx}{dt} + Sx = Fe^{i\omega t} .$$

Here, x is a coordinate of a physical quantity under oscillations, t is time, F and ω are the amplitude and angular frequency of the driving force, and other notations are the same as those in the equation for one-dimensional *damped oscillations*. The solution of the equation in ques-

tion is the sum of two terms. The first one corresponds to free oscillations and exponentially decays in the course of time t. The second term is the steady state solution given by

$$x = \frac{F\exp(i\omega t - i\theta)}{\sqrt{(S - \omega^2 M)^2 + R^2\omega^2}} ,$$

where $\tan\theta = R\omega/(S - \omega^2 M)$. The amplitude of the steady state oscillations $|x|$ reaches a maximum value at a given frequency $\omega = (S/M - R^2/2M^2)^{1/2}$. This phenomenon is called *resonance*.

format (**1**) Different software applications save data in certain ways. There are several standard file formats; some examples of graphical formats are *.gif* or *.jpg*.
  (**2**) Hard drives need to be formatted after partitioning so that they can be used by the operating system. It allows the hard disk to be ready for use. Floppy disks also need to be formatted. There is also the SCSI low-level format that makes the drive ready to be used by a SCSI controller.

form factor The ratio of the effective value (root mean square value) of an alternating quantity to the average value during half of an interval (cycle). It compares the various kinds of periodic waveforms. The effective value V_{eff} of alternating voltage v of interval T is

$$V_{eff} = \sqrt{\left[\frac{1}{T}\int_0^T v^2\right]} .$$

When the alternating voltage has a sinusoidal waveform, the effective value is

$$V_{eff} = \sqrt{\left[\frac{1}{T}\int_0^T \left(V\cos\frac{2\pi t}{T}\right)^2\right]},$$

$$= \frac{V}{\sqrt{2}} .$$

The average value during a half of the interval of this sinusoidal waveform is $2V/\pi$. The form factor of sinusoidal waveform is $\pi/2\sqrt{2}$.

Forster critical distance The characteristic distance associated with the energy transfer between two chromophores. This distance depends on the spectral overlap between the donor emission and acceptor absorption bands, the refractive index of the medium, the quantum yield of the donor in the absence of the acceptor, and a geometric factor depending on the relative orientation of the donor and acceptor.

Forster dipole-dipole approximation The interaction between two chromophores is approximated according to the interactions of their associated dipole moments. The success of this approach was first shown by the experiments of Stryer and Haugland (L. Strayer and R.P. Haugland, *Proc. Natl. Acad. Sci. USA* **98** (1967) 719).

fountain effect Also called the *thermomechanical effect*, the *fountain effect* is a manifestation of zero viscosity in Helium II. Consider two containers of superfluid helium at a pressure, p, and temperature, T, connected by a narrow tube or a tube filled with closely packed particles. (*See figure.*) If we increase the pres-

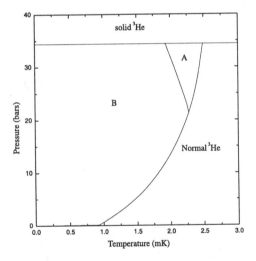

Fountain effect. The thin connecting tube is filled with packed powder.

sure in container A, the normal component of the helium will be unable to flow through the narrow channel due to the large viscosity of the normal fluid. The superfluid component, however, will flow unimpeded through the orifice into container B. This process will increase the temperature in container A and decrease it in

container B such that

$$\frac{\Delta p}{\Delta T} = \rho S \,,$$

where ρ is the density of the fluid and S is the entropy of the normal fluid. Conversely, if we begin by increasing the temperature in container B, superfluid flows into container B, producing the hydrostatic pressure difference given above. These effects are quite large – at 1.5 K, a temperature difference of 0.001 K produces a 2 cm pressure head in the liquid helium.

Fraunhofer doublet One of the three types of small aperture objective lenses used in most prismatic binocular telescopes, low-power microscopes and telescopes. The field of view in these objectives is very small and the aperture is usually smaller than f/5. In this case the most important aberrations are the primary spherical, coma, and chromatic. Cemented achromatic aplanatic doublets are commonly used. Image quality deteriorates rapidly away from the axis, but there is little that can be done to improve it, so either the crown or flint component of the cemented meniscus lens pair can be placed on the long conjugate side of the lens. If the crown is on the long conjugate side, the achromatic doublet is of the *Fraunhofer* type. If the flint is on the long conjugate side, it is a *Steinheil* type. Fraunhofer types usually have shallower surfaces and the advantage that the more stable crown glass is on the exposed side for a telescope objective. A third type is the *Gauss* type, which consists of uncemented meniscus lenses with an air gap, so it is possible to achieve better correction for spherical aberration at the expense of coma.

Fraunhofer lines The thousands of fine dark absorption lines that cross the spectrum of the photosphere (the glowing outer layer) of the sun. A few of the lines are due to absorption in the terrestrial atmosphere. The most prominent 600 of these lines were observed and named by J. Fraunhofer in 1814, who recognized that the D lines coincided with the emission lines of sodium. G.R. Kirchoff in 1859 gave the modern interpretation of these lines. They are useful spectral benchmarks, used in the specification and measurement of refractive indices, for example. Some of these benchmarks, together with their origin and approximate wavelength in angstroms, are:

A	terrestrial O2	7594-7621
B	terrestrial O2	6867-6884
C	H-alpha	6562.816
Alpha	terrestrial O2	6276-6287
D1	Sodium	5895.923
D2	Sodium	5889.953
D3	He	5875.618
E2	Fe	5269.541
B1	Mg	5183.618
B2	Mg	5172.699
B3	Fe	5168.901
B4	Fe	5167.491
F	H	4861.327
D	Fe	4668.140
E	Fe	4383.547
F	H	4340.465
G	Fe	4307.906
G	Ca	4307.741
G	Ca	4226.728
H	H	4101.735
H	Ca+	3968.468
K	Ca+	3933.666

The lines A, B, and alpha are oxygen bands caused by absorption in the earth's atmosphere.

free vibrations   Vibrations that continue after eliminating a driving force.

freeze resistance   The ability of certain organisms to function below the normal freezing temperature by the use of specific molecules that depress the freezing point of their bodily liquids.

frequency   The number of cycles completed by a periodic quantity, the acoustic disturbance, in a unit time. The units of frequency f are hertz (Hz); 1 Hz corresponds to 1 cycle per second. Frequencies audible to the average human ear are in the range between 20 and 20,000 Hz. The frequency is the inverse of the period T, $f = 1/T$, the time interval during which one oscillation cycle is executed.

frequency allocation   Bands of frequencies for specified services are allocated by international agreement and refer to the frequency on which a transmitter has to operate within specified tolerance.

frequency, angular   For any oscillation, the number of vibrations per unit time f, multiplied by 2π, $\omega = 2\pi f$. Also known as *angular velocity* or *radian frequency*. The units of angular frequency are radians per second (rad/s).

frequency, audible   The range of frequencies that the average human ear can hear, from 20 to 20,000 Hz. Disturbances below the useful frequency range for human hearing (below 20 Hz) are classified as *infrasound*, and those with a frequency too high (above 20,000 Hz) are called *ultrasound*.

frequency band   A continuous range of frequencies extending between two limiting frequencies, the nth band having a lower frequency $f_{min}(n)$ and upper frequency $f_{max}(n)$. The frequency scale is divided into contiguous bands, and is said to be proportional if $f_{max}(n)/f_{min}(n)$ is the same for each band. The center frequency of a band $f_c(n)$ is defined as the geometric mean, $f_c(n) \equiv \sqrt{f_{min}(n) f_{max}(n)}$. In an octave band the relation $f_{max}(n) = 2 f_{min}(n)$ holds. In noise control the $\frac{1}{3}$-octave frequency partitioning scheme is used, for which $f_{max}(n) = 2^{1/3} f_{min}(n)$. The principle of fixed frequency ratios is also applied in the theory of musical temperament. Frequency ratios corresponding to classic music intervals that sound harmonious are 2:1 (octave), 3:2 (perfect fifth), 4:3 (perfect fourth) and 5:4 (major third).

frequency compatibility   This refers to the ability of television sets to receive color broadcast in black and white without special adaption.

frequency conversion   The shift from one frequency band to another using multiplication by a sine wave. *See* mixer.

frequency, cut-off   A frequency at which axially decaying modes develop (for example in waveguides or horns). These modes decay exponentially with distance from the sound source or the reflection location. For modes with frequencies above the cut-off frequency, axial propagation takes place, and the corresponding modes are termed *cut-on*.

frequency discriminator   A device used to provide frequency – amplitude conversion when frequency demodulation occurs. The process requires that output voltage or current amplitude is linearly proportional to the frequency of the input signal.

frequency, distress   With respect to the S.O.S. signal, the international code signal used by aircraft or ships at sea when in distress.

frequency divider   A unit or device that divides (reduces) an input frequency by a preset number. If the input frequency is f_{in}, then the output of the circuit will be

$$f_{out} = f_{in} \div N \, ,$$

where N is the preset divisor. Frequency dividers are usually based on binary counters, hence they will operate on and generate square waves. Programmable counters are often used, allowing a user to change the value of N. *See also* frequency synthesizer.

frequency doubler   A circuit or device capable of providing power at twice the input AC frequency.

frequency, fundamental   The lowest resonant frequency of the system. The lowest frequency in a complex wave. Other resonant frequencies are called *overtones*.

frequency, maximum usable (MUF)   In radio wave propagation via the ionosphere, the highest frequency that can be used between two points at a particular time. It therefore gives the best frequency for long distance transmission.

frequency meter   A device capable of measuring the frequency of either a sinusoidal AC input voltage or a propagating electromagnetic wave. Depending on the magnitude of the frequency involved, the device may either count the number of waves in a predetermined dwell-time, or determine the frequency by means of a calibrated resonant cavity or circuit.

frequency multiplier   A unit or device that provides an output frequency at some preset

multiple of the input frequency. *See* frequency synthesizer.

frequency, resonant   A frequency at which some measure of a physical system subjected to periodic forcing develops a maximum. Three types are defined: *phase, amplitude* and *natural resonance*. They are nearly equal when dissipative effects are small. Also known as *resonance* or *natural frequency*.

frequency response   Frequency response expresses the relative gain of, for example, an amplifier as a function of frequency for a pure sinusoidal input signal. This information is usually presented in a graphical format. *See also* half-power bandwidth.

frequency selective amplifier   A frequency selective amplifier combines a frequency- or phase-dependent network with a broadband amplifying device to produce a narrow band-pass or band-stop filter. Depending on the network used and how it is connected, the frequency selective amplifier can be designed to either accept a very narrow range of frequencies (an *acceptor*) or reject a narrow range of frequencies (a *rejecter*).

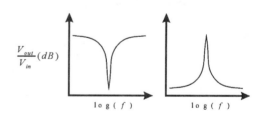

Rejecter and acceptor transfer functions of a frequency selective amplifier.

frequency swing   The total carrier frequency shift between the lower frequency extreme and the upper frequency extreme in telecommunication transmission.

frequency synthesizer   A frequency source whose output is a programmable integer multiple of an input reference source. A block diagram of a simple frequency synthesizer is shown below. A frequency reference, f_{ref}, is sent to one input of a phase difference detector whose output voltage controls a voltage control oscillator (VCO). The output of the VCO is feedback to the other input of the phase difference detector via a divide–by–N counter. This counter generates a pulse for every N pulse of the VCO; hence the output frequency of the VCO is $f_{out} = N \times f_{ref}$. The counter's N value can be digitally controlled from, for example, a microprocessor or a simple thumb wheel switch array and the reference frequency is usually derived from an accurate crystal controlled oscillator.

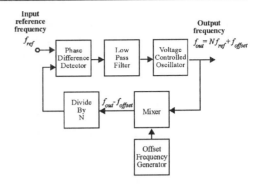

Block diagram showing the heterodyne–down conversion technique in a frequency synthesizer.

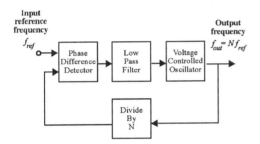

Operational block diagram of a basic frequency synthesizer.

Frequency synthesizers are commonly used in present day communications and radio systems. However, the output frequencies in these systems are too high for common TTL or CMOS divided–by–N counters. A technique called *heterodyne–down conversion* is used to handle, i.e., systhesize, higher frequencies. In this technique, a second, *offset* or *local* oscillator (f_{offset}) is mixed with the output frequency obtaining the frequency difference $f_{out} - f_{offset}$. This frequency is then processed by the divide–by–N counter as before. This additional step is shown in block form in the second illustration. The output frequency is thus determined by $f_{out} = N \times f_{ref} + f_{offset}$.

frequency time sharing   Feature of time division multiplexing in which some of the time interval between adjacent pulses is used by other independent message signals. It allows for the joint utilization of a common channel since several independent message signals can be sent without mutual interference. *See* multiplexing, time division.

Fresnel approximation   An approximation used in the theory of diffraction. Suppose that a point source is located at the center of the Cartesian coordinate system, and a receiver is located at the point (x, y, z). Then, the phase increment of a wave propagating from the source to the receiver is given by $\exp\left(ik\sqrt{x^2 + y^2 + z^2}\right)$, where k is a wave number. *Fresnel approximation* is a replacement of this phase increment by its approximate value $\exp\left[ik\left(x + \frac{y^2+z^2}{2x}\right)\right]$. This approximation is valid if $x \gg \sqrt{y^2 + z^2}$. Fresnel approximation for other problems (e.g., wave radiation by an aperture) employs an analogous replacement.

Fresnel integrals   These are the integrals that come from separating the error function into real and imaginary parts for a real variable x

$$\frac{1}{(1 + i)} erf(x) = C(x) - i S(x) ,$$

where the resulting real functions are the *Fresnel integrals* given by

$$C(x) = \int_0^x \cos\left(\pi t^2/2\right) dt$$

$$S(x) = \int_0^x \cos\left(\pi t^2/2\right) dt .$$

Fresnel integrals may be evaluated numerically or found graphically by use of the ingenious *Cornu spiral*, which is a double spiral curve formed by plotting C against S. As with any vibration curve, the amplitude of the diffraction pattern from Fresnel diffraction is computed from distances on this curve, and the square of this length then gives the intensity.

Fresnel lines   Fringed shadows that appear at the edges of the geometrical apertures in Fresnel diffraction. A *Fresnel fringe* is a single band in a group of these light and dark bands that can be viewed in the periphery of the Fresnel diffraction shadow. Unlike the Fraunhofer pattern, the minima in Fresnel diffraction do not go to zero.

Fresnel zones   An explanation of Fresnel diffraction effects obtained by dividing the wavefront falling on the obstacle into a number of concentric annular zones, such that the distances of the boundaries of the zones from the observation point increase by one half wavelength from zone to zone. By Huygen's principle, each point on the wave front can be considered the source of a secondary wave, each of which makes its own contribution to the light reaching the observation point. Each Fresnel zone contains approximately the same area so that each zone may be viewed as having the same number of secondary sources, and thus contribute the same amount of light at the observation point. However, the contributions at the observation are out of phase since each zone is a half wavelength farther than the next. The total contribution is the sum of a series of terms, one for each zone, which are alternately positive and negative. While all the zones have about the same area, the central zones point directly at the observation point, while the outer zones point more obliquely. This causes the magnitude of each term to decrease steadily from term to term. Then the sum of the series can be demonstrated to be half the first term, and the amplitude reaching the observation point from the entire unrestricted wavefront is half the amplitude that would result if all zones but the first were blocked off. The intensity (amplitude squared) is one fourth that due to the first zone alone. For the case of a circular aperture in which the wave is now blocked off by a screen that has a small circular aperture, the amount of light that reaches the central point of the diffraction pattern depends on the number of Fresnel zones that fit into the aperture. If the radius of the aperture is such that only the first Fresnel zone fills it, then the amplitude is twice and the intensity is four times that of the unscreened wave. Increasing the radius of the aperture so that two zones can fit, the amplitude is just the difference of the amplitude from these two zones, or practically zero (the contributions are equal but out of phase), and the intensity decreases even though the aperture has been made larger. Further increases of aperture radius cause the intensity to pass through maxima and minima each time the number of zones included becomes odd or even. If an odd number of zones fits the aperture, then the diffraction pattern has a central bright point; if an even number of zones fits the aperture, the pattern has a central dark point. The same effect occurs by moving the point of observation continuously to or away from the aperture along the perpendicular, thereby varying the size of the zones, and producing minima or maximum along the axis of the aperture. If the circular aperture problem is replaced by a circular obstacle problem, these methods lead to the surprising conclusion that there should be a bright spot in the center of the shadow. If the obstacle is made to consist of alternately opaque and transparent optical zones, it is possible to arrange for alternate Fresnel zones to be effective for a particular observation point. Then the result is high intensity at that observation point, since arrivals from alternate zones are in phase. This type of obstacle is the *Fresnel zone plate,* which produces a bright image of the source at the observation point, and thus acts as a lens. By this arrangement, Fresnel lenses produce images of any small bright source by diffraction.

friction   When the surface of one body slides over the surface of another, each body exerts a frictional force on the other. Frictional forces are oriented parallel to the surface in the direction opposite to the motion relative to the other body. Thus frictional forces always oppose the motion. They can exist when no motion is present. *Static friction* describes frictional forces acting between surfaces at rest with respect to each other. Its magnitude F_s is proportional to the normal force on the surface N, related through the coefficient of static friction μ_s, as $F_s \leq \mu_s N$. Forces of *kinetic friction* develop between surfaces in relative motion. The magnitude of the force of kinetic friction F_k is related to the normal force N through the coefficient of kinetic friction μ_k, $F_k = \mu_k N$. *Friction in fluids* that deform under the action of shear stress is described by viscous forces. A fluid element when subjected to a shear stress τ_{yx} experiences a rate of deformation (shear rate) described by du/dy, for Newtonian fluids $\tau_{yx} = \mu \frac{du}{dy}$. The coefficient of proportionality is the absolute (dynamic) viscosity μ.

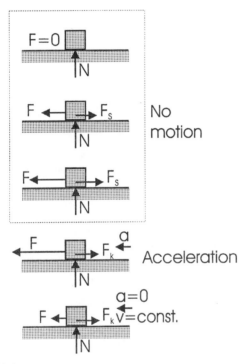

Friction.

frictional coefficient (electrophoresis)　The frictional coefficient $f f_o^{-1}$ relates the Stokes radius R_s of a macromolecule to its molecular mass M:

$$R_s = f f_o^{-1} (3vM)^{-1/3} (4\pi N_A)^{1/3}$$

where v is partial specific volume and N_A is Avogrado's number.

frictional electrification　An electric charge generated by rubbing a material against a dissimilar material. For example, an electric charge is generated on a glass rod rubbed against a silk cloth.

fringes　Interference and diffraction phenomena are characterized by patterns of maximum and minimum intensity called *fringe systems*. One of these alternate light and dark or color bands, in a circular or rectilinear pattern, is a bright or dark or color fringe. Depending on the means of separation of the original beam into interfering beams, interference fringes may be classified as fringes from division of wavefront, as from a diffraction grating, or fringes from division of amplitude, as in a semireflecting mirror. Fringes from plane-parallel plates as in the Fabry-Perot interferometer are fringes of equal inclination or *Haidinger fringes*. Fringes from other geometrical situations are fringes of equal thickness or *Fizeau fringes*. Since the visible spectrum occupies only one octave, fringes from path differences of one wavelength are possible in white light, and are called *white light fringes*. With longer path differences, fringes are seen only for nearly monochromatic light, and are called *monochromatic light fringes*.

fuel cell　A type of electrochemical cell where the reactants are not permanently contained in the cell but are fed into the cell when power is desired. The operation of a fuel cell is similar to a battery but no consummable electrodes are used.

full-duplex　Refers to a communications system or equipment capable of transmission in two directions, that is, transmitting and receiving simultaneously. In pure digital networks, this is achieved with two pairs of wires. In analog networks or in digital networks using carriers, it is achieved by dividing the bandwidth of the line into two frequencies, one for sending, and the other for receiving.

fullerphone　System of telegraphy that uses direct currents for actual transmission. Buzzer signals for keying and listening are also used, thus making the system more difficult to tap.

function generator　(1) A device capable of producing a desired time varying waveform (e.g., sine, square, or triangle) with adjustable frequency and amplitude.

(2) A circuit that accepts an input voltage as an independent variable, x, and generates a de-

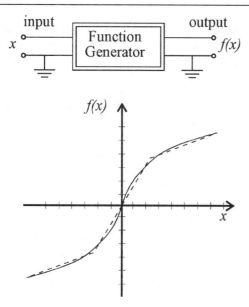

*A block diagram of a function generator accepting an
input voltage x and producing an output voltage $f(x)$.
Also shown is an arbitrary function with an example of
a piecewise linear approximation.*

pendent output voltage based on a prescribed
function, $f(x)$. Example functions include
$f(x) = \log(x)$, $f(x) = x^2$, or $f(x) = \sin(x)$.
Depending on the quality of the circuit in use,
the output may be a piecewise linear approximation to the desired function.

fuse   A device that is used to protect an electrical device or circuit from overloading. When
a loaded current exceeds a specified safe level,
a fuse heats itself and melts to open the circuit.
Some fuses consist of a wire enclosed in a small
glass or ceramic cartridge with metal terminals.

G

gain   The ratio of the output variable of a device to the input variable. For an electronic amplifier, it is the ratio of output voltage to the input voltage. The gain is equivalent to the transfer function for passive circuits.

gain–bandwidth product   A figure of merit that determines the useful frequency range of a transistor or amplifier in general. It is the mid-band gain times the bandwidth. For transistors, it is sometimes expressed as

$$\sqrt{\text{available } power \text{ gain}} \times (\text{bandwidth}) .$$

So, a high–gain, low–bandwidth amplifier is equivalent to a low–gain, wide–band amplifier by this algorithm.

gain, inverse   In most feedback systems, the inverse of the total gain is approximately equal to the transfer function of the feedback network. *See also* feedback.

Galvanic cell   It is believed that Luigi Galvani discovered that when two dissimilar metals are contacted by a moist substance, a direct current will flow through the metals. The early electrochemical cell designed using this principle is called a *Galvanic cell.*

galvanometer   An instrument used to determine the strength and direction of an electric current. It operates based on the fact that a magnetic field is created by the current in accordance with Ampere's circuit law. This field interacts with the field of a permanent magnet, causing the coil of the permanent magnet to deflect. The magnitude and direction of the deflection is then related to the magnitude and direction of the current in the coil.

galvanometer, ballistic   A moving coil galvanometer used to measure charge by detecting a burst of current passing through the moving coil. When an electric burst of current passes through the coil, the initial maximum deflection of the moving coil of the galvanometer is proportional to the total charge passed. The period of the movement of the coil should be longer than the duration of the transient current to be measured. *See also* galvanometer, moving coil.

galvanometer constant   A factor by which a galvanometer reading must be multiplied to obtain the current in a standard unit.

galvanometer, d'Arsonval   Instrument where the current carrying the coil is free to rotate in the magnetic field of a large permanent magnet. The deflection of the coil is related to the current it carries.

galvanometer, Einthoven   Also known as *string galvanometer.* A galvanometer that consists of a single conductive filament stretched tightly between the poles of a powerful magnet. The current causes deflection of the filament. The deflection, which is observed through a microscope, is proportional to the current strength.

galvanometer, Helmholtz   A galvanometer that uses a magnetic field generated by a Helmholtz coil. The generated magnetic field is relatively uniform. A Helmholtz coil consists of two coils. The two coils are identical and the distance between the coils is as long as the radius of the coil. The overlapped fields of the two coils cause a uniform magnetic field. The magnetic density flux B near the center of a Helmholtz

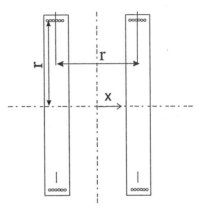

Helmholtz coil.

coil of radius r is

$$B = \frac{25\mu_0 I}{8\sqrt{5}r}\left(1 + O\left(x^4\right)\right),$$

where μ_0 is the permeability and I is the induced current and x is the distance from the center around the axes of a Helmholtz coil.

galvanometer, integrating   A galvanometer designed to measure very slow changes of the electric flux generated in a coil in an electric field.

galvanometer, moving coil   A galvanometer that measures a current passing through a light coil of many turns suspended or pivoted with a spiral spring in a fixed magnetic field produced by a magnet.

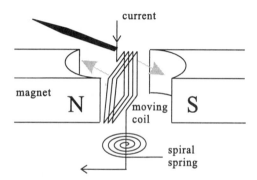

Moving coil galvanometer.

galvanometer, sine   A galvanometer that has a short magnetic needle suspended between two Helmholtz coils. The current in the coil deflects both the coil and the scale. As a result the needle keeps pointing zero scale when the current is loaded. The sine of the angle of deflection of coil (scale) is proportional to the current strength.

galvanometer, tangent   A galvanometer that has a short magnetic needle horizontally suspended at the center of a vertical coil. The current in the coil deflects the needle. The tangent of the angle of deflection is proportional to the strength of the current.

galvanometer, vibrational   A galvanometer that measures an alternating current. The natural oscillation frequency of the movable element of

the galvanometer is made equal to the frequency of the current.

Galvini's experiment   The demonstration that electrical charge causes a muscle to contract.

gamma cameras   A device for producing an image using gamma radiation, usually used to locate radioactive substances within a living body.

gamma ray detector   A device for detecting the presence of a photon of gamma radiation. Such devices are used in nuclear medicine, radiation therapy, high-energy physics, and astronomy.

gamma-ray spectrometry   The measurement of the energy of specific photons of gamma radiation. Such experiments can yield valuable information about the source of the gamma radiation.

gate   In digital electronics, a gate is an electronic circuit that performs an elementary Boolean function. The device is constructed such that it accepts one or more input voltages which are assumed to represent Boolean variables given a predefined logic convention. The gate then provides an output voltage (typically just one) which represents the result of the operation it was designed to implement. Examples include the **AND, OR,** and **EOR** gates.

A gate is a combinational circuit, meaning that the output state is the functional result of the present state of the input variables. *Compare with* flip-flop, whose output depends on the history of the input variables (*see* sequential logic).

The internal electronics of the device can be based on one of many different schemes: TTL (transistor – transistor logic), RTL (resistor – transistor logic), ECL (emitter – coupled logic), CMOS (complementary metal oxide semiconductor), and so on.

Gauss   This is the unit of magnetic flux density and is equal to 10^4 webers per square meter.

Gaussian channel   A Gaussian channel represents a communication channel model wherein the transmitted signal is corrupted and distorted only by additive Gaussian noise, a time-invariant amplitude attenuation and a time-invariant phase offset. As such, a Gaussian channel possesses infinite bandwidth. The Gaussian channel often represents an adequate channel model of a bandlimited channel for a limited time duration within the finite bandwidth over which the information-bearing signal is transmitted.

Gauss's law (electric)   Relationship between the total electric flux through a closed surface and the charge within that surface. In integral form, it is $\int_s E.da = Q/\epsilon_0$, where Q is the total charge enclosed by the surface s. If the charge is distributed over a volume, ρ being the charge density, then by application of the divergence.

generator, current   A device for generating constant currents. It is designed to make the output current independent of the impedance between the output terminals.

generator, electrostatic   A device for generating high-voltage electrical charges. A van der Graaff generator is a kind of electrostatic generator.

generator, Hartman   A device used to produce powerful ultrasonic sound waves. Shock waves induced by a supersonic gas jet at the edges of a nozzle resonate with the opening of a small cylindrical pipe placed opposite the nozzle.

generator, heteropolar   An electric generator whose conductors move through magnetic fields of opposite direction successively. Usually, a heteropolar generator has a conductor rotating in a non-uniform magnetic field. Most generators used are heteropolar generators.

generator, impulsive   Generator used to produce pulses of high voltage and short duration. Usually achieved via parallel charging and series discharging of capacitors.

generator, induction   An AC generator that consists of an induction motor connected to an AC source. Most induction generators in com-

mon use are three-phase AC generators. The rotor of an induction generator is driven mechanically above the synchronous speed corresponding to the frequency of AC from the source. The motor generates AC energy and sends the energy back to the source at the frequency of the source.

generator, reaction, AC   A kind of AC generator. Its rotor does not have any coils but has a salient rotor. It is used for low speed rotation usage, e.g., hydroelectric power generation. The salient pole is designed to make a sinusoidal change of the magnetic flux density. It requires an AC current supply from another AC generator. Therefore, it is connected in parallel with one or more synchronous generators. It is also called a *salient-pole synchronous generator.*

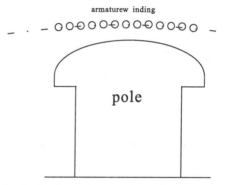

A salient-pole synchronous generator.

generators, acoustic   Transducers that convert electrical, mechanical or other forms of energy into sound; they can be either constant volume velocity or constant pressure. Both versions can be impedance-type or mobility-type. The concept of acoustical generators is of importance in acoustical circuits.

generators, electrical   A device for generating voltages and currents. It usually converts a particular mechanical power to electrical power to generate voltages and currents.

generators, tandem   A modified van der Graaff generator that is used as an elementary particle generator. It connects two generators in a series and a high voltage (positively charged) electrode between the pair of generators. Negative ions are accelerated from ground potential toward the electrode. The negative ions pass by the electrode and then pass through low pressure gas to remove their surplus electrons from the ions. The ions become positive and accelerated again to ground potential by the electrode in the same direction. The ions are twice accelerated in a tandem generator.

generator, van de Graaff Machine for generating high voltages at low currents via electrostatic charge separation. It consists of an insulating belt that is revolved at a high speed over separated rollers, one of which is usually insulating while the other is neutral metal. Adjacent to the rollers are metal combs, one of which is connected to the ground, while the other is connected to the inside of the generator electrical conducting dome. As the belt rotates, charge separation occurs between the insulating roller and the belt. The roller attracts oppositely charged ions from the comb via the plasma created in the air due to the high voltage between the comb and the roller. These are intercepted by the belt and carried away to the other roller, the charges being picked up by the other comb and distributed over the generator dome. In this way large potential differences can be achieved between the dome and the earth.

generic pump A device using various physical principles to convert mechanical energy into fluid energy, usually energy of motion of the fluid. These devices typically work either by compression or suction.

geophones A transducer used in seismic work that responds to motion of the ground at a location on or below the surface of the earth. It is used in exploration seismology to analyze acoustic waves reflected from different rock layers in the earth's subsurface.

germanium A group IV semiconducting element. In its natural crystalline form, it has a diamond-like tetrahedral structure with covalent bonds between neighbors. It is an excellent candidate for doping, thus it is technologically important in the construction of semiconducting electronic devices. *See also* doping; diode junction.

glow discharge A device used for the production of charge particles or monochromatic light. In electronic circuits, it can be used as a voltage reference.

The glow discharge consists of a discharge tube, two exposed conductors in the tube, and a means of supplying a moderately high voltage to the conductors. The discharge tube is an evacuated vessel that is partially filled with the intended gas. Operating pressures are of the order 100 microns. The conductors can be arranged in many different arrangements. The simplest is the parallel plate geometry, as shown in the following figure.

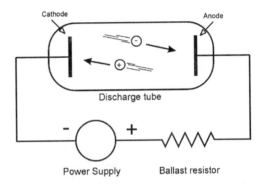

A basic arrangement for a glow discharge.

Considering only a DC voltage source, the conductor that is biased positively is called the *anode*. The negatively biased electrode is the *cathode*. The operation of the discharge produces many separated electrons and ions: a *plasma*. The electrons and negative ions are attracted toward the anode, while the positive ions are accelerated toward the cathode.

A typical I–V characteristic curve for the glow discharge is shown below. Normal operating conditions are 10s to 100s of milliamperes of current and 10s to 100s of volts. The actual values involved depend on many parameters, such as the type of gas and electrode material as well as gas pressure. Note that there is a region for which the voltage is constant for many different currents. This behavior can be exploited in the construction of a voltage regulator. A ballast resistor is shown in the figure below to limit the current drawn by the glow discharge.

The discharge is maintained from two main processes:

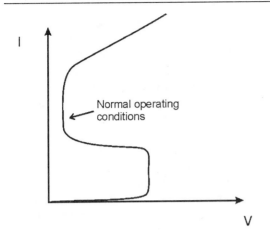

A typical I–V curve for a glow discharge in arbitrary units.

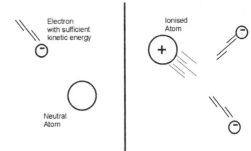

Ionization of the background atom by electron impact.

1. Positively charged ions colliding with the cathode will eject secondary electrons from the surface. The yield of electrons depends on the accelerating potential and the type of cathode material. These secondary electrons are accelerated toward the anode.

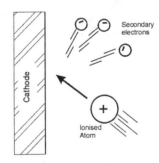

The collisional production of secondary electrons at the cathode surface.

2. As electrons are accelerated toward the anode they, with some degree of probability, will collide with the neutral background gas atoms. If the electron's energy is sufficient, then the atom can be ionized. This produces a positively charged ion that is accelerated toward the cathode. It also produces another free electron that, in turn, can ionize more atoms. If the electron collisional energy is insufficient, the atom will only become excited, which in turn emits photons. This gives the discharge its characteristic glow.

If the secondary electron yield at the cathode surface and the collisional probability of producing positive ions are sufficient, then the discharge will continue to "run" and current will flow. Otherwise, the discharge will "go out" and will not draw any current. Normally, the discharge will not start by itself. A means of producing a few separated charge particles is required. Commonly, the discharge process is started simply from background radiation, e.g., cosmic rays, ionizing a few of the atoms.

Goldman model   This model relates potential differences across a membrane in the presence of more than one ion. The ions are assumed to move independently of one another. The electric field within the membrane is assumed to be constant.

Gouy method   A technique for determining magnetic susceptibility that is based on the measurement of the force exerted on a sample by an inhomogeneous magnetic field. When any substance is placed in a magnetic field, the field produced within the sample either is greater than or less than the applied field, depending on whether the material is paramagnetic or diamagnetic. The method usually involves measuring the weight of a substance in the presence and absence of the magnetic field, making it one of the least expensive techniques for determining magnetic susceptibility.

gramophone   An instrument for reproducing acoustic signals, such as voice or music, by transmission of vibrations from a stylus that is in contact with a groove on a flat disk. Also known as a *phonograph*.

grating   Any arrangement of diffracting bodies that imposes on an incident wave a periodic variation of amplitude or phase or both. It is a device for producing spectra by diffraction and measuring the wavelengths of incident waves. One of the simplest kinds of grating consists of a number of identical equidistant slit apertures in an opaque screen, an idealization useful for mathematical analysis. Practical gratings may consist of equidistant diamond rulings on a plate or mirror or a replica made of these rulings. Making narrow slits the elements of a grating allows them to act as sources that radiate uniformly. Most of the important optical characteristics of diffraction gratings, such as resolving power and dispersion, are concerned with the interference effects from sources arising from each of the corresponding elements of the grating, and are associated with the periodicity of the diffracting elements rather than with their individual shape. Diffracted light from a grating may be either reflected or transmitted, and produces maxima of illumination or spectral lines according to the equation

$$d\left(\sin\iota + \sin\theta\right) = m\lambda \, ,$$

where d is the distance between corresponding elements $= 1/N$, where N is the number of grating lines per unit length of grating, ι is the angle of incidence, θ is the direction of the diffracted maximum with respect to the normal corresponding to order m ($m = 0$ for central image), and λ is the wavelength. The concave or Rowland grating removes the necessity of achromatic collimation and telescopic objectives. *See also* blazing of grating, echelette, echelon, grating, Rowland.

grating, holographic   The generation of a grating on a blank using the holographic process, in which a series of interference fringes are formed from the intersection of two coherent beams of light, corresponding to the grooves of the grating, are recorded in a photosensitive material. The spacing of these fringes is determined by the angle of intersection of the beams and by the wavelength of light. Subsequent chemical treatment, using the solubility dependence of the photoresist that in a suitable developer changes with exposure to light, forms a modulated profile on the surface of the blank

by selective dissolution. Since the grooves are determined by the interference of light, a grating made this way is free from the random and periodic errors present in gratings made with ruling engines. The idea of making gratings this way was considered by Michelson as early as 1915, but it was not until 1960 that Burch made gratings this way. Gratings suitable for general spectroscopic use did not appear until 1967, with the advent of high power lasers, coinciding with the practical realization of holography that was also made possible with the laser, and which was very much in vogue at that time, so the technique acquired the name *holographic grating,* although it is more correct to speak of them as *interference gratings.* The most serious shortcoming of this technique is that there is little control over the groove profile, usually producing gratings with a sinusoidal or a quasi-sinusoidal groove profile. It is also very difficult to control the blaze angle using this process. An advantage of holographic gratings is that they can be made with a periodicity smaller than with a ruling engine. The periodicity in a holographic grating is limited to 1/2 the exposing wavelength; e.g., using a He-Cd laser at 442 nm, this limit is .221 microns. Selecting holographic or ruled gratings depends on the requirements for grating period, blaze angle, and grating depth.

grating, radial   A nonspectroscopic grating used to measure angular displacement in which the wires or rods are set radially within a circular structure, like the spokes of a wheel. Usually radial gratings take the form of annulus between 10 and 20 mm wide on a wider annular blank. Conventional circular dividing engines can make such gratings with an accuracy of ±1 seconds of arc. Further reductions of residual error are possible with multiple printing of the whole grating at all orientations. These are available with up to 43,200 lines or one line per 0.5 min of arc.

grating, Rowland   A concave grating. These ratings are ruled on concave spherical mirrors of metal instead of plane surfaces. It diffracts and focuses the light at the same time and eliminates the necessity of using lenses, thereby eliminating chromatic aberration. It also has the advantage of being used for regions of the spectrum that are not transmitted by glass lenses, e.g., the ultraviolet. A mathematical treatment of the concave grating shows that if R is the radius of curvature of the spherical surface, then a circle of diameter R can be drawn tangent to the grating at its mid-point which gives the locus of points where the spectrum is in focus, provided the source slit is also on this circle. Most mountings for concave gratings are based on this Rowland circle condition for focus.

gratings, crossed   One of the techniques used for projecting a 1-D or 2-D grid of dots or lines. A series of dots is produced when the output of a collimated laser is passed through a diffraction grating. Using crossed diffraction gratings, the result will be a 2-D grid of dots. If a laser line generator is used, the result on passing the output through a single grating is a series of lines. Two such arrangements at right angles or a laser cross generator and crossed diffraction gratings will result in a 2-D grid of lines. The spread of the individual spots or lines is inversely related to the diffraction grating pitch. However, the brightness of the dots or lines may not be even close to uniform since the intensity decreases with the order of the diffracted beam. Lower density gratings (fewer lines/mm) will result in a larger number of more uniformly spaced higher order spots or lines of more nearly equal brightness, but they will be dimmer and more closely spaced (not deflected as much). Crossed gratings can be used to produce *Moiré fringes*. If the two gratings are identical transmission gratings that consist of alternating opaque and transparent elements, then when these gratings are placed face to face with a small inclination angle, no light will be transmitted where the opaque parts of one grating fall on top of the transparent parts of the other. This has the appearance of a set of dark fringes called Moiré fringes. If one grating is stationary, and the other moved perpendicular to its rulings, the Moiré fringes will cross a point each time the gratings advance by the grating interval. This allows measurement of the movement of the grating, by measuring the Moiré fringes, which magnify the movement of the gratings inversely as the angle between the rulings. This has useful applications in automation, where one grating may be attached to an object being processed and is moved across a stationary grating for controlled displacement of the object.

gratings, in series   Gratings may be placed consecutively one after the other in series. One advantage of this set-up is the easy adjustment even in the IR region. The law of diffraction suggests that by combining two grating orders, the beams with appropriate order numbers have the same entrance and exit angles. Thus the series set-up is invariant against tilt of the incident beams as long as both gratings are parallel.

grating spectrograph   An optical instrument that uses a grating to diffract light into specific wavelengths so as to form the spectrum of a light source and record it on film or with photodetectors situated in the spectrum at positions possibly corresponding to the lines of elements or compounds whose presence is of interest. Early spectrometers, such as those developed by Fraunhofer and others, used Newton's discovery of the dispersion of light by a prism instead of a diffraction grating. A concave grating requires no other means to form a sharp image of the slit on the film, but a plane grating or a prism requires additional achromatic lenses or concave mirrors for image-forming in addition to the diffracting element.

gravitational waves   A propagating gravitational field, a distortion of spacetime predicted by general relativity, that is produced by some change in the distribution of matter. It travels at the speed of light and alternately produces out-of-phase contractions and elongations of space along two axes perpendicular to the propagation direction. The strength of the field is characterized by its strain, the fractional changes in the lengths along the two axes. Also know as *gravitational radiation.*

grazing incidence   Light that strikes a surface at an angle nearly perpendicular to the normal. Similarly, *grazing emergence* occurs when an emergent ray is at an angle nearly perpendicular to the normal of the emergent surface of a medium.

ground   The point in a circuit that by definition has zero potential. In practice this is achieved by connecting the circuit directly to the earth via a conductor. This serves to create a fixed baseline potential from which other potentials can be measured.

ground glass   Glass that has been frosted by grinding or etching. It diffuses light by scattering in directions close to the incoming beam; at larger angles out from this direction, light falls off rapidly.

ground wave   In transmission of radio waves, the path traveled by wave along the earth's surface. It can interfere with sky waves to produce selective fading of radio signals.

group velocity   The concept of group velocity is used to describe the movement of acoustic waves in a moving fluid. Acoustic waves can, in their simplest form, be represented by plane waves, the direction of propagation $\vec{n}$ perpendicular to the wavefronts, that move with the speed of sound c. The velocity of the disturbance in a stagnant fluid is $c\vec{n}$. If the speed of the ambient fluid is $\vec{v}_0$, the velocity of the wave disturbance registered by an observer at rest, called the group velocity $\vec{v}_{gr}$, is the vector sum $\vec{v}_{gr} = c\vec{n} + \vec{v}_0$. Also describes the velocity of the envelope of a group of waves having slightly different frequencies and phase velocities. Also known as *(acoustic) modulation.*

Grove cell   A primary cell with a platinum electrode in a nitric (or sulfuric) acid electrolyte inside a porous cup. Outside this cup is a zinc electrode in a sulfuric acid electrolyte. It was invented by William Grove.

guard wires   A high-conductive connection to a large conducting body. It is usually connected to electrical equipment for safety and circuit completion.

Gunn effect   *See* diode, Gunn.

gun, sound from   The sound wave generated from the discharge of a gun, consisting of three separate signals. The first signal forms the envelope of the waves emitted by a projectile that moves at a speed higher than the speed of sound, and it is called the *head* or *bow wave.* This wave reaches the observer first, and is perceived as a sharp crack. The second wave is caused by the explosion of the shell, and the third one by the expanding gunpowder gases traveling at the speed of sound.

Truth table of the full–adder

C_{i-1}	X_i	Y_i	S_i	C_i
0	0	0	0	0
0	0	1	1	0
0	1	0	1	0
0	1	1	0	1
1	0	0	1	0
1	0	1	0	1
1	1	0	0	1
1	1	1	1	1

half–adder Forms the basis of multi–digit addition or subtraction of base–2 numbers. It is the smallest operational block that performs elementary binary addition or subtraction on a single digit. The half–adder has two inputs X and Y for the addend and augend. Normally a combinational circuit, it provides sum, S, and carry, C, outputs based on the immediate input conditions. The operation of the half–adder is defined by

$$S = (X \otimes \overline{Y}) \oplus (\overline{X} \otimes Y)$$

$$\equiv \overline{(X \oplus Y)} \oplus (X \otimes Y)$$

$$C = X \otimes Y,$$

and is illustrated in the accompanying truth table.

Truth table of the half–adder

X	Y	S	C
0	0	0	0
0	1	1	0
1	0	1	0
1	1	0	1

To perform basic addition of the ith digit of two binary numbers (*see* digital arithmetic), consideration must be given to the carry (or borrow) of the next lower significant digit. The *full–adder* has an extra input, C_{i-1} in addition to the addend and augend, X_i and Y_i respectively. The full–adder is constructed from two half–adders; the first adds X_i and Y_i and the second adds this subtotal with C_{i-1}. A truth table illustrating the operation of the full–adder is also presented.

half–duplexing Also known as *two way alternate*. Data can be transmitted in network communication in either direction but only in one direction at any given time. This is done by use of a circuit to provide transmission alternately in either direction.

half–power bandwidth The frequency difference between the two points in an amplifier's frequency response for which the power gain has finally dropped to one half the center frequency f_0 power gain. This is equivalent to the points where the relative gain has dropped by 3 dB. These two points f_a, f_b define the 3 dB bandwidth or pass–band $(f_a - f_b)$.

half–power frequency The frequency at which an amplifier's power gain has dropped by one half the center frequency power gain (i.e., by 3 dB). *See also* half–power bandwidth.

half tone A musical interval with a frequency ratio of $2^{1/12} = 1.0595$. In the theory of equal temperament, any two half tones approximate a major interval, any four a major third, any five a fourth, any seven a fifth, any nine a sixth and any eleven a seventh. Any twelve half tones form an octave. In just intonation (characterized by mathematically exact intervals) with reference to the major key of C, the frequencies for the keys D, E, F, G, A, B and C are tuned to 9/8 (major interval), 5/4 (major third), 4/3 (fourth), 3/2 (fifth), 5/3 (sixth), 15/8 (seventh) and 2 (octave) times the frequency of the first C. Also called half step or semitone. *See also* frequency band and octave.

half-wave plate One of the simplest devices for production and detection of circularly polarized light is the quarter-wave plate, which introduces a 90° phase shift between the ordinary and extraordinary vibrations. With the quarter wave plate oriented at 45° with the plane of polarization of the incident light, circularly polarized light is produced. A half-wave plate reflects the E-vector about its axis. Light passing through a quarter-wave plate twice is equivalent to passing once through a half-wave plate. Like the quarter-wave plate, these plates are usually made of thin sheets of mica or quartz cut parallel to the optic axis. The thickness is adjusted, depending on the wavelength (usually sodium D is selected) to introduce a phase difference of 180° between the ordinary and extraordinary vibrations. Plane polarized light passing through the plate has its plane of polarization rotated through 2θ where θ is the angle between the axis and the incident vibrations. In certain instruments, such as the Laurent Saccharimeter, it is desirable to compare two adjacent fields of light polarized at a certain angle with respect to each other; then half the field is covered with a half-wave plate, and the analyzer is rotated until two half fields are equally bright or dark.

half-width   The full width at half maximum (FWHM) or *half-width* expresses the extent of a function, $y = f(x)$, given by the difference between the two extreme values, $x2 - x1$, of the independent variable x at which the dependent variable y is equal to half of its maximum value, $ym/2$. For example, the half-width of the error function integrand $e - y2$ is 1.67. FWHM is frequently applied to spectral width of sources used for optical communications. When applied to pulse width where the independent variable is time, full duration at half maximum (FDHM) may be used.

haloes   (**1**) In meteorology, the short lived and sometimes faintly hued circles or arcs that are seen to surround a light source viewed through fog or light clouds. The theory attributing their formation to ice crystals was suggested by the 17th century philosopher Descartes. White haloes are formed by reflection from ice crystals, colored haloes from refraction. The size of scattering ice determines the size of the ring.

(**2**) The ring surrounding a photographic image of a bright source and resulting from the scattering of light in random directions.

(**3**) The broad rings that appear as a result of the diffraction of monoenergetic beams of electrons or X-rays from crystalline powder.

hamming distance   The number of bit positions by which two binary codewords differ. The error detecting and error correcting properties of a code depend on this distance.

hardware   Physical devices, generally for the interfacing of central processing units to the physical world. Each piece of hardware will perform some specific and unique task. The computer's hardware will handle chores such as inputing and encoding data, by way of a keyboard or scanner, for example. Output hardware include printers, display devices, and robotic equipment.

harmonic analyzer   Electronic device that measures the frequencies and relative amplitudes of harmonic components in a complex wave. Also known as *harmonic wave analyzer.*

harmonic motion   Motion in that the displacement of particles repeats itself in equal intervals of time, also called *periodic motion.* Periodic motion can be described in terms of sines and cosines. Since the term *harmonic* is used for expressions containing these functions, periodic motion is also called *harmonic motion.* Thus, harmonic motion along a line is given by the function $x = a\cos(kt + \theta)$, where t is the time parameter, and a, k and θ are constants. Also known as *harmonic vibration* or *simple harmonic motion.* When frictional forces that dissipate energy are present, the system will execute damped harmonic motion.

harmonics   A series of sounds which have frequencies that are integral multiples of some fundamental frequency. *See also* frequency, fundamental.

harp, sound from   Sound produced by plucking strings spanned on a triangular frame. The resulting vibration is a combination of several modes of vibration. When the string is plucked at the center, the resulting vibration will consist of *fundamental* and *odd harmonics.* When the string is plucked at a point other than its center, the spectrum of the constituent modes changes. For example, plucking the string at 1/4 of the distance from the end suppresses the 4th harmonic, etc. The string length determines the wavelength of the fundamental harmonic of the sound wave. The modern harp is equipped with a pedal mechanism that increases the span of tones that can be produced by the harp so that it can exceed the power of keyboard instruments.

hearing   The general perception and the specific response to acoustic stimuli. Two approaches are used to study and describe hearing.

1. The goal of auditory physiology is to understand the structure, organization, and functioning of various components of the auditory system at different stages of processing of auditory signals.

2. Auditory psychophysics and psychoacoustics deal with the way humans sense and perceive sound. Hearing involves the elements of intensity, frequency, pitch (sound rich in harmonics is perceived as having the pitch of the fundamental frequency independent of the ratio of energy carried by the fundamental frequency and the harmonics), localization of sound, and the perception of complex spectra.

hearing, abnormal   Impaired hearing is most commonly identified in pure tone audiometry by evaluating the auditory response of the individual to sinusoidal signals at octave intervals from 250 to 8,000 Hz sounded in a quiet room, using the audiogram. The impedance, or its inverse the admittance, of the middle ear is determined through immittance measurements that are useful in establishing the site of the lesion within the auditory system. An important aspect of hearing impairment is degradation in speech intelligibility, evaluated by speech audiometry, to determine the speech recognition threshold and word recognition score. The objectives of the tests are to establish the extent of hearing impairment and its cause (the site of the lesion).

hearing aids   Miniature, portable prosthetic devices for individuals with impaired hearing. Conventional electroacoustic hearing aids are head-worn sound amplifiers that consist of a microphone, audio amplifier, earphone, and battery. Assistive listening devices improve the speech-to-noise ratio of conventional hearing aids, responsible for poor speech recognition, by moving the detached microphone closer to the sound source. The microphone output signal

can be delivered to the amplifier either by wire or by means of radio frequency or infrared signals. Vibrotactile devices convert sound into an electrical signal to deliver it to the skin of the individual as a pattern of mechanical vibrations through vibrating mechanical contacts. Cochlear implants convert acoustical signals into electrical signals, process these, and deliver them to the nerve fibers in the inner ear by electrodes inserted surgically into the inner ear.

hearing loss   Impaired general perception or specific response to acoustic stimuli. Hearing loss can be congenital or caused by external factors or illness. Approximately 9% of the population in the United States is affected by hearing impairment of different levels of severity. Among the population above the age of 65, this percentage is between 30 and 40%. The focus of the discipline of audiology is the diagnosis and rehabilitation of individuals affected by hearing loss. The aim of the diagnosis is to establish the extent of hearing loss and its etiology using a battery of tests. Depending on the site of the lesion, hearing impairment can be classified as *conductive hearing loss, retrocochlear* (involving the lesion of the auditory nerve or neural pathways in the brain), or *middle ear pathology,* as well as *sensorineural hearing loss* (cochlear site lesion). *See also* hearing, abnormal; hearing aids.

heart-lung machine, artificial   A device consisting of a pump and an oxygenator used to oxygenate the blood and to pump it through the body. Usually used in heart or lung surgery.

heat exchangers   There are several types of heat exchangers used at low temperatures. In general, heat exchangers consist of a fluid at one temperature, T_1, and another fluid or body at a different temperature, T_2. In most situations, heat removal is the primary concern, so $T_1 > T_2$. A counterflow heat exchanger consists of two tubes with fluid flowing in opposite directions. Usually made of concentric tubes of stainless steel, brass or a copper-nickel alloy, this design allows heat flow across the wall of the inner tube. The temperature varies continuously along the length of the tube; therefore, this design is often called a *continuous counter-*

flow heat exchanger. At low temperatures, the Kapitza boundary resistance increases, making the use of simple concentric capillary tubes ineffective. A second type of heat exchanger, *a step heat exchanger,* is used to increase the surface area at low temperatures. This exchanger has a metal sinter to provide the necessary surface area. Often made of silver or copper powder, the sintered heat exchangers can provide many 10s of square meters of surface area. If the heat needs to be exchanged between a liquid (e.g., helium-3) and a solid, then sintered metals provide the best exchangers available. *See also* Kapitza boundary resistance.

heating effect of current   This is also called *Joule heating.* When a current I is flowing through a resistor with resistance R, the amount of energy loss per unit time (power loss) is $I^2 R$.

heat switches   A heat switch allows a thermal connection between two regions to be opened and closed upon demand by one of several techniques. Gas heat switches use the thermal conductivity of a gas, often helium or hydrogen, as the mechanism to open and close the switch. A closed volume is connected to the two regions of interest, and gas is introduced to close the switch, then removed to open it again. Helium-4 can be used, but at temperatures below 2.17 K, the helium-4 film on the inner surface of the closed volume becomes superfluid. The superfluid film will continue to conduct heat after the gas is removed, so care to remove all of the helium must be taken. This problem is not present if hydrogen is used, but hydrogen's flammability is a concern. In very low temperature cryostats, the remaining hydrogen can undergo ortho-para conversion causing substantial heat release. (Ortho-hydrogen is a spin $I=1$ molecule that converts into para-hydrogen, a spin 0 state, at low temperatures.) Helium-3, a rare isotope of helium, is an ideal choice, as it does not become superfluid until reaching temperatures below 2.5 mK, and does not have the problems hydrogen does. At temperatures above approximately 1 K, physical contact can be used. In this technique, a spring, screw, or motor forces a piston, attached to the first region, into contact with the second region. A common technique for temperatures below 1 K is to use a super-

conducting heat switch to provide the thermal contact. In the normal state, a superconducting metal carries heat through its conduction electrons. In the superconducting state, the electrons form Cooper pairs and cease to carry heat effectively. As a result, the thermal conductivity in the normal state is 1000 times (or more) that in the superconducting state. In practice, the switch is closed by producing a magnetic field which drives the heat switch into the normal phase, closing the switch. Foils of Al, Sn, Nb, and Ta are the most common metals used for superconducting heat switches due to their superconducting transition temperatures, critical magnetic fields, and ease of fabrication.

HeI   A term synonymous with normal liquid ^{4}He (in contrast to HeII). *See also* helium-4, normal; helium-4, superfluid.

HeII   A term synonymous with the superfluid phase of liquid ^{4}He. *See also* helium-4, superfluid.

helimagnetism   A property possessed by some metals, alloys, and salts of transition elements or rare earths in which the atomic magnetic moments, at sufficiently low temperatures, are arranged in ferromagnetic planes, the direction of the magnetism varying in a uniform way from plane to plane.

helium-3/helium-4 mixtures   Helium-3 and helium-4 are miscible in all proportions at high temperatures, but phase-separate at temperatures below 0.87 K. A schematic phase diagram is seen below.

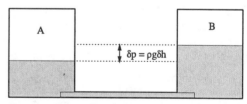

Phase diagram for ^{3}He/^{4}He mixtures.

The superfluid transition temperature can be seen to decrease with increasing ^{3}He concentration. Note that the ^{3}He dissolved in ^{4}He does not partake in the superfluidity of the surrounding ^{4}He. The ^{3}He-rich phase, located on the lower right of the figure is nearly pure, while the ^{4}He-rich phase contains some ^{3}He all the way to absolute zero, seen in the figure by the non-zero x-intercept. Near 0 K, the dilute phase (^{4}He-rich) has roughly 6.6% ^{3}He dissolved in it. This finite solubility is due to the large binding energy of ^{3}He atoms in ^{4}He, 2.8 K/atom (24 J/mol). The behavior of dilute solutions of ^{3}He in ^{4}He at low temperatures allows a cooling technique used in "dilution refrigerators" that can produce steady-state temperatures of 3–5 mK.

At temperatures below approximately 0.5 K, the liquid ^{4}He is deep in the superfluid phase and therefore acts only as a "dense vacuum". The ^{4}He is, under most circumstances, at thermal equilibrium due to its very high thermal conductivity. The entropy and specific heat of the ^{4}He is vastly smaller than that for the ^{3}He as well, so the ^{3}He atoms can ignore the ^{4}He atoms for thermal properties. The ^{3}He atoms do have to move the ^{4}He atoms around in order to flow, so the superfluid is said to increase the effective mass of the ^{3}He atoms. The effective mass, m_3^*, is about 2.4 times larger than the bare mass, m_3. With this change to the ^{3}He atoms, the theory for interactions between ^{3}He atoms and their collective behavior, Fermi liquid theory, provides an accurate description of the dilute gas of ^{3}He atoms.

As mentioned above, the specific heat of the ^{4}He is negligible compared to the ^{3}He, so the specific heat of the dilute phase is constant (roughly 3/2 k_B per atom as for an ideal gas) at moderate temperatures (0.25 K < T < 0.5 K) and varies linearly with temperature (as a Fermi liquid) at lower temperatures. At higher temperatures, the specific heat of the ^{4}He must be included for an accurate description.

The addition of ^{3}He atoms to a bath of superfluid ^{4}He changes the thermal flow properties by providing another path for heat conduction. At high temperatures, heat is carried by excitations of the superfluid, namely phonons and rotons, and the ^{3}He atoms are carried along for the ride. In addition, the thermal diffusion of the ^{3}He atoms and the heat flow through the normal fluid contribute, particularly at lower temperatures. The viscosity is altered in a similar manner. The presence of ^{3}He decreases the mean free path in the superfluid, thereby decreasing the viscosity.

The enthalpy of the dilute phase (low ^{3}He concentration) is higher than that in the concentrated (nearly pure ^{3}He) phase. In fact, for a 6.6% solution, the enthalpy of the dilute solution is roughly 8 times that in the concentrated solution. Thus, if ^{3}He atoms diffuse into the dilute phase from the concentrated phase, the concentrated phase will cool by the enthalpy difference per atom. This "evaporation" of the liquid ^{3}He in the concentrated phase to the "dense vacuum" of the dilute phase is the mechanism used in dilution refrigerators to cool down to several millikelvin. *See also* refrigerator, dilution.

helium-3, liquid   Helium-3, the lighter stable isotope of helium, liquefies at 3.2 K atmospheric pressure. Between liquefaction and approximately 0.1 K, liquid helium-3 behaves much like a dense classical gas. Thus, at temperatures above 1 K, the specific heat is constant, its viscosity is also constant (roughly 25 μPoise), and the speed of sound varies from 183 m/s at saturated vapor pressure to 422 m/s at melting pressure (34.4 bar).

At temperatures below 0.1 K, liquid helium-3 is well described by the Fermi liquid theory. This theory, described by Landau, replaces the bare ^{3}He atoms with quasiparticles that include the effects of interactions and the quantum nature of ^{3}He. The specific heat is roughly linear, with the actual dependence found to be

$$C_V/R = \gamma T + \Gamma T^3 \ln(T/\Theta_c) ,$$

where R is the gas constant, and γ, Γ, and Θ_c are temperature-independent constants. Liquid ^{3}He has a specific heat much larger than typical metals in this temperature range. The viscosity for ^{3}He varies roughly as T^{-2} in this regime, reaching nearly 1 Poise, the viscosity of machine oil.

At very low temperatures, below 0.003 K, liquid helium-3 goes through a second order phase transition into one of several superfluid phases. Due to the nuclear magnetic moment of ^{3}He (and hence its fermionic nature), superfluidity in helium-3 can be described by a somewhat altered version of BCS theory, originally developed to describe low temperature superconductors. (*See* helium-3, superfluid.) *See also* helium-4, liquid; helium-4, superfluid; helium-3, superfluid.

helium-3, melting curve   Solid ^{3}He melts at pressures of 29.3 bars (at 0.315 K) to 34.4 bars (below 0.001 K). The coexistence curve for liquid and solid ^{3}He, the melting curve, is known accurately below the minimum at 0.315 K and can, by measuring the melting pressure accurately, be used as a thermometer. The melting curve is shown in the figure below.

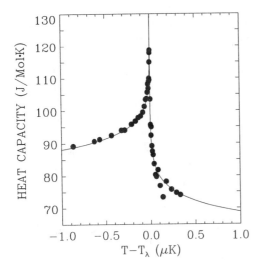

Heat Capacity of liquid ^{4}He.   The Lambda phenomenon.

There are two major features of the melting curve worth noting: first, the melting curve never reaches atmospheric pressure — ^{3}He is one of two substances that do not solidify upon cooling at atmospheric pressure (the other being ^{4}He). The second feature is the pronounced minimum in the melting curve visible at 0.315 K. The Clausius-Clapeyron equation:

$$\frac{dp_{\text{melt}}}{dT} = \frac{\Delta S}{\Delta V}$$

where ΔS is the entropy difference between the solid and liquid and ΔV is the difference in molar volumes. At the temperatures of interest, the molar volumes are roughly constant and ΔV is positive. The entropy of solid ^{3}He is $S_s = R \ln(2)$ to within 1% at temperatures above 10 mK while the entropy of the liquid is $S_\ell = 4.56RT$ where R is the gas constant. Thus, at temperatures below 0.315 K, the entropy of the solid is *larger* than that of the liquid, and the slope of the melting curve is negative. This remarkable fact means a compression of a solid-liquid mixture of ^{3}He below 0.315 K will produce *cooling*. This is in marked contrast to, for example, the response of a gas to being compressed. *See also* refrigerator Pomeranchuk.

helium-3, solid   Liquid helium-3 becomes a solid only at elevated pressures. The minimum pressure for solidification is 2.93 MPa (29.3 bars) at a temperature of 0.316 K. The solid at pressures below roughly 10 MPa is a body-centered-cubic structure, while at higher pressures the hexagonal-close-packed phase is stable. At very high pressures and temperatures (above roughly a kbar and 20 K), a face-centered-cubic structure is observed.

At temperatures above approximately 1 mK, solid helium-3 is a nuclear paramagnet with an entropy of roughly $Nk_B \ln(2)$ due to the N disordered spins. This entropy is actually larger than that in liquid helium-3 at temperatures below 0.3 K. The Clausius-Clapeyron equation therefore indicates that the slope of the melting (liquid-solid coexistence) curve is negative at low temperatures; a mixture of liquid and solid helium-3 will actually cool as the pressure increases. At a temperature of 0.93 mK, the nuclear spins go through a strongly first-order phase transition into an antiferromagnetic phase. This phase, known as the U2D2 phase, does not consist of alternating spins as in a simple antiferromagnet. Instead, the system forms planes of spins which are ferromagnetic – the structure consisting of two such planes with the spins up, then two planes with spins down, etc.: hence U2D2 represents up-two, down-two. The entropy of the U2D2 phase varies as T^3, analogous to phonons. (In any system that has elementary excitations with a linear dispersion relation, the entropy is expected to depend cubically on temperature.) The U2D2 phase is stable at low magnetic fields, below approximately 0.45 Tesla. Above this field, the system goes through another first-order phase transition into a different antiferromagnet, the canted-normal antiferromagnet (CNAF). This is a phase with an antiferromagnetic signature and a ferromagnetic signature at right angles to one another. A one-dimensional representation of the U2D2 and the CNAF phases is shown below, respectively:

$$\uparrow\uparrow\downarrow\downarrow\uparrow\uparrow\downarrow\downarrow \cdots$$

$$\nearrow\nwarrow\diagup\nearrow\nwarrow\diagup\nearrow\nwarrow\diagup\nearrow\nwarrow \cdots$$

The primary interactions between ^{3}He atoms in the low temperature solid is due to atomic exchange processes. There are many possible exchange configurations, several of which contribute to the properties of solid ^{3}He. The simplest exchange is simply nearest-neighbor exchange, but this is suppressed by the steric hindrance of nearby atoms. (*Steric hindrance* describes the fact that the atoms are not point particles and must move past one another.) Multiple spin exchange in rings is therefore favored over two particle exchange. Three, four, and six particle exchange are all relevant for solid ^{3}He. This exchange produces an effective spin interaction through the Pauli exclusion principle requiring the total wavefunction to be antisymmetric.

Above the transition temperature, the magnetic susceptibility per unit volume behaves according to a Curie-Weiss law,

$$\chi = \lambda/(T - \Theta)$$

where λ is the Curie constant per unit volume and Θ is the Weiss temperature. The susceptibility changes discontinuously at the (field driven) transition between U2D2 and CNAF, increasing by roughly a factor of six. *See also* helium-3, melting curve.

helium-3, superfluid   Liquid helium-3 becomes a superfluid at 2.5 mK (at 34.4 bar). Since ^{3}He has one unpaired neutron in the nucleus, it has a net spin angular momentum of 1/2 (in units of $\hbar$) and is therefore a fermion. As such, it cannot have more than one atom in a given quantum state. This is in marked contrast to the case for liquid ^{4}He which forms a superfluid when a large number of atoms form a interacting Bose condensate in the ground state. Superfluidity in liquid ^{3}He can be described by the formation of Cooper pairs, similar to the process in superconductors. Liquid ^{3}He can

be described as a "nearly ferromagnetic" liquid. One ^{3}He atom will tend to polarize nearby atoms through the nuclear dipolar interactions. This cloud of polarized atoms will then "prefer" another atom of the same spin as the first one (e.g., spin up) over a spin of the opposite state (spin down). This produces a net interaction between these two like spins that is attractive. At sufficiently low temperatures, these atoms form Cooper pairs of spin 1. Since the wavefunction for fermions must be antisymmetric, the orbital wave function must therefore be in an odd angular momentum state and is found to be in an $\ell = 1$ state. (In contrast, electrons in low temperature superconductors form Cooper pairs of total spin 0 and angular momentum $\ell = 0$.) Thus the spin state of the wavefunction can be $|\uparrow\uparrow>, |\downarrow\downarrow>, or |\uparrow\downarrow> + |\downarrow\uparrow>$. This internal structure of the Cooper pair leads to complex behavior. At high pressures and relatively high temperatures, superfluid ^{3}He-A is stable. The A-phase consists of equal spin pairs (those pairs with $S_z = 1$). This pairing leads to ^{3}He-A being very anisotropic, having preferred orientations in both spin-space and real-space. This anisotropy produces complex magnetic behavior and liquid-crystal-like behavior. At lower temperatures and pressures, there is a first order phase transition to the B-phase. The B-phase consists of all three possible spin states available for the Cooper pairs. The B-phase also has complex magnetic and orbital behavior. As the magnetic field increases, the volume of phase space in which the B phase is stable decreases until, at approximately 0.6 T, the A phase becomes the stable phase. A third phase, the A$_1$ phase, exists in magnetic fields between the A phase and normal ^{3}He. The A$_1$ phase consists of Cooper pairs of only one spin orientation (e.g., $|\uparrow\uparrow>$). This phase occupies a very small region of phase space which is not visible on the scale of the figure, but which grows linearly with magnetic field to be roughly 0.5 mK in a 10 T field. The specific heat of superfluid helium-3 shows a finite discontinuity at the transition temperature, indicative of a second order phase transition. There is a slight kink at the transition between A and B phases.

The specific heat varies roughly as T^3 with deviations due to strong-coupling effects. The viscosity falls rapidly below T_c, with the limit-

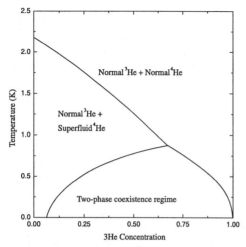

Phase diagram for superfluid ^{3}He.

ing value being difficult to determine due to the anisotropic nature of the superfluid phases. Unlike ^{4}He, liquid ^{3}He has a nuclear magnetic moment and a corresponding response to magnetic probes. Nuclear magnetic resonance (NMR) measurements first indicated the complex order present in superfluid helium-3. In typical NMR experiments, the net spin, **S**, is tipped away from its equilibrium direction along an external field, **B**$_0$. The precession of **S** is subsequently measured. The frequency for superfluid ^{3}He-A differs from that for normal ^{3}He, ω_0, by a temperature dependent term Ω_A. If, instead of tipping the spin away from the magnetic field, the magnitude of the magnetic field is changed suddenly, the spin parallel to **B**$_0$ oscillates at a frequency equal to Ω_A. This parallel resonance does not occur for normal NMR systems. Similar frequency shifts are seen in the B-phase under certain circumstances. In addition, superfluid helium-3 also exhibits sound modes (first sound, second sound, etc.) similar to those found in superfluid ^{4}He. There are also textures in superfluid ^{3}He similar to those found in (room temperature) liquid crystals. *See also* helium-4, superfluid; helium-3, liquid; helium-3, melting curve; superconductivity, BCS theory; Cooper pair; liquid crystal.

helium-4, liquid   Helium-4 has a critical temperature of 5.2 K and a normal boiling point of 4.21 K. Due to its large zero-point motion, ^{4}He remains a liquid under its own vapor pressure all the way to absolute zero (0 K). In fact, it does

not solidify until the pressure exceeds 25.4 bar. At temperatures well above 2.17 K, liquid ^{4}He behaves much like a dense, classical gas. The specific heat is therefore roughly 3/2R where R is the ideal gas constant. (Of course, *liquid* helium-4 is not exactly a non-interacting gas, so the comparison is only approximately true.) The velocity of sound is low in liquid ^{4}He, approximately 230 m/s. The thermal conductivity near its boiling point is quite poor (roughly 10,000 times worse than copper at similar temperatures). Upon approaching 2.17 K, however, liquid helium-4 goes through a dramatic change of character. The specific heat diverges logarithmically as a critical temperature, $T_c \simeq$ 2.1768 K, is approached. This divergence is a manifestation of a phase transition to an exotic phase, the *superfluid* phase. The shape of the specific heat, shown below, gives the transition temperature its common name, the *lambda point*. The properties of the superfluid phase are discussed in detail in a separate entry. *See also* helium-4, superfluid; helium-4, solid; bubbles, suppression of; lambda point.

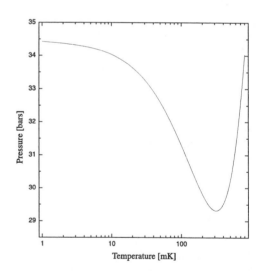

Helium-3 melting curve. The minimum is located at 0.316 K and 29.32 bars, and the zero temperature pressure is 34.39 bars.

helium-4, solid   Unlike all other substances, liquid helium does not solidify at atmospheric pressure at *any* temperature. Liquid helium-4 does not solidify at absolute zero until the pressure approaches 2.5 MPa (25 bars) and solidifies at its normal boiling point, 4.2 K, when the pressure is 14 MPa (140 bars). Under most temperatures and pressures for which the solid is stable, helium forms a hexagonal-close-packed (hcp) structure with an in-plane lattice constant of approximately 3.5°. At very high temperatures, the stable phase is face-centered cubic, while a body-centered cubic structure is stable over a limited range of temperatures and pressures.

helium-4, superfluid Liquid ^{4}He goes through a second order phase transition at a temperature of 2.1768 K into a superfluid state. Below this temperature, the viscosity is vanishingly small, the thermal conductivity is quite large, and the helium will support several new sound-like phenomena. Since the ^{4}He atoms are bosons, it is tempting to claim the superfluidity is due to Bose-Einstein condensation, but the truth is somewhat more subtle than that. It is true that the general concepts of Bose-Einstein condensation can be used in discussing superfluidity in ^{4}He, but the interactions in liquid ^{4}He are much too strong to claim the bosons are *noninteracting*. Thus, the term "Bose-condensed helium" is not, precisely speaking, an accurate description of superfluid helium-4.

Superfluid helium is often discussed in terms of the two-fluid model which assumes the helium is composed of two interpenetrating, noninteracting fluids – the normal component and the superfluid component. The superfluid component consists of the atoms that occupy the ground state and therefore have zero viscosity and carry no entropy. The normal fluid consists of any excited atoms, and these atoms carry all of the entropy and have finite, non-zero viscosity. Note that this model is a *description* of superfluid helium and should not be taken too literally.

The thermodynamic and transport properties of superfluid ^{4}He differ drastically from those of the normal fluid. The specific heat diverges logarithmically (see the figure in the entry for liquid helium-4) as it approaches the phase transition. Superfluid helium-4 supports several wave phenomena not found in other states of matter. Ordinary acoustic waves are known as "first sound" in superfluid helium. The velocity of first sound at low temperatures is roughly 240 m/s and shows a cusp at the lambda point.

First sound is attenuated at temperatures above the lambda transition by the same processes as occur in normal fluids, viscosity and thermal conductivity. At the lambda temperature, there is a discontinuity in the attenuation due to the phase transition. At approximately 1 K, there is a large peak due to a process specific to superfluids, second viscosity. *Second viscosity* is caused by irreversible processes in superfluid ^{4}He involving multi-phonon and phonon-roton scattering. (*Rotons* are excitations of the superfluid not present in normal fluids.)

In addition to first sound, superfluid ^{4}He also supports temperatures waves, *second sound*. If a sinusoidally varying amount of heat is dissipated at one end of a container of superfluid helium, it will propagate nearly unchanged to the other end as measured with a thermometer. Unlike *first sound,* which consists of variations of pressure with density, second sound is due to the variation of the temperature with entropy while the density remains constant. In confined geometries, superfluid ^{4}He can also support *third* and *fourth* sound. Third sound occurs in films of superfluid and is effectively a second sound wave which propagates on the surface of superfluid helium. If superfluid helium is confined to small tubes or tubes packed with powder, the normal fluid is held in place by viscous forces while the superfluid component is free to move. Thermal waves in such a geometry, fourth sound, propagate at a speed between that for first and second sound.

The viscosity of liquid helium below the lambda point is the physical property most commonly associated with the "super" in superfluidity. If a torus of liquid helium above T_λ is rotated slowly, then cooled below the transition, the fluid will continue to rotate even if the container is slowed to a stop. In fact, the superfluid helium will flow *ceaselessly* as long as a limiting velocity, the critical velocity, is not surpassed. If superfluid helium is set into motion with speeds greater than this velocity, the system becomes dissipative. In practice, there are several critical velocities, each one signaling the onset of different physical phenomena. The values of these critical velocities depend on the type of experiment performed and the geometry of the experimental cell. Numerically, these velocities can vary between a few millimeters per second and several tens of meters per second. In most cases, the critical flow velocity is the speed at which the flow becomes turbulent and dissipation begins. This turbulence is caused by some form of vortex creation in the superfluid.

The thermal conductivity of superfluid helium is also quite spectacular in its deviation from that of normal fluids. In normal helium-4, the thermal conductivity is roughly that of a high-density, ideal gas. Upon cooling below T_λ, the thermal conductivity increases by six orders of magnitude or more. This incredibly high thermal conduction results in a lack of bubbles in superfluid helium. Another aspect of superfluidity which affects heat flow is the fountain effect — the superfluid component will flow towards a heat source and, in a suitable apparatus, will produce a pressure gradient that can be quite large. *See also* helium-4, liquid; helium-4, solid; bubbles, suppression of; lambda point; fountain effect.

helium, liquid   *See* helium-4, liquid; helium-3, liquid.

helium, liquid, cooling power of   Liquid helium has a latent heat of vaporization of 20.90 kJ/kg (2.6kJ/ℓ). This translates into the ability to cool 1 kg of copper from 300 K to 4 K using 31 ℓ liquid helium in the process. If the enthalpy of the cold helium gas (roughly 200 kJ/ℓ) is used to precool the copper, the amount of helium necessary drops to 0.8 ℓ per kilogram of copper. The latent heat of liquid helium is nearly an order of magnitude lower than that of liquid nitrogen, so it is commonplace to cool most cryostats to 77 K to decrease the volume of liquid helium needed.

helium, liquid, transfer tube   A transfer tube for liquid helium must be a more complicated device than a liquid nitrogen transfer tube due to the small latent heat of vaporization of liquid helium (23.9 kJ/kg compared to 199 kJ/kg for liquid nitrogen). Such a transfer tube consists of an inner tube made of a low thermal conductivity metal (often steel) surrounded by another metal tube with a vacuum space in between. The liquid helium flows inside the inner tube while the vacuum space thermally isolates the inner tube and the liquid helium from the room temperature environment a few centimeters away. This vacuum-jacketed transfer tube allows the transfer of liquid helium into a cryostat with only small losses.

helium, superfluid   *See* helium-4, superfluid; helium-3, superfluid.

helmholtz coil   (1) An apparatus for providing a nearly constant magnetic field in a small region for experimentation or measurement. Unlike a solenoid, the helmholtz coils are two many-turn loops, coaxial, and separated by a distance the same order of magnitude as the coil diameters. The magnetic field generated when current flows in the loops has low first and second derivatives in the center of the coils, and thus is well behaved for many experimental uses. The geometry allows insertion of nonmagnetic equipment (such as evacuated tubes and phosphor-impregnated glass) in the central region of nearly constant magnetic field strength that would be impossible in a solenoid.

(2) These comprise two parallel coils carrying equal currents separated by a distance equal to their radius. The magnetic field in the center of the coils is uniform to within few percent.

hemodynamics   The study of blood circulation and the forces involved.

henry   An SI unit of inductance or mutual inductance, normally denoted by a symbol H. One henry is the inductance of a circuit in which a current changes at a rate of one ampere per second, inducing an electromotive force of one volt. Permeability is measured in henry per meter.

Henry's function   The mathematical formula used in electrophoresis to account for the retarding force on the macromolecular ion of interest due to counterions. Being of opposite charge, the counterions move in the opposite direction of the macromolecular ion. The interactions between the macromolecular ion and the counterions with their associated solvent impedes the movement of the macromolecular ion.

high speed switching   *See* circuit, switching.

high tension   (1) High voltages, typically, 100 kilo-volts (100 kV) or more.

(2) Anode voltages, typically in the range of 60 to 250 volts.

Hodgkin-Huxley model   A model describing the electrochemical processes associated with nerve-cell discharges. The three basic steps of the model are:

1. activation of sodium ion conductance,
2. subsequent inactivation of sodium ion conductance, and
3. activation of potassium ion conductance.

hole   A vacant electron state that behaves like an electron with positive charge.

Electrons in semiconductors are only allowed to have certain specific energies. The allowed energies are grouped into bands. The lowest band is the valence band and in semiconductors the available electron energy states are mostly filled. The next highest band is the conduction band and it is mostly empty. It is the electrons at the top of the valence band and the bottom of the conduction band that are the most interesting as these electrons are sufficiently close to vacant levels and can readily change states.

The quintessential dispersion relationship (energy E vs. momentum k) for electrons in a bulk semiconductor is shown below. The group velocity of an electron is given by

$$v = \frac{dE}{dk} \frac{1}{\hbar}.$$

If one considers an electron in the semiconductor under the influence of an applied electric field $\mathcal{E}$, then it is possible to show that

$$-e\mathcal{E} = \frac{\hbar^2}{d^2 E/dk^2} \frac{dv}{dt}.$$

Comparing this with Newton's equation $F = ma$, then

$$\frac{\hbar^2}{d^2 E/dk^2} = m*$$

is the effective mass. Note, it is at the top of the valence band, $d^2 E/dk^2$, and hence the mass of the electron in this state, is *negative.*

The concept of an electron with negative mass can be simplified if one considers the small number of vacant states in the valence band. First consider a completely filled band with n electrons with velocities $v_1, v_2, \ldots, v_n$. Since

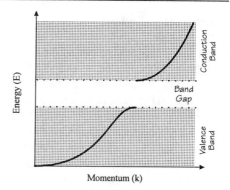

Typical dispersion relationship for an electron in a bulk semiconductor.

the band is full and there is a net zero current flow, there are as many electrons with velocity $-v$ as there are with velocity $+v$. Hence,

$$-|e| \sum_n v_n \equiv 0.$$

If the ith electron is missing, there will be a net current flow

$$-|e| \sum_{n \neq i} v_n.$$

But since

$$-|e| \sum_{n \neq i} v_n + -|e| v_i \equiv 0,$$

the net current flow is equal to $+|e| v_i$ and is considered to be due to a positive charge. It is possible to show that the effective mass of this missing electron, or hole, is positive at the top of the valence band. This idea of hole current is illustrated below as if there is an applied electric field moving the electrons, successively filling the vacancy.

Therefore, holes can be considered as positive charges with positive masses and hole conduction can be thought of similarly as for electron conduction. When an electric potential is applied across a semiconductor, there are two contributions to the conduction current:

1. electrons in the conduction band, and
2. holes in the valence band.

Energy level diagrams are usually defined for electrons. However, if desired, they could be defined for holes as well provided the energy is measured downward from the top of the valence band.

hole current   *See* hole.

hologram   Image formation by the method of wave-front reconstruction, as proposed by Gabor in 1947. A hologram is a special diffraction screen that reconstructs in detail the wave field emitted by the subject. Output from a laser is split into two beams one of which illuminates the subject, the other is a reference beam. The two beams form an interference pattern that is recorded on a high-resolution photographic plate. The plate contains all the information needed to reconstruct the wave field of the subject. After development, the plate is viewed with a single beam from a laser, and part of the diffracted wave field is a precise, three-dimensional copy of the original wave reflected from the subject. The imagine is seen in depth, and moving the point of view changes the perspective of the view.

homeostasis   The ability of an organism to maintain a stable internal condition (such as body temperature) by regulating its physiology in response to external environmental conditions.

homing adapter   In automatic switching, this device enables the automatic return of a sequential selector to a predetermined unoperated position upon its release.

homochronous   Term applied to digital signals whose corresponding significant instants have a constant, but uncontrolled, phase relationship.

Hooke's law   Used, together with Newton's law, to describe vibrations in linear systems. The elastic force F acting on a spring is proportional to the length change of the spring x due to the force F, $F = -kx$, where k is the force constant of the spring. This relation is known as *Hooke's law* and holds for moderate displacement amplitudes. The direction of the force is opposite to the displacement of the endpoint from the origin and it is a restoring force that points toward the origin. For a stretched spring $x > 0$ and the force F is negative, for a compressed spring $x < 0$ and the force F is positive.

hopping   (1) Method employed to force a jammer to cover a wider spectrum by randomly hopping the data-modulated carrier from one frequency to a next. Examples are *slow frequency* and *fast frequency hopping*.

(2) It also refers to a small jump and, with reference to the internet, indicates the route that a computer takes in order to relay its information from one point to the next.

horizontal scanning   The scanning of an electron beam over a phosphor surface to produce a television image. The image is formed by many horizontal scan lines. The number of lines per image depends on the signal encoding (525 lines in American broadcast, 625 for European). The electron beam, therefore, must sequentially scan each line to form the image.

horns, sound from   Sound radiated by acoustic transducers consisting of a tube of varying cross-sectional areas employed to match the impedance of a relatively heavy vibrating diaphragm to the light medium used to propagate the sound. Horns have different shapes (parabolic, conical, exponential, etc.) and their diameter is smaller than the wavelength of sound passing through them. Horns are characterized by their acoustic impedance as a function of frequency, cut-off frequency, resistance to reactance ratio, and directional characteristics.

hot-wire ammeter   An instrument that uses the thermal expansion of a wire or bi-strip to measure the current passing through it. Certain mechanical devices are used to magnify the actual increase in the length of the wire.

h–parameters   Used in the circuit description of a transistor and the inter-relationship between the various currents and voltages in a transistor. The transistor can be considered as a two-port, four-terminal network circuit. Since the transistor has only three connections, there will be one common to the two ports appearing in the figure. Hence, depending on the model circuit, we can have a common base, common emitter, or common collector and the scheme is designated as CB, CE, or CC, respectively; an example is given later.

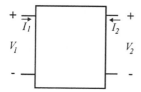

A two-port, four-terminal network representation of a transistor.

There are four different variables I_1, I_2, V_1, and V_2 external to the network as shown in the figure. Any two of these can be chosen as independent variables and the other two expressed in terms of them. For example, the currents I_1 and I_2 can be chosen as the independent variables. Then, the voltages V_1 and V_2 can be solved in terms of the currents

$$V_1 = f(I_1, I_2) \ ,$$

and

$$V_2 = g(I_1, I_2) \ .$$

If these expressions are expanded in a Taylor series, ignoring higher order terms;

$$\delta V_1 = \frac{\partial f}{\partial I_1}\delta I_1 + \frac{\partial f}{\partial I_2}\delta I_2 \ ,$$

and

$$\delta V_2 = \frac{\partial g}{\partial I_1}\delta I_1 + \frac{\partial g}{\partial I_2}\delta I_2 \ .$$

The quantities δV_1, δV_2, δI_1, and δI_2 are the small signal or incremental voltages and currents. These are usually written in the lower case v_1, v_2, i_1, and i_2. The AC component is assumed to be small compared with the DC values. Also, f and g are assumed to be linear functions over the small range of the AC signals. Then, the above becomes

$$v_1 = \frac{\partial f}{\partial I_1}i_1 + \frac{\partial f}{\partial I_2}i_2 = z_{11}i_1 + z_{12}i_2 \ ,$$

and

$$v_2 = \frac{\partial g}{\partial I_1}i_1 + \frac{\partial g}{\partial I_2}i_2 = z_{21}i_1 + z_{22}i_2 \ .$$

Since z_{ij} have units of impedance, it is said that these relations are in the impedance representation.

If we had instead assumed that V_1 and V_2 were the independent variables, then we would have arrived at the admittance representation,

$$i_1 = y_{11}v_1 + y_{12}v_2 \ ,$$

and

$$i_2 = y_{21}v_1 + y_{22}v_2 \ .$$

There are a total of six different representations corresponding to different choices of two independent variables from the set i_1, i_2, v_1, and v_2. A very useful representation is the *hybrid* representation based on i_1 and v_2 as the independent variables.

$$v_1 = h_{11}i_1 + h_{12}v_2 = h_i i_1 + h_r v_2 \ ,$$

and

$$i_2 = h_{21}i_1 + h_{22}v_2 = h_f i_1 + h_o v_2 \ .$$

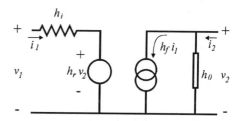

Hybrid circuit representation of a transistor illustrating the meaning of the h–parameters.

These *h–parameters* have especially useful meanings:

$h_i = h_{11}$ is the input impedance with the output short–circuited,

$h_r = h_{12}$ is the reverse voltage ratio with input open–circuited,

$h_f = h_{21}$ is the forward current ratio with output short–circuited,

$h_o = h_{22}$ is the output admittance with input open–circuited.

A circuit model of a transistor based on the hybrid scheme is shown in the second figure.

This parameter formalism and circuit model is generally applicable to all schemes of modeling the transistor (i.e., CE, CB, or CC); however, the h–parameters will be different for each case. A second subscript is used to indicate the scheme used.

Symbol	Probability
a	0.51
c	0.26
d	0.11
b	0.06
e	0.06

Model		Hybrid equation
	CE	$v_b = h_{ie}i_b + h_{re}v_c$ $i_c = h_{fe}i_b + h_{oe}v_c$
	CC	$v_b = h_{ic}i_b + h_{rc}v_e$ $i_e = h_{fc}i_b + h_{oc}v_e$
	CB	$v_e = h_{ib}i_e + h_{rb}v_c$ $i_c = h_{fb}i_e + h_{ob}v_c$

For example, using the above model of a transistor and assuming a common emitter scheme, $h_{fe} = I_C/I_E$ refers to the forward current gain and h_{ie} is the input impedance at the base of the transistor. The numbered parameters or labels in the circuit model can be replaced with the appropriate lettered labels once the scheme is determined; e.g., $h_r v_2$ becomes $h_{rb}v_c$ in the CB scheme.

Huffman code   The Huffman code represents a very common variable-length memoryless code that attempts to match the average code word length to the source entropy. Suppose the Huffman encoder is to encode a five stream of information data symbols drawn from the symbol set {a, b, c, d, e} appearing with the respective probabilities of 0.51, 0.06, 0.26, 0.11 and 0.06. This source's entropy equals: -0.51 log(0.51) - 0.06 log(0.06) - 0.26 log(0.26) - 0.11 log(0.11) - 0.06 log(0.06) = 1.838. The Huffman encoder would proceed as follows:

Step 1: Arrange the information symbols in descending order of probability. If two or more symbols occur with equal probability, the order among them is immaterial.

Step 2: Combine the bottom two entries into a new entry, with a new probability equal to the sum of the two entries combined. Re-order the new list in descending order of probability.

Symbol	Prob.			Symb.	Prob.
a	0.51			a	0.51
c	0.26			c	0.26
d	0.11			{b,e}	0.12
b	0.06	>	{ b,e } 0.12	d	0.11
e	0.06	>			

Step 3: Repeat Step 2 until only two entries are left.

Symb.	Prob.	Symb.	Prob.
a	0.51	a	0.51
c	0.26	c	0.26
d	0.11	d	0.11
b	0.06 >	{b,e}	0.12
c	0.06 >		

Symb.	Prob.	Symb.	Prob.
a	0.51	a	0.51
c	0.26	c	0.26
{b,e}	0.12 >	{b,d,e}	0.23
d	0.11 >		

Symb.	Prob.
a	0.51
{b,c,d,e}	0.49

Step 4: Starting with the right of the table, assign the most significant bit of the code word. Then move left in the table and assign another bit if a split occurs. The assigned codewords are written in parenthesis below:

Symb.	Prob.	Symb.	Prob.
a	0.51	a	0.51
c	0.26	c	0.26
d	0.11	d	0.11
b	0.06 (1110) >	{b,e}	0.12
e	0.06 (1111) >		

Symb.	Prob.	Symb.	Prob.
a	0.51	a	0.51 (0)
c	0.26	c	0.26 > (10)
{b,e}	0.12 >	{b,d,e}	0.23
d	0.11 (110) >		

Symb.	Prob.
a	0.51
{b,c,d,e}	0.49

Thus, $a \rightarrow 0$, $b \rightarrow 1110$, $c \rightarrow 10$, $d \rightarrow 110$, $e \rightarrow 1111$. The average length equals: $1 \times 0.51 + 4x0.06 + 2 \times 0.26 + 3 \times 0.11 + 4 \times 0.06 = 1.940$, only 5.5% above the theoretical minimum specified by the source entropy of 1.838. If block coding is used, the average length would have equaled 3.

The Huffman code self-punctuates — it needs no explicit punctuating symbols as does the Morse Code. Successive blocks of the Huffman encoder output may be directly concatenated and uniquely recoverable by a Huffman decoder.

human voice   Sound generated by the human vocal system, including the organs for human speech. Vocal sound is produced by forcing air from the lungs (by contracting the chest muscles) through the larynx and the vocal chords. The vocal chord is a muscular organ shaped like a diaphragm with a slit-like opening that allows modulating the air stream and controlling the fundamental frequency of sound by changing the tension of the lips of the slit. The cavities and orifices of the throat, mouth and nose together with the tongue form an acoustic network and contribute to the formation of different sounds by varying the shape of the oral cavity. In this way some of the harmonics formed by the vocal chords are emphasized while others suppressed. The sound intensity at a distance of two meters away from the speaker in a normal conversation is 2.10^{-8} W/m^2 and the frequency range of the human voice is 40 to 12,000 Hz.

Huygens' principle   A method of analysis used for problems of wave propagation that avoids the difficulties of a rigorous mathematical computation of waves in inhomogeneous media or near obstacles. The principle makes only general assumptions about wave propagation and states that each point of an advancing wavefront is itself the center of a new disturbance, the source of a new train of waves. It also notes that the entire advancing wave as a whole can be considered to be the result of the secondary waves that arise from points in the medium already traversed. This view of wave propagation facilitates the study of various phenomena such as diffraction, where the light on reaching a slit or edge of an obstacle is regarded as being the source of wave fronts that proceed from there so as to interfere and give the maxima and minima observable as diffraction fringes.

hybrid coil   Used in telephony to interface a 2–wire communication system with combined receive-and-transmit signals to a 4–wire system with independent receive-and-transmit circuits.

Most local telephone exchanges use a 2–wire communication system to service the common telephone. However, physical separation of the transmit and receive signals are required in the switching networks. Separation of the two signals is also desirable since it is easier to transmit various types of data (e.g., digital or voice) and it is possible to insert amplifiers in the transmission circuit for enhanced performance. Therefore, a circuit that interfaces the 2–wire and 4–wire systems is needed; referred to generally as a *hybrid circuit*. The hybrid coil is a simple and conventional way to accomplish this task.

hybrid communication network   A hybrid communication network represents a communication system capable of transmitting and receiving both analog and digital signals.

hydrogen, liquid   Hydrogen molecules, H_2, liquefy at 20.3 K at 1 bar. Hydrogen remains a liquid down to 14.0 K, at which point it solidifies. In the past, liquid hydrogen was used as a refrigerant. Since hydrogen combines with oxygen explosively if present in sufficient concentrations, most laboratories use liquid helium and or cryocoolers instead of liquid hydrogen.

hydrophones   Devices that receive underwater sound waves (underwater microphones) and convert them into electric signals. Hydrophones take advantage of the piezoelectric properties of crystals such as quartz or the magnetostrictive properties of materials such as nickel. They can detect sound emitted by a sound source and the direction of the sound source (for example, an underwater vehicle) by passive listening. Hydrophones are also used to measure the distance of underwater objects by measuring the time needed for the emitted sound beam, reflected by the object, to return in the form of an echo. They can operate in the sonic or ultrasonic frequency range. Hydrophones are characterized by their directivity and the frequency range.

hyperchromism   An increased absorption of electromagnetic radiation (usually in the ultraviolet region) caused by geometry-dependent interaction between two or more parts of a molecule.

hyperfine structure   A set of very closely spaced lines making up a spectral line or paramagnetic resonance line. There are at least two types of hyperfine structure:

1. The splitting of an element's spectral line into doublets, triplets, etc., can be the result of the interactions between the electron spin and the spins of adjacent magnetic nuclei via the coupling of the total angular momentum of the orbital electron with the nuclear spin.

2. The presence of several isotopes in the sample being tested, in which each isotope contributes one or more components of the spectral line (this type of hyperfine structure is termed *isotope structure*).

hyperopia   *See* eye, far-sighted.

hypersonic waves   Sound waves of frequencies above 500 megahertz.

hypochromism   A decreased absorption of electromagnetic radiation (usually in the ultraviolet region) caused by geometry-dependent interaction between two or more parts of a molecule.

hysteresis   Occurs in ferromagnetic materials. When a changing magnetic field H produced by a changing electric current I is applied to ferromagnetic material it exhibits a magnetization B that is not a simple linear function of the applied field. B will eventually show saturation for high values of H. When H is reduced the magnetization is less than what occurred as H was increased, i.e., the magnetization lags behind the magnetizing field. On reducing, reversing and then increasing H, the B field goes through a complete loop. The phenomenon is known as *hysteresis*.

hysteresis loop   The hysteresis loop is the complete functional relationship between H and B which forms a closed loop. *See* hysteresis.

hysteresis loss   The area inside the hysteresis loop representing the loss of energy occurring in each cycle of the changing current I is known as the *hysteresis loss*.

hysteresis tester   An instrument for the rapid determination of the hysteresis loss of a given specimen of magnetic material. The version attributed to Ewing comprises a specimen built up out of strips of a prescribed shape and size and is rotated by hand between the poles of a horseshoe magnet which is suspended and balanced so as to be free to turn on an axis in the same straight line as the axis of rotation of the specimen. The latter is rotated sufficiently fast to produce a steady deflection of the horseshoe magnet, which is read off. The hysteresis loss in the specimen is proportional to the sine of the angle of deflection.

I

ignitron   A three terminal device capable of switching extremely large currents (hundreds of amperes) at high voltages. It has electrical characteristics similar to an SCR. It works by controlling a gas discharge between an anode and cathode. The anode and cathode are well insulated, so when the device is off, it can separate several thousand volts. The device is turned on by initiating a gas discharge. This is accomplished with a minute arc to the cathode with the ignitor terminal. Once the discharge is initiated, the current flow between cathode and anode maintains the discharge. Ignitrons are used in high current regulated voltage supplies or for controlling automated spot welding equipment.

illuminance   Luminous flux density on a surface, i.e., luminous flux incident per unit area of a surface, when the latter is uniformly illuminated. Synonymous with illumination, a more general term. Also synonymous with the intensity of illumination. Practical units of measurement are lumen per square meter or meter-candle or lux, lumens per square foot or foot-candle, lumens per square centimeter or phot. 1 phot = 10,000 lux = 929.03 foot-candles. Note that luminance of a source is the number of lumens emitted per solid angle (steradian) by a unit source area. Visual acuity and other properties of vision depend on the illumination, and minimum values are tabulated for various occupations in lighting codes. The eye has maximum efficiency between 10 and 100 foot-candles.

illumination   Synonymous with illuminance (*illuminance* is the preferred usage, since *illumination* has more a general meaning). A surface is illuminated when a luminous flux is incident on it. The flux per unit area at any point on the surface is the illumination (or *illuminance* or *intensity of illumination*) at that point.

image analysis   The extraction of scientifically useful information from an image.

image converter   A device that converts television video information between different encoding formats. This device is sometimes referred to as an image buffer.

There are many different video encoding formats for the transmission and display of images, e.g., scanning frequency and number of scan lines constituting the image. This is because development of television occurred simultaneously in different countries and horizontal scanning in the display of the television image is syncronized with the local power frequency. So, an image converter would be used to convert a 625–line, 50 Hz signal (European) to a 525–line, 60 Hz signal (American).

image dissector   An early video tube for the encoding of light images to electrical signals that formed the basis of early television cameras. The device is obsolete and no longer used in modern cameras.

image distance   The distance from the vertex to the image. The vertex is the point on the surface of the lens where the optical axis crosses. From the thin-lens equation, the image distance s for a thin-lens, whose focal length is f, can be calculated from the object distance s' as

$$\frac{1}{s} = \frac{1}{f} + \frac{1}{s'},$$

$$\frac{1}{f} = \frac{(n'-n)}{n}\left(\frac{1}{R} + \frac{1}{R'}\right),$$

where the radius surface of the lens is R and R'.

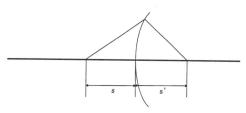

Image distance.

Focal length, image distance, and object distance of a thick lens should be measured from the principal planes of the lens.

image duplication   A 1:1 reproduction of an object by light rays. An image forming optical duplication gathers light from an object point and transforms it into a beam that converges toward another point (real image) or diverges from another point (virtual image).

image enhancement electron micrograph A signal-processing operation performed to increase the quality of an image.

image, primary   A real image formed by the objective of optical instruments such as a microscope or telescope. The primary image is also known as the first real image. (*See* image, real). The eyepiece (ocular) magnifies the primary image. It can be defined as an image that includes the point where each ray of light from the objective passes through. If astigmatism exists, the primary image of the point source can not be a perfect point. In a compound microscope, the objective lens forms a real, inverted image of the object. The image is formed in space on the plane of the field stop of the eyepiece within the microscope tube. The eyepiece observes this real image. In telescopes, the primary image is formed by the objective lens or the objective mirror. For projection with a microscope, the primary image is formed ahead of the first focal length of the ocular, which forms a real image again. The magnification of the eyepiece should be treated as a lateral magnification. For visual observation, the primary image is generated inside the focal length of the ocular as a magnifier. The magnification of the eyepiece should be treated as an angular magnification.

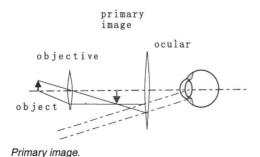

Primary image.

image, real   An optical image such as that formed by the light from an object that actually passes through the image. The real image is luminous and is generally visible. It can be projected on a screen and the projected image is sharp.

images, acoustic   Geometric figures formed in space by acoustic mirrors, lenses or other acoustic equivalents of optical system components. Also know as *imaging*. Acoustic imaging deals with the generation of real-time images of the internal structure of metallic or biological objects that are opaque to light by irradiating them with sound. Also known as *sonography* or *ultrasonic imaging*.

image, ultrasonic   The use of ultrasound to produce an image of the internal structures of the body.

image, virtual   An optical image where rays of light only appear to diverge. A virtual image cannot be projected on a screen. The image of an object in a plane mirror is a virtual image. The virtual image is not luminous actually. The rays of light do not actually focus on a virtual image. With a converging lens, the object closer to the lens than the first focal point will form a virtual image, not a real image. The image formed by a concave lens is also a virtual image. The image formed by (viewed in) a plane mirror is a virtual image. A stereoscopic image generated by a hologram is a virtual image.

imaging, dark-field   An imaging technique in which illumination conditions are regulated by apertures and stops to permit only certain electrons or light photons to reach the lens. This allows for the illumination of certain parts of the sample while the remaining parts of the sample remain dark.

imaging, medical   The use of X-rays, ultrasound, and other techniques to make images of internal structures of the body.

imaging, NMR   The use of nuclear magnetic resonance (NMR) to produce images of internal structures of the body.

imaging, radionucleotide   The use of radioactive decay to visualize internal structures of the body.

immunofluorescence   A process in which fluorescently labeled antibodies are attached to cells. These cells are then examined and separated by an optical microscope while shining ultraviolet radiation on the sample.

impedance   The relationship between a sinusoidally varying quantity, e.g., force, voltage, electric field strength, to a second quantity that measures the response of the system to the sinusoidal quantity, e.g., velocity, current, magnetic field strength. The response of the system will often depend on the frequency of the applied disturbance. The most common usage of the term is in alternating current circuits where the impedance Z relates the voltage V (expressed as a complex number) to the current I $Z = V/I = V_0/I_0 e^{i\phi}$, where V_0 and I_0 are the voltage and current amplitudes, and ϕ is the phase difference between them. Similar relationships are found in other systems, e.g., mechanical systems, where Z relates an applied sinusoidal force to the resultant velocity of a body.

impedance, acoustic   The complex ratio of the sound pressure $\hat{p}$ on a given surface to the sound flux through that surface, the volume velocity of the fluid $\hat{U}$, $\hat{Z}_A(\omega) = \hat{p}/\hat{U}$, expressed in units kg/(m^{4}s). It was introduced as the analogy to mechanics where the impedance is the ratio of the force amplitude to the velocity amplitude. The real and imaginary parts of Z, R and X, respectively, are *acoustic resistance* (associated with the dissipation of acoustic energy) and *reactance* (resulting from the effective mass and stiffness of the medium).

impedance, characteristic   The impedance of a waveguide/line to a transmitted wave of a specific frequency assuming the guide is of infinite length. For a lossless line, it is equal to $\sqrt{CL}$, where C and L are the capacitance and inductance per unit length. If a guide is terminated with an impedance equal to the characteristic impedance, there will be no reflection from its far end, thus simulating a guide of infinite length.

impedance diagram   A diagram used to analyze the steady-state behavior of a circuit or to analyze the response of a circuit to a particular input. It shows the relationship between resistance, reactance and impedance in a circuit or a portion of a circuit. The real axis indicates resistance R and the imaginary axis indicates reactance X of a circuit.

$$Z^2 = R^2 + X^2 ,$$
$$Z = R + iX ,$$
$$X = X_L - X_C ,$$

where X_L and X_C are reactance caused by inductance and capacitance, respectively.

The impedance diagram of a circuit can be obtained by dividing each of the voltage phasor diagrams by the magnitude of the current. The impedance diagram applies linear circuit theorems to the AC circuits. It means the ratio of complex voltage to complex current for each resistive, inductive, and capacitive element of the circuit is constant. It comes from Ohm's law. The phasor diagram is also used for similar purposes. The phasor diagram of each branch of the circuit is used to create the impedance diagram of the circuit. The phasor diagram is based on the equation of complex voltage V and complex current I:

$$V = RI + j\omega LI - j\frac{1}{\omega C}I ,$$

where *omega*, R, L, and C are the frequency of the input sinusoidal signal, resistance, inductance, and capacitance, respectively. When the phase angle of the current of a branch is ϕ to the current of another place of the circuit, it can be described as

$$V = \left(RI + j\omega LI - j\frac{1}{\omega C}I \right) e^{j\phi} .$$

In such case, the phasor diagram is used as rotated by phase angle ϕ.

impedance, driving point   The ratio of the complex component V_i of the applied alternating voltage to the complex component I_i of the alternating current.

183

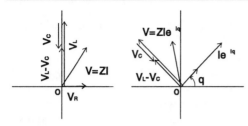

Phasor diagram.

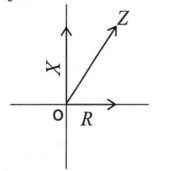

Impedance diagram.

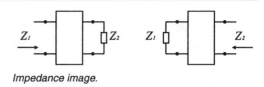

Impedance image.

impedance, dynamic   The impedance of an electric component or an electric device with an AC signal input.

The dynamic impedance Z includes not only resistance R but also reactance X:

$$Z^2 = R^2 + X^2 ,$$
$$Z = R + iX .$$

The resistance (the real part of the impedance) indicates the loss of power, The reactance (the imaginary part of the impedance), which is caused by capacitance and/or inductance, indicates the phase difference between the voltage and the current. Dynamic impedance caused by capacitance and/or inductance indicates frequency dependency. *See* impedance diagram.

impedance image   A pair of impedances of a quadripole that satisfies two conditions:
1. when the first pair of terminals of the quadripole is terminated with an impedance Z_1, the impedance of the second pair of terminals is Z_2, and
2. when the second pair of terminals of the quadripole is terminated with an impedance Z_2, the impedance of the first pair of terminals is Z_1.

impedance, input   The impedance presented to an input source of a device by the device.

impedance, iterative   The impedance which, when connected to two terminals of a four-terminal network, produces the equivalent impedance across the other two terminals.

impedance matching   The method of maximizing the power transmitted from a source to a sink. Maximum power transmission occurs when the resistances of the source R_o and sink R_i are equal $R_o = R_i$, and the reactances, X_o and X_i, are equal but opposite: $X_o = -X_i$. The standing wave ratio is a minimum when the impedances are matched in this way.

impedance, membrane   The degree of difficulty of the flow of current (usually in the form of ions) through a membrane.

impedance, membrane, measurements
Techniques used to determine quantitatively the difficulty with which current flows through a cellular membrane.

impedance, output   The impedance presented to the load of a device by the device.

impedance, reflected   The input impedance of a transformer is given by the sum of two terms The impedance of the primary winding and the reflected impedance due to the secondary winding. The reflected impedance is equal to $\omega^2 * M^2/Z$, where ω is the angular frequency of the applied voltage, M is the mutual inductance between the windings, and Z is the impedance of the secondary winding.

impedance, source   The impedance of the source (be it a supplier of charge, heat, light, etc.) as seen by the rest of the device.

impedance, synchronous   The impedance of an alternator while it is operating at a fixed frequency. It depends on the stationary impedance of the alternator and its armature reactance.

impedance, transfer   The complex ratio of the applied sinusoidal quantity, such as voltage, force, etc., to the corresponding response (current, velocity etc.) between any two points of the device.

impulse function   The electrical potential along an axon during the generation of a nerve signal.

impulsive sound   Sound caused by a short duration disturbance. An impulse excitation can be caused by a force that is applied for a very short or infinitesimal length of time and is nonperiodic. Impulsive sound is of importance in architectural acoustics, for example, in determining the reverberation time of a room. The reverberation time of a room for impulsive sound (such as a hand-clap) can be considerably greater than that computed using methods for sustained sound.

impulsive voltage/current   A waveform that rises rapidly and reaches a short duration of voltage/current plateau and then falls rapidly to zero. In practical usage, the ringing of the impulse should be considered.

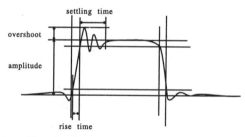

Impulsive signal.

impurity   A substance intentionally added to a semiconductor to change the available number of charge carriers in the crystalline lattice. *See* doping.

inactivation   The decline or stoppage in current of a particular ion through the cellular wall.

inactivation, kinetic interpretation   The conductance of ions through the cellular membrane is dependent on the potential difference across the membrane. In the case of sodium ions involved in nerve pulses, the movement of sufficient numbers of ions causes the potential difference to change to the state of inactivation.

inactivation, voltage dependence   The process of stopping the flow of ions through the cellular membrane because of the potential difference across the membrane.

incandescence   The emission of visible light radiated by a substance with a high temperature (> 3000 K). The radiation itself is sometimes called *incandescence.* Electric lights are incandescent.

inclination, magnetic   The angle between the magnetic field vector of the earth's magnetic field and the horizontal plane. Same as *magnetic dip.*

incoherence   Absence of a fixed phase difference between two sinusoidal waves. The phase difference from two incoherent sources may vary rapidly and irregularly with time. For optical sources, the interference fringes of the resultant disturbance changes as the phase difference changes, and at any given instant the maxima and minima change positions faster than can be resolved and the result appears to be a uniform illumination — i.e., superposition of incoherent light waves gives an intensity equal to the sum of the intensities of the individual waves. When fluctuations in phase of beams from different sources are completely uncorrelated, the beams are incoherent. When the beams are from the same source, the fluctuations are correlated and the beams are completely coherent or partially coherent depending on how complete the correlation is. The degree of correlation is measured by the degree of distinctness of the interference fringes when the beams are superimposed.

incubator   A device in which the environmental conditions can be carefully controlled for the purpose of sustaining living organisms, such as premature babies, developing eggs, and cultured microorganisms.

index of refraction   (*refractive index, refractive constant*) A dimensionless quantity with the symbol n. The refractive index of media is equal to the ratio of the speed of an electromagnetic wave in two mediums. The absolute refractive index of a medium n is the ratio of the wave speed in free space to c wave speed in the medium v:

$$n = \frac{c}{v}.$$

The index of refraction depends on the wavelength. Usually, the index of refraction for yellow light (sodium D-lines, wavelength 589.3 nm) is used. Optical length s is a product of the index of refraction and the length l of the path in the free space: $s = nl$. The reduced distance l' is equal to the optical path length in the air l divided by n.

induced charge   When a charge is brought near an uncharged conductor, charges of the opposite sign in the conductor will move to the parts nearer the charge and those of the same sign will move away from it. These charges on the conductor are known as induced charges.

induced magnetic fields   Magnetic fields acquired by some magnetic materials when placed in an external magnetic field.

inductance   General term given to the creation of an inducted potential difference in a circuit due to a changing magnetic field which threads the circuit. It is a result of Faraday's induction law.

inductance, leakage   The inductance in a transformer that results in leakage reactance, and is a result of flux linking only one coil of the transformer.

inductance, mutual   The mutual inductance M between two circuits is defined as $M = \Gamma/I$, where Γ is the flux linking one circuit as a result of the current I flowing in the other. From Faraday's induction law, one determines that the voltage V induced in the first circuit is given by $V = -M dI/dt$.

inductance, self   The self inductance L of a circuit is defined as $L = \Gamma/I$, where Γ is the flux linking the circuit as a result of the current I flowing in it. From Faraday's induction law,

one determines that the voltage V induced in the circuit is given by $V = -L dI/dt$.

induction balance   Invented by A.G. Bell. Two coils (a transmit coil and a receive coil) work on the principle of eddy current generation and the inductive imbalance between the two coils. A very low frequency current causes the transmitting coil to create an electromagnetic field where polarity is pointed to an object. If the object is metallic or ferromagnetic, an eddy current induced inside of it creates its own magnetic field. This field is then detected by the receive coil. Some metal detectors use an induction balance.

induction coil   A coil used to produce an intermittent high voltage from a source of low, constant voltage. The low voltage source is connected to a primary coil of few turns that surrounds an iron core via a switch which interrupts the current. Around this coil is a secondary coil of many turns. The rapid variations of the flux linking the primary coil cause a correspondingly large voltage to be induced in the secondary coil.

induction, electromagnetic   The setting up of an electric field by reason of the variation in magnetic flux density with time. Any current so induced is in such a direction as to oppose the change in magnetic flux.

induction heating   Method of heating a conductor via the application of an alternating magnetic field. This creates circulating eddy currents in the conductor, as a result of Faraday's induction law, which heat the conducting material via the Joule effect. With this method of heating, the material is not contaminated with combustion gases. It also allows one to primarily heat only the surface by the use of a high frequency field which, due to the skin effect, will produce heating currents only in the surface.

induction machine   A device that produces high-voltage electrical charges by electrostatic induction. It is also known as an *electrostatic generator*. The van der Graaf generator and Wimshurst machines are well-known examples.

induction meter   A motor meter that uses a kind of induction motor. It is also used as a watt-hour meter for AC current because the loss of power is during measurement and it is reliable.

inductive reactance   A part of the reactance X normally denoted by symbol X_L, measured S (Siemens) or Ohm. It is caused by the existence of inductance. It is considered as a positive imaginary number $X_L = j2\pi f L$, where f is the frequency of input signal, L is the inductance of the component, and j is the unit imaginary number. In a purely inductive reactance circuit, the current lags the applied voltage in its phase by $\pi/2$.

inductor   A coil (turns of wire) introduces electromagnetic inductance. It is measured in henry. Usually the symbol L is used to indicate an inductor. The eddy current induced in an inductor causes the power loss of the inductor to increase with frequency. The effective inductance is affected by the stray capacitances between the turns of the existing coil. The inductance of inductors in a series is an algebraic sum of each inductance of the inductor. The inverse of the inductance of the inductors in parallel is the algebraic sum of the inverse of the inductance of each inductor.

inductor, stored energy in   With a varying current I passing in an inductor of inductance L, it is necessary to provide energy to drive the current against the induced electromotive force. The electromotive force V_{emf} is equal to $L dI/dt$. Therefore, the electromagnetic energy stored in the inductor U is:

$$\frac{dU}{dt} = V_{\text{emf}} I \, ,$$
$$= LI\frac{dI}{dt} \, ,$$
$$dU = L dI/dt dt$$
$$U = \frac{1}{2}LI^2 \, .$$

U is stored in the magnetic field of the inductor.

information   Evaluated facts and judgments with application; the summarization of data. Technically, data are raw facts and figures that are processed into information. *Data* and *infor-*

mation are terms often used synonymously and interchangeably. Information can be said to be structured in the following manner: data, text, spreadsheets, pictures, voice and video. Data are discretely defined fields. Text is a collection of words. Spreadsheets are data in matrix (row and column) form. Pictures are lists of vectors or frames of bits. Voice is a continuous stream of sound waves. Video is a sequence of frames. Databases can store all kinds of information.

information channel   That by which data is transmitted from input point to output point. It also refers to the communication link connecting a PC or server to a hub in the wiring closet.

information entropy   This provides a quantitative measure of the degree of randomness of a system and is a measure of the average information content per source symbol; it can be quantified by the probabilities of the source symbols.

information, mutual   When noise is introduced into the channel, the symbol at the channel output will not always be identical with the state at the channel input. A measure of the average information rate at the receiver output, and thus the average information rate through the channel, is given by the log of the ratio of the final and initial uncertainties regarding the source. If there is no noise in the channel, the mutual information is 1 and if the noise is so great that the output states are independent of the input states, then the mutual information is zero.

information transmission, substrate for
The medium through which information is transmitted, such as electromagnetic waves for radio.

infrasonic waves   Sound waves below the frequency range of human hearing, covering the part of the acoustic spectrum below 20 Hz. The compressibility of air is responsible for acoustic oscillations above the frequency of about 0.003 Hz. Below this frequency, transverse oscillation can develop, and buoyancy acts as a restoring force in a stratified medium. At very low frequencies (characteristic periods of the order of hours) the transverse oscillation component is dominant, and such waves are called *gravity waves.* Infrasound accompanies natural phenomena (avalanches, powerful storms, eruptions of volcanoes, earthquakes, wind, ocean waves) and can also be generated artificially (explosions, rocket launches). Infrasound can propagate to very large distances due to refraction in the atmosphere and low absorption.

injection loss   The loss of light that results when two fibers are joined at a connection point.

input   The signal, current or voltage applied to a circuit or device or the terminal at which they are applied.

input characteristics   The electrical properties (impedance, capacitance, etc.) associated with the input channel of a device; in a FET (field effect transistor), the dependence of the input current on the voltages between the source and gate and between the drain and gate.

insects, sound from   Sound produced for communicative purposes by many insect species using a wide variety of mechanisms. Frictional methods are predominant, and sound production can occur during locomotory, cleaning or feeding movements. Stridulation, the rubbing of one body part against another (a file with a series of pegs or teeth can be rubbed against the scraper formed by a single edge or ridge), is the most common. Another common method of sound generation is by a vibrating membrane driven by muscles. Sound frequency and patterns vary and their true nature cannot be appreciated by humans because of the narrow frequency response and the long time-constant of the human auditory system.

insert (in antenna)   Also known as *aerial insert.* In a buried cable run, it refers to the raising of the cable followed by an overhead run usually on poles. The cable is subsequently returned to the ground. This procedure becomes necessary where it is not practical to run cables underground such as in areas with rivers.

insertion gain   In transmission line theory, this refers to a negative insertion loss where the insertion of a line or network between a generator and a load introduces dissipative elements.

If the impedance matching is reduced, the power delivered to the load is reduced. The insertion of a line or network in this case causes an increase in load current representing a negative loss. *See* insertion loss.

insertion loss   The insertion of a line or network between a generator and a load may improve or diminish the impedance match between the source and load and introduce dissipative elements. Increased power delivered to the load at the receiver leads to this type of loss. It can be quantified by the number of decibels by which the current in the load causes a change, by the insertion. *See* insertion gain.

insulated conductor   A conductor that surrounds a non-conducting material itself and is separated from other conductors.

insulation, acoustic   Materials used to diminish the energy of sound that passes through them or strikes a surface, which is of considerable importance, for example, in architectural acoustics. It is customary to differentiate between airborne (human speech, music) and structure-borne (mechanical excitations of structures by machines or people walking) sound excitation. The sound transmission coefficient τ, defined as the ratio of the transmitted and incident sound, and the sound reduction index, $R = 10\log(1/\tau)$, expressed in decibels, provide a measure of sound insulation.

insulator   A substance that is electrically non-conductive. One or more energy bands of an insulator are full and other bands are empty. Only an electron having an energy of enough electron volts can jump to a conduction band. The energy is higher than room temperature (2.6×10^{-2} eV) and sufficient to disrupt. *See* semiconductors.

integrated communication system (ICS)
An integrated communication system (ICS) represents a communication system that joins and interoperates two or more originally autonomous communication systems, which have consequently lost their initial independence but have become one single interdependent system.

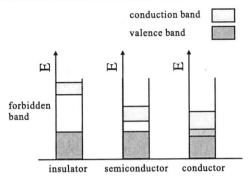

Insulator.

integrator   A circuit that takes a single input and gives the integral of the input signal as the output signal.

intensity   The radiant energy per unit time (flux) or the number of photons per unit time, flowing through a unit area, through a surface normal to the direction of propagation. For mechanical waves, the intensity is proportional to the square of the amplitude of the wave. For a traveling light wave, the intensity is proportional to the average energy flux per unit time, or the mean square value of the optical disturbance. The optical disturbance varies with time too rapidly to be observed directly, so it is the light intensity that measures the observable effects of light.

intensity modulation   Image reproduction by varying the intensity of an electron beam and thus the light output of a cathode-ray tube in accordance with the magnitude of the signals it receives.

intensity of sound   Average rate at which acoustic energy is transmitted in a specified direction through a unit area of a surface perpendicular to the direction of propagation $\vec{n}$, $\vec{I} = \frac{\bar{n}p^2}{\rho c}$. The unit of I is watt per square meter (W/m^2). Acoustic energy travels with the speed c in the direction $\vec{n}$; p is the pressure and ρ the density. Also called *acoustic energy flux* or *acoustic intensity*.

interaction, exchange   Since electrons are fermions, they obey the *Pauli exclusion principle* which disallows more than one fermion in a given state. Examining a two electron system,

we find the singlet state, an anti-symmetric spin state with $S = 0$, has a lower energy than any of the triplet states, those states with symmetric spins and $S = 1$. This difference leads to an effective spin interaction, the so-called exchange interaction. It is possible to recast the exchange interaction into a spin Hamiltonian. For the two electron case, the spin Hamiltonian has the form

$$H_{\text{spin}} = --J\mathbf{S}_1 \dot{\mathbf{S}}_2$$

where J is the exchange coupling constant (the energy difference between spin states) and the $\mathbf{S}_i$, the spin of the i-th electron. In an N-electron (or N-spin) system, the spin Hamiltonian generalizes to include all pair interactions (represented above) plus higher order interactions. In many systems, only the two-particle interactions are relevant, and the spin Hamiltonian is the Heisenberg Hamiltonian,

$$H_{\text{Heisenberg}} = --\Sigma J_{ij}\mathbf{S}_i \dot{\mathbf{S}}_j$$

where the sum is over all pairs of spins.

interfacial tension   The force exerted on molecules at the interface between two boundaries, such as surface tension.

interference (light)   The systematic attenuation and reinforcement of the amplitudes over distance and time of two or more overlapping light waves that have the same or nearly the same frequency. The maxima and minima of light wave interference cannot be described by the ray approximation of the wave equation. From Huygen's principle, interference occurs when there are two or more paths of different lengths from a source to the observation point. The interference is constructive (destructive) if the phases and amplitudes increase (decrease) the resultant amplitude squared relative to the sum of the squares of the amplitudes. Since energy is conserved, the energy that is missing from the destructive interference zones or dark spots in the interference pattern is found at the constructive zones or bright spots. Interference can also occur when there is more than a single source, provided there is a fixed phase relationship between the sources, i.e., the sources are coherent. The interference of light was first discovered by Thomas Young in 1801 using a single light source and twin pinholes or slits. Fresnel and Young explained the resulting fringes using the wave theory of light.

interference, acoustic   The variation over distance or time of the amplitude of a wave that results from a superposition (algebraic or vector addition) of two or more waves. A medium can simultaneously transmit any number of waves, which propagate independently of the other. The displacement of the medium at any point and any instant of time is the algebraic sum of the displacements caused by the individual waves at that instant of time. The term *interference* is commonly used to describe this effect, although the term *superposition* would be more accurate. Unlike light that requires coherent light for interference, in acoustics, separate sources will be coherent and can give rise to interference effects. *See also* modulation, acoustic.

interference, constructive   The case of superposition of waves arriving simultaneously at a point so as to give a resultant intensity that is greater than the sum of the squares of the amplitudes of individual disturbances. Since the equations of the amplitude wave motion is a linear equation, the sum of any number of solutions is also a solution. The intensity, which is the observable quantity for light phenomena, is the square of the amplitude. Thus the resultant intensity is not merely the sum of the intensities of the individual waves, but can be greater or less than the sum of individual intensities. *See* interference.

interference, destructive   The case of superposition of waves arriving simultaneously at a point so as to give a resultant intensity that is less than the sum of the squares of the amplitudes of the individual disturbances. *See* interference, constructive.

interference fringes   The maxima and minima of intensities seen during the optical interference of light waves. The maxima occur as a result of constructive interference of light waves, which arrive in phase, to give a bright spot, while the minima occur as a result of the destructive interference of light waves, which arrive out of phase, to give a dark spot. *See* interference.

interference, heterodyne   The mixing of two signals of different frequencies in a nonlinear device resulting in the generation of two new frequencies that are the sum of and the difference between the two original frequencies. The effect is used in the heterodyne receiver where two signals having slightly different frequencies are combined to form an audio-frequency beat signal that can be heard with a loudspeaker. Under certain conditions, a steady, high-pitched audio tone, known as *heterodyne interference* or *heterodyne whistle,* can be heard in the amplitude-modulation radio receiver as a result of the heterodyne action.

interference, radio frequency   An unwanted signal that enters the transmission line from radio and television transmitters at a level sufficient to degrade the performance of the channel by a significant amount. With this type of interference, the cable acts as an antenna.

interference, thin film   Interference phenomena that occur from the reflection of light from the two surfaces of a thin transparent film. The maxima in the easily observed interference occur when the thickness of the film is an odd multiple of a quarter of the wavelength of the incident light, and the minima occur when the thickness is an even multiple. If the thickness varies across the film, then an image of the film using a lens will show a different brightness in different places. Using the eye to form an image of the thin film directly on the retina, instead of a lens and screen, shows a system of interference fringes, with brightness varying according to whether the quarter wavelength is an even or odd multiple of the thickness. Lines of equal thickness appear as lines of equal brightness. These results apply equally well to films with an index of refraction greater than the surrounding medium (e.g., a thin plate of glass) or with an index of refraction less than the surrounding medium (e.g., an air gap between two thick plates of glass).

interference with diffraction   *Interference* is the modulation of wave amplitude into reinforcing maxima and canceling minima produced by the superposition of a finite and usually small number of beams. *Diffraction* is the modification of amplitude determined by a superposition via integration of infinitesimal elements of a wave front. The double slit pattern is thus a combination of interference (superposition of beams from each of the two slits, yielding narrow maxima and minima) and diffraction (integration over the wave front from each slit yielding a modulation of the interference maxima and minima).

interferogram   The pattern of interference that results when two waves of the same wavelength are brought together. Extremely accurate measurements of distance can be made using this technique.

interferometer, acoustic   A device for measuring the velocity and attenuation (absorption coefficient) of sound waves in a fluid by the interference method. In the acoustic interferometer an electrically driven crystal oscillator is used to induce longitudinal vibrations in a column of fluid. A movable reflector plate is placed parallel to the radiating surface, to allow standing waves to form in the fluid column. Varying the spacing l between the reflector and the source causes various modes of resonance of the fluid column, leading to changes of the driving current of the crystal oscillator circuit. The periodic maxima of the driving current (that resemble patterns registered by optical interferometers) correspond to the resonance patterns in the fluid that are one half acoustic wavelength $\lambda/2$ apart. Plotting the crystal driving current vs. the reflector position yields information regarding the acoustic wavelength λ. If the frequency of the sound waves f is known, the velocity of sound can be determined as $c = \lambda f$. In an absorbing fluid the relative amplitude of the reflected wave decreases as the spacing between the oscillator and the reflector increases, and this is manifested as the decrease of amplitude of the current peaks of the interferometer pattern. The absorption coefficient is measured by measuring the shape of the current peaks as a function of spacing l.

interferometer, Fabry-Perot   Optical instrument that utilizes the interference fringes produced by multiple reflections of light from a broad source in the air gap between two plane parallel plates that are thinly silvered. A lens, which may be the lens of the eye, is used to bring together for interference the parallel transmitted rays from break-up by reflection of the incident ray on the first silvered surface. The condition for reinforcement of the transmitted rays is the same as that for the Michelson interferometer, $2d \cos\theta = m\lambda$, which is satisfied by the points on a circle with the center given by the intersection of the axis of the lens with the screen, so maxima are a series of concentric rings, which are not images of the source, with spacing that changes with the air gap distance, d. To vary d, one plate is fixed, and the other can be moved via a slow-motion screw attached to an accurately machined carriage arrangement. The rings are very narrow, so light from a source that consists of two closely spaced wavelengths produces two clearly separated sets of rings. Fabry-Perot interferometers are thus useful as spectrometers with high resolution for nearly monochromatic light. If the light is not nearly monochromatic, the interference pattern becomes too difficult to interpret, unless another instrument is used to do a preliminary wavelength separation.

interferometer, Michelson-Morely   An optical interference apparatus based on amplitude division of a wave front into two beams that are sent in different directions against plane mirrors. When the beams are recombined, interference fringes are formed. The main optical parts consist of two plane mirrors and two plane plates of glass, one of which is sometimes slightly silvered. Light from an extended (not a point or slit) source encounters this slightly silvered glass and is divided into a reflected and transmitted beam of equal intensity. A mirror is used to reflect this beam back to the glass plate, while a second mirror reflects the transmitted beam back to the glass plate where they are recombined. The secondary glass plate is inserted along the transmitted beam in the compensating plate used to give equal pathlengths in glass to the reflected and transmitted beams. The compensating plate is not necessary for fringes from monochromatic beams, but it is required for fringes in white light. The mirror reflecting back the reflected beam from the first glass plate is mounted on a moving carriage that rides on an accurately machined track and is attached to a slowly turning screw that is calibrated for distance measurements of the mirror's motion. Adjustments on the two mirrors allow them to be made accurately perpendicular to each other. When the pathlengths are the same for each beam and the images coincide, interference fringes will be seen. The fringes will be circular if the mirrors are exactly in adjustment, with maxima given by angles relative to the axis given by $2d \cos \theta = m\lambda$, where d is the separation of the virtual image of the fixed mirror from the moving mirror. Since the fringes are determined by a phase difference determined by an angle of incidence, these are fringes of equal inclination. Unlike other types of fringes, these fringes may retain their visibility over very large path differences. Also, unlike many other types of interferometers, the two beams traverse widely separated path lenses, which make possible many applications where a beam is required to traverse a substance to be compared with the reference beam. Small differences in the index of refraction of two substances can be accurately measured this way. The Michelson interferometer is also used to set the standard length of one meter in terms of the wavelength of the red line of cadmium. The Michelson interferometer was used in the Michelson-Morely experiment to establish the absence of an ether drift, where the distance d was made as large as 11 m by reflecting the light back and forth between 16 pairs of mirrors. *See* Michelson-Morley experiment.

interferometers   An optical instrument that uses the interference of light waves originating from a common source, with various practical applications depending on its design. The prevalent interferometer designs are Michelson, Twyman-Green, Fabry-Perot, Lummer-Gehreke, Jamin, Mach-Zehnder, and Fizeau. The *Michelson interferometer* has widely separated beams and a path difference that is readily varied, with refracting material commonly inserted into the path of one of the beams, so as to measure distances in wavelengths, and refractive indices of the inserted material. *Twyman-Green* is used to test the accuracy of optical surfaces. *Fabry-Perot* is used to accurately measure wavelengths and hyperfine structures. *Lummer-Gehreke* is used in the *UV*. *Jamin* is used to measure the refraction of gases. *Mach-Zehnder*

is used to study slight changes of refractive index over a considerable area, as in the flow pattern in wind tunnels. *Fizeau* is used to test the uniformity of optical thickness of plane-parallel transparent plates.

intermediate state   The intermediate state of a superconductor occurs when the magnetic field nears the critical field. At fields $H < H_c$, the magnetic flux is expelled from the superconductor via the Meissner effect. This exclusion of the flux enhances the magnetic field near the surfaces of the superconductor, however, and near H_c, this is sufficient to drive a portion of the superconductor normal. The stable form for this state is a state in which there are alternating strips of superconducting and normal metal aligned parallel to the magnetic field.

intermodulation   Process in nonlinear device or system whereby the components of a complex wave modulate each other to produce new waves having frequencies equal to the sum and differences of the frequencies of the various components of the harmonics of the input wave. This causes distortion in nonlinear devices.

intermodulation, acoustic   Modulation of the components of a complex wave by each other, generating new waves whose frequencies are equal to the sums and differences of integral multiples of the frequencies of the original waves. *See also* modulation, acoustic; interference, acoustic; interference, heterodyne.

International Commission on Radiation Protection   The international body that provides guidance about all issues related to safety of ionizing radiation. First organized in 1928 as the *International X-ray and Radium Protection Committee,* its name was changed to the *International Commission on Radiation Protection* in 1950. This body makes recommendations about the basic principles of radiation safety and leaves the detailed recommendations to the various national regulatory bodies.

International Telecommunication Union (ITU)   An agency of United Nations that is responsible for standardizing international telecommunications. Its sectors are concerned with allocating radio frequencies worldwide to competing interest groups and with telephone and data communication systems.

internodal segment, electrical characteristics Large nerve cells in vertebrates are covered with a membrane wrapped around the axon (myelin). Gaps between the myelin are called nodes and the myelin-wrapped regions are internodal segments. The myelin greatly impedes the flow of ions in this region of the axon, resulting in little current flow.

internodal segment, equivalent electrical network The presence of the myelin modifies the electrical properties of the internodal segments. In an electrical analysis of the nervous system, the internodal segments have both resistive and capacitive aspects of their electrical properties.

interocular distance The separation of the two eye pupils when the observer is viewing distant objects, approximately 65 mm. For the case of two photographs taken by identical cameras from positions representing an observer's eyes, a stereoscopic effect is seen if the camera lenses are separated by the correct interocular distance, and correct viewing distance.

intervals, musical The spacing in pitch or frequency between two sounds. Two notes forming the musical interval of one octave have their frequencies as the ratio 2:1. The frequency interval is expressed as the ratio of the frequencies or the logarithm of this ratio. *See also* octave; frequency band; pitch, acoustic.

intrinsic conductivity The conductivity of a pure semiconductor material as opposed to the extrinsic conductivity due to the presence of impurities in the semiconductor material.

invariance of charge The charge of an electron or proton that appeared in the equations governing electrostatics and electrodynamics is invariant under Lorentz transformation. In other words, the charge of a particle is independent of its speed.

inverse square law The law stating that for any propagating wave, the rate of flow of energy across a unit area perpendicular to the direction of propagation (intensity) from a point source varies as the inverse square of the distance between the source and the receiver. The inverse square law is also one of the two fundamental laws of photometry, and states that illuminance or irradiance falls off at the inverse square of distance from the source.

inverter An electronic device that inverts the input, i.e., produces a high output for a low input and vice versa.

ion channels (cell) Molecular structures on the surface of cells that regulate the flow of particular ions through the cell membrane.

ionic current (cell) The movement of charged species both inside and outside of living cells.

ionic current (cell), measurements The determination of the movement of charged species inside and outside a living cell.

ionization chamber An instrument used to detect the presence of ionizing radiation by measuring the current due to the ionization of the medium inside the chamber (usually a gas) by the ionizing radiation.

ionography The study of the ions that move across cellular membranes.

ionophore Any molecule that transports a specific ion across a cellular membrane.

ion pump A vacuum pump in which the remaining gas molecules are ionized and drawn out by electric fields (it is usually used after a roughing pump).

ion transport (cell membrane) The movement of charged species through the membrane of cells.

irradiance A measure of the time rate of transfer of radiant energy (radiant power) per unit area that flows onto or across a surface. Radiant energy is any energy transferred by electromagnetic waves without a corresponding transfer of mass. Also called *radiant flux density*.

irradiation   Application of radiation (anything that propagates as a ray, such as electromagnetic waves, and the emission of radioactive substances) to a material body.

isochromatic line   Lines that are of the same color, as in the interference fringes produced in birefringent materials. For a pencil beam of plane-polarized rays incident obliquely on a uniaxial crystal cut perpendicular to the axis, there will be two emergent pencils, plane-polarized at right angles. For a thin crystal, the two beams are not separated on emergence and will interfere, giving a series of concentric light and dark rings. If white light is used, the rings will be colored, and due to symmetry about the axis, the color is constant around any circle centered on the axis; such lines are termed *isochromatic lines*.

isochromatic surface   Surfaces that give the locus of points of constant phase difference between the ordinary and extraordinary rays in uniaxial and biaxial media, which depend on the direction of the light. Sometimes referred to as *Bertin's surfaces*. Observed isochromatic lines superficially resemble the intersections of isochromatic surfaces with the crystal face, and these intersections can provide qualitative explanation of the isochromatic line forms.

isoelectric point   The pH value (hydrogen ion concentration) of a solution at which the colloid particles in the solution have zero net charge. The solution has minimum viscosity, conductivity and osmotic pressure at this pH value.

isomagnetic lines   Lines connecting points of equality in magnetic properties within magnetic materials.

isoplanatism   In the diffraction theory of aberrations, this refers to regions that are free of coma, which is one of the five aberrations of a lens with spherical surfaces. Skew rays from a point object meet at the same point on the image plane, instead of a pear shaped spot (coma).

J

jammers   Units producing specific types of jamming waveforms, e.g., a single-tone or pulsed noise jammer. *See* jamming.

jamming   Waveforms that are used on some fraction of transmitted symbols creating bursts of errors at the receiver output. A spread spectrum system is particularly susceptible to it, and relies on error-correcting codes combined with interleaving to combat it.

jitter   Type of analog communication line distortion caused by the variation of a signal from its reference timing positions. This can cause data transmission errors, particularly at high speeds. It also can create a short time line or circuit instability.

Josephson junction   A thin insulator separating two superconducting materials through which electron hole pairs tunnel.

junction   The interface where two types of materials meet. Two different bandgap materials give a heterojunction (e.g., GaAs/AlGaAs), and different dopings in the same material coming in contact give a homojunction. *See also* p-n junction.

K

kaleidophone   A thin metal bar of rectangular cross section carrying a bead at the upper end and clamped in a vice used to generate vibrations of prescribed frequencies in the different planes of vibration, thus forming characteristic patterns. The frequency of vibrations is the same for the two planes when the cross section of the bar is square or circular. The stiffness of the bar is greater in the plane of greater thickness in case of a rectangular cross section, leading to higher vibration frequency in this plane when compared to the side of smaller thickness. The ratio of vibration frequencies in the two planes can be adjusted to the desired value by appropriate selection of the dimensions of the cross section. The kaleidophone was invented by Wheatstone. A modification of the original design, with the bar divided into two parts, allows the continuous variation of the frequency ratio by changing the location where the bar is clamped.

Kaleidoscope   An optical toy consisting of a tube and between two to four plane mirrors. Most kaleidoscopes consist of two or three mirrors, and the mirrors are placed at an angle of 45° or 60°. It produces symmetrical patterns by multiple reflection by the mirrors. Front-surface-mirrors are used to generate a clear image. Objects are illuminated at one end of the tube and the image is observed through a small hole at the other end.

Kalman filter   A method of recursively estimating a state from a series of measurements. The Kalman filter is not a filter in the sense of a circuit, but rather an estimator algorithm based upon linear relationships of measurements to the state being estimated, as linear transitions in the state from measurement to measurement. The power of the kalman approach is seen when an individual measurement does not allow full observability of the state, but when all measurements are sufficient to observe the state being estimated.

Kapitza boundary resistance   The Kapitza boundary resistance is the resistance that occurs between liquid helium (either ^{3}He or ^{4}He) and solids. Acoustic mismatch theory, taking into account the different velocities of sound, predicts the thermal resistance should be $R_K \propto (AT^3)^{-1}$ where A is the area of the interface. This result is in reasonable agreement over a limited temperature range, but in marked contrast to the boundary resistance between two dielectric solids; acoustic mismatch theory does not work very well for solid-helium interfaces above roughly 1 K and below roughly 10 mK. In fact, depending on surface treatment, the thermal resistance can be several orders of magnitude smaller than the theory predicts. It is also found that the resistances for liquid ^{4}He, liquid ^{3}He, and solid ^{3}He are all roughly the same at $T = 1$ K. The Kapitza resistance between ^{3}He and metals with magnetic moments (even as impurities) is even smaller than that of ultra-pure metals and $R_K \propto T^{-1} or T^{-2}$. There is evidence that this is due to dipole-dipole coupling between the ^{3}He nuclei and the magnetic moments in the metal, but the question is still not completely resolved.

Kerr effect   A nonlinear electro-optic effect that makes certain substances behave like a uniaxial crystal (doubly refracting with a single optic axis) when placed in an electric field. The optic axis is parallel to the lines of force, and the magnitude of the effect is proportional to the electric field squared. The effect was first observed by Kerr in 1895 for glass, and is also seen for gases and liquids (nitrobenzene). The Kerr cell consists of a glass cell containing a liquid for which the effect is strong, located between the plates of a capacitor. Polarized light passing through the medium across the field can be interrupted at high frequency, and is useful as an electro-optic shutter.

keying   (**1**) Entering data by typing on a keyboard.

(**2**) The process of changing some characteristic of a direct current or other carrier between a set of discrete values in order to carry information.

(**3**) The process that causes modulation at a telegraph or radiotelegraph transmitter.

keying frequency, maximum The highest rate at which it is possible to key, and can be dependent on the response time of the system.

keying, frequency shift (FSK) A commonly used method of frequency modulation in which a one and a zero (the two possible states) are each transmitted on separate frequencies for the transmission of binary messages. The amplitude and phase are held constant in this technique.

keying, multiple frequency shift (MFSK) The same process as for frequency shift keying (FSK) but applied to the scenario where a modulator in a coded digital communication system provides a one-to-one mapping of the channel symbols into a set of signals.

keying, multiple phase shift (MPSK) A modulation technique to convert binary data into an analog form comprising a single sinusoidal wave. The phase of a carrier changes by π radians or 180°. In this modulation technique, frequency and amplitude are held constant. The modulator in a coded digital communication system provides a one-to-one mapping of the channel symbols into a set of signals.

keying, phase shift (PSK) *See* keying, multiple phase shift.

kidneys, artificial A device for filtering water and wastes from the blood and producing urine.

Kirchhoff's rules The rules found by G.R. Kirchhoff (1824–87). The first law (the *current law*) states that the algebraic sum of the currents that meet at a point is zero. It is useful in parallel circuits.

The second law (the *voltage law*) states that the algebraic sum of the electromotive force in any closed path is equal to the sum of the products of current and resistance in the path. The sum of the voltage drop is equal to the sum of the voltage sources, which is called the *effective voltage*. It is useful in series circuits.

Kirchoff's law (emission and absorption) The *Kirchoff radiation law,* as derived from thermodynamics, states that the ratio between "ab-sorbtivity" (absorption) and emissive power is the same for each kind of ray for all thermal radiators in thermal equilibrium, and is equal to the emissive power of a perfectly black body at the same temperature. The ratio depends only on the wavelength and the temperature. The absorption determines the loss of light on transmission through the material and there is not a simple relation with the absorptance that measures the loss of light on a single reflection. Kirchoff's law is a very general relation between the absorption and emission of radiation by surfaces of different bodies. If the absorption is large, the emission must also be large. Black bodies absorb all wavelengths completely, and also give the largest amount of radiation at a given temperature.

klystron An electron tube in which the velocity of the electrons is regulated/modulated. Discovered by Russel and Sigurd Varian in 1937, it is composed of an electron gun, a modulating cavity (buncher), and a collecting cavity (catcher).

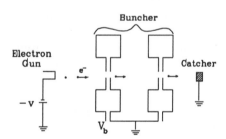

Block diagram of a klystron. The electrons are emitted at the electron gun. Their velocity is modified by the field generated by V_b in the buncher to collect them into bunches and they are then collected in the catcher.

Knight shift When measuring the nuclear magnetic resonance frequency of a metal, the frequency is found to differ from that of a free atom of the same metal or that of a salt of the same metal. This frequency shift, the *Knight shift,* is due to a change in the local magnetic field at the nucleus. The conduction electrons in the metal are polarized by the external magnetic field, B_0. These polarized electronic spins interact with the nucleus (the hyperfine interaction), producing a magnetic field at the nucleus, thereby changing the resonance frequency. The magnitude of the Knight shift is generally a few percent or less of the unshifted frequency.

Kondo effect   The presence of magnetic impurities can have drastic effects on the resistance of metals at low temperatures. Near such an impurity, the spin of conduction electrons becomes polarized by the impurity magnetic moment, and other conduction electrons inelastically scatter off the cloud of electrons surrounding the impurity. This scattering flips the spin of the electrons involved. This increased scattering rate produces an increase in the resistivity of the metal at low temperatures. The *Kondo effect resistivity, $-\rho_K \ln(T)$,* dominates at low temperatures and produces a minimum in the total resistivity at the temperature when the effect becomes important.

Korringa relation   The spin-lattice relaxation time, τ_1, is related to the temperature of the lattice and electrons in a nuclear paramagnet according to the Korringa relation:

$$\tau_1 T \approx \kappa \,,$$

where κ is the Korringa constant. This relationship is only approximate, as the Korringa constant actually varies slowly with the magnetic field and with temperature at extremely low temperatures. *See also* Bloch's equations; nuclear magnetic resonance.

Kundt's tube   A tube used to measure the speed of sound. It is a wide tube closed at one end by a piston and at the opposite end by a diaphragm attached to a rod clamped at its center. The tube is filled with air or other gas and contains a light powder. When longitudinal vibrations are excited in the rod, they are transferred to the gas in the tube through the diaphragm. The position of the piston is adjusted so that a certain number of standing waves forms in the tube. These waves are visualized by the powder in the tube that becomes lumped at nodes, giving the length of standing waves generated in the tube. By knowing the frequency of the sound generated in the rod and the wavelength, the speed of sound can be determined.

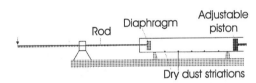

Kundt's tube.

L

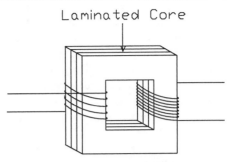

A laminated core used in a transformer.

lagging current   In a series RLC circuit, if the capacitive reactance X_C is less than the inductive reactance X_L, the current is lagging behind the voltage by a phase angle $\phi = \tan^{-1}[(X_L - X_C)/Z]$.

Lalande cell   The zinc-copper oxide cell invented by F. de Lalande and G. Chaperson in 1881 is known as the *Lalande cell.* It consists of a zinc anode and copper oxide cathode in a KOH electrolyte. It is a primary cell employing the following chemical reaction:

$$Zn + CuO \text{——} Cu + ZnO \ .$$

lambda leaks   The bane of scientists working at low temperatures, lambda leaks are leaks that only appear when the apparatus is immersed in liquid helium below the lambda temperature, 2.17 K. Such a leak occurs when the source of the leak (a crack, hole, etc.) is small enough to prevent gases and normal liquids from entering due to viscous drag on the gas or liquid. The viscosity falls to zero below the lambda temperature, allowing lambda leaks to occur.

lambda phenomenon   *See* helium-4, superfluid; lambda point.

lambda point   The lambda point is the temperature at which pure ^{4}He becomes a superfluid, 2.1768 K. The name is due to the particular shape of the specific heat vs. temperature curve at the phase transition. *See also* helium-4, liquid; helium-4, superfluid.

lambert   A unit of luminance, equal to $1/\pi$ candle/cm^2.

laminated core   An iron core made up by the lamination of sheet iron or steel generally used in transformers.

lamp, arc   An electric lamp in which light is generated by an arc. The arc is a spark produced when current flows through ionized gas between the two electrodes. Carbon electrodes are often used. The electrodes are vaporized by the heat of the arc.

lamp, fluorescent   A lamp emits light by fluorescence (luminescence). It contains gas (sodium vapor, mercury vapor, and so on) at a low pressure. The gas is excited by collisions with electrons emitted by a cathode. When gas that is raised to an excited state returns to a ground state, ultraviolet light is emitted. Fluorescent substances are coated on the inner surface of a fluorescent lamp and emit visible light when they absorb the ultraviolet light emitted by the gas in the lamp.

lamp, incandescent   A lamp uses incandescence raised by an electrically heated filament. Incandescence is an emission of the light by substances heated to a high temperature ($> 3000K$).

lamp, tungsten   An incandescent lamp uses tungsten as its filament. It is used as a standard lighting.

LAN   Acronym for *local area network.* A communications network that serves users within a confined geographical area in the same building or group of adjacent buildings. It is made up of an interconnection of servers, workstations, and a network operating system using a communications link. *Servers* are high-speed machines that hold programs and data shared by network users. The *workstations* or *clients* are the users' personal computers, which can access the network servers. They can retrieve all software and data from the server. In small LANs, which are easier to install and manage, the workstations can be used as servers, and users can therefore access data on another computer. A printer can be attached to a workstation or to a server and be shared by network users. For large networks, dedicated servers are required. LANs run on a network operating system such as NetWare, UNIX, or Windows NT. The message transfer is managed by a transport protocol such as TCP/IP or IPX. The physical transmission of data is performed by an access method such as Ethernet or Token Ring, which is implemented in network adapters that are plugged into the machines. Network adapters are interconnected by twisted pair or optical fiber cable.

Langevin function   This is a mathematical function that is important in the theory of paramagnetism. Analytically it is given by:

$$L(x) = \coth x - 1/x \ ,$$

$x \ll !, L(x) \cong x/3$, the polarizability of molecules having a permanent electric dipole moment and the paramagnetic susceptibility of a classical collection of magnetic dipoles given by a function of this type.

language, programming/machine   These describe a problem solution as a source program. Each language has a translator or compiler to convert the source program to an object program. Machine language programming involves writing instructions directly in object code with only binary symbols 0 and 1.

laser   An acronym that denotes the process of *light amplification by stimulated emission of radiation.* In this process, stimulated emission from an atom or molecule is used to amplify an electromagnetic field, and this amplifier may be combined with a resonator to make an oscillator. The two-mirrors of the Fabry-Perot interferometer provide the optical resonator and atoms or molecules in a metastable excited state, forming an inverted population of energy levels, and provide the gain medium. The excited state is attained with an optical pump, or an electrical discharge may provide the energy source. The resonator mirrors feed photons belonging to the laser modes back into the resonance cavity so that their number can grow through repeated

interaction with the gain medium and provide stimulated emission. One of the cavity mirrors is the output mirror and is allowed to transmit 1 to 2% of the light produced. This is a loss mechanism that must be overcome if the gain medium is to lase. The amplitude of the laser field will then grow until a steady state is reached where the laser radiation rate is the same as the net rate at which energy is supplied. Many different kinds of lasers have been developed covering the spectrum from IR to UV, some of which are tunable with a selectable wavelength, and some of which operate at more than one wavelength. Unlike ordinary light sources, such as glowing wires, the radiation from lasers is highly coherent, highly collimated, extremely monochromatic, and intense. The degree of control of visible light with lasers approaches that of radio frequency oscillators and microwave sources. Since the light from lasers is concentrated in one or a small number of modes, the photon occupation number in each mode is very large, making the radiation field more classical than conventional light sources.

laser beam, directionality of   Laser beams are exceptionally narrow. The width of a laser beam is determined by the size of the opening provided by the partially silvered mirror through which the beam exits. The main source of spreading out of the beam as it exits is from diffraction around the edges of the opening, so very little spreading-out occurs. Photons emitted at an angle relative to the laser tube axis are quickly reflected out the sides of the tube via the silvered ends, which are carefully arranged to be perpendicular to the tube axis. Since the emitted photon in the stimulated emission process travels in the same direction as the stimulating incoming photon, the net effect of arranging the cavity mirrors this way is to make the beam from the laser highly directional.

laser beam, monochromaticity of   In stimulated emission, an incoming photon induces an electron in an excited state to change energy levels, but only if the incoming photon has an energy that exactly matches the difference of energy between the two states. This makes stimulated emission similar to a resonance process, in which the incoming photon triggers an electron to change energy states only if energies, and thus photon wavelengths, exactly match. When the stimulated emission only involves a single pair of energy levels, the output beam has a single wavelength and the radiation in the laser beam is monochromatic. The early lasers each had their own characteristic wavelength, depending on the material used, and could not be tuned very far, just the width of the laser line. The output wavelength is determined only roughly by the amplifying medium. It is more precisely determined by the tuning of the laser resonator. By applying a magnetic field or changing the temperature of a solid laser material, more tuning could be obtained. With the development of dye lasers, it is possible to cover the range of wavelengths from 350 to 950 nm continuously, throughout the entire visible spectrum.

laser, cavity of   In a laser, the two-mirror Fabry-Perot interferometer serves as the optical resonator or cavity. Laser processes begin when an atom or molecule in the metastable excited state spontaneously emits a photon parallel to the axis of the laser cavity. This photon causes other excited atoms or molecules to emit a photon via the stimulated emission process such that the emitted photon travels in the same direction as the original one and exactly in phase. To ensure that more photons are created in an avalanche of stimulated emission from a multiple reflection process, both ends of the tube are silvered to form mirrors that reflect the photons back and forth. So that some of the photons may escape the tube and form the laser beam, one of the mirrors is only partially silvered and serves as the output window for the laser beam.

laser communications   Laser communications represent a form of optical communication with a laser as the light source. A laser represents a single-frequency phase-coherent light beam. Telecommunications using laser beams is distinguished by lasers' very wide frequency spectrum, lasers' efficient use of transmission power, and the laser beam's precise spatial directivity. The visible spectral band over which lasers operate spans over half a trillion megahertz; thus, one laser beam may theoretically offer a transmission rate surpassing that of the entire radio-frequency spectrum. Because of the

precise spatial directivity and spectral coherency of laser beams, they are highly power-efficient and useful for long distance communications such as satellite or spacecraft telecommunications, where the satellites or spacecrafts typically have very limited power resources. However, laser beams are highly vulnerable to obstruction by fog, rain or snow in outdoor atmospheric channels. Laser beams are often transmitted within protective pipes for earth-bound telecommunications. Laser fiber optics transmission is typically in baseband, with the information signal represented as a sequence of on-and-off light pulses. Semiconductor photodiodes represent the most common optical communication receivers.

laser, effect on biological tissues   The absorption of electromagnetic energy by biological tissues means that energy is deposited in tissues when irradiated. For the specific example of lasers, the physiological response depends on the wavelength and intensity of the laser light as well as the degree of focusing of the light.

laser, efficiency of   Ratio of the output power to the input power. The overall efficiency is a product of efficiency factors, which are individually defined, depending on the mechanisms of energy transfer in the laser. The *pumping efficiency* is the ratio of the total pump power absorbed by the gain medium to the electrical input power into the pump source, and this efficiency is composed of efficiency factors, such as the ratio of lamp radiation within absorption bands of the gain medium to electrical input power. Another factor is the fraction of the electrical input power that results in potentially useful radiation. Still another is the efficiency of transfer of the useful radiation from the pump source to the gain medium, and so on. The *energy extraction efficiency* is the ratio of actually extracted power to available power.

laser, gain in   Increase in signal power in transmission from one point to another. Units for power gain in common engineering usage are decibels or db. An active medium exhibits gain rather than absorption at a certain frequency. The bandwidth of an active medium is the full distance between frequencies at which the power gain has fallen to 1/2 its peak value, corresponding to 3 dB down from peak gain. This amplification bandwidth is considerably smaller than the atomic linewidth, especially at higher gains — an effect called *gain narrowing,* which reduces useful bandwidth of high gain amplifiers. Gain is synonymous with *amplification.*

laser induced fluorescence   The spontaneous re-emission of radiation that occurs after a gas is put into an excited state by illumination with a high-power laser. The amount of spontaneous emission or fluorescence from an atomic transition is proportional to the upper-level population of the transition. Since the upper levels are initially sparsely populated, the observed laser induced fluorescence is a measure of the lower level number density. This provides a useful technique for measuring the population of metastable levels, which cannot decay by giving off a photon. Laser induced fluorescence (LIF) diagnostics operate by pumping electrons out of the metastable state to a higher excited state which can decay radioactively; measurement of the LIF can be used to determine the cross-section for excitation into the metastable state.

laser materials, in general   The active laser medium that emits radiation from stimulated electronic transitions to lower energy states. Laser operation has been demonstrated in a wide variety of media, but only a few types of materials have been developed for commercial use. For a material to be useful as a laser there are certain optical, chemical, thermal, and mechanical properties it must have. The ions providing the optical emission must be able to efficiently absorb pump energy, and emit efficiently at the desired frequency. Often, there is not a single ion species that can do both efficiently, and combinations of ions are used in the same host material: one to absorb pump energy (sensitizer ion) and one to provide the lasing (activator ions), with a strong overlap in spectra of both types of ion. For a laser material to be useful commercially it must also be economically produced in sufficient quantity at high quality. To be useful outside the laboratory, the material must be stable and robust in its environment. It should be

chemically stable and preserve the ion valence state and resist ion diffusion out of the optical path, while resisting internal stresses that may be either thermally or optically generated. The requirements for lasing materials are often mutually contradictory and no single material can meet all of these criteria simultaneously. Thus for successful design of a laser system, a wide variety of materials must be considered and selection made based on a thorough understanding of their optical and other physical properties.

laser, medical application   The use of laser radiation for a therapeutic purpose, such as surgery on the retina.

laser mode   The field pattern of light in an optical resonator. The optical resonator of most lasers consists of two reflectors facing each other, aligned so that multiple reflections take place. An analysis of interference by multiple reflections of light in this resonator reveals the nature of the axial and transverse modes. As in the Fabry-Perot interferometer, there is full transmittance at a series of discrete wavelengths separated by a free spectral range determined by the wavelength and the separation of the mirrors. Each of these frequencies is a spectral mode or axial mode. These modes represent resonant frequencies that are exactly an integral number of half-wavelengths along the resonator axis between the mirrors. In a real cavity, any wave as it bounces between mirrors will also spread transversely due to diffraction, and will also distort in transverse amplitude and acquire diffraction ripples in even a single pass through the laser cavity. Analytical or computer calculations may be carried out to find the change of the transverse field pattern, with repeated passes through the laser cavity, as pioneered by Fox and Li. Usually these are carried out with the laser gain omitted for simplicity; for any given cavity with finite diameter and mirrors, there will be a distinct set of transverse amplitude and phase patterns, which self-replicate in form, though are reduced in amplitude after each round trip through the cavity. These are termed *transverse eigenmodes* or *transverse cavity modes* and depend on the detailed shape and curvature of the end mirrors. They are analogous to the transverse modes of electromagnetic waves in closed waveguides.

laser, pulsed   A laser with an emission waveform that consists of short duration bursts, each characterized by a rise and decay time. Between pulses, the laser is inactive, in contrast with the *continuous wave* or *CW* laser, which has a constant output. A widely used technique to shorten the pulse duration and allow the laser pumping process to build up a larger than usual population inversion is *Q-switching,* in which one of the end mirrors is effectively blocked to remove feedback, which is subsequently restored to its usual large value, dumping the entire population inversion in a single short laser pulse.

laser, safety   The knowledge of operating a laser in a manner that results in no injury to the user or any bystanders.

laser surgery   The use of laser light to accomplish a surgical procedure, such as removing growths on the skin.

laser therapy   The use of laser light for a healing purpose.

lattice vibration   Periodic oscillation of the atoms in a crystal lattice about their equilibrium positions.

law of mass action   The law of mass action describes the equilibrium behavior of a variety of chemical systems in solution and gas phases by stating that for a reaction of the type,

$$aA + bB \rightarrow cC + dD \, ,$$

the thermodynamic equilibrium constant is given by

$$K_{eq} = \frac{[C]^c [D]^d}{[A]^a [B]^b} \, .$$

The coefficients a, b, c, d are stoichiometric coefficients and A, B, C, D represent chemical species. The square brackets indicate that the concentration of the chemical inside is at equilibrium. Therefore, the value of the equilibrium constant at a given temperature can be calculated only when the equilibrium concentrations of the reaction components are known. Notice that the unit for K_{eq} depends on the reaction being considered because it is determined by the powers of the various concentration terms.

Knowledge of the equilibrium constant for a reaction allows for the prediction of several important features of the reaction. Namely, the tendency of the reaction to occur (but not the speed of the reaction), whether or not a given set of concentrations represents an equilibrium condition, and the equilibrium position that will be achieved from a given set of initial concentrations.

The tendency of a reaction to occur is indicated by the magnitude of the equilibrium constant. If K_{eq} is much larger that 1, the equilibrium lies to the right and the reaction system would mostly consist of products. If K_{eq} is less than 1, the equilibrium lies to the left and the reaction system would consist of mostly reactants. This would mean that the given reaction does not occur to a very significant extent.

See van't Hoff's law.

leader stroke   A thin, highly ionized, and highly conducting channel that grows from one electrode toward another of opposite polarity in a gas (usually air) during the initial stage of a spark discharge. It is neutralized when the tip of the leader reaches the second electrode, triggering the next phase of the discharge known as the return stroke. The leader stroke is clearly detectable in spark discharges through large gaps such as lightning flashes which begin with the growth of a leader from cloud to earth or earth to cloud. *See also* lightning flash; lightning stroke.

leading current   In an RLC circuit, if the capacitive reactance X_C is more than the inductive reactance X_L, the phase angle $\phi = \tan^{-1}[(X_L - X_C)/Z]$ is a negative quantity. In this case, the current reaches its maximum before the voltage does and the current is called the leading current.

leakage current   The undesirable current that "leaks" through an insulator.

leakage fields   Fields that extend beyond the region over which they were designed to exist.

least distance of distinct vision   Conventionally, the near point of a normal eye is a distance (*nearest distance of distinct vision*) of about 25 cm (10 in). This is called the *least distance of distinct vision*. It is the average of the least distance between the eye and an object such that the object can be seen in focus by an unaided eye. It is used to compare the magnifying powers of microscopes. For a simple microscope (magnifier), the angular magnification M is described by the least distance of distinct vision L and the object distance s:

$$M = \frac{L}{l} \left(= \frac{25}{l} \right) ,$$

where distances are measured in cm. Consequently, if the image is viewed at infinity,

$$M = \frac{L}{f} ,$$

where f is the focus length. If the image is viewed at the near point of the eye (image distance $s' = -L$),

$$s = \frac{Lf}{L + f} ,$$
$$M = \frac{L}{f} + 1 .$$

These angular magnifications are estimated by the least distance of distinct vision. However, actual magnifications depend on the particular observer. It is also known as the *distance of most distinct vision.*

lecher wires   (**1**) Two parallel wires used to measure high radio frequencies. The two wires, which are a few wavelengths long for the frequency to be measured, are either adjusted by sliding a shortening bar along them, or terminated at their far end and varied electrically by tuning a capacitor that is in series with the wires. When connected to the high frequency source, standing waves will be generated in the wires when their lengths are multiples of half-wavelengths. Measurement of the corresponding node or anti-node positions allows the frequency to be calculated.

(**2**) Two parallel straight wires, as in a two-wire transmission line, with a sliding short circuit copper strip between them. The wires can be tuned to a specific frequency of an oscillatory electrical wave by moving the strip along

the wires. Generally, lecher wires are used in the microwave frequency range as part of a wavemeter for determining wavelength. They can also be used as a tuned circuit or an impedance matching device.

Leclanche cell   The zinc-carbon cell originally invented by Georges Leclanche is known as a *Leclanche cell.* It consists of a zinc anode and a manganese dioxide cathode. The Leclanche cells are noted for their low cost and good shelf life. It is a primary cell employing the following chemical reaction:

$$Zn + 2MnO_2 \rightarrow ZnO \cdot Mn_2\,O_3 \;\; .$$

Lenard spiral   A spiral of bismuth wire, mounted between mica plates, that is used to measure magnetic field strength. It has small or negligible inductance while its resistance is strongly dependent on the strength of the magnetic field directed orthogonally to the axis of the spiral. An increase in the magnetic field results in an increase in the resistance and vice versa. Thus a measurement of resistance can be related to a magnetic field strength.

lens   A piece of isotropic, transparent material that has two surfaces with a common axis. The common axis is called the *optical axis.* The point on the surface of the lens where the optical axis crosses is called the *vertex of the lens.* The geometrical center of the lens is called the *optical center.* A light ray that passes through the optical center will not be deviated. A lens is used for refraction of light. Usually, polished glass or molded polymer is used as the material. There are various kinds of lenses and their surfaces are usually spherical (*see* lens, spherical). The surface of others are aspherical, cylindrical, parabolic, toroidal and so on. The curvature of the surfaces affect the functions of the lens: focus length, aberration, astigmatism, and so on. The surface of a lens, called *concave* or *convex,* curves inward or outward, respectively. Lenses are divided into positive or negative (converging or diverging). A positive lens causes parallel rays of light to converge and a negative lens causes them to diverge. The *shape* of the lens refers to the shape of the periphery of the lens.

The *form* of the lens indicates the relative allocation of the curvatures. Lenses are classed as thin or thick. A *thin lens* is a lens whose thickness is small enough to be neglected in the calculation of optical quantities of the lens. A lens should be treated as a *thick lens* when a precise solution of a lens problem is required. A typical problem is the design of a camera lens. The *focal length* of a thin lens is the distance between the optical center and the focal point of the lens. The image distance and the object distance of a thin lens are measured from the vertices. Solving a lens problem, the thin-lens focal length, the image distance, and the object distance of a thick lens should be measured from the principal planes of the lens.

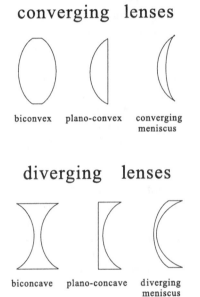

converging lenses

biconvex   plano-convex   converging meniscus

diverging lenses

biconcave   plano-concave   diverging meniscus

Lens.

lens, achromatic   An optical system that is corrected for chromatic aberration. Usually, a combination of positive lenses and negative lenses of different refractive indices is used. An achromatic lens is designed so that the dispersions of the two lenses neutralize each other, and their refractions do not neutralize each other. The simplest achromatic lens consists of a positive lens and a negative lens, and is called an *achromatic doublet*. It is used as an objective of a telescope and is also known as an *achromat*.

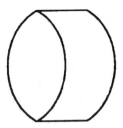

An achromatic lens (achromatic doublet).

lens, antiflex   An antiflex objective is used to visualize structures that would be invisible in normal bright–field microscopy. An example is the observation of cells grown on the bottom of a Petri dish visualized by incident-light-reflection contrast techniques.

lens, apochromatic   An optical system that is highly corrected for both spherical and chromatic aberration for two or more colors and is used as a microscope objective.

lens, astigmatic   A toroidal (toric) lens that is used in eyeglasses to correct astigmatism. It is also known as an *astigmat*.

lens, Barlow   A lens system of one or more negative lenses that are placed just ahead of the focal plane of the objective, between the objective and eyepiece, in a telescope. It is used to increase the effective focal length and thereby increase magnification. Plano-concave lenses are usually used.

lens, blooming of   The process of coating a transparent, thin film on a lens to reduce the reflection of light at the surface of the lens. A substance, such as a magnesium fluoride, is deposited on the lens to form a thin film with a thickness t of one quarter wave length $\lambda_c = \lambda/n_c$. The thickness of the coating t is

$$t = \frac{\lambda_c}{4},$$
$$= \frac{\lambda}{4n_c},$$

where the refractive index of the substance is n_c. Usually evaporation is used for the deposit. The substance should have a lower refractive index than the lens to be bloomed. The reflectivity R of the interface of the different media with indices n_1 and n_2 is

$$R = \left(\frac{n_2 - n_1}{n_2 + n_1}\right)^2 .$$

The desirable refractive index of the coating material is dependent on this relationship and the number of layers to be deposited on the lens. For example, with one layer coating, the desirable refractive index of the coating n_c is

$$n_c = \sqrt{n_2 n_1} ,$$

where n_2 and n_1 are the refractive index of free space and one of the lens, respectively.

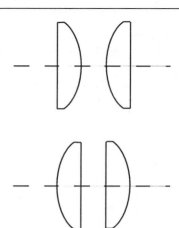

A condenser lens.

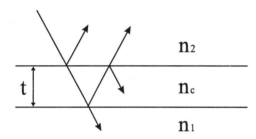

Lens blooming.

lens, concentric   A lens with two spherical surfaces. The spherical surfaces have the same center.

lens, condenser   A lens system that is used to concentrate as much light from a source as possible. Also known as a *condensing lens,* it is used in an instrument of illumination for various kinds of optical systems. A condenser lens should be free from aberration. Usually a double plano-convex condenser lens is used. It is also used in an projection system. An *Abbe condenser* is well known. It consists of a pair of lenses and has a variable large-aperture. An Abbe condenser is used as a microscope objective.

lens, converging   (lens, positive) A lens that causes parallel rays of light to converge on a principal focus on the axis of the lens. The center of a converging lens is thicker than the edge of the lens. A converging lens has a positive focal length. In a converging lens, the second focal point lies on the opposite side of the light

source. (*See also* lens.) With a converging lens, the object closer to the lens than the first focal point will form a virtual image, not a real image. A biconvex lens, a plano-convex lens, and a converging meniscus are well known as converging lenses.

lens, crossed   A kind of spherical lens designed with particular radii of the curvature of the two surfaces in order to realize a minimum spherical aberration for parallel incident rays. The radii are dependent on the refractive index of the lens.

lens, crystalline   An elastic, jelly-like lens of the eye that is elastic biconvex. The refractive index of the crystalline lens is high (about 1.4), and it is highest in the center and lowest at the equator. It lies between the anterior chamber, which is filled with aqueous humor, and the vitreous, filled with vitreous humor. Its high refractive power can be altered by varying its thickness which is altered by the ciliary muscle.

lens, cylindrical   A lens in which one or both of its curved surfaces are a portion of a cylinder. It has axial astigmatism and is used in the correction of visual deficiencies. A planar cylindrical lens is used for correcting astigmatism of the eye.

lens, decentered   A lens whose optical center is different from the geometrical center of the rim of the lens. A decentered lens works as a lens combined with a weak prism.

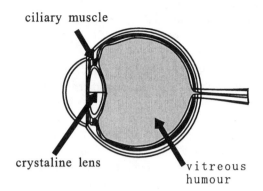

ciliary muscle

crystaline lens

vitreous humour

Crystalline lens.

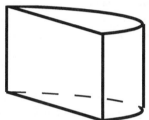

Cylindrical lens.

lens, diverging (lens, negative) A lens whose focal length is negative. A diverging lens causes rays of parallel light to diverge. Biconcave, plano-concave, and diverging meniscus are known as a diverging lens. *See also* lens.

lens, equation, Newtonian form One of the *lens-users equations,* this form uses image distance and object distance measured not from vertices but from the focal points of a lens. An equation that treats the relationship between the distances between two conjugate points and their respective foci. The product of the distances is equal to the square of the focal length of the lens. The distance between an object and a focal point x_o, the distance between an image and another focal point x_i, object size y_o, and image size y_i, and lateral magnification m, satisfy the equation;

$$m = \frac{y_i}{y_o} = \frac{f}{x_o} = \frac{x_i}{f} ,$$

where the focal length is f. These relations derive from the Newtonian form lens equation as

$$x_o x_i = f^2 .$$

It can be used with both thin and thick lenses.

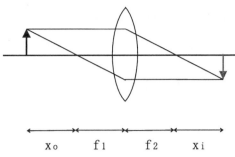

Newtonian form lens equation.

lens, equivalent A lens or a system of lenses that forms almost the same image as a given lens or a given system of lenses.

lens, landscape A simple meniscus lens or an achromatic doublet (*see* lens, achromatic) with its stop rate greater than f/11. The angle of the field is limited to about 40° because of the oblique astigmatism of a landscape. It is also called an *achromatic meniscus*. A meniscus of which the concave side facing the object is corrected for both astigmatism and coma still has spherical and chromatic aberration and distortion. A combination of two landscape lenses will improve its aberration.

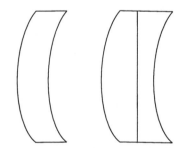

Landscape lens.

lens maker's equation Also called the *lens equation* or *lens maker's formula,* this equation is used with thin lenses. The focal power P and the focus length f of a thin lens can be calculated by the equation. It uses the fact that the refractive power P of a thin lens is the algebraic sum of the pair of surfaces of the lens P_1, P_2; $P = P_1 + P_2$. When a thin lens of index n_l is surrounded by a medium of index n_m, the powers of the surfaces, whose radii are R_1 and R_2, are calculated by the surface power equation

$$P_1 = \frac{n_l - n_m}{R_1} .$$

Therefore, the power of the lens P is obtained as

$$P = \frac{1}{f} = (n_l - n_m) \left(\frac{1}{R_1} - \frac{1}{R_2} \right) .$$

It is used to decide the two radii of the surfaces of the lens. It is important for lens makers to decide how to grind the lens.

lens, negative *See* lens, diverging.

lens, objective The lens of an optical system that is nearest to the object to be observed, it usually has a short focal length. The objective lens of a microscope forms a real air image (*see* image, primary). For projection with a microscope, the primary image is formed ahead of the first focal length of the ocular which forms a real image again. For visual observation, the primary image is generated inside the focal length of the ocular as a magnifier.

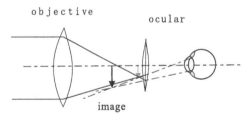

Image formation in a telescope.

lens, positive *See* lens, converging.

lens, power of (focal power) The power of a lens shows the ability of a lens to converge the parallel rays of light. It is equal to the reciprocal of the focal length, the distance between the optical center of a lens and its principal focus. The power P of a thin lens is the algebraic sum of the pair of surfaces of the lens P_1, P_2; $P = P_1 + P_2$. The angular magnification of a lens M_θ and power of the lens P have a relationship $M_\theta = \frac{1}{4} P$. This equation is known as the *quarter-power equation* and it applies to the lens that is used without accommodation; an image

is seen at infinity and the lens is set close to the eye. With accommodation, $M_\theta = \frac{1}{4} P + 1$.

lens, speed of The speed of a lens indicates how much light the lens can gather and transmit. The extent of the energy of light gathered by the lens from a light source is proportional to the area of the lens. The gathered energy is inversely proportional to the area of corresponding image through which the gathered light passes. The energy flux density passing through the image is proportional to the $(D/f)^2$, for a lens of focal length f and diameter D. The *f-stop* (*f-number*) is described as $f/\#(= f/D)$. D/f is called the *relative aperture* of the lens. The *speed* of a lens is measured by the f-number of the lens. The *f-number* of a lens is equal to the ratio of the focal length to the diameter of aperture of the lens. A lens with a lower f-number (more rapid lens) can gather and transmit more rays of light. The *depth of field* of a lens is equal to the ratio of the focal length to the speed of the lens.

lens, spherical A lens whose surfaces form portions of spheres. It is easier to make spherical lenses than other kinds of lenses. Therefore, this kind of lens is widely used. Aspherical or deformed spherical lenses are used, because a spherical lens has aberrations. A spherical lens sometimes refers to a lens of complete sphere form.

lens, split (a billet split lens) A lens that is cut into two parts along the optical axis of the lens. It is used in an interferometer.

lens, Stokes A variable-power compound lens made up of cylindrical lenses of equal power mounted so that the angle between their axes can be varied.

lens, telephoto The telephoto lens is a kind of camera lens, especially for a *single-lens reflex* (SLR) camera. It has a long focal length, usually longer than 80 mm and is used for taking a photograph of an object far away from the camera. Its field of view is very narrow. To avoid making a telephoto lens too long, a negative lens — of which the focal length is shorter than the other positive lens — is used as the second lens.

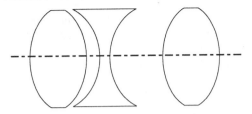

Telephoto lens.

lens, thick   A lens whose thickness is such that the effect of the thickness cannot be neglected to consider the optics of the lens or the lens system. Whether a lens is regarded as thin or thick also depends on the precision required. The principal planes of a lens are a pair of planes that are perpendicular to the optical axis of the lens. The image of any object on another plane is formed on the other plane with unity lateral magnification.

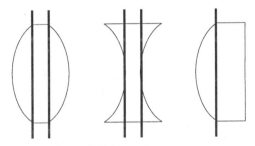

Principal planes of thick lenses.

With a thick lens, the focal length, image distance, and object distance of the lens should be measured from the principal planes of the lens.

lens, thin   A lens whose thickness is insignificant enough to be neglected in the calculation of the optical quantities of the lens. The thickness of a thin lens is sufficiently smaller than the focal length of the lens, radii of the pair of the curvatures, the image distance, and the object distance. Whether a lens is regarded as thin or thick also depends on the precision required. The focal length of a thin lens is the distance between the optical center and the focal point of the lens. The image distance and the object distance of a thin lens are measured from the vertices. The rays of light, which pass through the first focal point of a lens, will form parallel rays just after the rays pass through the lens. The second focal point is formed by a set of parallel rays of light which pass through a thin lens. With a positive lens, the second focal point lies on the opposite side of the light source. Usually, "the focal length" of a lens means the second focal length of the lens. With a converging lens (positive lens), the object closer to the lens than the first focal point will form a virtual image, not a real image.

lens, thin, combination of   The focal length and power of the combination of thin lenses f, P can be calculated from the focal length of the lenses f_i, P_i, $i = 1, \ldots, n$ as:

$$\frac{1}{f} = \sum_{1}^{n} \frac{1}{f_i},$$

$$P = \sum_{1}^{n} P_i.$$

The combination of thin lenses can be described as a product of the optical matrix of each thin lenses.

lens, toroidal   Also known as a *toric lens*, this is a lens whose surface forms parts of toric surfaces. A toroidal lens is used for the correction of astigmatism on both of the perpendicular meridian planes. The *plano-cylindrical, spherocylindrical,* and *spherotonic* are troidal lens.

Lenz's law of induction   This law states that when a conductor moves with respect to a magnetic field, the currents induced in the conductor are in such a direction that the reaction between them and the magnetic field opposes the motion.

Leyden jar   An early form of capacitor that was first thoroughly investigated by Pieter van Musschenbroek of the University of Leyden in 1746. It consisted of a stoppered glass jar filled with water and a nail piercing the stopper and dipping into the water. Holding the jar in one hand then touching the nail to an electrode of an electrostatic machine and finally disconnecting it caused the jar to acquire and store charge. An electric shock was experienced when the free hand touched the nail. John Bevis modified the Leyden jar so that inner and outer surfaces were covered with metal foil. This was a closer arrangement to modern capacitors.

lidar Acronym for *light detection and ranging*. Method for determining positions of distant objects by use of a laser beam. The process involves reflection of laser light from an object and the determination of time required for a beam to reach the object and return to the emitter. Distance is then computed from the product of light speed and time. Similar to radar, but based on electromagnetic waves in the visible part of the spectrum.

light, absorption of The process whereby light is absorbed by the atoms or molecules of a material. The energy of absorbed light in general serves to excite electrons into higher energy states in the material, from which they can subsequently decay by emission of electromagnetic radiation with less energy (frequency) than that of the absorbed radiation. Energy retained in the material appears as thermal energy in the atoms and molecules of the matter.

light, corpuscular theory Early theory of light which asserted that light consisted of a stream of particles referred to as *corpuscles*. The theory (alternatively known as "emission theory of light") was most successfully expounded by Issac Newton (1642–1727), who postulated the existence of different kinds of "light-particles" that could excite a material ether filling all of space into different types of vibrations so as to produce the sensation of distinct colors. The notion of light as a stream of localized entities, inherent in the corpuscular theory, was revived in the quantum theory of light, which interprets light on a submicroscopic scale in terms of "particles" of energy called *photons*.

light, emission of Process whereby light is emitted by the atoms or molecules of a material. Sources of light emission are

1. accelerated charges (electrons or ions), and

2. excited atoms and molecules. Light emitted by accelerated charges results from the conversion of *kinetic energy* of charges into radiation energy. Light emitted by excited atoms and molecules is produced (primarily) by the conversion of *potential energy* (in excited states) into radiation energy.

light emitting diode A semiconductor device converting electrical energy into light, which is heavily used as a light emitter in displays, etc. Commonly, an LED is a forward biased p-n junction diode in which the light is emitted when a hole and electron recombine. The most recent LEDs in use are GaN, giving blue/green light, and SiC LEDs.

light, energy of Energy associated with light on the basis of its capacity to exert forces (which can do work) on electric charges. In the electromagnetic wave theory of light, the energy transported across a unit area per second by a light wave is represented by the magnitude of a vector quantity, $\mathbf{S}$, referred to as the *Poynting vector*, and defined in terms of the electric and magnetic fields of the wave, $\mathbf{E}$ and $\mathbf{B}$, via the relation

$$\mathbf{S} = \frac{1}{\mu}(\mathbf{E} \times \mathbf{B}) \, ,$$

where μ is the *permeability* constant of the medium. Alternatively, in the quantum theory of radiation, the energy of light is said to reside in discrete entities termed *photons*, each of which is associated with a quantized amount of energy hf, where f is identified with the frequency of the radiation and h is a fundamental constant of nature known as Planck's constant, with the (approximate) value $h = 6.626075 \cdot 10^{-34}$ J.s.

light, momentum of Linear momentum associated with light on the basis of its capacity to exert forces that transfer momentum to (the electrons, atoms and molecules of) material media. In the electromagnetic wave theory of light, the momentum transported across a unit area per second by a light wave is represented by the ratio of the Poynting vector, $\mathbf{S}$, divided by the speed of light, c. Alternatively, in the quantum theory of radiation, the momentum of light is associated with discrete entities termed *photons*, each of which is characterized by a quantized momentum equal to hf/c, where f is the "frequency" of the light and h is Planck's constant, (approximately) equal to 6.626075×10^{-34} J.s. *See* light, energy of.

light, monochromatic Light consisting of electromagnetic waves (or photons) characterized by a single frequency f. The term, signifying *single color,* derives from connection between different frequencies of light and distinct colors in the visible spectrum. It contrasts with non-monochromatic light consisting of a combination of electromagnetic waves of different frequencies. Since in practice, all physical light beams (of non-infinite extent) must consist of mixtures of electromagnetic waves with frequencies within some non-zero range, monochromatic light is in practice *quasi-monochromatic* — meaning that it consists of electromagnetic waves having only a small spread of frequencies about a central (dominant) frequency.

light, monochromatic, biological action Certain monochromatic single–wavelength light seems to have a biostimulation and sometimes therapeutic effect when applied to cells in tissue. Accelerated tissue repair, stimulated by the monochromatic light, is useful in fields of dentistry, dermatology, and neurology. These effects are sometimes called photo–stimulation.

Wavelengths used are usually in the 630 to 950 nm range. The character of the light source, continuous or pulsating, seems to have an effect on cell functioning. For example, with continuous light, relief of pain, relaxation of muscle fibers, and reduced swelling may be observed. This may be due to an increase in cellular blood flow. On the other hand, a pulsating light source may stimulate protein production and calcium accumulation, which might result in accelerated healing of damaged tissue. *See* red light, healing effect.

lightning arrester (surge protector) A device used to protect electronic and electrical equipment from damage due to voltage spikes in power lines, communication lines, as well as other long wires during lightning surges. The arrester is a shunt device with an impedance that changes from high to low during the spike and thus provides a bypass circuit preventing damage to the attached apparatus. Examples of lightning arresters are short air gaps with a specific breakdown voltage, or semiconducting elements with a resistance that becomes very low above a predetermined voltage.

lightning conductor (lightning rod) A rod, usually made from copper, mounted as high as possible above buildings, trees, power lines and other structures to be protected against lightning strokes. The upper end of the rod consists of one or more sharp points while the lower end is firmly connected to conductors embedded in the ground. Lightning rods primarily serve to neutralize the charge on a nearby cloud by conducting negative charge to or from the ground and through the atmosphere at a relatively slow rate. As a result, the probability of a direct lightning stroke is reduced. As an approximate rule, the rod acts as a shield over a cone with a radius at ground level equal to the height of the rod.

lightning flash (or discharge) Refers to all the phenomena associated with electrical discharges between clouds, and between clouds and the earth. Its main constituents are the leader stroke and the return stroke. The latter produces the highly luminous part of the discharge. On average, a lightning flash lasts for 200 milliseconds and consists of several pulses each of some 10 milliseconds separated by 40 milliseconds. *See also* lightning stroke.

lightning stroke (return stroke) A highly luminous channel triggered by the leader stroke of a lightning flash. The luminosity travels from ground to cloud as a large current travels from ground up to neutralize the highly charged leader channel at a velocity of 0.1 to 0.3 of the speed of light. The most dangerous effects of the lightning stroke are connected with the current peak which can reach values of 100 kA.

lightning surge Large voltage transient in transmission lines, communication lines, and other long wires induced by lightning strokes. The damage from a lightning surge to any apparatus is minimized by an installed lightning arrester. *See also* lightning stroke; lightning arrester.

light pipe A flexible transparent polymer rod that transmits light from one end to the other even when it is bent. It is used to guide a light beam. The principle of total internal reflection is applied to light pipes. The angle of the incidence of the rays of light is larger than a specific critical angle, and the interface between two different media acts a mirror. The refractive index of the light pipe n should be higher than the refractive index of the surrounding media n_m. The critical angle ϕ_c is

$$\phi_c = \arcsin\left(\frac{n_m}{n}\right) ,$$

where n_m and n is the refractive indices of the surrounding media and one of the light pipe, respectively. Usually, $\phi_c = \arcsin\left(\frac{1}{n}\right)$. It is used in various fields, as an instrument for internal human organs, communications system, and so on.

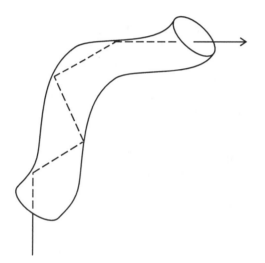

Optical fiber.

light, pressure of   Pressure associated with light on the basis of its capacity to exert forces on electric charges that produce pressure (defined as *force per unit area*) on a material object. In the electromagnetic wave theory of light, pressure exerted by light on an object is directly proportional to the magnitude of the Poynting vector of the light wave, **S**, divided by the speed of light c. The relationship between pressure and the Poynting vector follows from equivalence between rate of transfer of linear momentum across the unit area of a surface and the force per unit area on surface. *See* light, energy of; light, momentum of.

light pulse   Propagating an electromagnetic wave (with a frequency in the visible part of

spectrum) that has a finite (usually short) limit in time, Δt, and a corresponding finite limit along the direction of propagation. A light pulse corresponding to a length in time Δt must consist of a combination of "wavelets" with a range of frequencies, Δf, approximately equal to $1/\Delta t$.

light, speed of   The rate of propagation of light (and other electromagnetic radiation). The speed of light in a vacuum is a fundamental constant of nature (denoted c) with a value (approximately) equal to 2.997925×10^8 m/s. The speed of light in a macroscopic medium, v, can never exceed c, and is expressed in terms of c and the *index of refraction* of the medium, n, via the relation

$$v = \frac{c}{n} ,$$

where the value of n in general depends on both the properties of the medium and the frequency of the light.

light, unpolarized   Light consisting of an (effectively) uniform mixture of light waves having all possible polarizations. Corresponds in practice to light in which the state of polarization varies randomly within a time interval less than the minimum time required for a measurement, such that no (resultant) state of polarization can be detected. *See* polarization of light.

light, wave theory of   Theory in which light (and other electromagnetic radiation) is interpreted to be a type of wave. Current theory of electromagnetism interprets light to consist of oscillating (time varying) electric and magnetic fields that propagate through space as waves with a speed dependent on the properties of the material medium that occupies the space.

light-year   Length (abbreviated ly) equal to the distance that light travels through a vacuum in one year. It represents a convenient unit of length for the specification of astronomical distances, and has a value given by the product of the speed of light in vacuum c times the number of seconds in a year, (approximately) equal to 9.461×10^{15} m ($= 1$ ly).

limbs, artificial   Artificial body parts that function like their natural counterparts. Used in the substitution of limbs for people who have suffered accidents leading to the loss of such extremities.

The artificial limbs form a subset of the general term *prothesis.*

limiter   A circuit or component used to limit the amplitude of a signal to a predetermined level. For example, a resistor R is used as a current limiter in a circuit. The current limit for a bias voltage V is V/R.

limit of resolution   The smallest angular separation between two point sources of light (subtended at the position of the detector) for which the images of the two sources are seen to be separated. The limit of resolution of light sources is most commonly determined by the so-called *Rayleigh criterion,* which states that the light from separate sources S_1 and S_2 will be resolved as distinct images only if the peak of the diffraction pattern produced by one source (at the observing screen or retina) is displaced from the peak of the diffraction pattern produced by the second source by a distance equal to or greater than the half-width of the diffraction patterns.

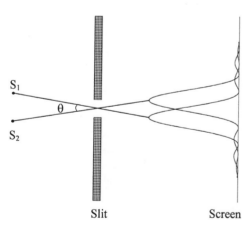

Limit of resolution.

It follows from this criterion that the limit of resolution of light from two sources can be shown to be approximately equal to the ratio of the wavelength of the light divided by the dimension of the aperture through which the light passes en route to the surface on which the light is detected.

linear circuit   A linear circuit is a circuit to which the superposition theorem can be applied. Namely, if there are N independent sources presented in the circuit, any branch voltage or current is composed of the sum of N contributions each of which is due to each of the independent sources acting individually when all others are set equal to zero.

linear code   A code is a linear code only if for any two elements a_1 and a_2 chosen from the alphabet and any two valid codewords $\mathbf{C}_1$ and $\mathbf{C}_2$, $a_1\mathbf{C}_1 + a_2\mathbf{C}_2$ also represents a valid codeword. A linear code must thus contain an all-zero codeword.

line, artificial   A network that simulates the electrical characteristics of a line over a given frequency range. For example, a transmission line can be simulated with a network of inductors, capacitors, and resistors.

line broadening   An increase in the natural range of frequencies of the radiation emitted or absorbed by a source or absorber of radiation. Corresponds to a "broadening" of the graph of the intensity of the emitted or absorbed radiation plotted versus the frequency of the radiation. Primary mechanisms resulting in line broadening are:

1. *Doppler-broadening,* produced by the spread in the velocities of the emitting or absorbing atoms or molecules, and

2. *collisional broadening,* resulting from collisions between the emitting or absorbing atoms or molecules that produce a change in the allowed energy states of the atoms or molecules, giving rise to an increase in the possible frequencies of the emitted or absorbed radiation.

line profile   Plot of the intensity of emitted or absorbed radiation as a function of the frequency (or wavelength) of the radiation. The "natural" line profiles associated with noninteracting atoms or molecules are well represented by Lorentzian functions of the frequency of the forms shown in the figures below, where Figs. (a) and (b) correspond to emission and absorption line profiles, respectively.

lines of force (electric)   Continuous lines drawn to represent the direction of an electric field. The number of lines passing through a small area of fixed size, oriented at right angles to the lines, gives the magnitude of the electric field.

line width   Width, at half maximum intensity, of the peak in a graph of intensity of emitted or absorbed radiation vs. frequency (or wavelength) of radiation. (*See* line profile and accompanying figures, in which the line width is denoted Δ f.)

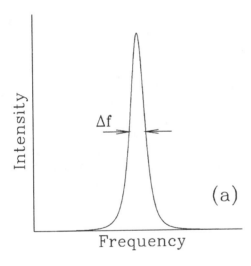

(a)

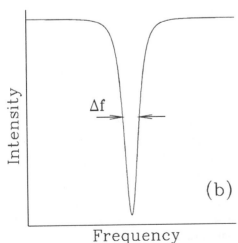

(b)

Line profile.

linkage magnetic   The product of the number of turns in a coil and the magnetic flux passing through the coil.

liquefaction coefficient   The liquefaction coefficient is a measure of the efficiency of the liquefaction of a (usually) cryogenic liquid. For isenthalpic processes (in Joule-Thompson liquefiers, for example), the efficiency is

$$E = \frac{H_f - H_i}{H_f - H_\ell} ,$$

where H_i, H_f are the enthalpy at the beginning and end of the process and H_ℓ is the enthalpy of the liquid state.

liquefier, Claude   This is associated with expansion engine liquefiers. Claude's first machines used isentropic expansion with liquid air being produced in the engine. Joule-Thompson expansion for the final liquefaction stage was later used. The operating efficiency is similar to the Linde system. This liquefier is often used in conjunction with Philips-Stirling and turbine expanders.

liquefier, Collins   This is a Joule-Kelvin combination liquefier that does external work. There are two pistons in cylinder expansion engines whose respective working temperatures are about 60 K/30 K and 15 K/9 K. The incoming helium gas enters via a series of heat exchangers and is distributed in the following manner: hotter engine, 30%; colder engine, 55%; Joule-Kelvin stage, 15%. The mechanical components have to be carefully designed so as to enable them to operate at temperatures where conventional lubrication methods are impossible. Gas purification is essential since even small traces of air can solidify and cause seizure of the piston.

liquefier, Hampson, air   This is a single stage air liquefaction process utilizing a heat exchanger that makes it possible to liquefy air by Joule-Kelvin expansion alone starting from room temperature. At first some cooling occurs but no liquid is produced as the high pressure incoming air expands at the Joule-Kelvin expansion valve. The resultant colder, low pressure air flows back through the heat exchanger and cools the incoming air stream. As a result the whole system gradually cools. Eventually steady operating conditions are reached and a certain fraction of the incoming air then liquefies as it expands. This liquefier is also popularly known as a *Linde-Hampson system.*

liquefier, Kapitza, helium   This system employs the Heylandt principle. Helium at a pressure of 30 atm is cooled to 65 to 70 K with liquid nitrogen. This then passes through the heat interchangers, expands to 2.2 atm in an expansion engine, and leaves the liquefier via the heat interchangers. A further amount of helium at 15 to 18 atm is cooled by the gas from the expansion engine and then expands at a valve where part of it liquefies. The final temperature drop to 4.2 K is obtained by the Joule-Thompson effect.

liquefier, Kapitza, hydrogen   Hydrogen is liquefied in a similar manner as for helium in the Kapitza liquefier. *See* liquefier, Kapitza, helium.

liquefier, Linde   The main components of this liquefier are a compressor, heat exchanger, and expansion throttle used for liquefying air. It is similar to the Hampson which only employs a different design of a heat exchanger. The efficiency is improved by letting the gas expand in two stages; further improvement occurred when Linde introduced a liquid ammonia pre-cooling stage so that the compressed gas entered the exchanger at about −48° C. *See also* liquefier, Hampson air.

liquefier, Philips, air   This is a single stage air liquefaction process that uses the Stirling cycle. The working fluid is operated on a closed cycle with the regenerator performing the functions of a heat exchanger by separating the colder and hotter ends of the system. Air condensation surfaces are attached to the outside of what corresponds to the colder cylinder.

liquefier, Simon (expansion)   This is a single expansion helium liquefier. Helium gas is compressed isothermally into a chamber to a pressure of 100 atm. A temperature of 15 K is maintained. This temperature can be reduced to 10 K by reducing the pressure above the evaporating hydrogen to well below its triple point pressure.

The gas is eventually adiabatically cooled to the final temperature.

liquid crystal   States of organization of atoms or molecules showing less symmetry than a crystal, but not as disordered as a liquid. The scientific term for this is *mesomorphic phase.* There are at least 9000 known molecules giving rise to mesomorphic phases. They are used in some types of displays (e.g., 7-segment display). Application of an electric field causes a large increase in light scattering creating a bright region.

liquid refrigerant level, surface detection
There are several techniques for detecting the level of a liquid cryogen. One common solution is to immerse a superconducting wire in the liquid and to drive current through the wire. If the current and wire are chosen carefully, the current will drive any of the superconductor above the liquid normal while leaving the portion of the wire immersed in the liquid superconducting. The resistance of the wire is therefore a measure of the distance to the surface of the liquid. A technique that does not dissipate so much heat in the refrigerant is to use a capacitor to measure the liquid level. Often, a capacitor made of two long concentric cylinders will be partially immersed in the liquid and attached to an AC bridge. As the liquid level rises and falls, the amount of dielectric (i.e., refrigerant) in between the plates of the capacitor changes, and the bridge measures the concomitant capacitance changes. In another technique, a thin tube with a rubber diaphragm on one end is inserted, open end first, into a storage dewar of liquid helium. As the tube is cooled, oscillations will begin in the diaphragm. These oscillations, called *Taconis oscillations,* are described by a complicated, non-linear set of hydrodynamic equations which must include the heat deposited into the helium gas and liquid by the tube, the heat conducted along the tube, along the gas, and other effects. If these equations are solved numerically or approximately, the frequency of these oscillations depends on whether the bottom of the tube is in contact with liquid helium or gaseous helium. Experimentally, the surface of the liquid is found by finding the point at which the frequency of oscillations changes discontinuously. Of course, visual location of the fluid

and or a float can also be used if the apparatus allows direct viewing of the fluid.

Lissajous' figures   Stable patterns showing the path of a particle moving in a plane forced by two harmonic oscillations whose frequencies ω_1 and ω_2 are related as $m\omega_1 = n\omega_2$, where m and n are integers. Lissajous' figures are obtained when $\cos \omega_2 t$ is displayed against $\cos \omega_1 t$ for different values of the integers m and n. They are called *Lissajous' figures* in honor of Jules Antoine Lissajous.

Lloyd's mirror   A mirror arranged relative to a (point) source of light and a viewing screen so as to produce an interference pattern on screen resulting from the superposition on the viewing screen of part of the wavefront of an incident light beam which reaches the screen directly from the source and part of the wavefront that arrives at the screen after reflection from the mirror surface.

load, electric   The electrical impedance connected to the output of an electrical energy source such as an active electrical circuit, electrical generator, battery or amplifier. It is also used in reference to the electrical power consumption of a device such as the power used by an electric motor, heater, etc.

load, lagging   A load in which the voltage reaches its maximum value before the current. A load that has a higher inductive than capacitive value.

load, leading   A load in which the current reaches its maximum value before the voltage. A load that has a higher capacitive value than inductive value.

load-line   A line representing the current in the load vs. voltage drop across a device in series with the load for a given input voltage. Its intersection with the device current vs. voltage curve is the operating current and voltage for the given device-load combination at the given input voltage.

loadstone   Also known as *magnetite*, it was discovered to be a natural magnet several cen-

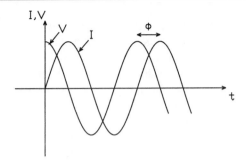

Load, lagging.

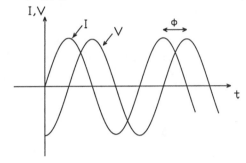

Load, leading.

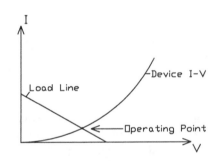

The operating point determined from the load line and the device I-V.

turies BC. It is basically Fe_3O_4 and was found originally in Magnesia of Asia Minor.

lobe   In the directivity pattern of an antenna or array, an area of the pattern bounded by directions of minimum radiation.

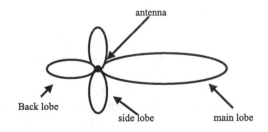

lobe length   In a Token Ring network, the length of cable between the *multi-station access unit,* i.e., the central hub and the workstation.

lobe switching   A method used for obtaining optimum signal from downcoming waves. This is done by switching into circuit networks that accept a chosen direction for optimum signal.

logic gates   Electronic devices in which the low or high voltage states of inputs and outputs determine a binary value of 0 or 1. They are used for logic operations, mathematical computations, storage, etc. (e.g., AND, OR, NAND gates).

logic operations   The operations NOT, AND, OR, NAND, NOR, XOR that act on Boolean arguments to produce a Boolean result.

longitudinal waves   A wave in which the direction of some vector characteristic of the wave — for example, the displacement of the particles — is along the direction of propagation. Also called *compressional* or *dilatational waves.* Longitudinal waves always have the highest propagation speed, $c_l = \left[\frac{E(1-\mu)}{\rho(1+\mu)(1-2\mu)}\right]^{\frac{1}{2}}$, with ρ as the density, μ the Poisson's ratio, and E as the elastic modulus of the medium.

long sight   (hypermetropia, hyperopia) A vision defect of the eye. The shortness of the eyeball makes it difficult for the lens of the eye to accommodate to project the image of near objects onto the retina. The correction for long sight is a converging lens that focuses parallel rays to the far point. This correction is treated by using a combination of thin lenses.

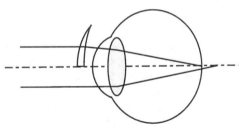

Correction for long sight.

loop gain   The product of the amplifier gain and the feedback fraction in a series voltage feedback loop. The input voltage to the amplifier will be given by $V_i = V_e + \lambda V_i$ where V_e is the externally supplied input voltage and λ is the loop gain.

loss, acoustic energy in liquids and solids
Loss of acoustic energy that occurs during the propagation of acoustic waves through a liquid or solid medium. The sources of these losses are either associated with losses at the boundary of the medium (viscous shear, heat conduction between medium and walls) or by those due to dissipation within the medium (viscous and thermal losses, losses due to molecular energy exchange conversion of the compressional energy of the medium into internal energy of molecular vibration). For small damping, such losses along the x direction are described by the attenuation constant α, and the pressure amplitude p_A at any point in the medium is $p_A = p_0 e^{-\alpha x}$.

loudness   The magnitude of the subjective physiological sensation produced by a sound. The measurement of loudness involves relating the loudness of a sequence of sounds to the sound pressure level, as well as determining its variation as a function of a single parameter, such as frequency content or duration (to determine loudness functions and equal loudness contours), or differentiating between sounds that vary only in level (loudness discrimination). Loudness functions are influenced by both physical characteristics of the sound (frequency, temporal properties, duration, sound pressure levels, background noise levels) and the auditory system of the listener. Loudness functions describe loudness (expressed in sones) as a function of the sound pressure level (in decibels) for binaural single frequency tones. Equal loudness contours for a set of loudness levels (expressed in phons) are plotted in the frequency-sound pressure level diagram.

loudspeakers   A system of interdependent electromechanical components with a transducer that converts electrical signal energy into acoustical energy that it radiates into a bounded space, such as a room, or into open space. Also known as a *speaker.* A loud speaker system consists of drive element(s), radiation aid(s) (such as baffle, enclosure, horn), and crossover. The drive element or driver is the electro-mechano-acoustic transducer. *Electrodynamic drivers* are the actuator counterpart of electrodynamic microphones. The *enclosure* is the cabinet in which the loud speaker is mounted and the *baffle* is the support structure for the driver. The electronic dividing network that filters and distributes the electrical signal to the drive elements of the loud speaker system is called the *crossover.*

love waves   Horizontally polarized shear (transversal) waves in plates. A horizontal dispersive surface wave in geophysics, multireflected between internal boundaries of an elastic body, applied chiefly in the study of seismic waves in the earth's crust.

lumen   The SI unit of luminous flux (with dimensions of energy/time). It is equal to the amount of light (measured with respect to the visual response of the eye) emitted per second in a unit solid angle of one steradian from a uniform source of one candela. This is approximately equivalent to 0.00146 W of light energy at a wavelength of 555 nm.

luminance   A measure of the brightness of the surface of a light source as measured with respect to the visual response of the eye. More precisely, it is the luminous intensity of the surface of a light source divided by the projection of the area of the surface perpendicular to the line between the surface and the observer. Luminance is measured in SI units of candela per meter^2. *See* luminous intensity.

luminescence   The emission of light by a material substance other than as a result of thermal energy (or incandescence). Luminescent emission in general follows a process in which a substance either absorbs radiation or is bombarded by charged particles, causing excitation of electrons into higher energy states from which they decay by the emission of radiation in the form of light.

luminosity   The magnitude of the visual sensation produced by a source of light, measured in terms of luminous flux emitted or reflected from the source. In astronomy, the term refers to the total output of radiation from a celestial object. *See* luminous flux.

luminous efficiency   A dimensionless ratio defined by the luminous flux emitted by a source of light divided by the total rate of electromagnetic energy emitted by the source; it is measured in SI units of lumens per watt. Luminous efficiency measures the relative effectiveness of a source of electromagnetic energy in evoking the visual sensation of brightness. Quantity is alternatively defined as the luminous flux emitted by a light source divided by the total power supplied to the source.

luminous emittance   Luminous flux emitted from a light source per unit area, measured in SI units of lumens per meter^2 (lm/m^2). This contrasts with *radiant emittance,* defined to equal the total radiant energy per second emitted from a source per unit area. *See* luminous flux.

luminous exitance   Luminous flux emitted from a light source per unit area per unit wavelength interval, and measured in SI units of lumens per meter^3 (lm/m^3). This contrasts with *radiant* (or *spectral*) exitance, defined to equal the total radiant energy per second emitted from a source per unit area per unit wavelength interval. *See* luminous flux.

luminous flux   The amount of light passing a given point per second measured in terms of the capacity of light to evoke the visual sensation of brightness. It is measured in SI units of lumen (lm), equal to 1 candela times 1 steradian. (*See* luminous intensity and accompanying definition of *candela.*) The capacity of light to evoke the sensation of brightness in the visual response of the eye depends on frequency or wavelength of radiation as determined by the *relative luminosity* curve shown below.

For a given quantity of radiant energy, the visual response of the eye is a maximum for radiation of wavelength 555 nm. The definition of *candela* equates 1 watt of light of wavelength 555 nm to luminous flux of 685 lm; as a consequence, 1 watt of radiation, with a wave-

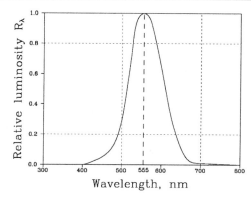

Luminous flux.

length corresponding to a relative luminosity R_λ, equates to a luminous flux of $685 \times R_\lambda$ lm.

luminous intensity   The rate of light emission from a (point) source of light per unit solid angle centered on the direction between the source and the observer, as measured with respect to the visual response of the eye. It is equivalent to the luminous flux per steradian, and measured in SI units of candela (cd), equivalent to 1 lumen per steradian. The candela (in practice) defines the SI unit of the lumen, and is approximately equivalent to the luminous intensity of one candle. More precisely, one candela is defined to equal 1/60-th of the luminous intensity of a square centimeter of the surface of a blackbody at the melting temperature of platinum, 2042 K, measured in the direction perpendicular to the surface.

Lummer-Gehrcke plate   Instrument used for the study of spectral lines. It consists of a few-millimeter thick plate of accurately plane-parallel glass or quartz with an attached prism at one end, through which light enters the plate in a manner such that its angle of incidence at the inner surfaces of the plate is slightly less than the critical angle for total internal reflection. Transmitted rays resulting from internal reflections then leave the surface at nearly grazing angles and are brought to a focus on a viewing screen by a lens so as to produce identical sets of fringes on either side of the plate. *See* accompanying diagram.

Multiple reflections at the interior surfaces of Lummer-Gehrcke plate.

This design has the advantage of a high reflection coefficient near the critical angle even for ultraviolet radiation.

lumped parameter   In circuit analysis, any component, such as inductance, capacitance, or resistance, that can be treated as a single parameter concentrated at a point in an electric circuit. This treatment is only valid for a certain frequency range where the wavelength of the alternating current in the conductors is larger than the dimensions of the component.

Lyman series   A series of lines in the emission or absorption spectrum of hydrogen (or hydrogenic) atoms corresponding to wavelengths $\lambda(= 1/f)$ in the ultraviolet part of the electromagnetic spectrum. Wavelengths of successive lines in series are given by the formula

$$\frac{1}{\lambda} = R\left(1 - \frac{1}{n^2}\right), n = 2, 3, 4, \dots \,,$$

where R is the Rydberg constant with the value

$$R = 1.0973732 \cdot 10^7 m^{-1} \,.$$

Lines in the series correspond to radiation emitted or absorbed in transitions of single electrons in an atom between the (allowed) outer orbits in the atom and the *innermost* orbit in the atom.

M

macromolecules, biological Biological macromolecules are molecules formed by hundreds and even thousands of individual atoms that work together and react as a unit. Their function is prescribed by their particular components.

Biomolecules do not behave as static structures but as dynamic ensembles of charge and shape. Macromolecular assembly and interactions are dictated largely by the electrostatic potential surfaces around the macromolecules and by the innate dynamic behavior and flexibility of specific components. Understanding these features is central to the study of protein and RNA folding, in which the pathways and macromolecular forces that drive folding are being studied.

Examples of macromolecules are polymers built from monomers by condensation reactions. The reverse effect, i.e., polymers broken down into monomers, can be achieved by hydrolysis reactions. In condensation reactions water is removed as monomers are joined, while in hydrolysis reactions water is added to break polymers down into monomer units. Among the most important polymers are the carbohydrates whose monomers are monosaccharides. The general molecular formula for carbohydrates is $C_n(H_2O)_m$. The most common ones have 5 or 6 carbons.

Other examples of macromolecules found in biology include lipids (such as phospholipids, fatty acids, and triacylglycerols), proteins, and amino acids that form DNA and RNA.

macroscopic transport parameters (cell) Transport through the cell membrane is determined by the diffusion constant of molecules and ions that diffuse across the membrane (passive diffusion), the reaction time between molecules and transmembrane proteins (facilitated diffusion), and the rate of energy conversion (active transport).

In *passive diffusion,* molecules directly cross the phospholipid bilayer dissolving in the aqueous solution at the other side of the membrane. The direction of transport is from the high to the low concentration side of the membrane. Only small, relatively hydrophobic molecules are capable of this (e.g., O_2, CO_2, benzene, H_2O, and ethanol). On the other hand, *facilitated diffusion* is carried out by the assisted passage of molecules through the membrane by proteins that enable the crossing without allowing the molecule to interact with the hydrophobic interior of the bilayer. Molecules participating in this diffusion include polar and charged molecules (e.g., carbohydrates, amino acids, nucleosides, and ions). Because *active transport* (ATP) is usually against the concentration gradient of the molecule, it requires the use of energy as provided by the hydrolysis of ATP.

magnetic amplifier This comprises an iron-core transformer with an extra winding to which a control signal can be applied. The amplifier modulates the voltage across the load in an AC circuit.

magnetic analysis The determination of the magnetic characteristics, under either direct or alternating fields, of ferromagnetic alloys can throw light on their phase-structure. Magnetic analysis has been used in identifying components in alloy systems, and in studying the effects of heat-treatment and other physical and mechanical variations.

magnetic axis A line that passes through the effective poles of a magnet.

magnetic balance A type of fluxmeter in which the force required to prevent the movement of a current-carrying coil in a magnetic field is measured.

magnetic blowout A coil used in circuit breakers to deflect any electrical arc formed to lengthen the arc or apply it to a cool surface and thus extinguish it.

magnetic bottle A configuration of magnetic fields used to confine a plasma for long enough for the plasma to react.

magnetic braking   A type of brake in which the brake is activated and deactivated by the action of an electromagnet.

magnetic circuits   These constitute the complete paths for magnetic flux lines. A magnetic circuit is analogous to an electric circuit with magnetomotive forces as the equivalent of voltage, flux density as the equivalent of current, and reluctance as the equivalent of resistance.

magnetic damping   The slowing down of the motion of a conductor when it passes through a magnetic field due to the production of eddy currents.

magnetic disk   Information storage device which encodes information in magnetic bits in a thin layer of magnetic material on the surface of the disk. Data can be read from and written to this storage device. It has faster access time than tape storage since the read-write head can move directly to the position on the disk where the data is stored. Magnetic disks come in two categories: floppy disks that can be inserted into and removed from a disk drive attached to the computer, and hard disks that are permanently installed in a computer. Floppy disks are usually 3.5 inches in diameter and can hold 1.44 megabytes of information after formatting. Hard disks are usually installed in a computer and have capacities in the gigabyte range. Zip disks, which are removable and hold approximately 100 megabytes (or more), of information, are also used for information storage.

magnetic drum   Information storage device in the shape of a drum that is coated with a thin layer of magnetic material and is rapidly rotating. Information is encoded on the outer surface of the drum. Fixed read-write heads which sense the stray field of magnetic bits are used to write and retrieve data. *See* magnetic recording.

magnetic effect of current   Hans C. Oersted discovered in 1820 that electric currents produce magnetic fields. The exact mathematical expression for the magnetic field produced by a line element of an arbitrary current distribution is known as the *Bio-Savart law:*

$$dB = \frac{\mu_o}{4\pi} \frac{i\,ds\,\sin\theta}{r^2} .$$

magnetic element   (1) Small part of a larger magnetic circuit or magnetic material.

(2) A small section of a magnetic circuit broken into elements in a computer model so that magnetic properties can be modeled.

(3) A part of an instrument that a magnetic field acts on.

(4) The three magnetic elements that describe the earth's magnetic field: the *horizontal component,* the *angle of dip* (inclination), and the *angle of declination. See* inclination, magnetic; declination, magnetic.

magnetic equator   A circle on the earth's surface oriented perpendicular to a line joining the magnetic north and magnetic south poles; also referred to as the *aclinic line.* The magnetic equator lies approximately halfway between these two poles. The magnetic equator marks the points on the earth's surface where the vertical component of the earth's magnetic field is zero. The position of this line changes slightly each year as a result of the slow drift of the earth's north and south magnetic poles.

magnetic field   The field of a magnet (or current carrying wire) is the region surrounding the magnet (or current carrying wire) where magnetic forces occur.

magnetic field, given by Faraday's induction law $\oint E.dl = -d\Phi/dt$, where Φ is the magnetic flux linking the closed circuit. Alternatively, $\nabla \times E = -\partial B/\partial t$ by Stoke's theorem. This relationship forms the basis of electrical generators.

magnetic field lines   A set of lines that describe the strength and direction of a magnetic field; also called *lines of magnetic force.* Arrows on the lines point in the direction of the magnetic field. The spacing of lines is inversely proportional to the magnetic field strength. Closely spaced lines indicate a strong magnetic field. Lines never cross. For permanent magnets, the lines of magnetic field emerge from the north pole and enter the south pole of the magnet. For a straight current-carrying wire, the lines form circles about the wire perpendicular to the length of the wire. *See* magnetic field of long straight conductor.

magnetic field of a conductor   The magnetic field of a conductor is determined by the electric current flowing in it. It can be calculated using Biot-Savart's law or Ampere's law in cases of high symmetry. *See* Ampere's law; Biot-Savart's law; magnetic field of a long straight conductor.

magnetic field of circular loop   For a circular loop of wire of radius a, carrying current i, lying in the x-y plane with its center at the origin, the magnitude of the magnetic field along the z-axis is given by

$$\mu_o i a^2 / 2 \left(a^2 + z^2 \right)^{3/2} .$$

μ_0 is $4\pi \times 10^{-7} T.m/A$. If the electric current circulates anticlockwise as seen from above, the magnetic field on the z-axis points in the $+z$ direction. *See* right-hand screw rule; magnetic field of conductor.

magnetic field of displacement current
Displacement current arises from a changing electric field and gives rise to a magnetic field. The value of the displacement current through a certain area is given by $\epsilon_o d\Phi_E/dt$ where ϵ_0 is 8.85×10^{-12} farad/meter and the second term is the rate of change of electric flux through the area of interest. This magnetic field may be calculated using Biot-Savart's law or Ampere's law (cases of high symmetry) once the value of the displacement current is known. *See* Biot-Savart's law; Ampere's law; displacement current.

magnetic field of long straight conductor
The magnetic field strength a distance r from a long straight wire carrying current i is

$$\mu_o i / 2\pi r .$$

Its direction is given by the *right-hand screw rule* with the magnetic field lines circulating anticlockwise and forming circles when the wire

is viewed end-on with the current coming out of the page. *See* right-hand screw rule; magnetic field of conductor.

magnetic field, vector nature   The magnetic field is a vector. It has a direction as well as a magnitude. The magnetic field vector points away from any nearby north magnetic poles and toward any nearby south magnetic poles. The vector is also along magnetic field lines at any point in space. *See* magnetic field lines.

magnetic field, work done by   The work done by a magnetic field on a magnetic material in a complete cycle is given by the area of the hysteresis loop. This energy is dissipated in the magnet and is referred to as *hysteresis loss*. Mathematically it is the integral of HdM. This quantity tends to be small for soft magnets such as those used in transformers and large for permanent magnets such as those used in electric motors. These losses come about due mostly to pinning of domain walls which hinder reversal of the magnetization. *See* hysteresis loss.

magnetic force between parallel conductors   The magnitude of the magnetic force per unit length (F/L) between two parallel wires separated by a distance d and carrying currents i_1 and i_2 is $\mu_0 i_1 i_2 / 2\pi d$. The force is attractive if the currents are parallel and repulsive if the currents are antiparallel. The source of this force can be thought of in the following way: One of the currents creates a magnetic field at the position of the second current. This magnetic field then exerts a force on the second current with the result that the first current exerts a force on the second current.

magnetic force on a conductor   The magnetic force on a conductor-carrying current i can be determined from

$$dF = i d\mathbf{L} x \mathbf{B}$$

where $d\mathbf{L}$ is a small length of conductor whose direction is along the current, and $\mathbf{B}$ is the applied magnetic field. For a long straight wire of length L in magnetic field $\mathbf{B}$ the force is

$$i \mathbf{L} x \mathbf{B} .$$

magnetic force on complete circuit A magnetic field exerts a force on parts of a circuit carrying a current. For a complete circuit in the presence of a uniform magnetic field, these forces add up vectorially to give zero. If a part of the circuit is movable, the magnetic field may cause that part of the circuit to move. *See* magnetic force on a conductor.

magnetic force on moving charge The magnetic force on a moving charge q is perpendicular to both the magnetic field **B** and the particle velocity **v** and is given by

$$q\mathbf{v}x\mathbf{B} .$$

Thus if the magnetic field is parallel or antiparallel to the velocity no force is exerted. If the magnetic field is uniform and perpendicular to the particle velocity, the particle will execute circular motion. If the magnetic field is not perpendicular to the particle velocity, then the particle will trace out a spiral pattern. Note that the force will reverse direction if the sign of the charge q is reversed.

magnetic induction Also called magnetic flux density. Commonly denoted by **B**, has CGS units of Gauss and MKS units Tesla or Webers/meter^2, and is a vector quantity. One Tesla is equal to 10^4 Gauss. It is determined by the applied magnetic field (or magnetic intensity) H and the magnetic moment per unit volume M of the medium. In CGS units

$$B = H + 4\pi M$$

and in MKS units

$$B = \mu_o(H + M) .$$

See Tesla.

magnetic intensity Also known as magnetic field strength. Commonly denoted by H, has CGS units of Oersteds and MKS units amps/meter and is a vector quantity. It is determined by source of the magnetic field only, and does not depend on the medium. One Oersted is equal to $1000/4\pi$ amps/meter. *See* Oersted.

magnetic leakage Phenomenon by which magnetic field lines leak out of a magnetic material. Can occur at cracks or imperfections in the surface of a magnetic material. Also refers to a magnetic field in the airgap in a magnet where magnetic field lines bow outward weakening the magnetic field in the airgap. Magnetic leakage generally reduces the efficiency of operation of a device. It also can be used to detect flaws such as cracks in materials that are magnetized, since imperfection changes the distribution of magnetic flux. *See* magnetic field lines.

magnetic lens A magnetic field arrangement designed to focus or guide moving charged particles. This magnetic field arrangement usually has axial symmetry. The magnetic field can be generated by current-carrying coils and permanent magnets. The path followed by the charged particles depends on their velocity and charge as well as on the magnetic field configuration of the magnetic lens. Thus only charged particles of a pre-selected charge and energy are guided or focused by a selected magnetic field configuration. *See* magnetic force on moving charge.

magnetic meridian An imaginary line that passes overhead from the magnetic south pole to the magnetic north pole of the earth. This line follows the direction of the horizontal component of the earth's magnetic field.

magnetic mirror A configuration of magnetic fields that reflects a moving charged particle impinging on it. It can be used to confine a plasma and is used to confine the plasma in controlled fusion experiments. A simple mirror would consist of a magnetic field along the z-axis that got stronger with increasing values of z. A particle moving with a component of its velocity along the $+z$ direction, will spiral along the field line due to the magnetic force on it and can be reflected. This is less effective for particles whose direction of travel lies close to the $+z$ direction and some of these may not be reflected. More complex magnetic field configurations are possible and are used to more efficiently reflect and confine charged particles. *See* magnetic lens. *See also* magnetic force on a moving charge.

magnetic moment A property of a permanent magnet, a current-carrying circuit, or material that has an induced magnetic moment in the presence of a magnetic field (a diamagnetic or paramagnetic material). Commonly denoted by m, has CGS unit of EMU and MKS units of amp.m^2. One EMU is equal to 10^3 amp.m^2. It is the sum of the moments of the atoms or molecules making up the material in the case of permanent magnets where all the atomic moments are aligned. Equal to (current $\times$ area) for a current-carrying circular coil.

magnetic monopole   Isolated north or south magnetic pole analogous to isolated positive and negative electric charge. Predicted to exist by some symmetry elementary particle theories but so far have not been found to exist in nature despite a number of experimental searches. If these particles exist the equations describing electric phenomena would have the same structure as those describing magnetic phenomena. Maxwell's equations, which describe electromagnetism, would be symmetric with respect to electric and magnetic fields.

magnetic needle   A needle made up of a magnetic material that can be thought of as a simple bar magnet. This needle is pivoted at its center so that it is free to rotate and line up with a magnetic field. Commonly used in a compass which is used to determine the direction of magnetic north on earth.

magnetic poles   North and south poles of a magnet. Always found to exist in pairs. Magnetic field lines emerge from the north pole and enter the south pole of the magnet. Like poles are repelled and unlike poles are attracted to each other. If a bar magnet is cut in half, a new north and south pole appear on either side of the break with the two new magnets each having one north and one south pole. *See* magnetic field lines.

magnetic potential   A scalar magnetic potential useful in the area of magnetostatics. The difference in magnetostatic potential between an initial (i) and final (f) position for a magnetic pole is defined as the work done in moving a unit magnetic pole from i to f. Mathematically, it is given by the integral of $-Hdl$ between these two points. H is the magnetic field intensity and dl is an infinitesimal path length.

magnetic quantization   In the presence of a magnetic field a moving charged particle has only certain allowable energies, referred to as *Landau levels*. Effects of quantization are only observable at low temperatures. It can lead to de Haas-van Alphen oscillations of diamagnetic moment of a material, and Shubnikov-de Haas oscillations of the electrical resistance of a material in the presence of a changing magnetic field. These oscillations are related to electronic structure — in particular, the size and shape of the Fermi surface — and are a useful probe of this structure.

magnetic recording   Recording of information in the magnetization of a material. The magnetic material is usually made up of gamma-Fe$_2$O$_3$ particles in a binder that is coated onto a support structure such as a flexible plastic tape or a solid plastic disk. The recorded magnetization is usually in the plane of the material. This magnetic information is read using a read head whose primary component is a small sensing coil that senses the stray field from the recorded magnetization. It can also be read using a laser beam of polarized light with the polarization of the reflected light being rotated depending on the magnetization of the medium. The magnetic information can be written using a small applied field that can be created by the sensing coil. Information can be in analog form such as is used in a simple tape recorder where the magnitude of the recorded magnetization is proportional to the signal generating it. It can also be digital such as with computer magnetic disks and computer tapes. In this latter case, the information is stored as magnetic bits, corresponding to ones and zeros. *See* magnetic disk; magnetic tape; magneto-optic.

magnetic resistance   Reluctance of a magnetic circuit. *See* reluctance.

magnetic saturation   State of a magnetically ordered material in which all the atomic magnetic moments are aligned. No domain walls are present and the material can be thought of as one large domain. An external magnetic field is often applied to get complete alignment. Application of a larger magnetic field than is required for saturation may only increase the magnetic moment by a small amount due to induced magnetic contributions.

magnetic screening   The screening of a region of space so that no magnetic field enters this region from outside. Materials with a high magnetic permeability are used to screen the region. The most widely used is mumetal, an alloy consisting mostly of Ni and Fe. *See* permeability, magnetic.

magnetic shell   A thin sheet of magnetic material in the shape of a sphere whose inner and outer surfaces have equal pole densities of opposite sign. It can also refer to the partially filled electron shell of an atom or ion where the spin and/or orbital angular momenta of electrons may not be zero so that a net atomic magnetic moment exists. In the rare-earths, the 4f sub-shell is the magnetic shell. It is partially filled and the electrons of this shell have a net spin and orbital magnetic moment that add to give the total moment of the rare-earth atom. In the special case of Gd there is only a spin magnetic moment. This idea does not apply to transition metal elements since the 3d electrons giving rise to the magnetism are not localized in an atomic shell but rather are distributed in the conduction band.

magnetic susceptibility, measurement below 1°   For paramagnetic salts, the magnetic susceptibility is temperature sensitive and provides a thermometric parameter. However, discrepancies between the thermodynamic temperature and the "magnetic" temperature occur at temperatures below 1 K. For a widely used thermometer salt such as CMN, the thermodynamic and magnetic temperature differ by less than 1% down to 6 mK for single crystals and powder specimen. For salts such as CPA, CMA, and FAA the differences in the two temperatures become significant at temperatures of 0.4 to 1 K.

Magnetic susceptibility is the ratio of the intensity of magnetization produced in a material to the intensity of the magnetic field to which the material is subjected; it measures the amount of magnetization of a substance by an applied magnetic field. Diamagnetic materials have small negative susceptibilities while paramagnetic materials have small positive susceptibilities, and ferromagnetic materials have large positive susceptibilities. The magnetic susceptibility varies linearly with $1/T$ in accordance with Curie's law above 1 K; however, it becomes almost temperature independent at very low temperatures since it behaves like a Fermi gas.

magnetic tape   A flexible plastic tape (often mylar) coated with a magnetic material, usually particles of $\alpha-Fe_2O_3$, on which information may be stored in the form of magnetic bits. Each magnetic bit contains many $\gamma-Fe_2O_3$ particles and the magnetization of the magnetic bit lies along the length of the tape. *See* magnetic recording.

magnetic torque   Torque τ exerted by a magnetic field **B** on a magnetic moment μ. Given by

$$\tau = \mu x \mathbf{B} .$$

Tends to align magnetic moments along an applied magnetic field. The earth's magnetic field exerts a torque on a compass needle since the compass needle is magnetized. This aligns the compass needle along a magnetic north-south direction.

magnetic viscosity   Phenomenon by which magnetization of a material changes with time. This usually occurs when the applied magnetic field is changed to a new value and held at that value. In some materials — for example, Tb-Co amorphous alloys — energy barriers are hindering magnetic reversal and a magnetic viscosity results when the magnetic system overcomes these barriers by thermal activation. At very low temperatures (below 10 K), mesoscopic quantum tunneling may also lead to a magnetic viscosity in this and similar systems. This type of magnetic viscosity is always accompanied by magnetic hysteresis. Magnetic viscosity may also be observed in materials where the structure is changing with time, such as in Fe with a small amount of carbon.

magnetic well   A configuration of magnetic fields designed to contain a plasma. Two magnetic mirrors designed to reflect moving charged particles would serve as a simple magnetic well. A magnetic well is used in fusion experiments to contain a plasma that has a very high temperature. *See* magnetic lens; magnetic mirror.

magnetism, molecular theory of   In the molecular theory of magnetism (also called the *Weiss theory of magnetism* or *mean field theory*) it is assumed that the exchange interaction usually described by $-J\mu_i \cdot \mu_j$ can be replaced by its average value. (J is the exchange constant and the μs represent neighboring magnetic moments.) When summed over nearest neighbors ($\mu_i's$), this average value is proportional to the magnetization M of the material, and the term is written as $\lambda M \mu_i$. This theory qualitatively describes many aspects of magnetism in three dimensions but ignores the effects of thermal fluctuations of magnetic moment below the magnetic ordering temperature. It leads to incorrect predictions for details of the magnetic transition, such as the critical exponents associated with magnetization as a function of applied field and magnetic susceptibility as a function of temperature. In lower dimensions where thermal effects on magnetization are stronger, the failures of this theory are more apparent.

magnetization   Magnetic moment per unit volume within a material as a result of the magnetic polarization of the material. This results from the alignment of permanent atomic magnetic moments in the case of Fe, Co and Ni. Contributions from induced magnetization such as diamagnetism are also present and are usually much smaller than the atomic moment contribution. *See* magnetic moment; diamagnetism.

magnetization curve   A plot of magnetization (or magnetic induction) as a function of applied magnetic field. Used to find important magnetic parameters describing a magnetic material. These include the saturation magnetization and the coercive force. Differences in these curves when the magnetic field is increased and then decreased indicate the presence of magnetic hysteresis. *See* coercivity; magnetic saturation; hysteresis.

magnetization, intensity of   Given by $\mu_o M$ where M is the magnetization per unit volume and has a unit of Tesla. Only defined for MKS system of units.

magnetizing current   The electric current that flows through a coil surrounding a core, usually made of a soft Fe alloy, and which establishes an applied magnetic field in the core. This applied magnetic field magnetizes the core. The coil and core make up the main part of an electromagnet. *See* electromagnets.

magneto-accoustic emission   This involves the study of the propagation of sound waves in metals in the presence of a magnetic field. At low temperatures, the interaction of sound waves with electrons is the primary source of attenuation of the sound wave. This attenuation is modified by the presence of a magnetic field and this effect can be used to probe electronic structure. It can also refer to accoustic energy generated by changes in magnetization, and can be associated with strains and magnetostriction. *See* magnetostriction.

magnetohydrodynamic wave   Electromagnetic waves in a plasma coupled to an oscillation in the plasma density in the presence of a magnetic field. Frequency is usually low — less than the cyclotron frequency of the charged ions in the plasma. Important in the earth's ionosphere and in the various layers of the sun.

magnetometer   A device used to measure a magnetic field such as the earth's field, or the magnetic field created by the magnetic moment of a magnetic material. Various devices exist to perform this function and operate using a number of different principles. It can measure the magnetic moment of a material in a vibrating sample magnetometer. In this device, a piece of the material is mechanically vibrated in a coil and the induced voltage from the stray field of the material is measured and is proportional to the magnetic moment. It must be calibrated using a known magnetic standard and can sense magnetic moments as small as 10^{-5} EMU. It can also measure a magnetic field by an induction method through the movement of a coil, by rotating it, for example in a magnetic field. It can also monitor a property that changes with the magnetic field such as the Hall voltage across a material induced by the magnetic field.

magnetometer, impedance   The ratio of induced voltage to current in pick-up coils of a vibrating sample magnetometer. In general, it is a complex quantity. If the capacitive and inductive parts of the impedance cancel, then the total impedance is real and the induced voltage and current are in phase. It is an important quantity since detection electronics must be matched in impedance to the pick-up coils in the magnetometer. *See* magnetometer.

magnetometer, Q   A measure of losses. Larger Q corresponds to smaller losses. *See* magnetometer.

magnetomotive force   The magnetomotive force around a complete magnetic circuit is defined as the work required to move a magnetic pole of unit strength once around the magnetic circuit. *See* reluctance, magnetic.

magneto-optic   Interaction of light with a magnetic material. In the *Faraday effect,* the polarization direction of polarized light is rotated when the light passes through a transparent material in the presence of a magnetic field along the direction of propagation. In the *Kerr effect,* polarized light has its direction of polarization rotated on reflection from a ferromagnetic material. This latter effect forms the basis for magneto-optical recording. A number of other magneto-optic effects exist including the *Zeeman* (normal and anomalous) *effect,* where spectral lines are split by a magnetic field, the *Voight effect,* where an anisotropic substance placed in a magnetic field becomes birefringent, and the *Cotton-Mouton effect* in which double refraction of light in a liquid in the presence of a magnetic field occurs. *See* Paschen-Back effect; magnetic recording.

magnetoresistors   Material whose resistance changes when subjected to a magnetic field. Observed with the magnetic field parallel or perpendicular to the electric current. Observed in many magnetic materials. The increase in resistance is a few percent or less in most alloys studied when fields up to 5 Tesla are applied. It can be large in semiconductors and in metals. High field magnetoresistance yields information on the electronic structure of metals, in particular the fermi surface shape. At low temperatures quantum oscillations of the magnetoresistance, called *Shubnikov-de Haas oscillations,* are observable in single crystal metals and yield more detailed information on the fermi surface. More recently nanostructured magnetic materials (particles or layers with dimensions of tens of nanometers) have been discovered that show a giant magnetoresistance — the resistance can change by more than 100% when the material is magnetized. This forms the physical basis for a read head based on a magnetoresistance for magnetic recording.

magnetostatic energy   Energy stored in a magnetic field or energy required to create a magnetic field. Energy density (energy per unit volume) is

$$B^2/2\mu_o \, .$$

To get the magnetostatic energy, this energy density must be summed (integrated) over the volume containing the magnetic field.

magnetostriction   Compressive or extensive stress in a magnetic material when its magnetization is changed by, for example, placing it in a magnetic field. This leads to a change in a dimension of a material when its magnetization is changed. The strain (the fractional change in a dimension of a material) is typically small — of the order of 10^{-5} or less in going from the unmagnetized state to the saturated state. Magnetostriction significantly influences the type of domain pattern in a material. *See* magnetostriction oscillator.

magnetostriction oscillator   A device in which one or more dimensions of a magnetostrictive material oscillate. An oscillating applied magnetic field leads to oscillations of the magnetic moment of the material. One or more dimensions of the material oscillate in response to the oscillating magnetic moment of the material converting magnetic energy to mechanical energy. This device is easily realized in principle by placing a coil with an alternating current around a soft magnetostrictive magnet. The alternating current provides the oscillating magnetic field. It can be used to generate a sound wave in the audible to ultrasonic frequency range in any medium that is physically in contact with the magnetic material. *See* magnetostriction.

magnetron   A device that converts DC electrical energy to microwave energy with high efficiency. Electrons emitted by a cathode toward an anode interact with crossed electric and magnetic fields in a cavity to produce these microwaves. The shape and size of the cavity determines the frequency of operation. It can operate in a pulsed mode generating microwave power for radar applications, or continuous mode for microwave cooking.

magnification   The magnification of an optical system indicates the effectiveness of enlarging or reducing an image. There are several kinds of magnification: *lateral magnification* of an image, *axial magnification* of an image, or *magnification of the magnifying power* of an optical instrument. It is important which magnification should be considered for use to treat optical magnification. The term *magnification* is sometimes used simply to mean lateral magnification or the power of a lens without qualification.

magnification, angular   The symbol used is M or γ. Angular magnification of an optical system is the ratio of the angles subtended at the eye by the image *theta'* and object *theta*. It can be obtained approximately by the ratio of the tangent function value of the angles for smaller angles. The angular magnification of a lens M_θ and power of the lens P have a relationship $M_\theta = \frac{1}{4}P$. This equation is known as the *quarter-power equation.* It applies to the lens that is used without accommodation; an image is seen at infinity and the lens is set close to the eye. With accommodation, $M_\theta = \frac{1}{4}P + 1$. It is also known as the *converging ratio.*

magnification, axial   Also called *longitudinal magnification.*   For an object with depth (e.g., three-dimensional), the magnification along the optical axis should be considered. The axial magnification is the ratio of length along the optical axis to the conjugate length in the object. The axial magnification M_x is defined as the ratio of a short length in the image measured along the optical axis to the conjugate

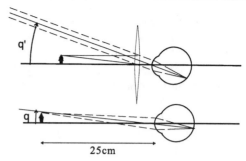
Angular magnification.

length in the object:

$$M_x = \frac{dx_i}{dx_o},$$

where x_i and x_o are the distance between the image and the focal point and the distance between the object from the focal point, respectively.

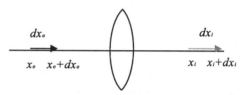
Axial magnification.

The axial magnification M_x for small distances from the focal plane is equal to the square of the lateral magnification M_l : $M_x = M_l^2$.

magnification, lateral   The ratio of the size of an image perpendicular to the optical axis y' to the size of the object perpendicular to the axis y. It is often simply called *magnification.*

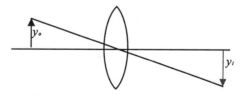
Lateral magnification.

The lateral magnification can be obtained as the ratio of object distance s' to the image distance s.

magnification, longitudinal   *See* magnification, axial.

magnification, normal   The multiplication of the primary magnification and the magnification of ocular. The normal magnification multiplied by the tube factor is the general magnification of the instrument.

magnification of lens   For the refraction of a thin lens, the relationship of the image distance s, object distance s', and the focal length f, can be written as

$$\frac{1}{s} + \frac{1}{s'} = \frac{1}{f},$$

$$\frac{1}{f} = (n' - n)/n \left(\frac{1}{R} + \frac{1}{R'} \right),$$

where R and R' indicate the radii of the surfaces. For a concave lens, f is less than zero. The lateral magnification m can be calculated as

$$m = -\frac{s'}{s}.$$

The magnifying power P of a thin lens is the algebraic sum of the pair of surfaces of the lens $P_1, P_2; P = P_1 + P_2$. The angular magnification of a lens M_θ and power of the lens P have a relationship $M_\theta = \frac{1}{4}P$. This equation is known as the *quarter-power equation*. It applies to the lens used without accommodation; the image is seen at infinity and the lens is set close to the eye. With accommodation, $M_\theta = \frac{1}{4}P + 1$.

magnification of mirror   For the reflection of a mirror, the relationship of the image distance s, the object distance s', the radius of the surface of mirror R, and the focal length f, can be written as

$$\frac{1}{s} + \frac{1}{s'} = \frac{1}{f},$$

$$f = -\frac{R}{2}.$$

For a concave mirror, f is greater than zero. The lateral magnification m can be calculated as

$$m = -\frac{s'}{s}.$$

Where the refractive index of a given medium is n, the refractive power of a mirror P is defined as

$$P = \frac{-2}{R}.$$

magnification of optical instruments   A measure that indicates how much an optical instrument enlarges or reduces the image of an object. *See* magnifying power.

magnification, primary   The transverse magnification provided by the objective lens. The magnified real image formed by the objective lens is called the primary image. In a compound microscope, the image formed by the objective is an inverted image of the object. The primary magnification m_p is a lateral magnification:

$$M_p = \frac{-L}{f},$$

where the distance of the image and the focal point of the objective lens is L, and the focal length of the objective lens is f.

magnifications in vibrations   The magnitude of the transfer function of a periodically forced vibrating system. It describes the magnitude of a measured output signal, such as displacement, velocity or force, transmitted to the base for the system excited with a sinusoidal direct-force excitation of magnitude M and frequency ω. For a damped, second order, one degree of freedom system, for example, the peak response of the transfer function is usually at the resonant frequency $\omega_r = \omega_n \sqrt{1 - 2\zeta^2}$ with ω_r as the resonant frequency in rad/sec, ω_n as the undamped natural frequency in rad/sec, and ζ as the non-dimensional damping ratio.

magnification, transverse   *See* magnification, lateral.

magnifier   *See* microscope, simple.

magnifying power   *See* magnification, angular. It is also known as instrument magnification. This is the ratio of the size of the retinal image of an object formed by an optical system to the size of the retinal image of the object seen with the unaided eye at the least distance of distinct vision (normal viewing distance). The size of the retinal image can be measured as the angle subtended at the eye by the image. The image of the object seen with the object means the *in situ* image for telescopes, and the image at the conventional distance of distinct vision, i.e., 25 cm from the eye, for a microscope. In a compound microscope, the magnification power is the ratio of the angle subtended at the eye to the real image that is formed in space on the plane of the field stop of the eyepiece within the microscope tube. For projection with a microscope, the magnification by the eyepiece should be treated as a lateral magnification. For visual observation, it should be an angular magnification, in a telescope, the real image formed ahead of the first focal length of the eyepiece. *See* magnification of optical instruments.

Malus' law   A law (first published in 1809 by Malus) determining the intensity of radiation transmitted through a polarizer in terms of the incident intensity. Malus' law equates the transmitted intensity, I, to the product of the incident intensity, I_0, and the square of the cosine of the angle θ between the polarization vector of the incident radiation and the direction of the plane of transmission of the polarizer, as expressed by the equation

$$I = I_0 \cos^2 \theta \,.$$

mammography   A type of X-ray examination of the breasts specifically tailored for the detection of tumors. A *mammogram* is an X-ray image of the breast that comes out of the exam.

manostats   Pressure-control devices used for controlling the vapor pressure of a boiling liquid to within certain limits. A stable manostat uses a capacitance pressure sensor and operational amplifiers that activate solenoid valves to control the vapor flow. Pressure stability of better than 0.1% is achieved with ^{4}HE from 2 to 1000 Torr, corresponding to variations of < 1 mK.

maser   Acronym for *microwave amplification by stimulated emission of radiation*. More precisely, a device (invented in 1955) that produces a narrowly directed beam of coherent monochromatic radiation with a (central) frequency in the microwave region of the electromagnetic spectrum between 10^9 and 10^{11} Hz. Analogous to laser (invented later) but operating at microwave rather than optical frequencies. The first maser operated at the "inversion transition" frequency of the ammonia molecule, 2.387×10^{10} Hz.

maser in communication   A maser designed to amplify microwave signals (as from artificial satellites) used in communications. It amplifies FM signals from a satellite by the stimulated emission of tunable Zeeman separated lines in ions of paramagnetic crystals (such as ruby). The maser has the advantage of amplification with extremely low noise.

masking, acoustic   A number of decibels by which the listener's threshold of audibility for a continuous sound is changed by the presence of another sound, called the *masking sound*. When the threshold of audibility of the original tone alone is a decibels, and the threshold of the same tone in the presence of the masking tone is b decibels, the masking of the original tone by the second one (characterized through its level in decibels) is $b - a$. Increasing the intensity of the masking sound to a level at which the original sound ceases to be audible results in the masking by the other sound. The masking effect is more pronounced above the frequency of the original tone than below. Noise can also cause masking. Masking will be perceived as beats when the frequencies of the two tones are close.

mass attenuation coefficient   In general, an attenuation coefficient is a measure of the rate of decrease of an average power with respect to a distance along the transmission path.

Given a material of thickness d, density ρ, a narrow beam of monoenergetic photons incident on one side with intensity I_0, and emerging on the other side with intensity I, and defining the mass thickness $l = \rho \cdot d$, the decay in intensity is given by the exponential attenuation law

$$I/I_0 = \exp[-(\mu/\rho) \cdot l] \,.$$

Rewriting the expression, we get the attenuation coefficient

$$\mu/\rho = l^{-1} \cdot \ln(I_0/I) \,.$$

Then the mass attenuation coefficient can be obtained from experimental measurements of I_0 and I.

Some values of μ/ρ rely on the theoretical values for the total cross section per atom, σ_{tot}, which is related to μ/ρ by

$$\mu/\rho = \sigma_{tot}/(u \cdot A) \,,$$

where u is the atomic mass unit, A is the relative atomic mass of the target element, and σ_{tot} is the total cross section for an interaction by the photon. In general, the attenuation coefficient, photon interaction cross sections, and related quantities are functions of the photon energy.

An experiment that relies on the mass attenuation coefficient is X–ray computed tomography (CT), where the reconstructed image represents the distribution of X-ray photon attenuation coefficients from the body under examination. An interesting problem arises because the X–rays used are not monochromatic. In this case the attenuation coefficient distribution has to be reconstructed at a certain effective energy of the beam. However, the highly non–linear dependence of the attenuation coefficient on photon energy results in systematic inaccuracy in the reconstructed image, known as the *beam hardening artifact*.

mass energy absorption coefficient   In the context of the scattering event described in the *mass energy transfer coefficient,* the mass energy absorption coefficient involves the further emission of radiation produced by the charged particles in traveling through the medium, and is defined as

$$\mu_{en}/\rho = (1-g)\mu_{tr}/\rho \,,$$

where μ_{tr}/ρ is the mass energy transfer coefficient. The factor g represents the average fraction of the kinetic energy of secondary charged particles (produced in all the types of interactions) that is subsequently lost in radiative energy–loss processes as the particles slow to rest in the medium. *See* mass energy transfer coefficient.

mass energy transfer coefficient   In the context of the interaction of photons with matter, the mass energy transfer coefficient μ_{tr}/ρ, when multiplied by the photon energy, is proportional to the sum of the kinetic energies of all the primary charged particles released by uncharged particles (here photons) per unit mass. In other words, μ_{tr}/ρ takes into account the transfer of

radiation coming from secondary photon radiations initially produced at the photon–atom interaction site, plus the quanta of radiation from the annihilation of positrons originating in the initial pair– and triplet–production interactions.

Hence μ_{tr}/ρ is defined as the sum of all the contributions coming from the total cross sections from photoelectric absorption, incoherent scattering, pair and triplet production. Then

$$\mu_{tr}/\rho = \left(f_{pe}\sigma_{pe} + f_{incoh}\sigma_{incoh} \right. $$
$$\left. + f_{pair}\sigma_{pair} + f_{trip}\sigma_{trip}\right)/(u \cdot A) \,,$$

where the factors f refer to the energy–transfer fractions, and σ to the individual cross sections. The factors f represent the average fractions of the photon energy E that is transferred to kinetic energy of charged particles in the remaining types of interactions. In particular,

$$f_{pe} = 1 - (X/E) \,,$$

where X is the average energy of fluorescence radiation emitted per absorbed photon;

$$f_{incoh} = 1 - \left(\langle E' \rangle + X\right)/E \,,$$

where $\langle E' \rangle$ is the average energy of the Compton–scattered photon;

$$f_{pair} = 1 - 2mc^2/E \,,$$

where mc^2 is the rest energy of the electron; and

$$f_{trip} = 1 - \left(2mc^2 + X\right)/E \,.$$

The fluorescence energy X depends on the distribution of atomic–electron vacancies produced in the process under consideration and is in general evaluated differently for photoelectric absorption, incoherent scattering, and triplet production. X should include the emission of "cascade" fluorescence X–rays associated with the complete atomic relaxation process initiated by the primary vacancy.

Because in calculating μ_{tr}/ρ, only the characteristics of the target atom are involved, for homogeneous mixtures and compounds, the total μ_{tr}/ρ can be obtained by

$$\mu_{tr}/\rho = \sum_i w_i \left(\mu_{tr}/\rho\right)_i$$

where w_i is the fraction by weight of the ith atomic constituent.

mass fragmentography   Analysis and identification of chemical substances by the study of parts or fragments of the whole compound. *See* mass spectrometry, medical applications.

mass spectrometer   A device used to separate out particles, usually atoms or molecules, according to mass. Particles are accelerated and enter a region where a magnetic field B perpendicular to the velocity of the particle is present. The particles follow a curved path due to the magnetic force on them. The radius of the curved path depends on the mass m, charge q, and velocity v of the particle according to

$$mv/qB .$$

Particles with larger mass follow paths with a larger radius allowing different masses to be selected. *See* magnetic force on a moving charge.

mass spectrometry, medical applications
By mass spectrometry, chemical substances can be identified by sorting gaseous ions in electric and magnetic fields. A *mass spectrometer* is a device that performs this type of sorting by using electrical means to detect the sorted ions. Devices that use photographic or other nonelectrical means are called *mass spectrographs*.

Mass spectrometry allows precise measurement of the mass of ions, to show the presence of different isotopes, and to measure the relative abundance of ions and isotopes in a mixture. Analysis may reveal that organic chemicals have produced a spectrum of ions from the fragmenting of the parent molecule. Then, by identifying the fragments according to their masses and relative abundances, the structure of the original molecule can be established.

Compounds such as H^2, C^{13}, N^{15}, O^{17}, and O^{18} have an enhanced proportion of isotopes that make them ideal to label substances involved in biological processes, and thus appropriate for mass spectrometry measurements. This tagging allows for precise chemical studies of such complex reactions as metabolism, photosynthesis, plant respiration, enzymatic reactions, phosphate-transfer reactions, and the direct application of oxygen in physiological ox-

idation. The details of the metabolic pathways involved in these reactions can be understood by analyzing the products from such processes by mass spectrometry.

matter waves   Waves associated with a particle of matter, as described by quantum mechanics. Also known as *de Broglie's waves*. It has been shown that particles with momentum p act like waves with a de Broglie wavelength λ, given by $\lambda = h/p$, where $h = 6.626 \times 10^{-34}$ J·s is the Planck constant. According to de Broglie's theory, particles of matter have wavelike properties that can give rise to interference effects, and electrons in an atom are associated with standing waves on a Bohr orbit.

maximum ratings   The maximum value of an input that a device can accept with no damage, or the maximum value of an output that a device will provide.

Maxwell bridge   An electric network designed for accurate inductance measurements. A schematic diagram is shown. Typically R_2 and C_3 are adjusted to achieve balance. An equation of balance is given as

$$L_x = R_1 R_3 C_3$$
$$R_x = R_1 R_3 / R_2 .$$

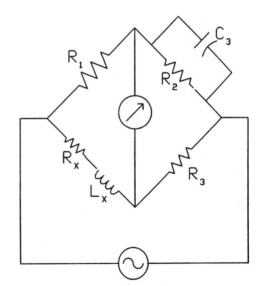

Maxwell bridge.

mean free path of sound   The average distance sound travels between successive reflections in an enclosure. Property that quantifies the propagation of sound in an enclosure.

mechanical properties, bone   Bones are a dynamic system where there is a continuous replacing of old cells with new ones. Bones regulate themselves so as to remove cells that are no longer functional and at the same time serve as the regulator of the amount of calcium in the blood. On the other hand, there are other parts of the bone that do not get renewed, like the bone's shaft. The cyclic discarding and renewal of cells is performed by two types of bone cells: the *osteoblasts,* which manufacture new bone tissue, and the *osteoclasts,* which dispose of old and worn cells.

The bones get their full strength and flexibility from the connective tissue that surrounds and intermingles with it, serving also as the connection between different bones. Vitamin D is known to be essential to the deposition of minerals in the bone structure. These minerals are chiefly calcium and phosphorus. Also, vitamin C, probably most effective as calcium, magnesium, manganese, and zinc ascorbates, is absolutely essential to the connective tissue's strength, flexibility, and endurance.

On a per weight basis, healthy bones are as strong as steel. The bone's interior is constructed somewhat like a bridge. Tiny strands of connective tissue, each strand capable of supporting a weight of 25 lbs, act like crossed wires to give great strength and flexibility. Thus, healthy bones will usually bend rather than break, as they so commonly do.

Of course, the total amount of force that a bone can sustain is dependent on the geometrical properties of the bone in consideration. For example, compression force properties of the femur indicate that human femur subjected to compression splinters at 1600 pounds/in^2. This is about twice the force exerted on the leg bones of a 160 pound runner and far more than current weightlifters can heft.

mechanical waves   Within the scope of acoustics, mechanical waves are defined as vibrations of rigid or elastic solid bodies, with the spectrum of vibration frequencies in the acoustic range. Typical examples are high-frequency elastic waves in delay lines or in nondestructive testing equipment, ground vibrations near factories, forge hammers, sound transmission through walls, ceilings, and enclosures, etc.

mechano-caloric effect   Refers to the transfer of heat and thereby the presence of a temperature difference from an imposed pressure difference. It is the inverse process of the *thermomechanical effect* where an imposed temperature difference causes a pressure difference.

medium, acoustic   Acoustic disturbances can be treated as small amplitude perturbations to the ambient state. The ambient state for a fluid, characterized by the pressure, density and velocity of the fluid when the perturbation is absent, defines the medium through which sound propagates. An acoustic medium is homogeneous when all ambient quantities are independent of position.

medium, homogeneous   A switching network is considered homogeneous if every connection between an inlet and an outlet uses the same number of crosspoints.

megaphone   A rectangular or conical horn used for amplifying or directing the sound of the speaker's voice. *See also* horns, sound from.

megger   (1) A test for measuring the resistance of the insulation in an electric motor. It is usually performed by passing a high voltage at low current through a motor's windings.

(2) The type of moving coil galvanometer used to measure high resistances. Part of the coil is in series with the unknown resistance, while the other part, which carries current directly from the source, is independent of it. The reading depends on the relative currents of the two parts of the coil, and is thus independent of the source voltage.

(3) The trade name of an instrument that is specifically designed to measure high electrical resistance. For example, it is used in testing the insulation resistance of power and communication lines, high tension insulators, wiring in buildings and moving craft.

Meissner effect (1) Named after German physicist Walther Meissner (1882–1974). The effect by which magnetic flux is excluded from a superconductor material when the temperature of the material is reduced below its superconducting transition temperature. It is often referred to as *perfect diamagnetism* since the applied magnetic field induces a magnetization in the superconductor that is opposite the applied field direction and exactly cancels the applied magnetic flux within the superconductor. Allows superconductors to be magnetically levitated. Flux is completely excluded in a type I superconductor and is partially excluded in a type II superconductor. *See* susceptibility, diamagnetic.

(2) A superconductor is highly diamagnetic. It is strongly repelled by and tends to expel a magnetic field regardless of whether the field was applied above or below the transition temperature. This effect shows that superconductivity is more complicated than simply being a state of zero resistance since that by itself would cause the flux to be trapped in a sample which was cooled through its transition temperature in a field less than its critical value.

membranes, vibration in Many sound generators in practical use take advantage of vibrations of membranes or diaphragms. Only a few diaphragms in practical applications are membranes in the strict sense; more often they are classified as plates. Vibrations in a membrane are described by the differential equation for the normal free vibration displacement ξ as $\nabla^2 \xi = \frac{1}{c^2}\frac{\partial^2 \xi}{\partial t^2}$, where the velocity c is $c = \sqrt{T/\rho}$ (T is the tension in the surface of the membrane and ρ the surface density). The solution of the governing equation for the motion of the membrane shows that on the nodal lines of the membrane no motion takes place. For a circular membrane, they can take the form of nodal lines and nodal circles. The number of natural frequencies of a circular membrane is large in a frequency range above the fundamental; thus membranes are often driven below their fundamental.

memory An electronic device or area in which information can be stored. Common types are random access memory (RAM), dy-namic random access memory (DRAM), read only memory (ROM), and programmable read only memory (PROM).

meridian plane Also known as *meridional plane,* this is a plane that includes or contains the axis of an optical system and the chief ray. The *sagittal plane* is the plane that contains the chief ray and is perpendicular to the median plane. The median plane and the sagittal plane are used to consider the astigmatism of a lens. The rays of light, which pass in the meridian plane, form a focal line. The focal line is called the *meridian focal line.* It is also known as *tangential plane.*

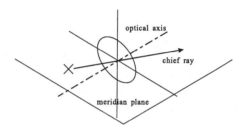

Meridian plane.

message Groups of characters or symbols processed and transmitted from one point to another and relayed over a communication system. For successful communication, the message at the destination should be identical to that emanated from the source.

metallic glasses Rapidly quenched combination of metals and semimetals that do not have long range structural (crystalline) order but do have some short range structural order. Many combinations of elements have been prepared as metallic glasses. Some combinations of Fe, Co and semimetals have extremely soft magnetic properties and are useful as cores in transformers where rapid reversal of the magnetization with low hysteresis losses is required. Combinations of rare-earths and transition metals in metallic glass form such as Tb-Co have extremely large coercivity. *See* hysteresis.

meteorological acoustics A branch of atmospheric physics or physical meteorology, in which the physical processes occurring in the atmosphere are described, modeled and explained.

meter, electric Any electrical measuring instrument. Chiefly, the term refers to an integrating meter such as the watt-hour meter used to measure the total energy consumed in an electrical circuit. *See also* meter, integrating.

meter, integrating Any instrument that measures the time integral of an electrical quantity. As an example, the domestic watt-hour meter is used to measure the total amount of electrical energy consumed.

Michelson-Morley experiment An experiment, first performed in the late 1880s by A.A. Michelson and E.W. Morley, designed to search for the effect on the speed of light produced by the motion of the earth through the proposed *ether,* conjectured to fill all of space. The experiment divided a light beam from a single source into two beams, which were directed along perpendicular paths and subsequently reflected back to the division point, as indicated in the diagram below.

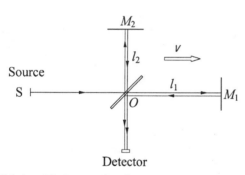

Michelson-Morley experiment.

Assumption that light traveled with a fixed speed in the ether (and adjustment of the lengths of the perpendicular paths to be equal) led to the expectation that the earth's motion with respect to the ether would cause relative speed of light and its consequent travel time along the two paths to differ. It was thought that the recombined light from the two paths would produce an interference pattern that could be altered by a change in the orientation of the source-mirror system relative to the earth's velocity vec-

tor through the ether. In contrast, the experiment found the orientation of the source-mirror system to have no effect on the interference pattern of the recombined light beams, and led to the eventual abandonment of the notion of ether. An explanation for the null result of the experiment was subsequently provided by Einstein's *special theory of relativity* in 1905, which recognized the speed of light in free space to be a fundamental constant of nature independent of the state of motion of the observer. *See* interferometer, Michelson-Morley.

micro-densitometry In micro–densitometry measurements, the optical density of materials can be measured at the microscopic level. From the output of such measurements the concentration of a substance can be attained.

In clinical uses, a densitometer measures concentrations of substances on surfaces of film or other supporting media by either a photocell measurement of the transmitted light through the medium, or by measurement of the distribution of a specific radioactive element on a radiochromatogram, as in a radiochromatogram scanner.

micro-dosimetry A micro–dosimetry apparatus yields accurate measurements of doses, down to the microscopic level. The technique acquires importance when microscopic doses of radiation have to be achieved in, for example, the treatment of cancer.

micro-electrophoresis By micro–electrophoresis the migration of charged colloidal particles or molecules through a solution under the influence of an applied electric field is studied. The applied electric field is usually provided by immersed electrodes. By examining the positions of the particles at different times during their migration toward one of the electrodes, their mobilities are calculated. Properties, like the mass and charge, of the moving particles can be determined from inputs such as the viscosity of the suspending media.

A related topic is *cataphoresis,* by which substances, especially proteins, are separated and molecular structures analyzed by measuring the rate of movement of each component in a colloidal suspension while under the influence of an electric field.

microfluorimetry   By microfluorimetry, fluorescence radiation emitted from a sample is measured.   The emitted radiation is in the ultraviolet-visible range. The apparatus used to make fluorescent measurements, the fluorometer, is similar to that used to make measurements of scattered radiation. The detector is usually placed perpendicular to the path of the incident radiation in order to avoid detecting the incident radiation. Because fluorescent intensity is, at low concentrations, linear in concentration, outputs from a microfluorometer can be callibrated to directly measure the concentration of the fluorescent substance under investigation. *See* microscopy, fluorescence.

microphone, carbon   A resistive sensor in that sound waves incident on a diaphragm apply a force on carbon particles in a container through a plunger attached to the diaphragm. The resistance of the carbon granules is proportional to the force applied to the diaphragm. Carbon microphones are characterized by high sensitivity, poor linearity and dynamic range, and are quite rugged. They are used in speech communications when high fidelity is not a requirement. In the past they were commonly used as telephone microphones.

microphone, hot wire   Consists of a fine platinum resistance wire grid mounted on a glass rod and attached to a container (that acts as a Helmholtz resonator) by means of a holder on the neck of the resonator. Electric current is passed through the grid causing resistive heating, so that the grid behaves as a hot wire. The sound wave passing across the opening of the container leads to an air surge, thus cooling the wire and causing its resistance to change. The resistance of the hot wire is sensed in a Wheatstone bridge arrangement. The grid is calibrated by varying the velocity of the air stream it is exposed to. High sensitivity and small inertia characterize hot wire microphones, so that their response is practically instantaneous.

microphone, moving coil   A sensor in that the pressure of sound waves incident on a diaphragm is converted into electrical signals through the motion of the coil attached rigidly to the diaphragm. Also referred to as an *elec-trodynamic microphone*. Under the influence of pressure oscillations the diaphragm is exposed to, the coil moves in the magnetic field generated by a permanent magnet. In this way electric current is induced in the coil by means of electromagnetic induction. The sensitivity of the microphone scales directly with the intensity of the magnetic field and the length of the magnetic path and inversely with the specific acoustic impedance. Moving coil microphones are characterized by low self-noise, relatively low impedance (suitable in applications requiring long cables); they are omnidirectional, insensitive to variations of ambient temperature and humidity, and are also quite rugged.

microphone, ribbon   An electrodynamic microphone, similar in operation to a moving coil microphone, with a corrugated ribbon suspended in the magnetic field replacing the coil and the diaphragm. Both sides of the ribbon are exposed to the sound field, and the ribbon deforms under the influence of the pressure difference. The motion of the ribbon in the magnetic field induces electric potential in the ribbon. Since it is a pressure-gradient sensor, its response is bidirectional, described by the directivity factor. The sensitivity of the ribbon microphone is similar to moving coil devices, it is less rugged, and the lower moving mass results in better frequency response.

microphones   Acoustic transducers that generate electrical signals proportional to the acoustic pressure or pressure gradient in the ambient in the vicinity of the transducer face. The properties of microphones that determine their feasibility in a specific application are the *electroacoustic performance* (sensitivity, directivity, frequency response, linearity . . .), *electrical characteristics* (output impedance), *sensitivity to external influences,* and *cost.* Microphones are used as communication devices (telephones, hearing aids), sound recording and broadcasting devices, and general-purpose and measurement devices. Microphones can be classified according to the mechanism that is employed to convert acoustic into electric energy: *moving coil* and *ribbon (electrodynamic) microphones, condenser, piezoelectric* and *electric microphones,* as well as *hot wire* and *carbon microphones.*

microscope, compound   A microscope that utilizes two lenses or lens systems. One lens forms an enlarged, real, inverted image of the object and is called an *objective lens*. The other lens magnifies the image formed by the objective and is called an *ocular* or *eyepiece*. The magnification of the objective lens is the transverse magnification. For visual observation, the magnification by the ocular is angular magnification. For projection with a microscope, the magnification by the ocular should be treated as a lateral magnification.

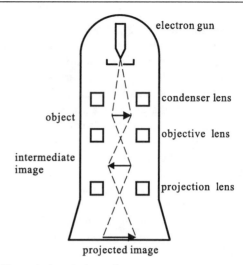

Transmission electron microscope.

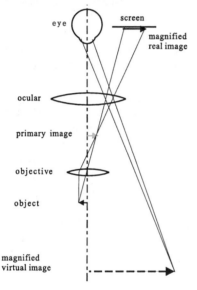

Compound microscope.

microscope, electron   An instrument that magnifies objects, it uses beams of electrons instead of rays of light. In an electron microscope, electrons that are emitted from a heated filament source are accelerated by electrostatic lens and have a very high energy. The electrons are focused to a very small point on the surface of a specimen by an electromagnetic lens. The image is projected onto a fluorescent screen or X-ray CCD plane. The non-destructive analysis of a sample can be made by measuring the back-scattered electrons and so on.

The resolution of an electron microscope depends on the wavelength of the electron which is related to the energy of the electron by de Broglie's equation.

microscope, field emission   A type of electron microscope. High positive voltage is applied to the metal tip surrounded by low-pressure gas (usually helium). The image is formed by field ionization at the surface of the specimen mounted at a very sharp and cooled (20 to 100 K) tip in an ultra-high vacuum micro chamber. A strong electric field creates positive ions by electron transfer from surrounding atoms or molecules. To produce a field ion image, carefully controlled amounts of image gas are introduced into the vacuum system. The type of image gas depends on the material to be investigated. Usually, neon, helium, hydrogen and argon are used as an image gas. The images are caused by image gas ions striking the fluorescent screen. Individual atoms on the surface of the tip can be resolved. The first observation of the atoms is realized by using FEM.

microscope, flying spot   A microscope in which a spot of light, produced in the lens system is scanned through an object. The spot of light passing through the object is detected by a photo cell. Use of a photo cell makes the quantum efficiency higher than a usual photograph. The image is produced on a cathode-ray tube.

microscope, simple   A diverting lens system that forms the enlarged image of a small object. The lens of a simple microscope is corrected for spherical and chromatic aberrations. Usually it is used to magnify the object at its focus to form an enlarged image at infinity. For a simple microscope, the angular magnification M is described by the least distance of distinct vision L and the object distance s

$$M = \frac{L}{l} \left(= \frac{25}{l} \right),$$

where the distances are measured in cm. Usually the image is viewed at infinity, therefore,

$$M = \frac{L}{f},$$

where f is the focus length. It is also known as a *magnifier,* or *magnifying glass. See* magnifying power; least distance of distinct vision.

microscope, stereoscopic   A microscope contains a pair of microscope systems. The two magnified images of the same field at different angles (a stereo pair) are observed simultaneously with a stereo viewer. Some kind of electron microscopes use the same technique to obtain the stereoscopic magnified image of the specimen.

microscope, traveling   A low magnifying-power (about $10 \times$) microscope equipped with a graticule in a plane of the ocular, it also has a rail. Therefore, it is possible to travel horizontally or vertically in order to make very accurate length determinations.

microscopy, fluorescence   Fluorescence microscopy detects and analyzes light emitted after a previously induced absorption event happened. At the moment of absorption of electromagnetic radiation (ultraviolet to visible range), excitation of the atoms being irradiated occurs. This results in one or more vacant orbitals nearer to the nucleus. Emission of radiation occurs when the excited electron returns to a lower energy electron orbital.

In general, after a photon is absorbed to an excited state, the decay of the electron can result in a release of heat, excitation of neighboring molecules, driving of a chemical reaction, or the emission of photons of lower energy. If the last case happens the emitted radiation is also termed luminescence. Luminescence is observed at energies that are equal to or less than the energy corresponding to the absorbed radiation.

Emission can occur by either of two mechanisms: fluorescence or phosphorescence. In *fluorescence,* the excited electron returns to the lower electron orbital immediately after absorption. When absorption ceases, fluorescence also immediately ceases. In *phosphorescence,* the excited electron decays to an intermediate orbital with an intermediate spin–flip, and then returns to the original orbital with a spin–flip that returns the electron to the original spin state. Phosphorescence occurs with low probability. Because inversion of the spinning electron during the last transition can require a relatively long time, the emission does not immediately cease when the absorption ceases. Therefore, fluorescence can be distinguished from phosphorescence by the time delay in the emission.

A standard method in the study of morphology is the use of fluorescence microscopy. Because the excitation and emission of light in the fluorescence process is typically done with specific radiation of determined wavelengths, very special fluorescent molecules are used to "tag" the tissue under consideration. In essence, the images obtained from fluorescence microscopy are produced by the emission from molecules of fluorescent dyes added to cells that attach to specific cellular components. Fluorescent antibodies are used to locate specific kinds of proteins and other materials in certain cells of a tissue or in certain regions of a cell. The antibodies can be prepared by, for example, injecting into a rabbit an antigen (myosin), which stimulates white blood cells to synthesize antibodies that react specifically with the antigen. After the antibodies are isolated and purified, the fluorescent dye, fluorescein, becomes attached to them by a chemical reaction. Once the fluorescent antibodies are spread over a tissue, they attach to the molecules that stimulated their formation (myosin). In this way, the image coming from the fluorescence microscope reveals the sites containing the antigen-antibody complex as bright areas in a dark background.

microscopy, ion   The field ion microscope is a development of the field emission microscope. A distinction between the field emission microscope and other electron microscopes is that the field emission microscope has a wire with a sharpened tip that is mounted in a cathode-ray tube. Electrons are then drawn from the tip toward a screen showing the image by the use of a high intensity electrical field. Because the high field at the tip exerts a large mechanical stress, only strong metals, such as tungsten, platinum, and molybdenum, can be examined in this way. The magnification of the field emission microscope is proportional to the ratio of the radius of curvature of the screen to the radius of the metal tip. A typical implementation may reach up to one million magnitudes in magnification.

In the *field ion microscope,* the tip is surrounded by helium gas at low pressure. The gas close to the atom planes on the tip is ionized and produces an image that can have a magnification of up to 10 million magnitudes. The field ion microscope has examined metals and semiconductors, as well as biological systems.

A further development of the field ion microscope is the atom probe. In this instrument, individual atoms can be removed from the tip and then passed through a time–of–flight spectrometer, which measures their energy and charge–to–mass ratio. In this way, the chemical nature of each atom in the field ion image may be determined.

microwave generator   A device that produces waveforms with a high frequency (usually from 1 GHz to 1 THz). Microwave ovens use these waves to heat the water molecules in different substances.

midband frequency   The central frequency of an amplifier's operating range. It is in the frequency region where the amplifier response is nearly independent of frequency. It is commonly taken as $\omega_0 = \sqrt{\omega_1 \omega_2}$ where ω_1 and ω_2 are the low and high frequency 3 dB points.

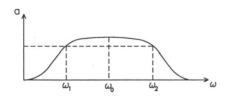

A plot of amplifier gain vs frequency showing the low and high frequency 3 dB points ω_1 and ω_2 and the midband frequency ω_0.

midband gain   The gain of an amplifier at the midband frequency. The gain is equal to the output signal divided by the input signal. *See also* loop gain.

midband loop gain   The loop gain of a series voltage feedback loop circuit at the midband frequency of the amplifier. *See also* loop gain.

miller coding   *See* modulation, delay.

Miller effect   The increase in the effective value of the base-collector capacitance in transistors due to the gain in a bipolar transistor. It is important for purposes such as rolloff frequency calculations.

mirage   An optical phenomenon caused by a bending of light rays in the atmosphere during abnormal vertical air density distribution.

mirror   An optical device for producing reflection, generally studied under plane, spherical, and various surfaces of revolution (e.g., paraboloid, ellipsoid, aspheric). A plane mirror reflects the light without either converging it or diverging it. The virtual image of the object is formed and the image is located behind the mirror at the same distance as the object is located in front of the mirror.

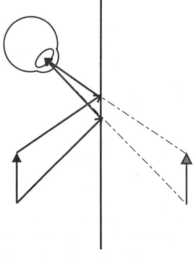

Virtual image formed by a plane mirror.

A *triple mirror,* which consists of three plane mirrors mounted at right angles to each other, is a kind of multiple plane mirror (*see* mirror, triple). Fresnel's mirror and Lloyd's mirror are also multiple plane mirrors. A concave mirror is a curved mirror. The curvature of this kind of mirror is concave to the direction of the object or light source. It forms a real image of the object and acts as a converging lens. The spherical mirror equation describes the relationship between the focal length f and the radius of the curvature of a spherical mirror R:

$$f = -\frac{R}{2} .$$

For a concave mirror, $R < 0$, $f > 0$. The lateral magnification m can be calculated as

$$m = -\frac{s'}{s} ,$$

where the image distance is s and the object distance is s'. The radius of its curvature is convex to the direction of the object and acts like a diverging lens.

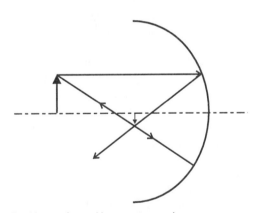

Real image formed by a concave mirror.

One of the applications of a convex mirror is the *keratometer.* The keratometer is a clinical device used for the measurement of the astigmatism. A spherical mirror can form a point image of a point object only when the object is put in the center of the curvature of the surface of the mirror. A mirror is free from chromatic aberration. However, a usual mirror has a spherical aberration. An aspherical mirror is used to reduce the spherical aberration. A paraboloidal mirror

and an ellipsoidal mirror are well known as aspherical mirrors. A paraboloidal mirror is used for applications in which an image or an object is at infinity. A paraboloidal radio antenna is a kind of parabolical mirror. Mirrors are used not only with visible light but with various kind of electromagnetic waves (e.g., microwave, X-ray, infrared, and ultraviolet).

mirror, aspherical A mirror of which the surfaces differ from a spherical surface. It is used to reduce spherical aberrations. A paraboloidal mirror and an ellipsoidal mirror are well known as aspherical mirrors. *See also* mirror.

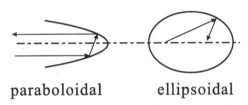

paraboloidal   ellipsoidal

Aspherical mirrors.

mirror, concave A curved surface mirror that has a concavely curved surface in the form of a part of a sphere. It can form either inverted real images or erect virtual images. For a spherical concave mirror, from the spherical mirror equation, focal length f is calculated from the radius of the curvature of a spherical mirror $R(< 0)$:

$$f = -\frac{R}{2} > 0 .$$

mirror, convex A mirror of which the surface is formed from the exterior surface of a sphere or paraboloid. It forms erect virtual images and gives a diminished wide image. For a spherical concave mirror, from the spherical mirror equation, focal length f is calculated from the radius of the curvature of a spherical mirror $R(> 0)$:

$$f = -\frac{R}{2} < 0 .$$

mirrors, acoustic A surface with a different specific acoustic impedance Z_S from the medium; acoustic waves propagating in a medium are reflected from this surface. The reflected plane wave is formed according to the law of mirrors, such that the angle of incidence θ_1 equals the angle of reflection (measured from the normal to the surface). The reflection coefficient is determined as $R(\theta_1, \omega) = \frac{\xi(\omega)\cos\theta_1 - 1}{\xi(\omega)\cos\theta_1 + 1}$, where the ratio of the specific acoustic impedance of the surface to the medium is $\xi(\omega) = Z_S/\rho c$. *See also* impedance, acoustic.

mirror, spherical   A spherical mirror of which the surface forms a portion of a sphere; it forms the images of real objects. The spherical mirror equation describes the relationship between the focal length f and the radius of the curvature of a spherical mirror R

$$f = -\frac{R}{2}.$$

The lateral magnification m can be calculated as:

$$m = -\frac{s'}{s},$$

where the image distance is s and the object distance is s'. A spherical mirror can form a point image of a point object only when the object is put in the center of the curvature of the surface of the mirror.

mirror, triple   Three plane mirrors are mounted at right angles to each other. A total reflecting prism cube, of which the corner is cut off by a plane crossing each cube face with equal angles. Any ray entering into a triple mirror will be returned parallel to the direction the light comes from. It is also known as a *cube-corner retro-reflector* or a *three plane mirror*. It is also used in some types of interferometers.

mixer   A device in a receiver system that carries out frequency conversion. *See* frequency conversion.

mobility   Mobility is the carrier drift velocity per unit electric field. It is related to the conductivity by $\sigma = ne\mu$ where σ is the conductivity, n is the carrier concentration, e is the electron charge, and μ is the mobility.

mobility coefficient (biophysical)   Consider a system of charged particles immersed in a liq-

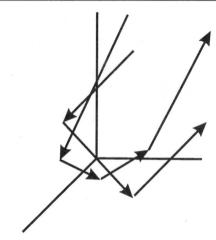

Triple mirror.

uid, subject to an external field E. The mobility μ is given by

$$\mu = D(e/k_B T),$$

where D is the diffusion constant, e the electronic charge, k_B the Boltzmann constant, and T the temperature. This relation is known as the *Einstein relation.*

The mobility of molecules under the influence of an external electric field forms the basis of the *electrophoresis technique.* In this technique the mobility of particles suspended in an electrolytic solution is determined for different particle types, because the mobility μ depends on the diffusion constant of each constituent. In essence, particle-type separation can be achieved from the dissimilar resulting speeds. *See* micro-electrophoresis; Raman scattering; electrophoretic.

modem   A single unit consisting of the two devices *mo*dulator and *dem*odulator used for transmission of digital data that is converted to an analog signal for transmission over a network, such as between terminals and a central computer over telephone lines. That is, it is a device used to convert one form of signal to another form for facility compatibility.

modes of vibration   Possible patterns of vibration of a system. For standing waves, in a vibrating string, for example, the lowest frequency f_1 is known as the fundamental frequency, and $\lambda_1 (= 2L)$ is the fundamental wavelength. The possible vibration patterns are characterized by the frequencies $f_n = \frac{n}{2L}\sqrt{\frac{T}{\mu}}$ and wavelengths $\lambda_n = (2/n)L$, where L is the length of the string, T, the tension and μ, the mass density of the string. The value labels modes or possible vibration patterns. The frequency $f_n = nf_1$ is the nth harmonic of the string. *See also* harmonics; frequency, fundamental.

modulation The variation in a reference signal's amplitude, frequency or phase used to encode a second signal.

modulation, acoustic Superposition of two traveling waves of equal amplitude and different frequencies ω_1 and ω_2. The resulting wave is $p = \cos(\omega_1 t - \gamma_1 x) + \cos(\omega_2 t - \gamma_2 x)$, which can be rewritten as $p = 2\cos\left[\frac{\omega_1 + \omega_2}{2}t - \frac{\gamma_1 + \gamma_2}{2}x\right]$ $\cos\left[\frac{\omega_1 - \omega_2}{2}t - \frac{\gamma_1 - \gamma_2}{2}x\right]$. The second equation describes the modulated wave, with the first term representing the high frequency carrier wave and the second one, the low frequency envelope. The velocity of the first term $c = (\omega_1 + \omega_2)/(\gamma_1 + \gamma_2)$ is called the phase velocity, and the velocity of the second term $v_g = (\omega_1 - \omega_2)/(\gamma_1 - \gamma_2)$, the group velocity. *See also* interference, acoustic.

modulation, amplitude A method of encoding a signal by changing a reference signal's amplitude. (Used to transmit certain radio signals: AM.)

modulation, delay Also known as *miller coding*. Signaling scheme used for magnetic tape recording and phase shift keyed signaling, since it utilizes a relatively narrow spectral bandwidth. The majority of the signaling energy lies in frequencies less than one half the symbol rate.

modulation, digital Modulation blends a data signal into a carrier for transmission over a network. Amplitude, duration, and position can be modulated. Carrier signals can be varied in the following manner. The most common methods are: (1) amplitude modulation (AM), which modulates the height of the carrier wave, (2) frequency modulation (FM), which modulates the frequency of the wave, and (3) phase modulation (PM), which modulates the polarity of the wave. The signals are produced by modu-

lating a baseband digital carrier or a pulse train. Such communication methods have been used for speech transmission, sampled data systems, and telemetry.

modulation, frequency A method of encoding a signal by changing a reference signal's frequency. (Used to transmit certain radio signals: FM.)

modulation index It is common in FM analysis to denote the transmission bandwidth on the relative magnitudes of the frequency deviation by this parameter, which is the ratio of the maximum allowable frequency deviation to the baseband bandwidth or actual modulation frequency.

modulation, phase A method of encoding a signal by changing the phase of a reference signal.

modulation, pulse A method of transmitting an analog signal using pulses. The analog signal may be represented by the amplitude (pulse-amplitude modulation, PAM), width (pulse-width modulation, PWM), or position (pulse-position modulation, PPM) of the pulses. Alternatively, the signal may be encoded in a sequence of binary numbers by pulse-code modulation, PCM.

modulation, pulse code (PCM) Modulation of a pulse train by coded representation of signal samples. It provides digital transmission systems that offer improved solutions to noise immunity and noise accumulation problems associated with analog transmission.

modulation, pulse frequency (PFM) In this type of modulation, the parameter varied is the frequency of the pulses with time.

modulation, pulse height Also known as *pulse amplitude modulation* (PAM). The blending of a signal into a carrier wave by varying the amplitude of the carrier. In this type of modulation, the parameter varied is the amplitude of the pulses with time. Broadcasting systems use this kind of modulation.

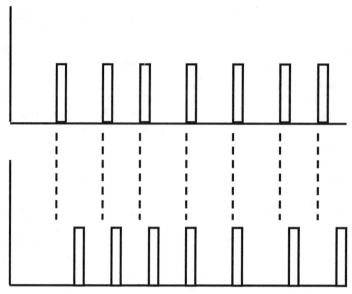

Modulation, pulse position (PPM).

pulse train

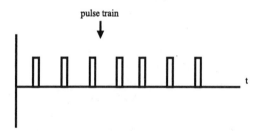

PAM.

modulation, pulse position (PPM)   In this type of modulation, the parameter varied is the position of the pulses with time. It is delayed by an amount depending on the amplitude of the modulating signal.

modulation, pulse rate (PRM)   In this type of modulation, the parameter varied is the rate at which the pulses are generated with time.

modulation, start-stop   This refers to a telegraph modulation technique that is isochronous for each character or block. However, it pos-

sesses an undefined interval between consecutive blocks or characters.

modulation, suppressed carrier   In the reception of an amplitude-modulated signal, an apparent reduction in the depth of modulation of a wanted signal caused by the presence at the detector of a stronger unwanted signal.

modulator   A device or circuit used to change a reference signal in order to encode an input signal.

Moiré fringes   Set of dark fringes created when two ruled gratings or periodic patterns are superimposed on one another with the angle of intersection between the patterns, θ, less than 45°. Fringes represent the loci of points of intersection between the superimposed patterns, and have in general a spatial separation, D, which increases as the angle θ decreases in accord with the equation $D = d/\theta$, where d represents the spatial period of the two patterns. Example of Moiré fringes produced by superposition of two gratings is shown in figure below. In general, the orientation of the fringes with respect to the lines of the original patterns approaches 90° as the angle θ goes to 0.

Moiré fringes allow the fidelity of the replication of a diffraction grating or ruled pattern to be checked to a high degree of precision, and

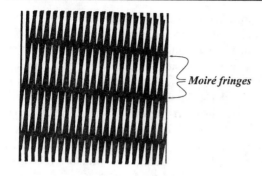

Moiré fringes

combine with measurements of D and d to make possible a determination of extremely small angles of intersection, θ, on the order of 1 second of arc.

molar conductivity   The conductivity of an electrolyte solution divided by the concentration of the electrolyte present.

molecular weight   The sum of the atomic weights of all the constituent atoms present in one molecule.

Mollier diagram   This is a plot of enthalpy (kJ/kg)versus entropy (kJ/(kg · K)). In this diagram, constant-pressure, constant-volume, and constant-temperature lines are indicated.

monochord   An instrument that uses the transverse vibrations of a string; an ancient method to generate musical sound. The monochord used by Pythagoras consisted of a sounding board and a box with a scale that allowed stretching one or more strings. Nowadays the monochord is used to compare the pitch of tones. The modern monochord consists of a thin metallic wire spanning two bridges.   The wire is stretched by a weight hanging over a pulley or by a spring tensioning device, and a movable bridge allows changing the vibrating length of the wire. Also called *sonometer.* Sonometers are usually equipped with a second wire of fixed length and tension that generates a reference frequency as comparison.

monochromatic wave   Wave with a harmonic variation in time (and/or space) characterized by a single frequency. *See* light, monochromatic.

monochromator   An instrument that is used to isolate a narrow band of wavelengths radiation (monochromatic radiation) from a wide band spectral source.

monopole, electric   A single point charge having either positive or negative polarity.

morphometrics   By morphometry, the measurement of the external form or shape (topology) of an object is understood.

Modern microscopy, e.g., confocal microscopy, allows for the study of the morphometry of objects. In some cases, a three–dimensional surface or tissue reconstruction can be achieved from two–dimensional cross sectional data.

Morse code   The Morse code, named after Samuel Finley Breese Morse, represents a binary valued variable-length memoryless code wherein each character is signified by a distinct group of dots and dashes. Each character group of dots and dashes are separated from each other by a space.

The Morse code may also be signaled by flagging — raising both arms vertically above the head signifies a dot; outstretching both arms horizontally to the two sides refers to a dash; lowering both arms to 45° vertical angles means space; circular arm motion overhead signifies erase or repeat; rapid vertical arm motion in front of the torso symbolizes end of message. The Morse code combines source encoding with channel coding. The Morse code needs 9.296 signaling time units to represent the average given information symbol, 24% over the optimum minimum. While a more efficient alternate but more complex coding scheme may be derived, the Morse code represents a sensible compromise between coding efficiency and ease for manual use.

MOS integrated circuit   *See* MOS logic circuit.

MOS logic circuit   (metal oxide semiconductor) One of two major categories of chip design, the other being known as bipolar. It uses metal, oxide and semiconductor layers. There are several varieties of MOS technologies, in-

Morse Code

Character	Morse Code	Occurr. Prob.	Signal. Time Units
Space		0.1859	6
A	.-	0.0642	9
B	-...	0.0127	13
C	-.-.	0.0218	15
D	-..	0.0317	11
E	.	0.1031	5
F	..-.	0.0208	13
G	--.	0.0152	13
H	....	0.0467	11
I	..	0.0575	7
J	.---	0.0008	17
K	-.-	0.0049	13
L	.-..	0.0321	13
M	--	0.0198	11
N	-.	0.0574	9
O	---	0.0632	15
P	.--.	0.0152	15
Q	--.-	0.0008	17
R	.-.	0.0484	11
S	...	0.0514	9
T	-	0.0796	7
U	..-	0.0228	11
V	...-	0.0083	13
W	.--	0.0175	13
X	-..-	0.0013	15
Y	-.--	0.0164	17
Z	--..	0.0005	15

cluding PMOS, NMOS and CMOS. Metal oxide semiconductor is a dominant main storage technology.

motor, asynchronous   An alternating-current motor in which the rotating magnetic field produced by the stator causes or aids the rotor in reaching and maintaining the operating rotational frequency, e.g., induction motors, AC commutator motors. *See also* motor, induction.

motor, electric   General term used for a wide variety of machines that convert electrical energy into mechanical energy.

motor, induction   An alternating-current motor in which the current in the rotor windings is induced by the alternating magnetic fields set up by currents in the stator windings. The interaction of the stator magnetic fields and those induced in the rotor sets the rotation frequency. Induction motors are classed as asynchronous motors. *See also* motor, asynchronous.

motor, repulsion   An alternating-current commutator motor in which the rotor is placed in an alternating magnetic field produced by a single phase stator winding, and in which the armature remains short-circuited in a line at a predetermined angle with respect to the stator field flux. The short-circuit is accomplished through brushes which rest on the commutator and are joined by a low resistance connector. It belongs to the class of asynchronous motors. *See also* motor, asynchronous.

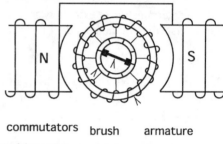

commutators   brush   armature

Repulsion motor.

motor, synchronous   An alternating-current motor, like induction motors, that operates on the principle of the revolving magnetic field, usually produced by the stator. However, the rotor of the synchronous motor is designed to produce a constant magnetic field. Consequently, it requires some means to spin the rotor at a rate that is in step with the revolving stator field so that the two fields will lock together and the rotor will be pulled around by the revolving field.

multi-channel analyzer multiplexing (MUX)   In such systems, many signals with overlapping frequency spectra are multiplexed, with the combined signal being used to phase-modulate the radio frequency (RF) carrier.

multiple access, code division   *See* multiplex.

multiplex   Combining two or more signals into a single bit stream that can be individually recovered. It is a way of combining two or more signals into a single signal for transmission via a telephone wire, television broadcast, microwave, or another medium. At the receiving end, the signals are separated again by a demultiplexer. Some examples of different multiplexing technologies are *time division multiplexing* (TDM), *frequency division multiplexing* (FDM), and *code division multiple access* (CDMA). CDMA multiplexing refers to the process that occurs in wireless communication with satellites in contrast to multiplexing utilizing cables.

multiplexing, code division (CDM) Different users employ signals that have very small cross correlation. Correlators can be used to extract individual signals from a mixture of signals even though they are transmitted simultaneously and in the same frequency band. Two ways of doing this are *frequency hopping* and *direct sequence.*

multiplexing, color division In optical communication systems, multiplexing of channels on a single transmission medium, since each color corresponds to a different frequency and a different wavelength. Each color in a transmitted polychromatic light beam on a channel is transmitted in one optical fiber or bundle of fibers. This occurs in the visible region of the electromagnetic radiation frequency spectrum and is the same process as *frequency division multiplexing* in the non-visible region of spectrum.

multiplexing, frequency division (FDM) A multiplex in which the multiplexed signals, for simultaneous transmission, occupy separate frequency ranges. It refers to the process of using several frequencies on the same channel to transmit several different streams of data simultaneously. This method is used for cable TV transmission. Filtering is the key operation used in the insertion or dropping of a signal or groups of signals that are multiplexed.

multiplexing, pulse mode Switching, for simultaneous transmission of signals over a single channel, in which connections between inlets and outlets of one or more switching stages are provided by a plurality of separate metallic paths.

multiplexing, space division (SDM) The combining of several independent and isolated fibers or wires in a single bundle or cable in order to be able to use each fiber, or bundle, as a separate communication channel over which several signals can be transmitted. Each spaced division multiplexed channel may be time division or frequency division multiplexed.

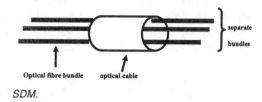

Optical fibre bundle   optical cable

separate bundles

SDM.

multiplexing, time division (TDM) A multiplexing technique, in which two or more signals are transmitted at the same time over the same communications channel. The individual signals are combined by interleaving bits. In this method of multiplexing, a channel is shared on a time basis rather than frequency. This form of multiplexing is a cheaper and simpler form than frequency division multiplexing.

multiplexing, trunk group A set of trunks treated as a unit from traffic point of view in which telephone companies multiplex many conversations over a single physical trunk. Economically, it costs about the same to install and maintain high and low bandwidth trunks, compared to permanent connection between any two switching stages in an exchange.

multiplier A resistor used with a voltmeter to allow the measurements of voltages not in the range of the voltmeter. An operational amplifier circuit whose output is the product of the two inputs.

multivibrator, astable A multivibrator with two quasi-stable states. It oscillates between the states spontaneously without reaching a steady state. Often used as a square wave or clocking waveform generator.

multivibrator, bistable A multivibrator that has two stable states that it switches between when triggered. It is very often used as a memory device in digital flip-flops.

multivibrator, monostable   A multivibrator that has one stable and one quasi-stable state. It changes to the quasi-stable state on being triggered and then returns spontaneously to the stable state. It can be used to gate other circuits.

multivibrators   Oscillating circuits, typically based on the saturation of transistors, which have two possible states used in applications such as timers and logic circuits.

multivibrator, stable states   Voltage levels in a multivibrator that will remain unchanged in the absence of a triggering signal.

muscle contraction, force-velocity relation
The force–velocity relationship during muscle contraction is experimentally determined by allowing a muscle to contract while doing force on a constant load. The resistive force of the muscle while shortening is then plotted as a function of its velocity (*see* figure below). The horizontal axis is plotted relative to a maximum velocity V_{max}, while the vertical axis is plotted relative to a maximum isometric force P_0. Point (A) in the figure illustrates where the sarcomeres are getting longer; negative relative velocity means that the actin and myosin are moving in opposite directions. Point (B) illustrates the shortening of the sarcomeres.

is determined experimentally by measuring the force a muscle can generate while being extended to a certain length (isometric conditions).

From these studies it was determined that the force generated by a muscle at a given length is a function of the magnitude of the overlap between the actin and myosin filaments (*see* muscle, mechanics). By the simple argument that the force performed by the muscle has to be linearly proportional to the number of cross bridges, then the length–tension curve has a peaked shape. The argument is the following: at short muscle lengths the thin and thick filaments in the sarcomeres overlap so much that the number of cross bridges is very small, thus a small force. As the muscle extends, there is a peak in the force right at the point that the thick and thin filaments do not overlap anymore (maximum number of cross bridges). With longer lengths, the number of cross bridges goes to zero as the overlap diminishes.

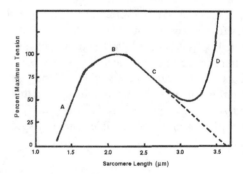

Muscle contraction.

In the figure above, the initial rise of the curve (A) indicates the overlap between the actin and myosin reaching a plateau (B) where the overlap is minimum. The curve then decreases (C) due to loss of cross bridges and would continue down to zero if it were not for the passive stiffness contribution of the muscle that then makes the curve increase rapidly (D).

There are additional contributions to the length–tension curve due to the spring–like response of the muscle that is linear in the displacement (*see* muscle, mechanics).

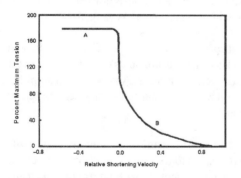

Muscle contraction.

muscle contraction, length-tension relation
The length–tension relationship of the muscle

muscle, mechanics   Muscle fibers in skeletal muscles consist of bundles of myofibrils that contract in response to neural or electrical stimuli. The myofibrils, in turn, consist of repeated cylindrical units called *sarcomeres* which are the smallest contractile unit in the muscle. The myofibrils are surrounded by a bag–like structure called the *sarcoplasmic reticulum.*

The sarcomeres are composed of thin and thick protein filaments that, upon movement of one relative to the other, achieve the contraction. The thin filament is composed of pairs of polymerized actin monomers arranged in a helix. The thick filaments are formed by myosin molecules, each one having two entwined tails (≈ 150 nm long) and a double globular head.

Contraction of the muscle is basically a result of the cyclical interactions between the thin and thick filaments. During contraction the globular heads of the myosin molecules attach to receptor sites on the actin molecules. These attachments form what are called *cross bridges.* Shortly after, the myosin undergoes a conformational change that exerts a pulling force on the actin filaments after which they detach from one another and the cycle starts again.

The whole process is triggered by a signal from a motor neuron that depolarizes the sarcoplasmic reticulum making it release Ca^{2+} into the sarcomeres. This constitutes the first part of the cycle where the Ca^{2+} then binds to the actin monomers producing a conformational change that exposes a receptor site for the myosin head to attach. After attachment of the myosin and actin, the actin molecule rotates, provoking the pulling that contracts the space in the sarcomeres and later detachment of the two filaments, at which point everything relaxes to its original position. The cycle continues as long as Ca^{2+} and ATP are present. When the depolarizing signal stops, Ca^{2+} is pumped back into the sarcoplasmic reticulum causing relaxation.

With respect to the mechanical response of muscles, muscles are like a spring. Muscles generate a restoring force when they are stretched beyond their resting length L_0 with a force of the form

$$F = k\,(L - L_0)\ ,$$

where k is the spring constant. When the muscle is stimulated, the contractile elements shorten and the muscle starts to develop tension that can be understood as if L_0 got shorter, i.e., L_0 gets smaller as contraction occurs.

If a weight were attached directly to the muscle, the weight would be pulled up progressively as the rate of stimulation increased until the muscle's restoring force precisely matched the weight.

musical instruments   Instruments that produce musical sound. They are classified according to the nature of the primary vibrator generating the sound as *string* instruments, *wind* instruments and *percussion* instruments. Acoustics studies the laws governing the action, design, and construction of musical instruments.

musical sound   A combination of sounds in rhythm, harmony and counterpoint, generated by musical instruments or human voice, as a medium of artistic expression. Musical acoustics considers the physical characteristics of sounds that might be perceived as music, the production of musical sound, and its transmission to the listener. The perceived qualities of musical sound, such as loudness, pitch and timbre, depend on physical parameters, such as sound pressure, frequency and spectrum.

mutual capacitance   Mutual capacitance, also known as the *coefficient of induction,* is defined as the ratio of the induced charge on the other conductor to the potential of a giving conductor when no other conductors are nearby.

mutual conductance   The derivative of the drain current with respect to the gate-source voltage in a common source FET circuit.

mutual intensity   Quantity defined to measure the coherence between fields at two separated space points. Equal to the average over the time t of a product of the "normalized" fields at the two separated space points evaluated at a common time t. The quantitative definition of mutual intensity requires introduction of a (scalar) field function of position and time, $V(\mathbf{r}, t)$, defined to determine the time averaged intensity, I, of the field at a space point $\mathbf{r}$ via the equation

$$I(\mathbf{r}) = \left\langle V^*(\mathbf{r}, t) V(\mathbf{r}, t) \right\rangle\ ,$$

where the angular brackets denote an average over t in a time interval large compared to one period of oscillation of the field. Latter definition allows mutual intensity of the fields at positions $\mathbf{r}_1$ and $\mathbf{r}_2$ to be defined by the quantity

$$\Gamma(\mathbf{r}_1, \mathbf{r}_2) \equiv \langle V^*(\mathbf{r}_1, t)\, V(\mathbf{r}_2, t) \rangle .$$

Quantity has the significance of determining the interference term in an expression for the field intensity resulting from the superposition of two fields derived from spatially separated point sources.

myoelectric activity   Myoelectric activity refers to the electrical properties and response of muscle tissue. Electrical impulses resulting from these myoelectric properties may be amplified and used in diverse applications such as the monitoring of the response of specific muscles to stimuli as well as the control or operation of prosthetic devices.

myopia   (short-sightedness) This results from the lenses of the eye refracting the parallel rays of light focused not on a retina. It is usually caused by an abnormally long eyeball. For correction, diverting spectacle lenses are used to move the image back to the retina. *See* eye, near-sighted.

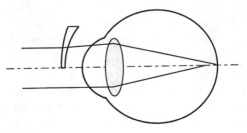

Myopia.

N

NAND   A Boolean operation in which the output is 0 if both inputs are 1 and the output is 1 if either of the two inputs is 0.

near point   The nearest point at which a human can focus on an object with accommodation. It is different from the least distance of distinct vision. The lenses of the human eye become harder with age which causes the nearest point to recede with age. *See also* least distance of distinct vision.

negative logic   Logic devices in which a high voltage determines a binary level 0 and a low voltage a binary level 1.

negative resistance   A resistance with a negative value. This occurs when the current in an electronic device decreases as the voltage across it increases, e.g., Esaki (tunnel) diodes, resonant tunneling diodes.

negentropy   In information theory, negentropy is related to the amount of information contained in a given system. It is the analog in information to what entropy is in thermodynamics. The analogy stems from the formal equivalence between the mathematical expression for entropy ($S = k \log W$) and Shannon–Wiener's information ($H = -\log_2 p_a$). The motivation for the negative sign in front is to yield the positive quantity H since probabilities p_a are less than 1.

neon liquefier   Any one of several machines that liquefy neon by causing it to undergo adiabatic expansion and to do external work.

neon tube   A tube in which neon gas at low pressure is ionized and heated to produce light.

nephelometer   An instrument that measures the size and concentration of particles suspended in liquids by analyzing the scattered light from the liquid. The nephelometer finds applications in the quantification of the turbidity in water caused by colloidally dispersed particles.

Nernst equilibrium potential   The Nernst potential defines the potential of the cell membrane due to a particular ion that is distributed both outside and inside the cell. The Nernst equilibrium potential for an ion is given by

$$V = \frac{RT}{zF} \cdot \ln \frac{Co}{Ci} ,$$

where R is the gas constant, T the absolute temperature, z the charge of the ion in consideration, F Faraday's constant, and C_o and C_i are the concentrations of the ion outside and inside the cell, respectively.

Nernst potentials exist separately for each ion. The total membrane potential is determined by the flow of all the ions that cross the plasma membrane.

nerve conduction velocity   Nerve conduction velocity refers to the velocity of propagation of an electrical impulse (action potential) through nerve fibers. The velocity of conduction along the nerve fiber is dependent upon several factors. The first factor is the outside diameter of the nerve fiber. The fastest conduction velocity occurs in the largest diameter nerve fibers. Another factor is the temperature of the nerve fiber. Conduction velocity increases at high temperature and decreases at low. Conduction velocity is also affected by myelination of the nerve fiber. Since ions cannot cross the lipid content of the myelin sheath, they spread passively down the nerve fiber until reaching the unmyelinated nodes of Ranvier. The nodes of Ranvier are packed with a high concentration of ion channels, which upon stimulation, propagate the nerve impulse to the next node. In this manner the action potential jumps quickly from node to node along the fiber in a process called *saltatory conduction* (from the Latin *saltare,* "to jump").

Because there is a definite relationship between the diameter of the nerve and the conduction velocity, the diameters of the nerves form a basis for classifying mammalian nerve fibers into groups in order of decreasing diameter and decreasing conduction velocity.

See node of Ranvier, electrical characteristics.

nerve impulses, propagation of   A nerve impulse has its origins when an external stimulus is applied to a neuron. The stimulus induces a change in the membrane potential of the neuron at the point of stimulation. In order for this nerve impulse, called *action potential,* to travel along axons or dendrites, the membrane needs to be sequentially depolarized all along the length of the path that the signal will follow. These membrane depolarizations are very localized in space and time and typically consist of a change in the membrane potential from -60 mV to approximately $+30$ mV, as measured in squid axons, in less than a millisecond. The depolarization is a result of a series of rapid and sequential opening and closing of voltage-gated Na^+ and K^+ channels. After the signal traverses a particular point in the membrane, the depolarization ceases soon after and returns to its resting value.

There are several steps in the depolarization process. This process starts with a relatively small initial change in the membrane potential (-60 to -40 mV) that leads to the rapid opening of Na^+ channels. The Na^+ then flows into the cell by diffusion due to the concentration gradient and also driven by the membrane potential. The large influx of Na^+ subsequently changes the membrane potential to $+30$ mV, approaching the Na^+ equilibrium potential of approximately $+50$ mV. At this time the Na^+ channels are inactivated and voltage–gated K^+ channels are opened, substantially increasing the permeability of the membrane to K+. Similarly to the Na^+, the K^+ flows rapidly out of the cell leading to a -75 mV in the membrane potential. At this membrane potential, approximately equal to the K^+ equilibrium potential, the K^+ channels are inactivated and the membrane returns to its resting level of -60 mV.

This process is repeated for every element of length that is adjacent to the initial stimulus thereby allowing the action potentials to travel down the length of nerve cell axons as electric signals.

network   An interconnection of basic components such as resistors, capacitors, and inductors in series, parallel, delta, π, etc, groupings to form a system that performs a specific function. The response of this system to an electrical input is dependent on the component type, value, and the manner of connection. The name given to a network may be described by:

- the types of components, such as resistive, resistance-capacitance (R-C), inductance-capacitance (L-C), inductance (L) networks, etc.

- the method of interconnection, such as series and parallel networks,

- the response, such as linear networks that have linear relationships between the voltages and the currents.

Networks are termed as active if they have an energy source or sink other than normal ohmic losses. Passive networks, however, do not have an energy source.

The term network may alternately be used to refer to the interconnection of communication facilities, e.g., computer terminals, telephones, etc.

network, distributed parameter   A circuit that behaves as if parameters such as resistances, capacitances, and inductances exist continuously over a physical length. For example, in a two wire transmission line, the distributed parameters are series resistance, series inductance, shunt conductance, and shunt capacitance per unit length of line.

network, ladder   A network that consists of H, L, T, or π networks connected in tandem. Ladder networks have been used as narrow bandpass filters and digital-to-analog converters.

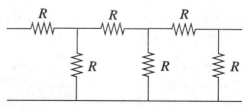

Ladder network.

network, linear A network in which the currents and voltages of the circuit elements have a linear relationship. This usually implies that the elements, such as capacitances, inductances or resistances, are constant in magnitude with varying currents.

network, lumped parameter In circuit analysis, any network in which distributed parameters such as inductances, capacitances, or resistances can be treated as single parameters concentrated at a point. This approach is usually valid for a specific frequency range.

network, nonlinear A network in which the currents and voltages of the circuit elements do not all have a linear relationship. For example, the voltage across a semiconductor diode is not directly proportional to the current through it. In signal transmission systems, a nonlinear network is one in which the signal transmission characteristics depend on the magnitude of the input signal.

neuron The neuron (nerve cells) forms the building block of the nervous system in the body. Neurons are highly specialized cells that transmit electrochemical signals throughout the body in response to internal and external stimulus. In humans, neurons may extend to more than a meter long, while in some invertebrates (e.g., the squid) the neuronal projections (axons) diameters can be as large as 1mm.

The neurons are composed of a cell body with nucleus, an axon (main signaling projection) and one or more smaller projections called dendrites. The axons carry signals over long distances and the dendrites, having a branch-like structure, serve to receive incoming signals from other neurons. The basic mechanism of the neuron is to respond (or not) by an action potential (*see* nerve impulses, propagation of) to several electrical inputs from the dendrites. If the neuron "fires" an action potential, then it propagates the signal to other neurons by releasing neuro-transmitters across the synapse with other neurons (*see* neuro-transmitters).

Major types of neurons include the *associative neurons,* found within the central nervous system (CNS), which link sensory and motor neurons, and *motor neurons,* which take impulses from the CNS to muscles, glands, or other effector tissues.

neuro-transmitters Neuro-transmitters are small hydrophilic molecules stored in the axon's bulbous end. There are more than 300 known neuro-transmitters of which the endorphins and acetylcholine are the most common examples.

The function of neuro-transmitters is to allow different neurons to communicate by electrical signals even when their membranes are not in direct contact. The way they work is the following: If an electrical signal is traveling through the membrane of a neuron, upon reaching the end of the axon the signal is carried from one neuron to the other via neuro-transmitters that are released in the space (synapse) in between the two neurons. The neuro-transmitters then travel from one (presynaptic) membrane to the other (postsynaptic) membrane, ultimately binding to the postsynapse neuron receptors in quantity that is in proportion to the strength of the electrical signal. The receptors, usually ligand-gated ion channels, then are opened (e.g., acetylcholine receptors of muscle cells).

In muscles, acetylcholine opens channels that are permeable to both Na^+ and K^+. The entrance of Na^+ to the muscle cell depolarizes the cell membrane and triggers an action potential. The action potential in turn opens the voltage-gated Ca^{2+} channels, leading to the increase in intracellular Ca^{2+} that results in contraction (*see* muscle, mechanics).

neutral point The common point of a Y-connection in polyphase circuits. Also refers to the point at zero voltage in a system consisting of a number of identical parts. Although both of these definitions generally refer to power and distribution transformers, they can also be used for systems where the circuit elements are non-reactive resistances. In this case, the number of resistances is two for direct-current or single-phase alternating-current, four for two-phase, and three for three-, six- or twelve-phase systems.

neutron therapy Neutron therapy falls into the larger category of *radiation therapies* by which energy is transported from a source to a target. The source can be ionizing radiation

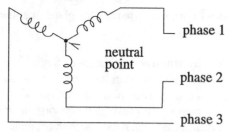

Neutral point for a Y-connection 3-phase circuit.

(e.g., X–rays, gamma rays) or particles (e.g., electrons, protons, neutrons). When the source, electromagnetic or particle, interacts with tissue, a production of free radicals and oxidants occur that subsequently damage or break cellular DNA, impairing cells from the capacity to divide and may even lead to cell death.

The therapy is widely used in cancer treatment. Damage to normal cells is diminished by the careful shielding of adjacent areas to the treated ones. When used properly, radiation may cause less damage than surgery and can often preserve organ structure and function.

Atomic elements used to some extent in cancer therapy as a source for neutrons in neutron therapy are the heavier actinides — those beyond plutonium in the periodic table.

Newton's rings   Interference pattern produced when light is reflected from the upper and lower surfaces of an air film of variable thickness formed in the space between a convex surface and a flat surface in which it is in contact. Typical geometry used to show Newton's rings interference pattern involves light incident on air gap between glass surfaces of different curvature. Diagram below shows example of interfering light rays in this geometry.

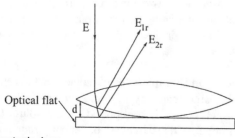

Newton's rings.

The maximum in the interference pattern for light of a particular wavelength occurs at positions where the film thickness matches an integer multiple of that wavelength. Resulting fringes follow lines of equal thickness corresponding to concentric rings centered at the contact point between the curved and flat surfaces.

A dark spot occurs in the center of the fringe pattern as a consequence of a phase change of π between rays reflected from glass to air and air to glass surfaces, respectively. Where incident radiation corresponds to white light, the interference pattern consists of colored rings with a central dark spot. The first observation of rings, credited to Isaac Newton (1642–1727), may have been made by Robert Hooke in the same era.

nitrogen, liquid   A single stage air liquefaction process using the Stirling cycle is usually the basis for Philips nitrogen liquefiers. Liquid nitrogen is usually preferred over liquid air on grounds of safety and constant temperature. It is obtained at a temperature of 77 K liquid, and nitrogen is usually used in the precooling stage when cooling with ^{4}He. The quality of vacua in the laboratory is improved by using liquid nitrogen cold traps. Liquid nitrogen is relatively inexpensive compared to liquid helium for cooling and is therefore preferred for cooling equipment using high temperature superconductors.

NMR nuclei in biological materials   A non-invasive diagnostic technique drawn from an application of *nuclear magnetic resonance* (NMR) is *magnetic resonance imaging* (MRI) where radio waves are beamed into a person who is under the influence of an external powerful magnetic field. Because different atoms in the body absorb radio waves at different frequencies of the radio waves, the absorption can be measured and specific data from specific atomic species can be reconstructed by a computer to render three–dimensional images of internal structures in the body. Molecules containing hydrogen (e.g., water molecules in body tissue) are especially suited to align magnetically and resonate, giving out enhanced absorbance.

Unhampered by bone and capable of producing images in a variety of planes, MRI is used in the diagnosis of brain tumors and disorders, spinal disorders, multiple sclerosis, and cardio-vascular disease. The procedure is considered to be without risk to the patient.

NMR probes Probes that can be imaged by nuclear magnetic resonance (NMR) include atoms of fluorine-19 and phosphorus-31, among others (not including hydrogen). These atoms can serve as probes for various tracer studies.

NMR zeugmatography In typical magnetic resonance imaging (MRI) studies, the nuclear signal, coming from the coupling of the external magnetic field and radio waves to the atomic nuclear spin, has a large wavelength (tens of meters). This means that any directional information is lost or very hard to determine.

One way to get spatial information from MRI studies is by using the dependence of the resonance of the nucleus to the external driving frequency of the radio waves. In this way spatial information is coded in the frequency part of the signal. This is accomplished by using a magnetic gradient instead of a constant field. Then, by changing the magnetic gradient spatial information is put into the frequency domain. By Fourier transformations the signal can be converted from k–space to real space. This gradient field method is called *zeugmatography*.

nodal planes/points A pair of conjugate points or planes on an axis of an optical system. An incident ray passing through one of them causes an emergent ray to pass through the other.

nodal slides The nodal slides are used for determination of nodal points or the nodal plane location for a lens system. The nodal slides are equipped with one collimating telescope that contains a distance marker. The collimator number coincides with the focal lengths of the lens combination.

node Locations in a standing wave at which the displacement is zero at all times. The displacement y_n for a vibration corresponding to the nth harmonic frequency ω_n in a standing wave, as a function of the longitudinal coordinate x and time t, is represented by $y_n(A_n \cos \omega_n t + B_n \sin \omega_n t) \sin \frac{\omega_n x}{c}$, where A_n

and B_n are constants and c the speed of sound. The frequency $f_n = nc/2\lambda$ is n times the fundamental frequency, and λ is the wavelength. By considering the term $\sin(\omega_n x/c) = (n\pi x/\lambda)$, it becomes obvious that the displacement y_n is zero at all times for x such that $(n\pi x)/\lambda = m\pi$, with $m = 0, 1, 2, 3, \ldots, n$. These positions are called the *nodal points* or *nodes of the standing wave pattern*.

node of Ranvier, electrical characteristics
The nodes of Ranvier are regularly spaced gaps in the myelin sheath along the length of a nerve fiber. Because these gaps are the only places where the membrane of the neuron is in contact with the extracellular plasma, this region has a high concentration of ion channels that permit the flow of ions in and out of the membrane.

The myelinated spacing is one of the key components so that the action potential does not decay rapidly with distance (high capacitance), and the high concentration of ion channels permit the repeat of the signal from node to node (rapid depolarization), thus giving rise to the "saltatory" transport of the action potential.

node of Ranvier, equivalent electrical network Under the assumptions of the Cable model, the circuit fragment models an axon, where the vertical RC components represent the axon membranes at the nodes of Ranvier, the top row of resistors corresponds to the resistance of the intracellular fluid, and the bottom row resistors correspond to the intracellular domain. The vertical resistors in the RC components represent the fact that the nodes of Ranvier act as leaky capacitors.

Because of the resting potential of the membrane, all the capacitors are charged. The action potential then corresponds to a discharge of a capacitor at a particular node.

Following further the assumptions of the Cable model, all of the horizontal resistances are the same. Using the "infinite axon" approximation, where both ends of the circuit are assumed the same, and the minimal current is flowing except in the node where the action potential is occurring, we can simplify the above circuit.

Typically R, the resistance of the fluid ($\approx 12 k\Omega$), is an order of magnitude smaller than the membrane resistance R_m ($\approx 200k\Omega$). R_t is the equivalent resistance of the rest of the axon.

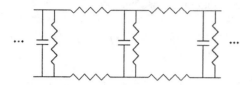

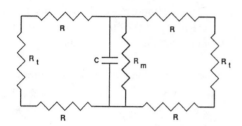

Ranvier.

noise   Sound in the acoustic spectrum that is unwanted, either because of its effect on humans, interference with the perception or detection of other sounds, its effect on fatigue, or malfunction of physical equipment. Sources of noise include industrial sources (machinery), transportation (moving sources, such as aircraft and road vehicles as well as airport or road noise), community noise, or noise generated by air conditioning systems.

noise, ambient, level   The surrounding background noise level caused by natural and man-made sources. The noise at radio frequencies that is generated by natural causes is referred to as static noise. Static noise normally originates from the atmospheric conditions caused by the presence of static electricity in the air through which radio waves are propagated. Manmade noise arises from human sources, usually electrical devices. At radio frequencies, ambient noise is usually heard as an unpleasant crackling sound at the output loudspeaker. Ambient noise should not be confused with noise generated by electrical circuits in receivers and transmitters.

noise, biological effects of   The effect of noise on humans depends on the intensity of noise and its duration, and ranges from annoyance, interference with speech, communications and sleep, to hearing damage (for example by rupturing the eardrum), and hearing loss in more severe cases. In addition to the acoustical factors nonacoustical factors such as attitude and environmental factors, influence the level of annoyance noise poses. Descriptors and measures, involving factors such as loudness and loudness level, noisiness and perceived noise level, sound levels, articulation index, speech interference level, indoor noise criteria, equivalent and percentile sound levels, day-night sound levels, etc. are available to find criteria regarding acceptable noise levels. Means to combat noise and its negative effects on humans include hearing protection devices (earplugs, earmuffs or helmets), passive measures of noise control (sound absorbing or damping materials, vibration isolation, sound barriers, enclosures), as well as active noise control (sound sensed by a microphone is processed and fed back through a loudspeaker to destructively interfere with the sound emitted by the primary source).

noise, density in space   The RMS noise-voltage density, v_N, is given by

$$V_N(\text{rms}) = v_N B^{\frac{1}{2}},$$

where V_N is the RMS noise voltage measure in a bandwidth B. In general, detected noise depends on the measurement bandwidth. A white-noise source has a v_N that does not depend on frequency. The squared noise density v_N^2 is also commonly encountered.

noise factor   The ratio of the product of the signal output S_o and input noise power N_i to the product of the input signal S_i and output noise power N_o, $= S_o N_i / S_i N_o$.

noise, Johnson   Noise arising from fluctuations of carrier velocities in a resistor at a finite temperature. This is also called *thermal* or *Nyquist noise*. These fluctuations give rise to a mean square noise voltage $V_{\text{noise}}^2 = 4kTRB$ where k is Boltzmann's constant, T is the temperature, R is the value of the resistance where the noise is generated, and B is the bandwidth.

noiseless coding   A source coding scheme is noiseless if all the encoder's input symbol values are uniquely related to the encoder's output symbol values. That is, no distortion or noise has been introduced in the coding process. Otherwise, the source code embodies a corrupted representation of the original data. The average number of bits necessary to encode a discrete memoryless source must equal or exceed the source's entropy.

noise level   The value, usually given in decibels, of a noise signal measured at a particular connection.

noise meter   The process of quantitatively determining one or more acoustical properties (such as duration, sound pressure level, variation as function of time, frequency content, presence of pure tones or background noise level, impulsive character) of acoustic noise. No single descriptor is available to uniquely characterize noise. Noise generated by moving sources is commonly characterized in terms of sound pressure levels at a certain defined location. Stationary sources are usually described by their sound power output that is measured either by moving a small source into a reverberation room or by measuring it on a hemispherical or rectangular measurement surface and applying a correction to these data. Data are also obtained in the normal environment in case of larger sources of noise and corrected appropriately. The response of the equipment used to measure noise is often weighted to match the way the human auditory system would respond to it. The sound level of noise is defined through the sound-level meter built and operated according to American Standard requirements. *See also* noise; noise, biological effects of.

noise, pseudo-noise   (1) Random noise is noise that arises from any randomly occurring transient disturbance. If the rate of occurrence of the disturbance is sufficiently high, it results in white noise, similar to thermal noise. If the rate of occurrence is low, random noise contributes to impulse noise. All electronic circuits and devices suffer from thermal and random noise.

(2) A very accessible and attractive technique for generating a source of digital noise is to generate a pseudo-random sequence. A pseudo-random sequence of binary digits has similar probability and correlation properties to an ideal string of random digits. The pseudo-random sequence is, however, totally predictable and repeatable although any portion of such a sequence looks for all intents and purposes just like a truly random sequence. It is easy to generate sequences of bits that have good randomness properties using standard deterministic logic elements such as shift registers.

noise, random   Random noise describes an acoustical quantity, such as acoustic pressure, whose variation as a function of time is described through the Gaussian (normal) distribution. Examples of such noise are the random noise caused by the random motion of air molecules. Electrical quantities can also be characterized by random noise (motion of electrons).

noise, shot   (Also known as Schottky noise, flicker noise.) Shot noise results from the statistical fluctuations of charge carriers across a junction and is given by

$$i_N^2 = 2eIB \,,$$

where i_N is the RMS noise current, e is the charge of an electron, I is the DC current, and B is the bandwidth of the measuring instrument. Shot noise is independent of frequency. The difference between thermal noise and shot noise is that the latter is related to the DC current through the junction.

no-load (electric circuit)   Operation of any electric circuit under rated operating conditions but in the absence of an impedance at the output.

non-inverter   A logic device that does not change the state of the input voltage.

nonlinear circuit   A nonlinear circuit is a circuit in which the amplitude of the current is no longer proportional to the amplitude of the voltage. A circuit containing a nonlinear device such as a diode is an example of nonlinear circuit.

non-linearity The type of relationship of two quantities x and y that cannot be expressed in the form $y = ax + b$ with a and b constants. A condition in an amplifier circuit in which the output is not proportional to the input. A property of circuits that cannot be analyzed with just series and parallel branches of linear elements.

nonlinear susceptibilities Tensor quantities that relate the electric polarization vector in a material medium, $\mathbf{P}(\mathbf{r}, t)$, to products of the components of the electric field vector $\mathbf{E}(\mathbf{r}, t)$, which induce the polarization. More precisely, functions $\chi_{jkl...}^{(N)}(\mathbf{r}, t - t_1, t - t_2, \ldots t - t_N)$, termed *electric susceptibilities,* are defined through an expansion of the j-th component of the electric polarization vector in products of the electric field components in a given medium in the form:

$$P_j(\mathbf{r}, t) = \int_{-\infty}^{\infty} dt_1 \, \chi_{jk}^{(1)}(\mathbf{r}, t - t_1) E_k(\mathbf{r}, t_1)$$

$$+ \int_{-\infty}^{\infty} dt_1 \int_{-\infty}^{\infty} dt_2 \, \chi_{jkl}^{(2)}(\mathbf{r}, t - t_1, t - t_2)$$

$$E_k(\mathbf{r}, t_1) E_l(\mathbf{r}, t_2)$$

$$+ \int_{-\infty}^{\infty} dt_1 \int_{-\infty}^{\infty} dt_2 \int_{-\infty}^{\infty} dt_3 \, \chi_{jklm}^{(3)}$$

$$(\mathbf{r}, t - t_1, t - t_2, t - t_3)$$
$$E_k(\mathbf{r}, t_1) E_l(\mathbf{r}, t_2) E_m(\mathbf{r}, t_3) + \cdots ,$$

where use is made of the summation convention (implying a sum over repeated indices from 1 to 3).

The quantities $\chi_{jkl...}^{(N)}(\mathbf{r}, t - t_1, t - t_2, \ldots t - t_N)$, for $N > 1$, are then defined to be the nonlinear susceptibilities, and $\chi_{jk}^{(1)}(\mathbf{r}, t - t_1)$ is defined to be the *linear susceptibility.* As defined, the magnitudes of the susceptibility functions are determined strictly by the properties of the material medium, while the time dependence of the functions is restricted by causality

through the equation

$$\chi_{jkl...}^{(N)}(\mathbf{r}, t - t_1, t - t_2, \ldots t - t_N) = 0 ,$$
$$t_s > t \quad (s = 1, 2, 3, \ldots) .$$

Fourier transformation of the functions $P_j(\mathbf{r}, t)$ and $E_j(\mathbf{r}, t)$, and replacement of the integration over the transform variable ω by a summation over discrete values of ω, make possible a definition of Fourier-transformed susceptibility functions via an equation of the form:

$$P_j(\mathbf{r}, t) = \sum_{N=1} \sum_{\pm \omega_n} e^{-i\omega_n t} \sum_{\substack{\pm \omega_1, \ldots \pm \omega_N \\ \sum_s^N \omega_s = \omega_n}}$$

$$\chi_{jkl...}^{(N)}(\omega_n; \omega_1, \omega_2, \ldots \omega_N)$$
$$\times E_k(\omega_1) E_l(\omega_2) \ldots E_m(\omega_N) ,$$

where the dependencies on the space coordinate $\mathbf{r}$ are suppressed on the right of the equality sign. In general, the magnitudes of the nonlinear susceptibilities are small compared to the magnitude of the linear susceptibility, and the terms in the expansion for $P_j(\mathbf{r}, t)$ beyond the first term are therefore small compared to the first term, except where the magnitude of the electric field $\mathbf{E}$ is large.

non-reciprocal device A device for which the reciprocity theorem does not apply. In these devices, I_o/V_i measured with the output terminals shorted is not equal to I_i/V_o with the input terminals shorted. Most such devices are active devices, although some passive devices involving gyromagnetic material may be non-reciprocal.

non-saturating circuit A circuit in which the output does not asymptotically approach a limiting value as the input is increased or decreased. Transistor-based switching circuits that do not operate in the saturation region of the transistor.

NOR A logic operation that gives 0 if either of the two inputs is 1 and gives 1 if both inputs are 0.

Norton equivalent A current source in parallel with a resistor that is equivalent to a circuit containing voltage sources and resistors. It is often used to simplify the analysis of complex circuits.

note A sign, defined by convention, that indicates the pitch of a musical sound by its position on the staff, and its duration by the shape of the sign.

n-type silicon Silicon (Si) that is doped with a donor impurity (pentavalent atoms such as arsenic (As), phosphorus (P), etc.) and in which electrons, which are negatively charged, are the majority carriers (electron concentration is much higher than hole concentration).

nuclear angiography Magnetic resonance imaging (MRI), of the same nature as nuclear magnetic resonance (NMR) phenomena, can be tailored to visualize flowing blood using magnetic resonance angiography, or *nuclear angiography*.

Because MRI uses the same principles as NMR, using radio waves coupled with high magnetic fields to detect structural and biochemical information about tissue in the body, it is a noninvasive procedure that is safer than imaging with X–rays or gamma rays.

Nuclear angiography takes advantage of the fact that MRI needs a longer scanning time than, for example, CT, which makes MRI more sensitive to motion studies. A direct application is the imaging of flowing blood by visualizing arteries and veins. Other areas that make use of this technique are in the examination of the bladder and a blood flow to the brain. Abnormal flow may indicate obstructions or other pathological conditions.

nuclear magnetic resonance (NMR) Resonant absorption of radio frequency radiation by the nucleus. A nucleus has discrete closely spaced energy levels corresponding to the orientation of the nuclear magnetic moment in an applied field. The nucleus is able to resonantly absorb radiation composed of photons whose energy corresponds to the difference between these energy levels. Can be useful in chemical identification since small changes in local magnetic field at the nucleus due to chemical bonding can be accurately measured in this technique. Forms the physical basis for magnetic resonance imaging (MRI) in medicine. *See* photon.

Nusselt equation Gives the relationship between the heat transfer coefficient to the thermal conductivity of the gas, the effective diameter of the tube, its dimensions, and the viscosity and heat capacity for the gas. Nusselt's equation for heat transfer by convective flow is given by

$$\mathrm{Nu} = \mathrm{const}(\mathrm{Re})^x (\mathrm{Pr})^y \, (D_e/L)^z \ ,$$

where Nu represents the Nusselt number, Re, Reynold's number and Pr, Prandtl's number. Values for the indices and the constant are found experimentally. The Nusselt number is obtained by the ratio of the density of the heat flux in the presence of natural convection to the density of the heat flux with non-moving interstitial fluid.

Nyquist criterion A condition for an amplifier to be stable, it states that if a polar plot of the loop gain as a function of the complex frequency encloses the point (1,0), the amplifier is unstable.

O

object distance The distance from the vertex to the object. From the thin-lens equation, the image distance s for a thin-lens, whose focal length is f, can be calculated from object distance s' as

$$\frac{1}{s} = \frac{1}{f} + \frac{1}{s'},$$

$$\frac{1}{f} = \frac{(n'-n)}{n}\left(\frac{1}{R} + \frac{1}{R'}\right)$$

where the radius surface of the lens is R and R'. The object distance of a thick lens should be measured from the principal planes of the lens. *See* image distance.

objective, immersion A high-power microscope objective in which the space between the lens and the cover glass is filled with an oil. The refractive index of the oil should be the same as the objective and the cover glass in order to reduce reflection losses and spherical aberration. The numerical aperture of a microscope objective indicates the brightness of the image (similar to the f-number), and it is proportional to the refractive index of the immersing medium. Because the refractive index of the immersing medium is higher than that of the air, the numerical aperture becomes higher. This causes an increase in the resolving power of the microscope.

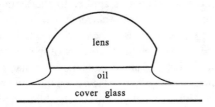

Immersion objective.

object, real An object that actually emits rays of light in an optical system. Sometimes used synonymously with real image.

object, virtual Sometimes used synonymously with *virtual image*. An object that appears to emit rays of light, but actually does not.

octave A band in the frequency scale for which the higher frequency f_{max} is twice the lower frequency f_{min}, $f_{max} = 2f_{min}$. The interval between any two frequencies having a ratio of 2 to 1. The concept of fixed frequency ratios, which define proportional frequency bands, is used in the theory of musical temperament. In the equal temperament with the 12 note per octave scale, the successive notes are 1/12 octave apart. An interval with the frequency ratio $2^{1/12} = 1.0595$ is called a half step.

ocular An ocular is a kind of magnifier. It enlarges the intermediate image of the object as formed by preceding lenses in the optical system and is not used to view an actual object. An ocular forms a virtual image. It is also known as an *eyepiece*.

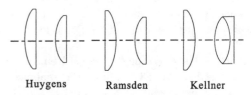

Huygens   Ramsden   Kellner

Ocular: Huygen's, Ramsden, and Kellner.

There are several kinds of oculars. A *Huygen's ocular* consists of a pair of plano-convex lenses of which the convex sides of both lenses are facing the object. A *Ramsden ocular* consists of two plano-convex lenses. The two lenses are almost the same as their convex sides are facing each other. A *Keller ocular* has a plano-convex lens and an achromatic doublet as a plano-convex lens for the eye-lens. It is an achromatic Ramsden ocular. The usage of an achromatic lens causes an increase in image quality, and is used for wide field telescopes.

An orthotropic ocular and a symmetrical ocular have long eye relief with wide-field and high magnification. A *Erfle ocular* consists of three achromatic doublets. It has well-corrected aberrations and is used for wide-field application.

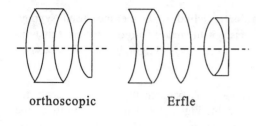

orthoscopic   Erfle

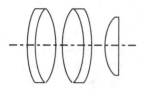

symmetrical

Ocular: orthotropic, symmetrical, Erfle.

Oersted   The CGS unit of magnetic field intensity abbreviated Oe. Named in honor of Danish physicist and chemist Hans C. Oersted (1777–1851) who is famous for pioneering work in electricity and magnetism. *See* magnetic intensity.

OFF   The state in which no power or input is given to a circuit or in which the circuit is producing no output signal. The condition in which the output of a digital flip-flop is low.

Offset   The deviation of a signal from zero. Often used for the DC portion of a combined AC/DC signal (a zero signal offset).

ohmic loss   The power dissipated by the resistance of an electrical circuit as a result of current flowing through it. The power dissipated, P, current, I, and resistance R are related by

$$P = I^2 R .$$

The total ohmic loss for a circuit in which there is a distribution of resistances with different currents flowing through each is obtained by summing the ohmic loss in each resistance.

Ohmic loss can occur when a current flows through an ionized gas as electrons dissipate their energy by collisions with ions, atoms, and molecules.

ohm, international   Symbol: Ω. The SI unit of electrical resistance. A current of 1 ampere through a resistance of 1 ohm requires a potential difference of 1 volt across it. Its definition is the resistance offered to a steady current by a column of mercury, 14.4521 gms in mass and 106.300 cm long, at 0° C.

ohmmeter   An instrument used for the direct measurement of electrical resistance. The component with an unknown resistance is usually connected across two terminals that have a potential difference between them. The magnitude of the current that flows is proportional to the resistance. A part of this current also passes through a galvanometer, and the value of the resistance is indicated by the deflection of the galvanometer, measured against a calibrated scale. However, the galvanometer, in many ohmmeters of this type have been replaced with a digital display. These use a variety of analog to digital conversion circuitry before the value of the resistance is displayed. Ohmmeters that rely on this type of two-terminal configuration have limited accuracy. Four-terminal or bridge type arrangements (e.g., Wheatstone bridge) must be used for greater accuracy.

Ohm's law   States that the electrical current, I, flowing through a metal conductor is directly proportional to the potential difference, V, across it. This relationship is expressed as

$$V = IR ,$$

where the constant of proportionality R is the resistance of the conductor. Ohm's law is only applicable to circuits carrying direct current through conductors with a constant resistance. Consequently, the temperature of the conductor must be held constant since resistance is usually temperature dependent.

Ohm's law cannot be applied to alternating-current circuits, since the current no longer simply depends on the resistance and potential difference, but also on the frequency of the source and the inductance and capacitance that may be contained in the circuit. However, the law has been modified to include the effects of these factors by substituting the impedance, Z, in place of the resistance, R, such that

$$V = IZ .$$

It is sometimes convenient to rewrite Ohm's law in terms of the current density J and the electric field strength E across the conductor as

$$J = \sigma E ,$$

where the constant of proportionality σ is the conductivity.

Ohm's law in acoustics   Describes the way individual partials of a musical sound are perceived by the human ear. The ear can separate a complex tone into its spectral components and perform a kind of Fourier spectral analysis.

ON   The state in which a circuit is receiving power and input and is producing an output. The state in which a digital flip-flop output is high.

online   An electrical device, such as a peripheral device of a computer, is online if it is directly connected to the computer and is ready to perform its function. On the other hand, offline denotes a peripheral device, possibly an online device, which is switched off, broken, or disconnected from a computer.

Onsager theory   Deals with the flow of both a heat current and an electric current in a wire simultaneously. Entropy is generated in the wire due to both processes. In the absence of a potential difference, a heat current depends only on the temperature difference but when there is a potential difference as well, the heat current depends on both the temperature difference and the potential difference. When both temperature and potential differences exist across a wire, the electric current depends on both of these differences. The heat flow, the entropy flow, and the electricity flow are irreversible, coupled flows that exist in a wire as a result of a departure from equilibrium conditions. If the departure from equilibrium is not too great, the heat and electric current I_s and I, respectively are linear functions of the temperature difference, ΔT, and potential differences, ΔE, and are given by the following Onsager equations:

$$I_s = L_{11}\Delta T/T + L_{12}\Delta E/T$$
$$I = L_{21}\Delta T/T + L_{22}\Delta E/T .$$

This expresses the linearity between the flows (or currents) and the generalized forces $\Delta T/T$ and $\Delta E/T$. They can also be represented by

Onsager's reciprocal relation $L_{12} = L_{21}$ where Ls are coefficients connected with electric resistance, thermal conductivity and the thermoelectric properties of the wire.

opacity   This measure indicates how a medium is opaque to electromagnetic radiation and is reciprocal to the transmittance.

opaque object   This is an object that does not transmit electromagnetic radiation, especially light.

open circuit   Term applied to part or all of an electrical circuit in which the impedance is infinite. In practice, this may be done by physically disconnecting a conductor or component necessary to complete that part of the circuit.

open circuit voltage   The voltage measured at a terminal when there is no load connected to the terminal (i.e., no current).

open-loop   An amplifier circuit in which there is no feedback connection.

opera glass   A very simple compact binocular Galilean telescope. Its magnification is low and its field of view not very wide.

operating point   The point that corresponds to the current and voltage values of a device under load. It is the intersection of the load-line and the device I-V curve. *See also* load-line.

operational amplifier   A very high gain DC differential amplifier with two inputs and a single output. It produces positive output when the noninverting input is higher than the inverting input and negative output when the inverting input is higher. It was originally used for mathematical purposes.

ophthalmometer   An instrument to measure curvature of the anterior corneal surface and astigmatism of the eye. It is also known as a *Keratometer.*

ophthalmoscope   An optical device to observe the inside of the eye (the retina, the fundus, and so on) through the pupil. It illuminates the eyeground through the pupil.

optical activity   Phenomenon in which the plane of polarization of a linearly polarized light wave is rotated in passage through a material medium. The effect occurs in materials in which the crystalline layers of the material or its constituent molecules exhibit a systematic twist, causing right- and left-handed circularly polarized light to propagate in the material with different speeds. The phenomenon occurs in quartz crystals and in certain sugar crystals as well as in sugar solutions. Rotation of the plane of linear polarization is explained by the fact that a linearly polarized light wave can be decomposed into two circularly polarized light waves, with oppositely directed polarizations, that propagate at (slightly) different speeds through optically active material, so as to produce a rotation of the plane of linear polarization at the exit surface of the material. The amount and direction of rotation of the polarization vector is in general dependent on both the wavelength of the light and the properties of the given material.

optical axial angle   The size of the angle between the two optic axes of a biaxial crystal. The angle θ can be calculated from the refractive indices α, β, and γ. α, β, and γ are the index of the fastest light, intermediate light, and the slowest light:

$$\tan^2\left(\frac{\theta}{2}\right) = \frac{\frac{1}{\alpha^2} - \frac{1}{\beta^2}}{\frac{1}{\beta^2} - \frac{1}{\gamma^2}} .$$

When $\theta > \pi/2$, the crystal is obtuse bisectrix and negative. When $\theta < \pi/2$, the crystal is acute bisectrix and positive. The optical axial angle is constant for each particular substance at a given temperature and pressure. The optical axes of the wave normals is obtained from the points of the two wave normal surfaces. *See* optical axial plane.

optical axial plane   Also called the *optical plane*. The plane that is defined by two optical axes of a biaxial crystal. The normal of the optical axial plane is called the *optic normal*. The optical axial angle is constant for each particular substance at a given temperature and pressure.

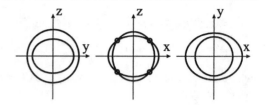

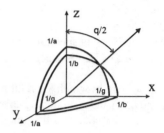

Optical axial angle.

The axis that bisects the angle between the optic axes is called the *bisectrix. See also* optical axis; optical axial angle.

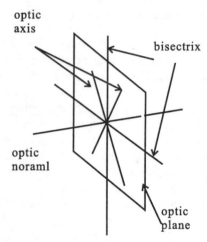

Optical plane.

optical axis   A straight line passing through the optical center, it is also known as the *principal axis. See also* meridian plane.

optical bench   A rigid but movable rod or track equipped with mountings for optical experiments. When an optical bench is used, it is possible to slide optical components along the bench and to determine the position precisely with attached scales. Some optical benches have internally damped honeycomb structures.

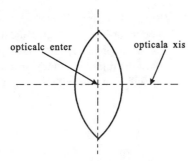

Optical axis and optical center of a lens.

optical bistability   Phenomenon in which the intensity of the light output from a material medium switches discontinuously between high and low values as the intensity of the input light is continuously varied. Effect in general occurs only where the interaction between the light and the material medium exhibits a nonlinear dependence on the electric field of the light wave, which causes the intensity of the light output from the medium, I_{out}, to be a multiple-valued function of the input intensity, I_{in} , corresponding to a graph of I_{out} vs. I_{in} of the form shown in the diagram.

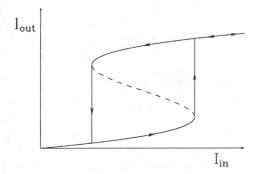

Optical bistability.

optical center   A point on the axis of optical components, in which each ray passing through the optical center does not deviate. It is also known as the *pole*.

optical channel   An optical channel may be free space, the atmosphere, or silica glass fiber optics, depending on the material medium of the optical communication system. The free space optical channel embodies an ideal communications channel because free space neither distorts

nor attenuates the transmitted signal. An atmospheric optical channel, with even minute temperature variations, may significantly broaden the spatial directivity or bend the path of optical signals. Fog or snow would sufficiently attenuate the optical signals to render outdoor atmospheric optical communication unfeasible for distances much farther than that between adjacent buildings. Indoor optical communication channels would not suffer from this fog and snow problem, but temperature variation would still affect the channel. Silica glass rods provide unparalleled transparency to optical signals. Signal distortion, however, occurs in long-distance transmission as the light wave reflects off the boundary between the two glass layers in the coaxial cylindrical step-index fiber optical fiber, wherein the core layer is made of glass with a slightly higher refraction index than the coaxial outer layer.

optical communication   Optical communication refers to the transmission and reception of information using electromagnetic waves in the visible and infrared parts of the electromagnetic frequency spectrum. Optical communication may also be through free space (in inter-satellite communications), through the earth's atmosphere, or through silica glass optic fibers. Free space or atmospheric optical communications differ from the more traditional radio communications primarily in the use of electromagnetic waves at much higher frequencies (on the order of 300,000 GHz), such that the signal wavelengths are much shorter than the dimension of the hardware devices themselves. This exceedingly high frequency provides very wide bandwidths for individual channels, facilitates the miniaturization of hardware components, exploits the unparalleled transparency of silica glass as a transmission wave guide via fiber optic communications. Lasers, with their superior spatial directivity relative to incoherent radiation, are often desirable optical light sources in free space, through the atmosphere, or in optical fibers. However, fog or snow would sufficiently attenuate the optical signals to render outdoor atmospheric optical communication unfeasible for distances much farther than that between adjacent buildings. Optical signals may be modulated using the conventional amplitude, frequency or phase techniques, but intensity modulation and polarization modulation are most common.

optical density   A measure of the light-stopping power of a transparent material, defined to equal the logarithm of the ratio of the intensity of the field in the incident light, I_0 , to the intensity of the field in the transmitted light, I .

optical glass   Glass used in the manufacture of optical parts, lenses, prisms, and so on. It should be free from defects such as bubbles and strain. In choosing a proper glass, refraction and dispersion of the glass are important. Practically, more than three refractive indices are used to characterize a glass. Usually, three Fraunhofer lines are used to specify the refractive index of the glass: the "Hydrogen C-line" ($\lambda = 656.2816$ nm, red), the "Helium D-line" ($\lambda = 587.5618$ nm, yellow), the "Hydrogen F-line" ($\lambda = 486.1327$ nm, blue). Usually, the index of refraction for yellow light ("Sodium D-lines"; wavelength 589.3 nm) is used as the main refractive index. There are more than 200 kinds of optical glass, which is divided into seven categories: crown, borosilicate crown, dense crown, light flint, flint, dense flint, and others.

Typical Value of Refractive Index of Optical Glass

Name	n_d
Borosilicate crown	1.51
Crown	1.52
Dense crown	1.6
Light flint	1.58
Flint	1.62
Dense flint	1.7

Ordinary crown glass has a refractive index within the range 1.51 to 1.54. Flint glass contains a refractive index between 1.58 and 1.72. Lanthanoid oxide is added to optical glass to make the refractive index higher.

optical path/path length   The optical path is simply the product of length/distance and the refractive index. More precisely, the optical path t is the integral of the refractive index ($n(s)$) over elements of length along the path (P) which rays pass through:

$$t = \int_P n(s)ds \ .$$

It is also known as *optical distance* or *optical length*. Fremat's principle says that light traverses the paths of which the optical pass-length has the smallest value.

optical pumping   Process whereby selected quantum energy levels in atomic or molecular systems are excited by optical radiation (called *pumping radiation*) so as to produce an inversion in the thermal distribution of the selected level(s) with respect to the lower (or ground) levels. Process is necessary for production of laser or maser radiation via enhanced stimulated emission.

optical switching   Process in which a nonlinear interaction between light and a material medium causes an optical signal to switch between two or more output modes as a function of the input intensity. Contrasts with *electro-optic switching* in which an optical signal is switched between output modes via application of an electric field to the medium. The phenomenon (in principle) makes possible signal routing processes used in optical communications and logical operations required in all-optical computing. High field intensity for enhancement of nonlinear interaction needed for switching is in general the result of either partial confinement of light inside a (nonlinear) etalon or concentration of light in a waveguide or directional coupler. Optical switching devices divide into types based on a single input beam and on both an input signal and a control beam.

optics, collecting   *See* optics, detection.

optics, detection   That part of a microscope of any kind, spectrometer, or other instrument that deals with the analysis of visible light, that gathers, collects, or detects the relevant light that is scattered or transmitted in any way from the body in observation. An example is given in *confocal detection optics*.

optometer Any of a large class of optical and mechanical instruments used to measure refractive errors of the eye, generally combining both subjective and objective elements in measurements; predecessors of current refractometers and automated refractors.

OR A logic operation that gives 1 if either of the two inputs is 1 and gives 0 if both inputs are 0.

organ pipes, vibration in Aerodynamically generated vibrations of air columns in pipes are the source of sound in the organ (flue pipe). Constant air flow enters the pipe through an opening and a lip strikes a knife edge above the mouth. The resonator, an open pipe, is connected to this lower structure. The air stream emerging through the lip undergoes unstable oscillation, and it can, at the appropriate stream velocity, excite the resonator. The fundamental frequency of the resonator determines the pitch of the pipe for self-excitation. The larger the cross section of the pipe (resonator) the more the fundamental dominates the tone produced by the pipe. *See also* pipes, sound from.

orthogonal code An orthogonal code refers to a code wherein various message bits are represented by distinct sequences of digits and any two such sequences are orthogonal to each other. That is, if each sequence is considered a vector, then the inner product of any two sequences in the code has an inner vector product equal to zero.

oscillation, parasitic Spontaneous oscillations in a circuit generated by lead inductances and inter-lead capacitances. They can be sources of noise in oscillator circuits.

oscillations, energy of Energy (E) associated with wave motion caused by the action of springlike forces and involving kinetic (K) and potential (u) energies. Waves can transport energy without the transport of mass. For example, the average energy density of transverse traveling waves on a stretched string with mass

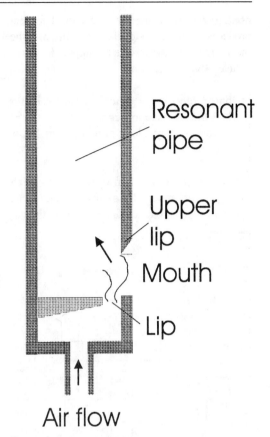

Organ pipe.

per unit length μ under tension T is $\left\langle \frac{dE}{dx} \right\rangle = \left\langle \frac{dK}{dx} \right\rangle + \langle u \rangle = \frac{1}{2}\mu\omega^2 y_0^2$, with ω as the angular frequency and y_0 the wave amplitude. The energy density is itself a traveling wave transporting energy at speed $v = \omega/k = \sqrt{T/\mu}$ and it is proportional to the squares of oscillation amplitude and frequency. *See also* power in wave motion.

oscillations, spontaneous Oscillations occurring naturally in unstable circuits or devices in the absence of a triggering signal, as opposed to those that occur only when triggered by an outside signal.

oscillator A device or circuit used to set up and maintain oscillations at a desired frequency. The most common form is an LC circuit giving an oscillator frequency of $\omega = \frac{1}{\sqrt{LC}}$ where L is the inductance and C is the capacitance.

oscillator, beat frequency   An oscillator that produces a reference signal to combine with the incoming signal to produce an output. It is used in heterodyne reception.

oscillator, blocking   A tube circuit oscillator in which the tube is highly conducting for a short period followed by a long period in which it is non-conducting.

oscillator, Colpitts   An oscillator in which two capacitors are used in series in the tank circuit with feedback taken from between the capacitors.

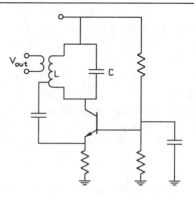

A Hartley oscillator, L and C are the tank circuit while the rest of the circuit provides the feedback to sustain the oscillations.

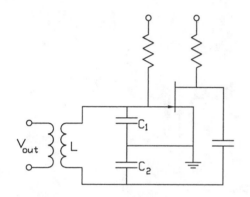

A Colpitts oscillator, L, C_1 and C_2 form the tank circuit, while the remainder of the components provide the feedback to sustain the oscillations.

oscillator, crystal   An oscillator in which the piezoelectric properties of a quartz crystal are used to produce the oscillations. The high Q value of the quartz leads to a very stable oscillator.

oscillator, Hartley   An oscillator in which the feedback connection to the tank circuit is made to the inductor.

oscillator, Hertzian   An oscillator consisting of two capacitors connected to a conducting rod with a spark gap. It produces highly damped oscillations at the frequency $\omega = \frac{1}{\sqrt{LC}}$ where L is the mutual inductance and C is the capacitance.

oscillator, local   An oscillator used in a superheterodyne receiver to give the reference frequency to the mixer used to convert the input frequency to the output frequency. Changing the local oscillator frequency changes the input frequency that will be converted to output, tuning the receiver.

oscillator, master   An oscillator used in amplifiers to establish the carrier frequency of the output.

oscillator, relaxation   An oscillator based on the charging and discharging of a capacitor. A relaxation oscillator can be quite stable, and normally produces a triangular or sawtooth output.

oscillator, squegging   A tube-based oscillator that reaches a high amplitude, and then is brought to zero by the blocking effects of leakage current on the grid.

oscillator, Wien bridge   A low distortion oscillator based on a feedback amplifier with a 180° phase shift at the desired frequency. A Wien bridge oscillator can operate at frequencies ranging from 20 Hz to 200 kHz with a very small distortion as long as the amplifier is kept in the linear regime.

osmocomformer   Refers to the way that some marine invertebrates adjust to changes to the salt concentration of their surroundings. At any given instant, these animals have the same osmotic pressure as the sea water that surrounds them in order to diminish osmotic flow from their bodies. When there occurs a change in the concentration of the surroundings, osmocomformers change their osmotic concentration to match that of the external environment and thus keep the osmotic balance. This behavior is in contrast to *osmoregulators,* animals that keep constant their own osmotic concentration in spite of external changes in salinity.

osmometer   An osmometer is a device for measuring osmotic pressure. The general scheme is given by the accompanying figure, where M is a membrane, I is the solvent with chemical potential μ_1, and II is the solvent and dissolved polymer with chemical potential μ_2. Both chemical potentials are at atmospheric pressure P_o. The difference in pressures, $P - Po$ gives the osmotic pressure π.

In this example, the membrane is permeable to molecules of the solvent but not to molecules of the solute (polymer in this case). The equilibrium condition requires the equality of the chemical potentials, $\mu_1 = \mu_3$, where

$$\mu_3 = \mu_2 + \int_{P_o}^{P_o+\pi} \left(\frac{\partial \mu_2}{\partial P}\right)_T dP .$$

Because $(\partial \mu_2/\partial P)_T = V_2$, $\overline{V_2}$ is the partial molar volume of the solvent which in practice is not pressure dependent. Hence $\mu_3 = \mu_2 + \overline{V}_2\pi$.

In the limit when the concentration of the solute c_2 in II tends to zero, the osmotic pressure is given by $\pi/c_2 = RT/M_2$, which expresses van't Hoff's law. M_2 is the molecular weight of a polydisperse polymer that in principle is the mean molecular weight from a mixture with M_i,

$$M_2 = \frac{\sum_i n_i M_i}{\sum_i n_i} .$$

osmoreceptors   In general, osmoreceptors react to minute changes in the osmolarity (concentration of particles) of some fluid in their vicinity by releasing some substance to counteract the change.

As an example, in the hypothalamus there are osmoreceptors (central receptors) that sense the osmolarity of extracellular fluid. When the osmolarity is high, they send a signal to the pituitary gland to release the hormone ADH (vasopressin, primary regulator of body water). ADH then acts on the distal tubules of the kidneys to

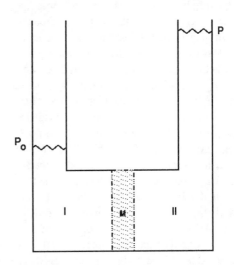

Osmometer.

increase reabsorption of water. This has the effect of increasing blood volume and blood pressure. Inhibition of production of ADH is conversely signaled when the osmolarity is low.

Another example of osmoreceptors are peripheral receptors. Peripheral receptors are osmoreceptors in the mouth and throat that signal the brain stem in the case of thirst.

osmoregulators   Osmoregulators and osmoregulatory mechanisms form one of the most important evolutionary innovations that enabled multicellular animals to carry around their own internal sea and stable internal fluid environment out of primordial waters. Osmoregulators maintain an internal environment that is constant in osmotic pressure and salt balance. Different species require specialized osmoregulatory mechanisms, which vary widely according to the nature of an organism's habitat.

In amphibians, the majority of the input of water occurs through the skin and across the wall of the bladder. This is realized by the production of a hormone that causes water to enter the body through the skin triggered from the brain when the animal is on a moist surface or immersed in water. In addition, before dehydration, hormonal stimulus can cause water to be "reclaimed" from the bladder and returned to the extracellular fluid. This is in contrast with other land vertebrates in which the bladder is largely a receptacle for urine that will be excreted.

In many terrestrial vertebrates, to counteract the loss of water from the lungs, the nasal cavities act as a "countercurrent exchange system". During inhalation, air passing through the nasal cavities is warmed by heat from adjacent tissues, but in the process, the temperature of these tissues falls. One result of this is a cold nose – a dog's nose is a good example. The inhaled air is further warmed and humidified in the lungs. Then, as it passes back out during the next exhalation, the warm, moist air flows over the cooler nasal surfaces, and the air gives up some of its heat. As the air cools, much of the water vapor condenses out on the nasal surfaces and so is not breathed out of the body. This process can save up to 20% of an animal's total need for water.

Also, in most terrestrial vertebrates, by controlling the amount of water and salt lost in the urine, the kidneys form the primary regulatory organs to keep a constant osmotic balance in the body.

osmosis, negative, anomalous   Negative osmosis, or *reverse osmosis* can be achieved by applying pressure to the side of a solution separated by a membrane in order to achieve the reverse flow of particles that would otherwise happen in normal osmosis.

It can be used as a separation technique in which a semipermeable membrane is placed between two solutions containing the same solvent. The membrane allows passage of the solution while preventing passage of larger molecules. Reverse osmosis occurs when pressure is applied to the solution on the side of the membrane that contains the lower solvent concentration. The pressure forces the solvent to flow from a region of low concentration to one of high concentration.

Negative osmosis is often used for water purification, concentrating impurities, recovering contaminated solvents, cleaning up polluted streams, and desalinizing sea water.

osmotic equilibrium (cell)   Cellular cytoplasm contains a high concentration of organic molecules including macromolecules, amino acids, sugars and nucleotides. In the absence of a counterbalance, this concentration of

molecules would drive water inward by osmosis, which if let undisturbed would result in swelling and eventual bursting of the cell.

Osmotic equilibrium across the cell membrane is then established by the proper diffusion of molecules through the membrane mediated by mechanisms such as active, passive, and facilitated diffusion. In passive diffusion small uncharged molecules diffuse across the cell membrane equilibrating the concentrations between out and in. Active transport requires ATP hydrolysis and is the main transport carried out by ion pumps. The ion pumps provide the required counterbalance to the osmotic gradient by providing counter gradients of Na^+ and K^+. In particular, the pumps establish a higher concentration of Na^+ outside the cell. Flow of K^+ through open channels further establishes an electric potential across the plasma membrane. This potential in turn drives Cl^- out of the cell. The differences in ion concentration balance the high concentration of organic molecules inside cells, equalizing the osmotic pressure and preventing the net influx of water.

In some simpler organisms (protists), membrane–bound contractile vacuoles pump fluid in a cyclical manner from within the cell to the outside by alternately filling and then contracting to release its contents at various points on the surface of the cell. This cyclic pumping functions in maintaining osmotic equilibrium.

osmotic pressure, Donnan   Such equilibrium as that found between a charged, immobile colloid (such as clay, ion exchange resin, or cytoplasm) and a solution of electrolyte is called *Donnan equilibrium.* In this system ions of like charge to the colloid tend to be expelled, and ions of opposite charge tend to be attracted by the colloid. The result is that the colloid compartment is electrically polarized relative to the solution in the same direction as the colloid charges, creating a *Donnan potential,* and the osmotic pressure is higher in the colloid compartment.

osmotic responses, kinetic theory, (cell)   An osmotic response is the result of the unbalanced pressures on both sides of a membrane that separates a solvent with different solute concentrations on both sides. The number of solvent particles hitting the side of the membrane with the most solute per unit time per unit area is less than the side with less solute. This means that the flow of particles from each side will be unequal, the side with the more solute gaining more solvent per unit time. The unequilibrated rate of exchange of solvent particles establishes then a flow from the solvent–rich to the solution–rich compartment creating an osmotic pressure that tends to equalize on both sides of the membrane. *See* osmotic transport.

osmotic transport   Osmotic transport refers to the transfer of a liquid solvent through a semipermeable membrane from a region with a low concentration of solute (high concentration of solvent) to one with a higher concentration of solute (low concentration of solvent). The membrane is only permeable to the solvent. Osmosis can be realized in practice by the following experiment: given a vessel separated into two compartments by a semipermeable membrane, if both compartments are filled to the same level with a solvent, and if a solute is added to one side, osmosis will occur, increasing the level of solvent at the side with the solute. The minimum pressure applied to the side with solute to stop the solvent transfer is called the osmotic pressure (*see* osmometer).

Dialysis is called the *transfer of solute* (rather than the solvent). The direction of transfer is from the area of higher to the area of lower concentration of the material transferred. In dialysis, dissolved salts are removed from solutions of proteins or other large molecules.

output   A signal (current or voltage) produced by a circuit and either measured directly or used as the input to a separate circuit. The terminal from which the output is taken.

output capacitor   A capacitor used to transmit the AC component of an output signal, while blocking the DC component of the signal.

output characteristics   The electrical properties (impedance, capacitance, etc.) associated with the output channel of a device. The dependence of the output current on the input and output voltages for a transistor circuit.

overload   A condition in which the desired input to or output from a circuit is larger than that which can be accommodated.

overtones   Components of the complex sound whose frequencies are integer multiples, greater than 1, of the fundamental frequency. Harmonics other than the fundamental component. *See also* frequency, fundamental; harmonics.

Owen's bridge   Owen's bridge is an AC variation of the Wheatstone bridge. Owen's bridge is used to measure inductance of an unknown inductor in terms of other known resistances and capacitances. A circuit diagram is shown below. Only R_3 and C_3 should be adjustable if you want the resistance and inductance balances to be independent of each other. The equations of balance are shown below:

$$R_x = R_1 \frac{C_2}{C_3}$$
$$L_x = R_1 R_3 C_2 .$$

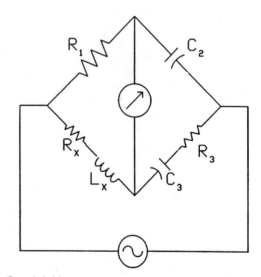

Owen's bridge.

oxygen, liquid   Oxygen liquefies at 90 K and can be produced in air liquefaction and oxygen separation plants. In liquid form it is very useful as a propellant for guided missiles and rockets. One of its advantages is that for a given mass, the volume of liquid is much less than in the gaseous state and it is therefore preferred for transportation and storage.

P

pacemakers, cardiac   Cardiac pacemakers are any of several electronic devices used to stimulate or regulate contractions of the heart muscle in people with cardiac problems. The devices are usually miniaturized and surgically implanted in the patient.

packet switching   A form of switching using packets to carry information. The packets are a combination of information (data) and a description of this information (metadata or header). The packet switch is a switch that examines a packet header to determine the packet's destination.

paleomagnetism   Study of the intensity and direction of the magnetic field of the earth in the present and the geological past, its origin and its change with time. This involves study of magnetized rocks in the earth's crust. When these rocks form from magma, a remanent magnetic moment is frozen in them along the direction of the earth's field. These rocks carry information about the intensity and orientation of the earth's magnetic field from that time.

panoramic   An optical instrument or a lens that takes a wide field of view.

pantograph   A sliding current collector bar used on top of electric trains to make contact with an overhead electric wire. The bar is supported by a four-bar parallel linkage, with no links fixed. The current collector bar is thrust upwards by powerful springs with sufficient pressure to follow variations in the height of the overhead wire and therefore maintain a low-resistance contact. The oscillations in the overhead wire and the pantograph, which may occur during high speeds, can lead to a break in the connection between them or cause excessive arcing.

parallax   Apparent change in the direction or position of an object when viewed from a different position or direction, often used to recognize the relative positions of the objects. In astronomy, the angle of parallax is used to determine distances of nearby stars.

parallel resonance   A condition in which the magnitude of the parallel inductance and capacitance in a load are equal.

parallel voltage feedback   A feedback condition in an operational amplifier circuit in which the feedback current is parallel to the input current and proportional to the output voltage of the amplifier.

paramagnetic materials   Materials whose atoms carry a permanent magnetic moment but are not magnetically ordered so that they do not have a spontaneous magnetic moment. They have a permeability smaller than a ferromagnet. Examples include all magnetic materials that show magnetic order when they are above their ordering temperature. *See* paramagnetism; permeability, magnetic.

paramagnetic probes   Probe used in *electron spin resonance* (ESR) studies, also called *electron paramagnetic resonance* (EPR). ESR is in general used to detect and measure unpaired electrons in the sample under study that ultimately leads to structural and dynamic information. Because the probe is paramagnetic (no net magnetic moment) it will not influence the sample or the measurements.

Ideally suited to study the effect of external magnetic fields within the surface of a sample, specifically over superconductors in the detection of induced electric currents when subjected to changing magnetic fields. The paramagnetic probes are usually molecules that in principle could also be attached, for example, in the myosin head for the study of muscle fiber activity. In addition, they could serve as probes of molecular motion of biomolecules and during experiments on phase transitions.

paramagnetic resonance   Also called *electron spin resonance* or *electron paramagnetic resonance*. Refers to resonant absorption of electromagnetic radiation, usually in the microwave range, by paramagnetic ions in a magnetic field. A magnetic atom has discrete energy levels corresponding to the orientation of the magnetic moment of the atom in an applied field. The atom is able to resonantly absorb radiation composed of photons whose energy corresponds to the difference between these energy levels. Gives information on the gyromagnetic ratio of magnetic atoms and ions.

paramagnetism   Describes a system of permanent magnetic moments, usually in a solid, that are not ordered but have a positive magnetic susceptibility. This susceptibility is independent of the applied field but is a strong function of temperature, thus:

$$\chi = Np^2/3k\,(T - T_0)$$

where N is the number of magnetic atoms per unit volume, p is the effective magnet moment per ion, k is Boltzmann's constant, T is temperature, and T_o is a constant called the *Curie-Weiss temperature*. For materials that do not order magnetically at low temperatures, T_0 is very small and may be zero. In the paramagnetic state the magnetic system is above its magnetic transition temperature and so magnetic moments are thermally agitated so that they point in random directions that change with time. *See* Curie Weiss law; paramagnetic materials; susceptibility, magnetic.

paraxial approximation   Only paraxial rays are considered in imaging by a lens/optical system and the small angle approximation $\sin\theta \approx \tan\theta \approx \theta$ is used.

Parseval theorem   Theorem relating the integral of the product of two functions, $f_1(x)$ and $f_2(x)$, to an integral over the product of the Fourier transforms of the respective functions, $g_1(k)$ and $g_2(k)$, or to a sum over a product of Fourier coefficients. The statement of theorem has distinct forms in the separate cases where functions f_1 and f_2 are either periodic functions of x in an interval L, or non-periodic functions of x in the interval between plus and minus infinity. The respective cases correspond to representations for $f_i(x)(i = 1, 2)$ in terms of Fourier coefficients a_{in} or transform functions $g_i(x)$ in

the forms:

$$f_i(x) = \sum_{n=-\infty}^{\infty} a_{in} e^{inKx}, \quad K = 2\pi/L,$$

$$f_i(x) = \int_{-\infty}^{\infty} dk\, g_i(k) e^{ikx},$$

from which the Parseval theorem can be derived in the respective forms:

$$\int_{-\frac{L}{2}}^{\frac{L}{2}} dx f_1^*(x) f_2(x) = L \sum_{n=-\infty}^{\infty} a_{1n}^* a_{2n},$$

or

$$\int_{-\infty}^{\infty} dx\, f_1^*(x) f_2(x) = 2\pi \int_{-\infty}^{\infty} dk\, g_1^*(k) g_2(k).$$

partially polarized light   Most simply, a mixture of polarized and unpolarized light. (*See* polarization of light.) A more precise definition derives from a general representation of light as a superposition of two linearly polarized components, characterized by polarization vectors vibrating in mutually orthogonal directions in the plane perpendicular to the direction of propagation of the light. Under the condition that the two components have equal amplitudes, and differ in phase by an amount that varies rapidly and irregularly with time, the light is said to be *unpolarized;* whereas, under the condition that the components have unequal amplitudes, and/or a phase difference that is not completely random, the light is said to be *wholly* or *partially polarized.* The representation of light as a superposition of linearly polarized components provides for an operational definition of partially polarized light in terms of measurements of the intensity of the light transmitted through a linear polarizer oriented so as to define a direction perpendicular to the direction of propagation of the light. Specifically, rotation of the polarizer through all orientations in the plane perpendicular to the propagation direction, and measurement of the intensities of the transmitted light, make possible (in principle) a determination of a maximum and a minimum in the transmitted intensity, I_{max} and I_{min} , in terms of which the *degree of polarization* of the light, P, can be defined by the ratio

$$P \equiv \frac{I_{max} - I_{min}}{I_{max} + I_{min}} .$$

The definition assigns P the values 1 and 0 in the cases of completely polarized light and (completely) unpolarized light, respectively, and allows partially polarized light to be defined in terms of values of P unequal to either 1 or 0.

partition, transmission of sound through
Transmission of sound energy (transmission loss) through single walls that are homogeneous and damped, dependent on the product of surface density and frequency. The thickness is not of importance for wall thicknesses below 30 cm. When walls are combined to form an enclosure, a small opening (thin windows, cracks around the door) can render noise reduction measures useless.

Paschen-Back effect   Named after German physicists Louis C.H.F. Paschen (1865–1947) and Ernst E.A. Back (1881–1959). An effect in which the spectral lines of a light source are split into multiplets when the light source is placed in a strong magnetic field. The strong magnetic field modifies the atomic structure of the atom breaking the coupling of the orbital angular momentum L and spin angular momentum S leading to the observed multiplets.

passive device   Devices such as resistors, inductors, or capacitors without a built-in power source, as opposed to active devices such as transistors.

pattern, acceptance   (**1**) The acceptance pattern of an antenna is the distribution of the off-axis power relative to the on-axis power as a function of angle or position. The acceptance pattern is the equivalent of a horizontal or vertical antenna pattern.

(**2**) The acceptance pattern of an optical fiber or fiber bundle, is the curve of total transmitted power plotted against the launch angle, where the launch angle is the angle with respect to the normal at which a light ray emerges from a fiber surface.

peak value (voltage/current)   The maximum positive or negative value of an alternating quantity (voltage or current). For example, a sinusoidally varying voltage with time has a peak value equal to the amplitude. The peak value of a sinusoidal voltage displayed on an oscilloscope is obtained by subtracting the lowest voltage from the highest voltage and dividing by two.

Peltier effect   (**1**) The heating or cooling that occurs at a junction of two dissimilar metals when an electric current is passed through it. The direction of the current determines whether the junction will be heated or cooled. The degree to which the junction is heated or cooled is determined by the magnitude of the current and the type of metals used. For a circuit consisting of two junctions of dissimilar metals (A and B, say) in series, one junction will be heated while the other cooled when a direct current is passed through the circuit. This is simply due to the current passing from metal A to metal B at one junction, and vice versa. The origin of the Peltier effect can be understood by thinking of electrons as being evaporated from metal A and condensing on metal B or vice versa. Heating or cooling will result if the energy required for evaporation is not equal to the energy required for condensation. Many Peltier devices are made from two dissimilar semiconductors and are used to cool miniature electronic components or provide temperature control for electronics for which the performance is critically dependent on temperature stability.

(**2**) The production or absorption of heat at a junction between two dissimilar conductors when a current is passed across the junction appears as power generated or absorbed at the junction. The latter is directly proportional to the current. $P = \pi_{ab} I$, where π_{ab} is the Peltier coefficient for current I passing from condutor b to conductor a.

penetration depth (low temperature)   Deals with the penetration of magnetic flux into the surface layer of a superconductor. The penetration depth is temperature dependent, decreasing from a value corresponding to total field penetration at the critical temperature T_c to an approximately constant value at $T_c/2$. Typically the penetration depth is about a few tens of nm for $T << T_c$.

periodic waves   A train of waves in which each particle exhibits periodic motion. The simplest special case of periodic waves is a simple harmonic wave, in that each particle is subjected to harmonic motion. *See also* harmonic motion. More complex waveforms can be described in terms of infinite sums of simple harmonic waves, whose frequencies are integral multiples of the fundamental frequency, using the Fourier series representation. *See also* harmonics.

peripheral vision   Act of seeing (or vision of) images produced by light falling on areas of the retina outside its central (macula) region. Makes possible visual awareness of objects located on either side of the column of space extending forward from the pupil of the eye.

periscope   An instrument that allows observation of objects not in direct line of sight. In the simplest form, it may consist of an optical L shaped tube with lenses, and a mirror or prism at the bend of L, to enable a person in a submerged submarine etc. to have a view of objects on the surface of water or over and around an obstacle.

Permalloy   An alloy made up of Fe and Ni, and heat treated in a prescribed way to have a high permeability and low hysteresis so that it is a good soft magnet. Useful as a transformer core material. *See* permeability, magnetic; hysteresis.

permanent magnet   Magnetic material with a permanent magnetic moment; i.e., it has a magnetic moment in the absence of an applied field. Possesses large coercivity and usually has a large magnetic moment. Useful in electric motors, generators and many other electrical devices. Best permanent magnet currently known is based on $Nd_2Fe_{14}B$. It is prepared in a nanocrystalline form and contains small crystallites of $\alpha-Fe$. Many other alloys are also used

as permanent magnets including Sm-Co and Alnico. *See* coercivity; alnico.

permeability, diffusive   During diffusive permeability, a membrane allows passage of certain, especially small, molecules or ions whose primary means of motion is of diffusive nature.

permeability, incremental (magnetic)   Defined at a particular bias field H. A small field ΔH is applied on top of the bias field leading to a small increase in the magnetic flux ΔB. The incremental permeability is then defined as $\Delta B/\Delta H$.

permeability, magnetic   The permeability μ is the proportionality constant between magnetic induction B and magnetic field intensity H and is defined by $\mathbf{B} = \mu\mathbf{H}$. If $\mathbf{B}$ is parallel to $\mathbf{H}$ then μ is a scalar quantity. For free space the permeablitiy is $\mu_0 = 4\pi x 10^{-7} H/m$. For paramagnetic materials the permeability can be an order of magnitude or more larger than μ_o. For ferromagnetic materials the permeability can be as much as six orders of magnitude larger than μ_o. If B is not parallel to H then the permeability is a tensor quantity. *See* susceptibility, magnetic.

permeability, osmotic   A membrane with osmotic permeability allows a rate of flow of a solvent to pass through during the case of an osmotic pressure difference between the two sides of the membrane (*see* osmotic transport).

permeance   The reciprocal of the reluctance of a magnetic circuit. *See* reluctance.

permendur   A magnetic alloy made of equal amounts of Fe and Co with a small amount of V (2%) added and annealed in a prescribed way. This material has a high permeability and so is a good soft magnet. *See* permeability, magnetic.

permittivity   Proportionality constant between electric displacement $\mathbf{D}$ within a medium and applied electric field thus $\mathbf{D} = \epsilon\mathbf{E}$. If $\mathbf{D}$ and $\mathbf{E}$ are parallel then ϵ is a scalar quantity. ϵ is related to the relative permittivity by $\epsilon = \epsilon_r\epsilon_0$ where ϵ_0 is $8.85x10^{-12}$ farads/meter. Also related to the electric susceptibility χ_e by

$$\epsilon = \epsilon_o(1 + \chi_e) \, .$$

persistence of vision   Ability of the eye to retain image on the retina for a brief time after termination of the optical excitation that creates the image. This allows the sequence of discrete images used in cinematography (for example) to produce an image on the eye that is continuous in time.

Petzval field curvature   A monochromatic aberration of an optical system in which the image surface is not planar but curved. This image surface is called the *Petzval surface* and the associated aberration is called curvature of field. If there are a number of thin lenses in air, the Petzval surface curvature R_P is given by:

$$\sum \frac{1}{n_i f_i} = \frac{1}{R_p} \, ,$$

where n_i and f_i are the index and the secondary focal lengths of the ith lens.

phantom circuit   Telecommunication circuit obtained by superimposing an additional circuit on two existing physical circuits by means of repeating coils. It enables the transmission of three messages with only two pairs of wiring.

phase   (1) The fraction of the period through which the time variable of a periodic quantity (such as the sound vibration) has moved, as measured from a point in time. It is commonly described in terms of angular measure, with one period equal to 360° or radians. *See also* phase angle.

(2) The type of state of the system, such as solid, liquid or gas, identified as having a distinct molecular arrangement that is homogeneous throughout. When more than one phase is present in a system, the phases are separated from each other by easily identifiable phase boundaries. *See also* phase changes.

phase angle   The argument $\omega\left(t \pm \frac{x}{c}\right) + \phi$ of the sinusoidal wave described by $p = p_A \sin\left[\omega\left(t \pm \frac{x}{c}\right) + \phi\right]$, obtained as a particular solution of the wave equation. The constant

ϕ is called the zero phase angle of the wave, the factor p_A is the peak value (amplitude) of the pressure, the coefficient ω is the angular frequency, t is time, x, the direction of propagation of the sound wave and c, the speed of sound in the medium. *See also* harmonic motion and periodic waves.

phase changes   Transitions of a substance from one phase to another, such as gas to liquid (condensation), solid to gas (sublimation), etc. Release or absorption of energy accompanies phase changes. *See also* phase.

phase conjugation   Process in which a given optical field is mixed with two counter-propagating electromagnetic beams in a nonlinear medium to produce radiation that propagates in a time-reversed manner with respect to the signal field. Process corresponds to *degenerate four-wave mixing* in which three fields of a single frequency mix in a nonlinear medium, characterized by a third order nonlinear susceptibility $\chi^{(3)}$, to produce a fourth field at the same frequency with an amplitude equal to the complex conjugate of the amplitude of one of the fields. Term *phase conjugation* derives from the fact that time-reversed form of signal field

$$\mathbf{E}(\mathbf{r}, t) = \mathbf{E}(\mathbf{r}, \omega)e^{-i\omega t} + \mathbf{E}^*(\mathbf{r}, \omega)e^{i\omega t} \, ,$$

represented by

$$\mathbf{E}(\mathbf{r}, -t) = \mathbf{E}^*(\mathbf{r}, \omega)e^{-i\omega t} + \mathbf{E}(\mathbf{r}, \omega)e^{i\omega t} \, ,$$

differs from field $\mathbf{E}(\mathbf{r}, t)$ only by replacement of the amplitude $\mathbf{E}(\mathbf{r}, \omega)$ by the complex conjugate amplitude $\mathbf{E}^*(\mathbf{r}, \omega)$ (corresponding to a field propagating in a direction opposite to the direction of propagation of the field with amplitude $\mathbf{E}(\mathbf{r}, \omega)$). With phase conjugation, when a signal field enters the four-wave mixing region after passage through an aberrating medium, the distortion produced in the signal field is removed from the time reversed (phase conjugate) field after passing through the same aberrating medium in the opposite direction.

phase constant   (1) A rating of a line or medium through which a plane wave of given frequency is being transmitted. It corresponds to the imaginary part of the propagation constant, and describes the rate of change of phase of a field component in the direction of propagation in radians per unit length.

(2) The constant ϕ in the argument of the sinusoidal wave described by $p = p_A \sin[\omega (t \pm \frac{x}{c}) + \phi]$. *See also* phase angle.

phase delay   Ratio of the total phase shift measured in radians of a sinusoidal signal propagating in the transmission line to the frequency measured in radians/second.

phase difference   The difference between the phases φ of two sinusoidally varying quantities $y_1(t) = \sin(\omega t + \varphi)$ and $y_2(t) = \sin(\omega t)$ that have the same frequency ω. Also called *phase angle*.

phase discriminator   An electronic device that generates an output signal that is proportional to the phase difference between an oscillator signal and a reference signal. It is used to control the oscillator and maintain it in synchronism with the reference signal. Also known as *phase detector.*

phase, in AC circuits   The displacement of a periodic waveform (usually sinusoidal) with respect to a specific reference time or another periodic waveform with the same angular frequency that does not rise and fall in unison. Usually expressed as an angle ranging from 0° to 360°. Two waveforms are said to be in phase, in quadrature, and in opposite phase (or antiphase) if the phase angle is 0°, 90°, and 180°, respectively. It is commonly used in many circuits where the current is not in step with the applied alternating potential difference.

phase lag   The negative of phase difference between a sinusoidally varying quantity $y_1(t) = \sin(\omega t + \varphi)$ and another quantity, which varies sinusoidally at the same frequency $y_2(t) = \sin(\omega t)$, when this phase difference is negative.

phase sensitive detector (PSD)   A detector that gives DC output proportional to the phase shift between a reference signal and the input signal. It is also known as a *synchronous rectifier, synchronous detector,* or *synchronous demodulator.*

phase shifter   A device used to create a phase shift between the input and output signals. A simple type consists of a resistor and capacitor in series with the output taken across the capacitor.

phase splitter   A device used to produce two outputs, one in phase and one 180° out of phase, with an input signal.

phase velocity   The velocity of a point that moves with a sound wave at constant phase. If the acoustic disturbance is represented as a set of harmonics, a phase angle can be assigned to any point of a component. The phase velocity is the distance covered per unit time of this point along the direction of propagation. In a nondispersive medium the phase velocity of all harmonics is the same, which is not the case in a dispersive medium. The phase velocity appears in the acoustic wave equation that describes the propagation of sound in a medium. In an unbounded, homogeneous medium the magnitude of the phase velocity corresponds to the speed of sound. The speed of sound is a function of the adiabatic bulk modulus and the density of the medium; its magnitude depends on temperature, pressure and material composition. *See also* modulation, acoustic.

phasor   A vector representation of sinusoidal waveforms. It is often used in the analysis of AC signals.

phonetics   Study of the sounds of human speech, their generation and the signs used to represent them. The phonetic transcription provides symbols for each phoneme of a language transcribed, as well as additional symbols to specify differences between variations of the same phoneme depending on the situation.

phonic chronometer   An electric chronometer driven by a phonic motor. Phonic motors allow the conversion of vibratory motion into rotary motion of constant speed using an electrically maintained tuning fork and a phonic wheel.

phonodeik   An instrument used to record the movement of air caused by sound. It consists of a horn that delivers the sound on a fine glass diaphragm. A fine wire attached to the diaphragm is wound around a vertical steel staff and maintained taut by a spring. A small mirror is fixed to the staff. Pressure oscillations caused by the sound waves set the diaphragm into motion, and its deflections are transferred to the mirror through the wire and the staff. A beam of light from a fixed source is reflected by the mirror and falls on a vertical film strip. In this way, the deflection of light caused by the motion of the mirror can be recorded on the film. This instrument is used to record curves representing speech and different musical instruments.

phonograph An instrument for recording or reproducing acoustic signals, such as voice or music, by transmission of vibrations from or to a stylus that is in contact with a groove on a rotating cylinder or disk. The early phonograph formed a groove of varying depth in a cylinder made of wax, while the stylus moved in and out of the wax following the motion of the diaphragm exposed to the sound waves. This motion was then magnified by retracing the groove on the cylinder or disk with a stylus attached to one end of the light lever, and allowing the other end of the lever to trace the curve on a smoked drum. *See also* gramophone.

phonometer Instrument that measures sound intensity. Webster's phonometer is based on the resonance method and consists of a tunable cylindrical resonator with a diaphragm mounted on the cylinder opening. The diaphragm is tuned to the sound whose intensity is measured by varying the tension of the wires supporting it. A small mirror attached to the diaphragm allows the experimenter to observe and measure its motion. The pressure amplitude of the sound impinging on the diaphragm is proportional to the displacement amplitude of the diaphragm.

phonon Quantum of sound representing excitations or energy levels in liquid Helium II in the form of longitudinal sound waves. The concept was explained by the peculiar behavior of liquid helium II by Landau in 1941. Thermal vibrations in a crystal lattice can be calculated by this quantum of thermal energy.

photodiode A commonly used diode that converts a photon input into a current output.

photoelasticity The effect in certain materials (also termed *stress birefringence*) wherein application of a mechanical stress causes materials to exhibit birefringence, made evident by the appearance of colored fringes when materials are observed through crossed polarizers. The phenomenon can be interpreted as the effect of stress-induced strain in altering density and polarizability of a material so as to produce an alteration in its dielectric tensor. Serves as the basis of technique for detection of internal stresses in mechanical structures by examination of structures through crossed polarizers for evidence of strain related birefringence.

photoelectric effect The effect wherein electrons (termed *photoelectrons*) are ejected from a material surface by incident electromagnetic radiation of sufficiently high frequency (usually in the visual part of the spectrum). The phenomenon is most pronounced where the material surface is metallic. The fact that maximum kinetic energy of photoelectrons is linearly dependent on the frequency of the incident radiation but independent of its intensity led Einstein to an explanation of the effect based on the interpretation of radiation in terms of localized packets of radiation energy called *photons*. An extended definition of the photoelectric effect refers to the broad category of radiation-induced changes in the electrical properties of a material, including changes in its conductivity. The effect is made use of in light beam-activated circuits of the type commonly used in *electric eye* devices.

photoemissive cell Device that detects and/or measures light or other electromagnetic radiation by the measurement of radiation-induced emission of electrons from the surface of a photocathode via the *photoelectric effect*. Represents the essential element of a *photomultiplier* tube.

photogrammetry The process of making maps or scale drawings from photographs. It has particular importance in the drawing of maps from aerial photographs.

It also relates to the general process of making precise measurements by means of photography. Typical fields that use this technique include archeology, architecture, medicine, and engineering.

photography   The process (and art) of producing images of objects on a photosensitive film by the collection of reflected radiation, usually in the form of light, from the surfaces of the objects.

photography, clinical   Use of photography to help in the diagnosis or treatment of diseases or other physiological conditions (*see* photogrammetry).

photography, color   Type of photography in which the images of the photographed objects reproduce the colors of the objects. Requires the use of color sensitive film.

photography of sound waves   Sound waves can be visualized by taking advantage of the dependence of the refractive index of light on the density of the medium. Optical measurement techniques, such as the shadowgraph, Schlieren methods, and optical interferometry, have been used to visualize compressible flow fields, such as supersonic flows and shock waves around airfoils, bullets and projectiles. Because of the high speed of the process, short exposure times are essential for successful photography. Short exposure times are accomplished using stroboscopic illumination or high-speed cinematography.

photography, spark   Type of photography in which illumination of objects to be imaged is provided by a spark to (severely) restrict the exposure time of the film. Allows for the production of sharp images of rapidly moving objects.

photolysis   Most generally, a process in which light (or other radiation) produces a chemical change in a substance. A common use of the term defines the process as one in which absorption of light causes decomposition of the molecules of a substance.

photomagnetism   Modification of magnetic properties, e.g., magnetic susceptibility, of a magnetic material by application of light (electromagnetic waves). Light modifies the elec-

tronic structure by exciting charge carrier. *See* susceptibility, magnetic.

photometer   An instrument for measuring the *luminous intensity* and/or flux produced by a source of visible light, usually in comparison with the luminous intensity of a reference light source. The original-type photometer compares luminous intensities, I_1 and I_2, of a source and reference source on an observing screen by varying distance d between reference and screen until two sources produce equal luminance on the screen's surface. The ratio of intensities of sources follows from the relation $I_1/d_1^2 = I_2/d_2^2$. The modern photometer in general measures the intensity of a light source via a calibrated *photoemissive cell.*

photometer, integrating   (Commonly in the form of an *integrating-sphere photometer.*) An instrument for measuring the total luminous flux emitted in all directions by a lamp or other light source. Allows for the determination of the *luminous efficiency* of a light source, given by the total luminous flux emitted by the source divided by the total power to the source. This integrating-sphere device makes use of a hollow sphere that can encompass the source to be measured and has a diffusely reflecting interior surface, the illumination of which from the reflected light is proportional to the total flux from the source.

photometry   The process (and/or science) of measuring the luminous intensity, flux, color, spectral or angular distribution, reflectance or transmittance of visible radiation (representing light). Contrasts with *radiometry,* defined to be the process of measuring the intensity of non-visible as well as visible radiation.

photometry, grease spot   Measurement of the intensity of a light source compared to a reference source by observing the effect produced when two sources illuminate *opposite* sides of an opaque white screen containing a central spot made translucent by treatment with a lower refractive index substance (such as oil or grease). The fact that unequal illumination of the screen results in the appearance of a dark central spot on bright surroundings, or the reverse, allows relative intensities of the sources to be determined from a "balance position" of the screen between sources for which the central spot disappears. *See* photometer.

photometry, heterochromatic   Branch of *photometry* concerned with a comparison of the illuminating effectiveness of light sources of different colors.

photomicrography   The process of making photographs of images of minute objects formed by a microscope. In general, this makes use of an instrument containing a *photomultiplier* for amplification of light from the separate segments of the image to be photographed.

photomultiplier   A device in which a radiation-induced photocurrent is amplified by focusing initial photoelectrons onto a succession of electrodes (called dynodes) so as to induce emission of secondary electrons. The acceleration of electrons between successive dynodes causes each secondary electron to produce additional secondaries which multiply the initial photocurrent. The resultant gain in electron current can equal (or exceed) 10^8. The device allows for detection of low levels of (visible or near-visible) radiation.

photon   The basic unit (or *quantum*) of light or other electromagnetic radiation. Plays the role of *carrier* of the electromagnetic field. An entity characterized by zero rest mass and a velocity of propagation along a particular direction with magnitude equal to the speed of light. Corresponds to a quantity of electromagnetic energy hf and angular momentum along the direction of propagation $h/2\pi$, where f is the frequency of the electromagnetic radiation and h is Planck's constant (6.626075×10^{-34} J·s). The existence of a photon is manifested by the fact that excitation and de-excitation of atoms and molecules takes place only via the absorption or emission of integer numbers of photons. Another (vision related) meaning of *photon* is the amount of light received by the retina of the eye from a surface with a luminance of 1 candela/m^2 when the area of the pupil is 1 mm^2.

photonics   The field concerned with the generation, propagation, processing, and detection of light or other radiant energy, often in connection with the transmission or detection of signals. Contrasts with *electronics* in that, control of electrons is replaced by control of photons, which displace electrons as the primary carriers of signals. The term emphasizes the quantized nature of a basic unit of radiation. The area defined by the term includes, for example, the production and amplification of radiation via lasers and other radiation sources, the design and fabrication of optical waveguides and interconnects, and the use of nonlinear optical effects in materials relevant to the generation and control of light and other electromagnetic radiation.

photosensitivity   Property of a material or organism whereby it is sensitive to visible (or near-visible) radiation. The term in general applies to materials or organisms readily affected by light. Photosensitivity in materials is exhibited via effects such as *photolysis,* increased conductivity, the emission of photoelectrons, or the photo-voltaic effect.

photosynthesis   In the most general sense, photosynthesis is the process in which green plants and certain other organisms use sunlight energy to manufacture carbohydrates from carbon dioxide and water with the help of chlorophyll. In most cases of photosynthesis, oxygen usually results as a by-product. In this sense, photosynthesis is the reverse of respiration in which carbohydrates are broken down to release energy.

In the first stage of photosynthesis, also called *light reaction,* direct light is required. Water is broken down into oxygen (released as gas) and hydrogen. Also, ATP molecules are produced. Next is the second stage called *dark reaction,* where the hydrogen and carbon dioxide (CO_2) are converted into intermediary compounds that ultimately yield the organic compound glucose ($C_6H_{12}O_6$) plus water. The chemical reaction involved is given by

$$6CO_2 + 12H_2O + \text{ energy } \rightarrow$$
$$C_6H_{12}O_6 + 6O_2 + 6H_2O .$$

The second stage does not require light to occur, using instead the energy released from the hydrolysis of ATP, where ATP reacts with water to yield ADP, inorganic phosphate, and energy.

Chlorophyll, exhibited mainly by the green pigment in plants, is contained in the *chloroplasts* (organelles contained in the cytoplasm of plant cells). Chlorophyll is the only substance in nature able to trap and store energy from sunlight. Because the red and blue-violet parts of the visible spectrum are the ones that get absorbed by chlorophyll, the rest of the spectrum, mainly green, gets reflected, thus giving the green color to plants.

phototherapy   Therapy of diseases or types of disorders, especially of the skin, using light. Ultraviolet and infrared radiation are of particular use in this kind of therapy. *See* red light, healing effect; light, monochromatic; biological action.

photovoltaic efficiency   Measure of the electric potential (or voltaic response) produced in a nonhomogeneous material by exposure to light or other electromagnetic radiation. Measures the process in which radiation absorbed by a material structure in the region of a potential barrier (such as at a *p-n* junction or metal-semiconductor contact) produces electric potential difference V_p in the region (e.g., by separation of electron-hole pairs). Photovoltaic efficiency of a process is defined by the ratio of power associated with (latent) voltage-induced current, I_p, expressed by $V_p I_p$, divided by power in input radiation, P_{in}.

pianoforte   The pianoforte action refers to communicating the entire energy to be radiated as sound by a vibrating string or wire during a very short time interval, 1/500 s, while the hammer is in contact with the string. In the pianoforte action the hammer is projected against the string, and it is free from the system of levers that have set it in motion. In addition to the point of impact and the velocity of the hammer head, which determine the loudness, the pianoforte performer can also influence the quality of the sound generated by the string. The way in which the motion has been initiated by the performer influences the vibrations of the hammer and the quality of the sound produced in this way.

piano, sound from   The piano is a stringed keyboard instrument in which a hammer striking the string is used to excite the sound; the hammer immediately rebounds after the action. The nonlinear elasticity of the hammer plays a key role in determining the character of the piano sound. An iron frame maintains the tension of the strings. The vibration of the string is transmitted to a soundboard that serves as the main source of acoustic radiation into the surrounding air. The frequencies generated by the piano span somewhat over seven octaves (from A_0 to C_8) with frequencies between 55 and 8360 Hz. When the piano key is struck and held, the sound begins to decay at one rate and then "breaks" to continue decaying with another, slower rate thus creating the prompt sound and the aftersound. This double decay characteristic is an important feature of the piano tone.

picture tube, color   A cathode ray tube used to display images in color by variation of the beam intensity. It uses a system of three different colored pixels in adjacent locations to produce the different colors.

piezoelectric effect   The piezoelectric effect takes advantage of the properties of crystalline quartz in that mechanical deformations can be generated along the mechanical axis of the crystal to induce longitudinal vibrations along this axis by applying an alternating electric field along the electric axis. As the frequency of the applied electric field approaches the natural frequency of any longitudinal vibration mode of the quartz crystal, the amplitude of the mechanical vibrations increases. Since the electroacoustic efficiency of these vibrations is high, the piezoelectric effect is frequently utilized to generate ultrasound in gases, liquids and solids. Conversely, when a stress applied to the quartz crystal results in a strain either along the optic or the mechanical axis, the crystal becomes electrically polarized and piezoelectric charges of opposite sign form on the two surfaces perpendicular to the optic axis. This effect is exploited in the operations of certain types of sensors. The most commonly applied vibration transducers take advantage of the piezoelectric effect. X-cut quartz crystals are typically used for piezoelectric energy conversion.

pigments   Finely divided particles of natural or synthetic substances used in suspension in materials to contribute to the optical (and/or physical) properties of the materials. Distinguished from dyes, used in solution, by insolubility in the host materials. Pigment particles are characterized by the ability to re-radiate light of particular wavelengths on absorption of light of other wavelengths so as to produce reflected and transmitted light of the same color. Commonly used to affect the color and opacity of paints and coatings.

pinch effect   Effect whereby electric charge carriers moving in the same direction are attracted toward each other by a magnetic force created by their movement. Results in a radially compressive force. In liquids and gases the flowing charge may pinch off so that no current will flow. *See* magnetic force on moving charge.

pinch-off voltage   The gate-source voltage in a JFET at which the drain current approaches zero due to a meeting of the depletion regions.

ping-pong mechanism   The ping-pong mechanism is a two-step chemical reaction in which a reactant reacts with a molecule that leads to a chemical resultant. Then the same molecule reacts again with the same kind of resultant to yield the original reactant. The importance of this mechanism lies in the importance of its by-products and of the fact that it is cyclic.

Examples of a ping-pong mechanism include mechanisms where membrane proteins exchange one charged molecule or atom for another, such as Cl^- for HCO_3^-, or Na^+ for H^+. These proteins play a role in cellular processes that include volume and pH regulation, and transport of ions across the membrane.

The protein that catalyzes the exchange of Cl^- for HCO_3^- functions by a ping-pong mechanism, in which the transport protein can exist either in a conformation with the transport site facing outward (E_o) or with the transport site facing inward (E_i). Conversion from E_o to E_i or vice versa occurs only when a suitable anion, such as Cl^-, is bound to the transport site. Thus, the system is confined to tightly coupled one-for-one exchange of ions.

Also CO_2, coming from metabolic processes in tissue, is a participant in a ping-pong mechanism when it diffuses into the red cell and intracellular carbonic anhydrase catalyzes its transformation into bicarbonate. After, intracellular bicarbonate is exchanged for extracellular chloride as bicarbonate flows passively down its concentration gradient. Once in the lungs, the process is reversed, and CO_2 diffuses into the atmosphere, thus completing the cycle.

pipes, sound from   Vibrations of air columns are the source of sound in the organ as well as in most wind instruments. In the theoretical analysis it is assumed that the walls of the pipe are rigid, the diameter is small compared to the length of the pipe and large enough to justify neglecting viscosity effects. The two types of pipes of practical importance are the open pipe (open at both ends) and the closed pipe (closed at one end). The modes of the organ pipe are characterized by displacement antinodes at the open ends and by an appropriate number of nodes (separated by antinodes depending on the vibration mode) in the middle region. A displacement node at the closed end and an antinode at the open end characterize the closed pipe. The closed pipe thus has a fundamental frequency that is one octave lower in pitch than the open pipe. *See also* organ pipes, vibrations in.

piston source   A vibrating piston mounted in an infinitely large rigid wall (baffle) or in the end of a long tube used to approximate sound sources.

pitch, acoustic   An aspect of the subjective sensation of sound that allows sounds to be ordered on a musical scale from "low" to "high". The variations in pitch lead to the sensation of melody. Pitch, measured in mels, corresponds to the frequency for a pure tone and to the fundamental frequency for a periodic complex tone. Pitch as a subjective attribute cannot be measured directly; a value is assigned to a sound by specifying the frequency of a sinusoidal sound vibration that has the same subjective pitch as the sound. In the perception of pure tones, two tones separated by an interval of one octave (the higher tone has twice the frequency of the lower tone) sound similar and have the same name on the musical scale. The concept of pitch is important in accounting for the perception of complex tones.

Pitot tube   An instrument that measures the static and/or stagnation pressure of a flowing fluid, consisting of a slender tube equipped with holes along the perimeter of the tube or one at the tip, pointing into the fluid and connected to a pressure indicating device. The static pressure probe is equipped with small measuring holes serving as pressure taps along the perimeter. The manometer connected to the tube indicates the static pressure p along the streamline at the location of the taps. The stagnation pressure p_0 is obtained when the fluid is decelerated to zero speed V_0 at the tip of the stagnation pressure probe. The pressure is sensed through the hole at the tip of the probe. The Bernoulli equation, $\frac{p_0}{\rho} + \frac{V_0^2}{2} = \frac{p}{\rho} + \frac{V^2}{2}$, relates the changes in speed to the changes in pressure along the streamline in incompressible flow of density ρ. The combination of the static and stagnation probes in the Pitot tube allows determining the flow velocity $V = \sqrt{\frac{2(p_0-p)}{\rho}}$ from the measured difference between stagnation and static pressures, $p_0 - p$.

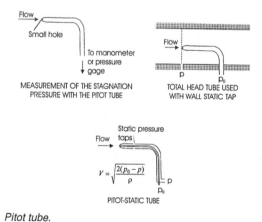

MEASUREMENT OF THE STAGNATION PRESSURE WITH THE PITOT TUBE

TOTAL HEAD TUBE USED WITH WALL STATIC TAP

PITOT-STATIC TUBE

Pitot tube.

plane wave   A wave in which the wavefront is a plane surface. The equiphase surfaces of plane waves form a family of parallel planes. Plane sound waves have the same acoustic properties at any position on the plane surface that is per-

pendicular to the direction of propagation of the sound wave. Plane sound waves can exist in a straight pipe or duct; they are one-dimensional and vary with time t and one Cartesian coordinate x. They can be described with the one-dimensional equation of motion, the acoustic wave equation, which relates the second derivative of sound pressure p along the direction of propagation x to the second derivative of the sound pressure with respect to time t through the square of the speed of sound c, $\frac{\partial^2 p}{\partial x^2} - \frac{1}{c^2}\frac{\partial^2 p}{\partial t^2} = 0$. The general solution of the acoustic wave equation is $p(x,t) = f(x - ct) + g(x + ct)$, where f and g are arbitrary functions.

Most generally, waves characterized by "planar wavefronts" having amplitudes and phases that have the same values for all points within any plane perpendicular to the direction of propagation. The less general, more usual definition defines plane wave to correspond to a solution of the scalar wave equation

$$\nabla^2 \Psi - \frac{1}{v^2}\frac{\partial^2 \Psi}{\partial t^2} = 0 \,,$$

of the form

$$\Psi(\mathbf{r}, t) = A e^{i(\mathbf{k}\cdot r - \omega t + \phi)} \,,$$

where A and $(\mathbf{k} \cdot r - \omega t + \phi)$ are referred to, respectively, as the "amplitude" and "phase" of the *wavefunction* Ψ, and $|\mathbf{k}| \equiv k = \omega/v$. The harmonic variation of a wavefunction with distance $\hat{\mathbf{k}} \cdot r$ and time t results in a wavefront that advances in the direction of vector $\mathbf{k}$ at the "phase velocity" ω/k. The planar character of wavefronts results from the dependence of function $e^{i(\mathbf{k}\cdot r - \omega t + \phi)}$ only on a component of coordinate vector $\mathbf{r}$ along vector $\mathbf{k}$, $(\hat{\mathbf{k}} \cdot r)$, and the absence of dependence on a component of $\mathbf{r}$ perpendicular to $\mathbf{k}$, $(\hat{\mathbf{k}} \times r)$. The resulting wave has infinite extent in the directions perpendicular to the propagation direction, infinite extension in time, and corresponding infinite length along the direction of propagation $\hat{\mathbf{k}}$.

plastic deformation   This is a permanent and irreversible deformation in crystals due to line defects.

plates, vibrations in   Objects in the class of plates, which can also include surfaces that are not flat, can emit sounds classified as "noises" with "metallic" character rather than "notes". Vibrations in a thin plate are described by the fundamental differential equation of motion for free vibrations (with the assumption that the middle layer of the plate is physically inextensible) as $\nabla^4\xi - k^4\xi = 0$, with $k^4 = \omega^2/c^4$. ω is the angular frequency and the parameter c is defined as $c^4 = \frac{qh^2}{3\rho_0(1-\mu^2)}$ (with q as Young's modulus, $2h$ the thickness of the plate, μ as Poisson's ratio and ρ_0, the volumetric density of the plate.

plethysmograph, capacitance The capacitance of a body part can be determined from a *plethysmograph*. In a plethysmograph, the capacitance of a body part can be measured when the reactance of this body part is determined as a function of the frequency of the external alternating current (*see* plethysmograph, impedance).

plethysmograph, impedance An impedance plethysmograph measures and records volumetric variations of an organ or body part on the basis of variations in the electrical resistance and reactance of the body or its segment as a function of the frequency of an input alternating current. The electrical variations in the output can be related to the amount of volume pulsing variations coming from blood and other intra- and extravascular fluids passing through the organ. With this technique, the flow of electrolytes injected in the blood flow can be easily followed by electrical conductivity measurements.

Considering the case where the input current is constant, the electrical resistance R of a portion of a conductor of uniform cross section a and length l is

$$R = \frac{\rho l}{a}$$
$$= \frac{\rho l^2}{l \cdot a}$$
$$= \frac{\rho l^2}{V}$$

where ρ is the resistivity of the material and V, the volume of the segment.

If the length of the conductor is constant and the volume is increased by increasing a, then δR

is proportional to δV as

$$\delta R = R_1 - R_2$$
$$= \rho l^2(\frac{1}{V1} - \frac{1}{V2})$$
$$= -\frac{\rho l^2 \delta V}{V1 \cdot V2}$$

or

$$\delta R \approx -\frac{\rho l^2 \delta V}{V^2}$$

for small changes, so that

$$\frac{\delta R}{R} = -\frac{\delta V}{V}.$$

This resulting equation is the basis for the volumetric variation measurements from an impedance plethysmograph. The equation results can be interpreted to mean that a small increase in volume brings a decrease in the electrical resistance.

p-n junction A junction formed by the contact of p and n doped semiconductors. The Fermi level difference in the p and n type materials produces a built-in potential in the device, which for silicon is ~ 1.1 V. *See n*-type silicon, *p*-type silicon.

P.O. Box (post-office box) One of the older forms of the Wheatstone bridge that consists of an arrangement of resistance coils, brass blocks, and tapered plugs that form three arms of the bridge. The resistance coils are inserted in the circuit by removing plugs from tapered holes between adjacent blocks. Each of the holes has a resistance associated with it. Each arm of the bridge contains several such holes; the resistance of an arm is read by adding the resistances of the unplugged holes.

Pockels effect Electro-optic effect in certain crystalline materials wherein the application of an electric field produces a change in the refractive properties of the materials proportional to the first power of the applied electric field strength. The effect occurs in crystals in which the applied electric field produces slight deformation of the ionic lattices (piezoelectric effect) and/or a redistribution of the bond charges, resulting in a change in the dielectric tensor of the material. First studied by F. Pockels in 1893.

polarimetry   The process (and/or science) of measuring the rotation of the plane of polarization of visible or near-visible radiation produced by passage of the radiation through a material medium. Serves as a basis for measurement of *optical activity* or circular dichroism of materials (such as sugar solutions).

polariscope   An instrument, consisting usually of a plane polarizer through which light passes and falls on a transparent sample material to reach a rotatable analyzer. It is used to investigate the effect of the sample material on the state of polarization of the emergent light and also to study strain in transparent material samples.

polarization, circular   Special type of *elliptical polarization* of light or other electromagnetic radiation in which the electric vector of the radiation rotates in time in the plane perpendicular to the direction of propagation of the radiant energy such that the tip of the vector describes a circular helix with its axis along the direction of propagation and with a period equal to the reciprocal of the frequency of the light. Circular polarization is referred to as "right handed" or "left handed" depending on whether the sense of rotation of the electric vector coincides with the curl of the fingers of the right or the left hand respectively when the thumb of the hand is pointed in the direction of the *Poynting vector.*

polarization, degree of   Measure of the extent to which light or other electromagnetic radiation can be said to be polarized. Representation of light as a superposition of linearly polarized components provides for an operational definition of degree of polarization in terms of measurements of the intensity of the light transmitted through a linear polarizer oriented so as to define a direction perpendicular to the direction of propagation of the light. Rotation of the polarizer through all orientations in the plane perpendicular to the propagation direction, and measurement of the intensities of the transmitted light, make possible (in principle) a determination of a maximum and a minimum in the transmitted intensity, I_{max} and I_{min}, in terms of which the *degree of polarization* of the light, P,

can be defined by the ratio

$$P \equiv \frac{I_{max} - I_{min}}{I_{max} + I_{min}}.$$

The definition assigns P the values 1 and 0 in the cases of completely polarized light and (completely) unpolarized light respectively. *See* partially polarized light.

polarization, electric   The electric dipole moment per unit volume of a dielectric. Electric polarization occurs when a dielectric is placed in an electric field that tries to align the electric dipoles parallel to each other. This results in a separation of electric charge in the assembly of dipole moments which in turn produces charges on the surface of the dielectric. The degree of polarization is dependent on temperature since thermal agitation tends to oppose the order produced by the electric field.

polarization, elliptical   State of polarization of light or other electromagnetic radiation in which the electric field vector of the radiation at a particular point in space rotates in time in the plane perpendicular to the direction of propagation of the radiant energy such that the tip of the vector traces out an ellipse. Elliptically polarized light can be represented as a superposition of two linearly polarized light waves of unequal amplitude, which have mutually orthogonal directions of polarization, and which are out of phase with respect to one another by a non-integer multiple of π.

polarization, linear   State of polarization of light or other electromagnetic radiation in which the electric vector of the radiation oscillates in time along a fixed direction in a plane perpendicular to the direction of propagation of the radiant energy. Also known as *plane polarization.*

polarization, membrane   *See* potential, resting; repolarization (cell); nerve impulses, propagation of.

polarization of light   Property of light and other electromagnetic radiation defined by a non-random orientation of the electric (and magnetic) vector of the radiation field in the plane perpendicular to the direction of propagation of the radiant energy. A specific state of polarization is in general specified in terms of the direction of the electric vector, **E**, in the plane perpendicular to the *Poynting vector.* Types of polarization can be categorized as *linear, elliptical* or *circular.*

polarization, plane Alternate term for *linear polarization.*

polarizer An optical device whose input is natural light and whose output is polarized light (attained usually with the help of a Nicol prism, Polaroid sheet, etc). Depending on the nature of polarization of the outcoming light (e.g., plane polarized, circularly polarized), the polarizer is called a *plane polarizer, circular polarizer,* etc.

Polaroid Trade name for a transparent sheet of dichroic material which transmits light that is linearly polarized along a particular direction. A common type of polaroid material consists of colorless plastic sheet treated with an iodine solution that creates parallel chains of polymeric molecules containing conductive iodine atoms which produce a plane of polarization by dichroism. Commonly used to reduce glare in optical and lighting devices. A generalization of the term relates to a range of photographic and optical products based on polymeric materials.

Polaroid camera Trade name for a camera that makes use of film containing its own developing and printing agents that make possible the production of a finished positive print within minutes after the photograph is taken. Developed by Edwin H. Land in 1948. Also known as *Land camera.*

pole piece Magnetic pieces that attach to opposite ends of a magnet to finish a magnetic circuit. Often have an air gap between them. Their size and shape determine the magnetic flux distribution in the gap. Can concentrate magnetic flux creating a large magnetic field in a small volume or spread flux lines out uniformly creating a small uniform magnetic field over a larger volume. *See* flux, magnetic.

pole strength A measure of the strength of a magnet. The pole strength p has MKS units

of $A.m$ and is defined by $m = pl$ where m the magnetic moment of a bar magnet with north and south poles of strength $+p$, and $-p$ and l is the separation of the poles.

polling A technique for coordinating access to a shared medium. A master checks whether each slave has data to send, and if it does, gives it a chance to use the medium.

potential, contact The potential difference that develops between two dissimilar metals placed in contact. It is given by the difference between work functions for the two metals and varies with the temperature of the junction. Two such junctions placed in series and kept at different temperatures produce a net electromotive force through the circuit which forms the basis of thermocouple thermometers. *See also* thermocouple.

potential, demarkation The demarkation potential is the threshold potential beyond which there is the initiation of an action potential. If a stimulus provided by sensory information or neurotransmitters changes the local membrane potential by as much as the demarcation potential of approximately -60 mV, then the initial signal leads to the rapid opening of Na^+ channels and to the initial steps toward creating an action potential. *See* nerve impulses, propagation of.

potential difference In electrical circuits, it is the work required to transfer unit charge between two points in the circuit. The SI unit is joules per coulomb but it is commonly referred to as volts. The potential difference across a resistance in an electrical circuit is obtained by applying Ohm's law. In an electric field, it is the work required to move a unit charge between two points A and B. It can be obtained by calculating the difference in electric potentials at points A and B, by

$$V = -\int_A^B \mathbf{E} \cdot d\mathbf{l}$$

where V is the potential difference, **E** is the electric field, and $d\mathbf{l}$ is a path element between A and B. *See also* Ohm's law; potential, electric.

potential divider   A circuit used to transmit a selected fraction of the input potential to the output. It typically consists of two resistors in series with the output taken across a single resistor.

potential, electric   The work required to bring a unit positive charge from infinity to a certain position. This is given by

$$V = \frac{W}{q},$$

where V is the electric potential, W is the work done by an external force, and q is the test charge brought from infinity at a constant speed to the required position. The SI unit of electric potential is joules per coulomb, otherwise known as volts i.e.,

$$1 \text{ volt } = 1 \text{ J/C}.$$

If the electric field $\mathbf{E}$ is known, the electric potential can be calculated from the line integral

$$V = -\int_{\infty}^{x} \mathbf{E} \cdot d\mathbf{l}$$

where x is the position at which the electric potential is required, $d\mathbf{l}$ is a length element along the path taken by the test charge coming from infinity to point x.

potential, evoked   When certain areas of the brain are driven toward electrical activity under stimulation of specific sensory pathways, the potentials coming from them are called *evoked potentials*. These evoked potentials are recorded by placing wires on the scalp over the areas of the brain being stimulated with a particular stimulus.

Common examples of evoked potential tests are the *visual evoked potentials* (VEP), the *brainstem auditory evoked potentials* (BAEP), and the *sensory evoked potentials* (SEP). In VEP the patient sits before a screen and responds to alternating visual patterns. In BAEP the auditory part of the brain is tested by presenting to the patient a series of clicks to each ear. In SEP a small electrical impulse is administered to the patient on an arm or leg. Not common but existing, are *motor evoked potential tests* that can detect lesions along motor neuron pathways of the central nervous system.

Evoked potential tests are often used to help make a diagnosis of multiple sclerosis (MS), because they can indicate dysfunction along neuronal pathways caused by demyelination of the axons of neurons.

potential, extracellular   Potential that arises when an action potential crosses the synapse and enters the post-synaptic membrane. The current that flows into post-synapse and into the membrane closes the current loop by flowing out of the cell along the length of the walls of the membrane and into the extracellular space, and subsequently re-enters the synapse.

Because the extracellular resistance R_{ex} is so small compared with the large resistance of the membrane R_m, the voltage across the membrane δV_m is effectively equal to the current I multiplied by R_m. Also, since $R_{ex} \ll R_m$, the extracellular potential drop δV_{ex} is going to be much smaller than δV_m.

By equating currents

$$\frac{\delta V_m}{R_m} = \frac{\delta V_{ex}}{R_{ex}},$$

we get typical values

$$\begin{aligned}
\delta V_{ex} &= \frac{\delta V_m}{R_m} \cdot R_{ex} \\
&= \frac{5 \times 10^{-3} V}{1 \times 10^5 \Omega} \cdot 50\Omega \\
&= 2.5\mu V.
\end{aligned}$$

The extracellular potentials of populations of neurons can be recorded and form the basis of the electroencephalographic measurements (EEG).

potential, graded (membrane)   A graded potential is a potential whose value depends on an external parameter. An example of a graded potential is the *postsynaptic potential* (PSP). The postsynaptic potential is a transient change in the electric polarization of the membrane caused by the influx of neurotransmitters as a response of an action potential at the presynaptic membrane. The PSP is a graded potential because its degree of hyperpolarization (increase in negative charge inside the cell membrane—inhibitory) or depolarization (decrease of the

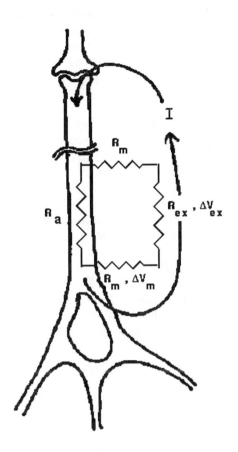

where HB is a weak acid and the reaction shows how it dissociates, the reaction rate is given by the constant K_a. The reaction constant is in turn given by the ratio $[H^+][B^-]/[HB]$. From the reaction it is evident that strong acids dissociate more readily and have high K_a while the opposite is true for weak acids.

In 1908 Henderson studied a metabolic reaction involving CO_2 and applied the law of mass action to get

$$[H^+]\left[HCO_3^-\right] = K_a\,[CO_2]\,[H_2O]\ .$$

Later, Sorensen (1909) introduced the "potential for hydrogen" notation where $pH = -\log[H^+]$. Using the pH notation also for the rate constant K_a ($pK_a = -\log K_a$), Hasselbalch in (1916) rewrote Henderson's equation to get the Henderson–Hasselbalch equilibrium equation

$$pH = pK_a + \log \frac{\left[HCO_3^-\right]}{[dCO_2]}\ ,$$

where dCO_2 stands for dissolved CO_2.

The pH and pK_a notations are particularly good since the hydrogen concentrations and the reaction constants K_a for weak biological acids are very small numbers. The acid-base balance equations above indicate then that strong acids have a low pK_a, and vice versa.

Extracellular potential.

negative charge—excitatory) varies on dependence of the activation of the ion channels by the neuro-transmitters (*see* neuro-transmitters).

potential gradient   The spatial rate of change of electric potential in a conductor, dielectric or free space. It is obtained by evaluating the potential difference per unit length along the direction of the electric field vector. In three dimensions, the electric field **E** is related to the spatial derivative of the potential V as follows:

$$\mathbf{E} = -\left(\mathbf{i}\frac{\partial V}{\partial x} + \mathbf{j}\frac{\partial V}{\partial y} + \mathbf{k}\frac{\partial V}{\partial z}\right)\ .$$

potential, Henderson   In the acid-base balance equation

$$HB \rightarrow H^+ + B^-\ ,$$

potential, interaction   (**1**) An interaction potential between two particles comes from their mutual interaction force. The interaction potential at point B, relative to point A, is proportional to the integrated mutual force if one would take one particle along a path that joins both points.

More specifically, the interaction potential is proportional to the work done W_{AB} in bringing a particle from point A to point B along a specific path,

$$W_{AB} = \int_A^B \vec{F}\cdot \vec{dl}\ ,$$

where F is the force between the particles, and l specifies the path of approach. For there to be a potential associated to a specific force, the work done has to be independent of the particular choice of the path. These forces are said to be conservative. Examples of conservative forces are the gravitational force and coulombic forces.

(2) The membrane potential arises from the charge separation that occurs between the extracellular and the intracellular regions. Because it requires electrical work to take an ion against the electric field created by the disbalance of charges, there is a corresponding potential difference across the membrane.

potential, intercellular (1) Potential difference between different cells.

(2) Before transmission of an action potential across a synapse, the pre- and post-synaptic membranes lie at different potentials. The release of neuro-transmitters propagates the signal by equalizing the potentials and provoking an action potential at post-synapse. *See* potential, extracellular; potential, membrane; neurotransmitters.

potential, junction The junction potential is the potential difference at the boundary between two solutions with different ionic compositions. It arises from the different diffusion constants of the ions in each particular solution.

The junction potential may give significant contributions in electrophysiological measurements of the order of mV. Namely, it may be important in measurements of ion permeation through ionic channels using the patch-clamp technique, in which extraneous ions with low mobilities are used in fairly high concentrations. *See* voltage clamp, ionic current in cell.

potential, liquid (1) The thermodynamics of simple molecular liquids are usually well described by pair interparticle potentials. An empirical potential for simple liquids is found to have a repulsive interaction at very small distances that is inversely proportional to the twelfth power of the distance between them. As they separate, an attractive interaction due to a weak charge-polarization or induced dipole-dipole interaction comes into effect. The interparticle potential can be written as

$$U(r) = 4\epsilon\left[\left(\frac{\sigma}{r}\right)^{12} - \left(\frac{\sigma}{r}\right)^{6}\right],$$

where ϵ and σ are the depth of the interaction well at equilibrium and the effective radius of the particles, respectively. These two parameters are experimentally obtained.

(2) Both sides of the interface between two ionic solutions may be at different potentials due to the differences in diffusion coefficients of the ions. The drop in potential gives rise to a junction potential. *See* potential, interaction; potential, junction.

potential, membrane The membrane potential, or transmembrane potential difference, is the electrical potential difference across a plasma membrane (*see* potential, resting; action potential).

The membrane potential arises from both the ion pumps and from the flow of ions through channels that are open in the cell plasma membrane (*see* osmotic equilibrium (cell)). The passive flow of ions through the membrane channels depends on the osmotic gradient of ions across the plasma membrane. The ion pumps use energy from ATP hydrolysis to actively transport ions across the plasma membrane in favor or against their osmotic gradient.

Because of the different mechanisms involved in distributing ionic composition in and out of the cytoplasm, the intracellular fluid is substantially different from that of extracellular fluids. The relationship between ion concentration and membrane potential is given by the *Nernst equation. See* Nernst equilibrium potential.

potential, miniature end plate (MEPP) The miniature end plate potential arises from small fluctuations (typically 0.5 mV) in the resting potential of post-synaptic cells. The profile of the potential is of the same shape, although much smaller than, the end plate potentials caused by the stimulation of the pre-synaptic cell. MEPPs are considered as evidence for the quantal release of neuro-transmitters at synapse, a single MEPP resulting from the release of the contents of a single synaptic vesicle. *See* neurotransmitters.

potential, receptor, hair cell In general, the dermis consists of several types of tissue: glands, nerve endings, fat cells, hair follicles, and muscles. The nerve endings, called *receptors,* perform an important sensory function, responding to various stimuli including: touch, pressure, pain, heat, and cold.

In particular, the receptor hair cells in the inner ear are responsible for the marked sensitivity of mammalian hearing. These hair cells are responsible for the transformation of the mechanical sound into a neural signal. Mechanical input, from acoustic pressure waves, deflect the hair cell's receptive organelle, the hair bundle. Bending of these hair cells (*stereocilia*) causes the opening of small channels in the cell membrane to which the hair cells are attached. Therefore, the movement of stereocilia controls the ion current flow into the cell. Bending the stereocilia in the direction away from (toward) the center of the cochlea leads to an increase (decrease) of the intracellular potential. Intracellular voltage changes as small as 0.1 mV are able to cause neuro-transmitter release at the synaptic contacts and in this way action potentials are evoked in the fibers of the auditory nerve.

potential, resting   The resting potential of a membrane is the electrical potential of the inside of a cell membrane relative to its surroundings. In almost all types of animal cells the inside of their membranes are negative with resting potentials in the range of −20 to −100 mV, with −70 mV being the typical value.

Although the membrane resting potentials are initially caused by the action of sodium ion pumps, their value comes primarily from the subsequent diffusion of potassium out of the cell through potassium leak channels. The resting potential is thus close to the Nernst potential for potassium. *See* action potential; osmotic equilibrium (cell).

potential well   A region in which a charged particle experiences a lower potential energy and consequently becomes trapped in that region. The particle can escape from the potential well provided it gains sufficient energy equal to the difference between its kinetic energy and the well depth. This is known as the *binding energy*. According to quantum physics, any particle trapped in a potential well can have only discrete energy levels; i.e., the levels are quantized. The concept of a potential well is mostly used to describe the physics of trapped atomic and subatomic particles such as electrons.

potentiometer   An instrument used to measure precisely an unknown potential difference of an EMF by comparing it to the known potential difference of a standard cell under the condition in which there is zero current from the unknown potential difference. The term *potentiometer* is alternately used to refer to a three-terminal voltage divider. Two of the terminals are connected by a fixed resistance while the third is a variable sliding contact that can be moved by a rotatable shaft. This forms a variable resistance between the sliding contact and the other two terminals. The changing resistance with changing position of the sliding contact may be linear, logarithmic, sinusoidal, etc. In everyday language, the term has been shortened to *pot*.

power density   The power density spectrum, $W(\omega)$, is defined as

$$W(\omega) = \lim_{T \to \infty} \frac{1}{T} |F_T(\omega)|^2 ,$$

where T is the time interval of interest and $F_T(\omega)$ is the Fourier transform of the given random function $f(t)$, given by

$$F_T(\omega) = \int_0^T dt f(t) \exp(-j\omega t) .$$

The function $W(\omega)$ is an even function of frequency:
$$W(-\omega) = W(\omega) ,$$
since for a real function $f(t)$,
$$F_T(-\omega) = F_T^*(\omega) .$$

Sometimes the spectrum is limited to positive frequencies by considering

$$W'(\omega) = 2W(\omega) \quad \text{for} \quad \omega > 0$$
$$= 0 \quad \text{for} \quad \omega < 0 .$$

The power in a band extending from ω_1 to ω_2 is

$$\int_{\omega_1}^{\omega_2} d\omega W'(\omega) .$$

power detection   A form of detection in which the demodulator supplies a substantial power output directly to the load without using an intermediate amplifying stage. A demodulator is a circuit, apparatus, or circuit element that is used in communication to demodulate the received signals, i.e., to extract the signal from a carrier with minimum distortion.

power factor   The ratio of the power, P (in watts) dissipated in a circuit to the effective volt-amperes, $V_{eff}I_{eff}$, applied to it. The power dissipated is given by

$$P = V_{eff}I_{eff} \cos \phi ,$$

where $\cos \phi$ is the power factor, ϕ is the phase angle between the alternating current and voltage, V_{eff} and I_{eff} are the root-mean-square values of the sinusoidal current and potential difference, respectively. The power dissipated in a pure inductance or pure capacitance is zero since $\phi = 90°$ in both cases, which makes the power factor equal zero. The power factor of a circuit is usually expressed in percent; $\cos \phi = 1$ is spoken of as *unit power factor* or *100 percent power factor.*

power gain   The ratio of the output power to the input power in a device or circuit.

power, in AC circuits   This can be divided into apparent power and actual power. The apparent power is the product of the effective voltage and effective current. Actual power is given by the product of apparent power and the power factor. Apparent power is expressed in volt-amperes while actual power is in watts. For a sinusoidal voltage and current, the effective current, I_{eff}, and voltage, V_{eff}, usually called the root-mean-square values, are given by

$$I_{eff} = \frac{I_0}{\sqrt{2}}$$

$$V_{eff} = \frac{V_0}{\sqrt{2}} ,$$

where I_0 and V_0 are the peak values of the current and voltage respectively. The power expended by a circuit is given by the actual power. It is common practice for an AC circuit or apparatus to be rated by the apparent power. *See also* power factor.

power in wave motion   The rate at which energy E, transported by the wave at speed v, is delivered by the wave. For a vibrating string, the average rate of energy (averaged over the time interval of motion) transported across a point on the string P is the average energy density that crosses the point in a unit time, multiplied by the length of the wave at that point (wave speed times the unit time), $P = v \left\langle \frac{dE}{dx} \right\rangle = \frac{1}{2} \mu \omega^2 y_0^2 v$, with μ as the mass density of the string, ω the angular frequency and y_0 the displacement amplitude. The power delivered by a wave is proportional to the square of both amplitude y_0 and frequency ω. *See also* oscillations, energy of.

power rating   The maximum power output available from a device, or the power it requires to operate.

power spectral density   For a specified bandwidth of radiation consisting of a continuous frequency spectrum, the total power in the specified bandwidth divided by the specified bandwidth is the *spectral density.*

Let $x(t)$ be a random process such that at time t_1, $x(t_1)$ is a random variable having a probability density, $p_{x_1}(\eta)$, which statistically describes the process at t_1. The probability that the process $x(t)$ will have a value in the range (a, b) at time t_1 is then given by

$$\text{Prob}\,[a \leq x\,(t_1) \leq b] = \int_a^b p_{x_1}(\eta)d\eta .$$

One simple measure of the degree of randomness of a process $x(t)$ is indicated by the process's autocorrelation function

$$R_x(t, \tau) = \mathcal{E}[x(t)x(t + \tau)] ,$$

where $\mathcal{E}$ denotes a statistical average over the joint density of the process $x(t)$ at time t and $x(t)$ at time $t + \tau$. The frequency characteristics of a stationary random process $x(t)$ are exhibited by its spectral density, $S_x(\omega)$, defined as the Fourier transform of the process's autocorrelation function

$$S_x(\omega) = \int_{-\infty}^{\infty} R_x(\tau) \exp^{-j\omega\tau} d\tau ,$$

where ω is the frequency. The function $S_x(\omega)$ is called the spectral density or power spectrum of the stationary random process $x(t)$.

Poynting vector Vector quantity, denoted S, the magnitude and direction of which determine the magnitude and direction of the energy transported by light or other electromagnetic radiation across a unit area in a unit of time. The vector is explicitly defined in terms of the cross product of the electric and magnetic vectors of the radiation, E and B, via the relation

$$S = \frac{1}{\mu}(E \times B) ,$$

where μ is the "permeability constant" of the medium. The significance of vector S derives from the fact that the integral of the normal component of S over a given surface equals the rate of flow of electromagnetic energy through the surface.

pre-amplifier An amplifier operating before the main amplifier to boost the signal. It is normally used before signal processing occurs to avoid amplifying the noise.

precooling This is a stage in an experiment used to speed up the overall cooling process. It is common practice to introduce a little air as exchange gas before the subsequent transfer of liquid helium into the cryostat for cooling. Cryostats in the temperature range above 1 K that use ^{4}He for cooling, usually use nitrogen for precooling.

presbyopia A reduction in the ability of the eye to accommodate to bring close objects to a focus on the retina. This is a naturally occurring consequence of aging.

primary cell An electrochemical cell that cannot be recharged is called a *primary cell*. They usually have high energy density and good shelf life. They are widely used in electronic equipment and are disposable. Typical examples are *zinc-carbon cell* and *alkaline cell*.

principal focus The point of intersection of a focal plane with the optic axis of the system. Corresponding to the first and second focal planes, we have the *first* and *second foci*.

principal maxima The peaks of largest intensity in the interference pattern produced by interfering beams of light or other electromagnetic radiation. Generally correspond to the zeroth and first-order maxima in an interference pattern, as, for example, the central and two neighboring maxima in the intensity pattern produced by the passage of coherent radiation through two slits.

principal planes/points Every lens/lens system has two principal points/planes and are two of six cardinal (or Gaussian) points/planes of the optical system. A ray entering a thick lens from the first focal point will emerge parallel and a ray parallel to the axis on the object side will pass through the second focal point. The extensions of the incident and emergent rays, in each case will intersect, by definition, the principal planes that cross the axis at the principal points. Principal planes in general do not coincide and may sometimes be located outside the optical system.

principle, convolution The convolution of two functions, $\psi_1(x)$ and $\psi_2(x)$, is by definition, the function $\psi(x)$ equal to:

$$\psi(x) = \int_{-\infty}^{+\infty} dy \, \psi_1(y) \cdot \psi_2(x - y) .$$

The Fourier transform of $\psi(x)$, $\psi(p)$ is the ordinary product of the respective transforms of ψ_1 and ψ_2. *See* principle, deconvolution.

principle, cryodyne *See* principle, Gifford-McMahon.

principle, deconvolution In general, the deconvolution principle is the inverse of the convolution principle (*see* principle, convolution). The goal in this procedure is, given a single function $\psi(x)$, two functions ψ_1 and ψ_2 (that when convoluted together yield $\psi(x)$) can be separately determined.

The method is of practical use to image restoration, enhancement, reconstruction, and signal filtering. Numerical algorithms exist that can reconstruct image objects by iterative coded-source image deconvolution. In many cases, a sharper image is sought when the source image has contributions from noise.

Novel techniques in the recording and imaging of X-ray and gamma rays use neural networks for the deconvolution phase mixed with nonlinear filtering for noise removal and edge enhancement.

principle, Gifford-McMahon   This is an example of expansion-cooling, and external work cycle. The cycle consists of four phases; in the pressurization phase, the warm volume is at a maximum. In the subsequent intake phase the valve remains open to enlarge the cold volume and reduce the warm volume. The expansion phase then takes over, when the exhaust valve is slowly opened and the cold volume is cooled by expansion. The final exhaust phase occurs when the displacer is moved downward to displace the remaining cold gas. This is also known as the *cryodyne*. The cryodyne is used to provide small refrigerators that cool microwave equipment or optical devices to 70, 20 or 4 K.

principle of complementarity   The wave nature (electromagnetic wave) and particle nature (collection of photons) of light are manifested in different experiments. Similarly, matter also exhibits both wavelike and particlelike properties. Niels Bohr proposed that matter has a dual nature and that the wave and particle aspects of matter complement each other. Einstein extended this duality concept to electromagnetic waves and photons.

principle of reversibility   Any ray in an optical system, if reversed in direction, will retrace the same path.

printed circuit   A pattern of conductors on a board of insulating material to which components are added to form a circuit. A printed circuit is often created by photolithography.

prism   A block of optical material with flat polished sides that are arranged at precise angles to each other. Prisms do not form images but can be used to deviate beams of light, invert or rotate an image, disperse light into its component wavelengths, or isolate separate states of polarization.

prism, achromatic   *See* aberration, chromatic.

prism binocular   Binoculars in which Porro or other prisms are used to produce erect final images and to reduce the length of the instrument. The distance between the objective lenses can thus be made greater than the interpupillary distance, thereby enhancing the stereoscopic effect produced by ordinary vision.

prism, combination of   A combination of prisms used to give a final erect image while simultaneously removing the left-right reversal produced by a typical telescope. Use of prisms avoids the problems of aberration and of increase in length of optical systems associated with lenses.

prism, Litrow   If a Brewster prism (a prism in which both a particular wavelength and linear polarization rays at minimum deviation enter and exit the prism at a Brewster angle) is cut in half along the plane in which there is a bisector between the two Brewster entrance/exit faces, a Litrow prism is used as the planar element of a hemispherical laser resonator.

prism, Nicol   A polarizing prism. The principal section is perpendicular to the entrance face, but the optic axis is neither parallel nor perpendicular to the face. It transmits the extraordinary ray, but the ordinary ray is totally internally reflected.

prism, Porro   A reflecting prism, commonly used in binoculars to provide erect images. It consists of two right angle prisms such that light entering perpendicular to the hypotenuse surface is totally reflected in turn by the two opposite surfaces, to emerge from the hypotenuse surface parallel to the incident light.

prism, Rochon   A common type of polarizing beam splitter when a ray of light, incident normally at the entrance face, travels along the optic axis in the first half of the prism. Both ordinary and extraordinary rays are undeviated and have the same refractive index. The second half of the prism has its optic axis at right angles to that in the first half, but the ordinary ray is undeviated since its refractive index is the same in both halves. The extraordinary ray has minimum index in the second half and is refracted at the cut.

prism, Wollaston   Wollaston prisms are used to provide double images of single sources and split a beam of light into two naturally orthogonal linearly polarized beams.

programming   The process of preparing a set of coded instructions that can be executed by a digital computer to yield the solution to a specific problem or perform a specific function.

progressive wave   A wave that transfers energy from one location in space to another (in contrast to a standing wave). Typical progressive waves are waves traveling along a stretched string and compression and rarefaction waves traveling along a tube. For example, a wave described by the shape $y = f(x)$ at the position x and at some time instant $t = 0$ traveling in the x direction with a constant velocity v is described by the equation $y = f(x - vt)$ at any time instant t. Also known as a *traveling wave.*

projection effect   Due to the three-dimensional nature of objects, only such points of objects that lie in the focal plane are imaged as sharp point images in the screen plane. Other points on the screen plane are depicted by small luminous areas that are sections cut out of this plane from the cone of image rays emanating from the off-focal plane points.

propagation constant   (1) A characteristic of a transmission line that summarizes the effects on the wave being transmitted by the line. It is a complex number, the real part of which (the attenuation constant) measures the signal loss while the complex part (the wavelength or phase constant) measures the shift in phase the wave undergoes.

(2) A complex quantity that measures the attenuation and phase change of a sinusoidal traveling wave along a transmission line. The real part of the propagation constant is the attenuation constant in nepers per unit length; the imaginary part is the phase change constant in radians per unit length. The propagation constant, γ, at a specific frequency is given by

$$\gamma = \log_e \mathbf{I}_1 / \mathbf{I}_2$$

where $\mathbf{I}_1$ and $\mathbf{I}_2$ is the current at two different points separated by unit length along the transmission line, $\mathbf{I}_1$ is closer to the signal source, and $\mathbf{I}_1/\mathbf{I}_2$ is the vector ratio of the currents. The propagation constant can also be written as

$$\gamma = \alpha + i\beta$$

where α is the attenuation constant, β is the phase change constant, and $i = \sqrt{-1}$. The amplitude of the vibration of a wave, E, at any distance, x, along the transmission line, is related to the initial amplitude, E_0 (i.e., at $x = 0$), of the wave by

$$E = E_0 e^{-\gamma x}.$$

The propagation constant is also known as *propagation coefficient.*

propagation constant, acoustic   The angular wave number k^*, which is a complex number in a porous material, multiplied by the imaginary unit $i = \sqrt{-1}$, $ik^* = \gamma = \alpha + i\beta$. α, the real part of the propagation constant γ, is called the *attenuation constant,* and it quantifies the exponential amplitude decay of the acoustic wave in the direction of propagation. The imaginary part of γ is the phase constant β. The complex wave number k^* is obtained from the solution of the wave equation.

propagation loss   The loss of energy from a beam of electromagnetic radiation due to absorption, scattering and spreading of the beam.

propeller sound   Noise generated by propellers — devices equipped with rotating blades mounted on a shaft — is of note as, for example, the major source of ship noise. Acoustic radiation is due to the complex dynamic interaction of propeller blades and nonuniform flow fields that can, in extreme cases, lead to cavitation. The sound characteristics depend on the blade form and operating conditions; noise control is an issue of continuing interest. Noncavitating propeller noise can be classified as (a) mechanical blade tonals that depend on angular velocity and blade number, (b) broadband noise caused by the interaction of propeller blades with turbulent structures and trailing edge vortices that cause a vibratory response, and (c) propeller-singing occurring at matching vortex shedding and blade resonant frequencies. Cavitation generates broadband noise, caused by growing and collapsing sheets of bubbles forming on the propeller blades.

prosthesis   An artificial replacement for a missing body part. Prostheses include artificial limbs, false teeth, hearing aids, artificial kidneys, and implanted pacemakers. In the construction of artificial organs, the use of biomaterials (materials that are biocompatible) are used.

   Modern devices include structural improvements that allow them to be of lighter materials, more realistic appearance, and greater flexibility. In some cases prostheses can permit the patient to participate in sports activities.

protocol   Two or more parties at the same level of communication are referred to as *peer entities* and a protocol is a set of rules and formats that govern the communication between peer entities.

proton pumps   Proton pumps are ion pumps that promote active transport of H^+ (protons). Because they involve active transport, they spend energy in their function by hydrolizing ATP. Proton pumps find their most use in areas of the cell where metabolic activity is being carried out, like intracellular organelles, endosomes, lysosomes, synaptosomes, chromaffin granules, golgi membranes, and endoplasmic reticulum where production of H^+ is the highest.

   Proton pumps are also of use in the production of ATP from glucose and other nutrients in the process called *cellular respiration*. An important step during cellular respiration is the electron transport chain (ETC), over which ATP is produced during the many steps in the process that involves oxidation of NADH and $FADH_2$ and reductions of O_2 into H_2O. The cytochromes, the typical protein carriers in the ETC, can accept electrons, pump protons, and generate H_2O from O_2 and H^+.

   In plant cells, vacuoles contain proton pumps in their membranes that, by transporting protons to one side, create a gradient in potential that is used later to move sugars and other ions in and out of the vacuole.

pseudo sound   Pressure fluctuations associated with flow fluctuations that propagate at the speed of the bulk flow (or a fraction of this speed) rather than at the speed of sound. Such fluctuations develop when, above a certain Reynolds number characteristic for the particular flow geometry, more or less steady vortices form at lower flow velocities. At higher flow velocities turbulence sets in, and the pressure fluctuations associated with turbulent flow are irregular. Small alternating pressures accompany the relatively large velocity amplitudes, and only a fraction of the energy carried by the flow is emitted as sound (noise). Pseudo sound is generated in heat exchangers in the flow over tube bundles, in blood vessels (blood flow), and fans (air flow).

p-type silicon   Silicon that has been doped with an acceptor type impurity (boron (B), gallium (Ga), etc.) and in which holes, positively charged, are the majority carriers (hole concentration is much higher than electron concentration).

pulse generator   A circuit used to generate pulses. A pulse generator will often allow variations in the repetition rate, amplitude, width, polarity, etc., of the pulses. They are commonly used to produce inputs for digital circuitry.

pulse height discriminator   A device that passes pulses higher than a predetermined pulse level, but blocks pulses lower than that level.

pulse, longitudinal   A pulse in a compressional zone (generated, for example, by giving a piston confined in a long tube a short, rapid inward stroke) traveling at a speed v, with particles of matter displaced in the direction of propagation (longitudinally). The fluid in contact with the piston subjected to the pulse is compressed; its density and pressure rise above the undisturbed values. The compressed fluid moves forward and compresses adjacent fluid layers, causing the compression pulse to advance down the tube. For a longitudinal sound wave in a gas, the velocity is $v = \sqrt{\frac{\gamma p_0}{\rho_0}}$, where γ is the ratio of specific heats for the gas, and p_0 and ρ_0, the undisturbed pressure and density, respectively.

pulse operation A method of operation in a circuit in which the signals are passed in the form of discrete pulses rather than as analog signals.

pulse rate (Also known as for pulse-repetition frequency.) The rate at which pulses are transmitted in a pulse train. It is the reciprocal of the period and is measured in units of Hertz.

pulse shaper A circuit used to shape a pulse to a desired form, or to modify its characteristics.

pulse train A series of pulses with similar characteristics occurring at regular intervals.

pulse, transverse A disturbance (created, for example, by applying a single sidewise movement to a stretched string) characterized by the motion of the particles of matter perpendicular to the direction of propagation of the disturbance (pulse). In a transverse pulse each particle remains at rest until the pulse reaches it, moves for the duration of the pulse and then returns to its initial state. For the stretched string, the speed of propagation of the disturbance is $v = \sqrt{F/\mu}$, where F is the elasticity measured by the tension in the string and μ the mass per unit length of the string.

puncture voltage The voltage gradient that leads to breakdown or "puncture" across an electrical insulator. This leads to a sudden decrease in the resistance of the material to such an extent that high-value currents may flow through local regions. In the case of solid insulation, puncture represents damage to the material, whereas in the case of liquid or gaseous insulation, the interruption of the current flow and the removal of the offending overvoltage restores the material close to its original condition. The breakdown of insulation is time dependent; the higher the voltage above the critical value, the shorter the time for breakdown.

pupil Pupils are derived from knowledge of the field stop in an optical system and are used to determine the limiting cone of rays from the object point to the conjugate image point. In the analysis of optical systems, two types of pupils are considered: the entrance pupil and the exit pupil.

1. *entrance pupil:* the image of the controlling aperture stop formed by all the imaging elements preceding it.

2. *exit pupil:* the image of the controlling aperture stop formed by all imaging elements following it.

In visual optics, the pupil is the aperture in the iris, normally circular and contractile, through which light enters the posterior portion of the eye.

pupil, entrance *See* pupil.

pupil, exit *See* pupil.

Purkinje effect Perceptual variation in relative lightness of different colors as illumination changes from daylight (photopic) to twilight/night (mesopic/scotopic). That is, during the day when vision is photopic, objects close to 550 nm wavelengths will tend to appear lighter than objects of 500 nm. When vision becomes scotopic, the situation is reversed. Also known as *Purkinje Shift*.

Q

Q-factor (**1**) Also known as the quality factor. It is a figure of merit for an electrical circuit (or any energy storage system), and is given by $Q = 2\pi \frac{\text{average energy stored}}{\text{average energy dissipated per half cycle}}$. For a resonance system with high Q values, it is also equal to $f_0/\Delta f$, where f_0 is the resonant frequency, and $\Delta f = f_2 - f_1$ is the frequency band, defined as those frequencies that give more than 50% of the total power delivered at resonance.

(**2**) Symbol: **Q**. Also known as the quality factor. A figure of merit for a tuned circuit. It determines the rate of decay of stored energy. The decay time becomes longer with increasing value of the Q-factor. For a tuned circuit, it is given by

$$Q = \frac{2\pi f_0 L}{R} = \frac{1}{2\pi f_0 C R}$$

where L = inductance, C = capacitance, R = resistance associated with a real inductance or capacitance, and f_0 = resonant frequency. The resistance may also be included as part of the tuned circuit to reduce the Q-factor. It is also given by $Q = f_0/\Delta f$, where Δf is the bandwidth at the -3dB points of the voltage vs. frequency response function of a resonant circuit.

Q-meter Laboratory instrument that measures the Q-factor of a circuit or circuit element by determining the ratio of reactance to resistance.

quadrature State of being separated by 90°, or one quarter cycle. Also known as *phase quadrature.*

quadripole The field that results from two magnetic dipoles arranged as a unit. If the dipolar axes point in opposite directions, then, at large distances compared to the physical dimensions of the quadripole, the magnetic field pattern has a quadrupolar symmetry.

quadrupole, electric In its simplest form, an arrangement of two equal electric dipoles in opposite orientation.

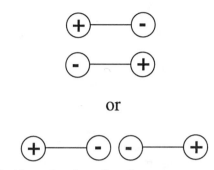

Electric quadrupole configurations.

The electric field falls off as $1/r^4$, where r is the distance from the center of mass.

quadrupole sound sources This refers to the arrangement of the sound sources of which there are two basic types: lateral and linear quadrupole sources. In the *lateral quadrupole arrangement,* the monopole sources with alternating phases are at the corners of a square. Sound is radiated in a cloverleaf pattern, with strong projection in front of each source and with the sound being canceled at points equidistant from adjacent opposite monopoles. In a *linear quadrupole arrangement,* the two opposite phase dipoles lie along the same line, as in a tuning fork, for example. In this arrangement, in the near field there are four maxima and four minima, while in the far field there are two maxima and two minima.

quality, sound *See also* timbre. A string can be plucked, bowed or struck, e.g., in a harp, violin or piano, respectively. Quality depends on the relative amplitude of various overtones to the fundamental tone. Fundamentals are indistinguishable; overtones introduce the necessary quality that is important for the ear. Notes of the same pitch and loudness produced by different instruments can be distinguished by the quality.

quantization of charge Charges, whether positive or negative, come in certain specific amounts, namely, $0, \pm q, \pm 2q, \pm 3q$ and so on. For example, the charge of one electron is equal to $-q$, while the charge of one proton is equal to $+q$. Robert A. Millikan used an oil-drop apparatus to demonstrate the quantization of charge and won a Nobel prize in 1923.

quarter-wave film   Film whose thickness, multiplied by the index of refraction of the film material, equals one fourth (or an odd integer multiple of one fourth) of the wavelength of the radiation incident on the film. The significance of quarter-wave film derives from the perfect constructive or destructive interference that occurs between parts of a wavetrain reflected at front and rear surfaces of the film (depending on dielectric constants of media on the two sides of the film), resulting in a maximum or a minimum in the reflectivity of the film.

quarter-wave line   A section of transmission line that is one quarter-wavelength long at the fundamental frequency being transmitted. When shorted at the far end, it has a high impedance at the fundamental frequency and for all odd harmonics, and a low impedance for all even harmonics. It is often used as an impedance matching device between two transmission lines with different impedances (Z_1, Z_2) such that $Z_0^2 = Z_1 Z_2$ where Z_0 is the impedance of the quarter-wave line.

quarter-wave plate   Plate of birefringent material with ordinary and extraordinary indices of refraction n_1 and n_2, cut with its surfaces parallel to the optic axis and with its thickness d adjusted so that the difference $(n_1 - n_2)d$ equals (exactly) one-quarter of the wavelength of the radiation of interest. Passage of this radiation through quarter-wave plate, perpendicular to the plate surfaces, produces a difference between the phases of the ordinary and extraordinary components of the radiation exactly equal to $\pi/2$, which serves to effect the conversion of linearly polarized light into circularly polarized light, and vice versa.

quiescent component   A device that has power applied to it, but is receiving no input signal.

quiescent current   A current in a circuit or device to which power has been applied, but no input signal has been applied. For example, the current in an AC amplifier circuit with no AC input.

quincke tube   A commonly used device for demonstrating the interference of sound waves and measuring their wavelength. Sound waves from a high frequency source enter the apparatus at S and the energy divides between A and B. The two waves reunite at C and are picked up by detector at D.

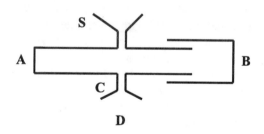

R

radiance   The radiant energy per second per unit solid angle passing through a unit area perpendicular to the bisector of the solid angle containing the measured radiation. Measured in units of watts/steradian/m^2.

radiation   (1) Propagating electromagnetic energy, *interpreted classically* to consist of oscillating (time-varying) electric and magnetic fields that propagate through space with a speed dependent on the properties of the matter that occupies the space, and *interpreted quantum-mechanically* to consist of units of electromagnetic energy known as photons with energies equal to hf, where f is the frequency of the radiation and h is Planck's constant. Classical interpretation of radiation applies where photon energies are small compared to the energy sensitivities of the measuring equipment such that the observed effects involve large numbers of photons that act on the average to produce classical fields. Types of electromagnetic radiation, classified by distinct ranges of frequencies, include radio, microwave and infrared waves, visible light, and ultraviolet, X- and gamma-rays.

(2) The term is alternatively used to refer to a stream of energetic nuclear particles or electrons.

radiation carcinogenesis   By definition, a carcinogen is any agent that causes cancer in animal and human tissue. Carcinogens may be inorganic (asbestos, arsenic) or organic (certain molds, viruses). Other types include X-rays, UV, and gamma rays. Contamination may be through air (radon, smoke), skin absorbed (pesticides), or ingested (nitrites).

Radiation carcinogenesis is then the development (from the beginning) of malignant cancer tumors exclusively from radiation exposure. Radiation carcinogenesis studies explore the relationship between radiation dose and response as well as dose rate effects. The studies provide assessment of dose response in individual organs and usually extrapolate results obtained from animal models to humans. *See* radiation effect, cumulative.

radiation damping, acoustic   Sound waves will undergo an exponential decay if there is absorbing material present in their path of propagation. The amplitude of the wave decreases with time due to the loss of mechanical energy to the medium or its surroundings. This effect is more significant, for example, in a room where standing waves are set up and absorbing material is present at the corners of the room. The acoustic resistance and the coefficient of absorption of the room need to be taken into account.

radiation effect, cumulative   Radiation effects are counted above the tolerance dosage of approximately 5 Rems per year (whole body dosage). The classification of doses range from mild and acute, to chronic, as a function of dosage. Effects may occur as a result of cumulative small doses of radiation or exposure to radiation from accidents. Mild radiation sickness is a common side-effect of radiation therapy for cancer.

Mild doses of radiation are 25 to 50 Rems in less than one week; these do not produce detectable clinical effects on the body. Doses of up to 150 Rems may show blood changes and may prove to be a longtime hazard. A dose up to 250 Rems is considered a moderate dose that will provoke nausea and vomiting within 24 hours. Injury may vary from slight to serious and the recovery is subject to good health and no complications. A dose of 250 to 350 Rems brings nausea and vomiting in under four hours with the possibility of mortality in two to four weeks. Higher doses, up to 600 Rems, are considered a median lethal dose with 50% probability of death or a very good chance of incapacitation. Doses greater than 600 Rems will provoke immediate nausea and vomiting with mortality in one to two weeks. The doses usually have the acute effects, the immediate effects within 48 hours, and delayed effects of injury or incapacity in the range of one to five weeks. There are chronic effects that result in changes to the skin and vascular changes. *See* radiation exposure; radiation standards; radiation effect, genetic.

radiation effect, genetic   Genetic effects of radiation are readily understood in terms of the effects that ionizing radiation has on living cells. Upon the irradiation of biological tissue, the radiation reacts with molecules and atoms causing the release of free electrons. This release comes from the interaction of high energy photons with the electron cloud of atoms and the transferal of the photon energy to the electrons, thus ionizing the atom and setting the electron free. Free electrons in turn react with water dissociating it and creating free radicals. Biological molecules, especially DNA, are very susceptible to radicals that cause strand breaks in them, thus altering the genetic code of cells.

The use of ionizing radiation to create free radicals is exploited in radiotherapy, where treatment and elimination of cancerous tumors is achieved by destroying the genetic code of malignant cells, thus causing a disruption of cell activity. The success of the therapy lies in the ratio of tumor cells vs. healthy cells that are affected. *See* radiation exposure; radiotherapy.

radiation exposure   Exposure to radiation emitting substances is becoming an increasingly complex problem due to increased exposure that exists both from natural as well as industrial sources.

Naturally occurring radiation sources come from the radioactive decay of radium and radon gas, found in some groundwater sources. Exposure dangers from radon, however, are mostly from breathing the gas after it is released into the atmosphere, rather than from drinking. Other sources are from thorium and uranium. There are also some biologically internal sources of radiation like potassium^{40} and carbon^{14} that occur naturally in living cells. Other types of radiation come from cosmic rays that have outer space origin.

The total annual dose received by a person from these naturally occurring radiation types at sea level is approximately 0.91 mSv. Twice this amount may be received by a person who lives at higher elevations where cosmic rays are more intense, or by people who live in a geographic place with soil with a high radium content. In this last case, radon from the decay of radium may accumulate indoors, and, if the house is not well-ventilated, the person may be exposed to doses as high as 100 mSv per year in their lungs.

Artificial radiation may come from nuclear power plants, uranium mining, and medical research, and may contribute substantially to water contamination. Nuclear weapons testing has been responsible for strontium^{90} and tritium contaminants in water. It has been estimated that the average annual radiation dose received from medical and dental irradiation in developed countries approaches in magnitude the dose received from natural background radiation.

Other artificial sources of radiation include radiation from radioactive minerals in crushed rocks, phosphate fertilizers, radiation from building materials, radiation-emitting components of television sets, and smoke detectors, among others. Total exposure of radiation from artificial sources amounts to approximately twice that of radiation from natural sources alone. *See* radiation standards.

radiation impedance, acoustic   Fundamental characteristic of medium in which sound is transmitted. It refers to the opposition of the transmission of sound in a particular medium. It is generally numerically given by ρc (g/cm^2 sec), where ρ is the density of the medium and c is the velocity of sound in that medium which is determined solely by the physical properties of the medium.

radiation, isotropic   Diffuse radiation such that the radiant energy is propagating in many different directions through a given small volume of space. The diffuse radiation has the same intensity in all directions.

radiation of sound   The propagation of sound waves from a source outward in a medium. There can be plane wave propagation or spherical wavefronts. Point sources produce spherical wavefronts.

radiation pressure   The pressure exerted by radiation on a given surface as a result of the rate of transfer of (radiant) linear momentum across the surface per unit area (producing a force per unit area on the surface). Electromagnetic radiation pressure can be interpreted to derive from the force of the magnetic field of the radiation in the direction of propagation. Pressure produced by the absorption (reflection) of electromagnetic radiation by a surface equals two times the time average of the *Poynting vector* of the radiation divided by *c*.

radiation pressure, acoustic   Traveling sound waves, on striking a surface, become absorbed or reflected. In doing so, they exert a force on the surface on which they fall. This can be used as a mechanism for measuring sound energy. *See* radiometers; rayleigh disk.

radiation protection   Protection against radiation has become increasingly important as more and more techniques and applications make use of radioactive materials in applications that involve materials quality testing to medical diagnostics and treatment of disease.

The most common type of radiation that people are exposed to is ultraviolet radiation from the sun. An increase of skin cancer, cataracts, and other effects from UV exposure have made UV exposure an important health issue. In efforts to provide meaningful ways of protection, the *UV Index* has been established. The index is meant to associate levels of exposure to risks of disease from radiation damage to tissue. The UV Index ranges between 0 and 10+, and is in proportion to the amount of UV radiation that reaches the earth's surface over a one-hour period at noon. Other useful indices used by suntan lotion and sunscreen companies is the sun protective factor (SPF). SPF numbers in these products indicate the amount of protection against the different types of UV.

Radiation protection against radioactive materials handling and managing is given by limiting personnel exposure to radiation from the substances. Storage is usually provided in lead sealed cabinets or cannisters. X-ray rooms are usually shielded as well as other facilities that expose doses of radiation to detectors. Pocket dosimeters provide a way of measuring the total accumulated dosage received, thereby monitoring radiation safety limits. *See* sun exposure and skin cancer, sunburn, radiation standards, radioactive waste, radioisotope storage.

radiation quality   *See* radiation standards; radioactive waste; radiation protection; radiation exposure.

radiation, resonance   Electromagnetic radiation with a frequency matched to a transition frequency of the atoms or molecules of a given material. Alternatively refers to the radiation emitted by a gaseous material when the atoms or molecules of the material are excited by incident radiation of the same or higher energy.

radiation, selected   Electromagnetic radiation of a particular frequency. Generally refers to radiation that is selected by reflection from a grating in a particular direction, or by transmission through an etalon via constructive interference at a selected wavelength.

radiation standards   Patients being treated for cancer are usually administered a dose of 50 Sv or more in daily exposures in periods that can last from four to six weeks. Protection for the normal tissue of the patient as well as protection to the medical personnel against excessive occupational exposure to stray radiation are taken in order to prevent or diminish damage and contamination. Comparable safeguards are utilized to minimize the exposure of workers employed in other activities involving radiation or radioactive material. Similarly, standards of safety have been developed for the handling and disposal of radioactive waste.

In the U.S., the FDA has regulatory power over radiation and radioactive materials management. This includes radiation control over consumer products that may be able to radiate excess radiation in, for example, X-ray machines, televisions, microwave ovens, etc.

For personnel who work in an environment that deals with radioactive materials usage and handling (with radiation doses that exceed 20 mR/hr at one meter), pocket dosimeters give an instantaneous measure of the total accumulated dosage. These dosimeters are able to detect X-rays as well as gamma radiation.

There are standards on the permissible dosage to which people may be exposed. Occupational exposures should not exceed 500 mSv per year, compared with public exposures of less than 5 mSv per year. *See* radioactive waste.

radio (1) An apparatus for receiving, broadcasting, or transmitting radio signals without connecting wires or waveguides.

(2) The process of transmission and reception of information by electromagnetic waves of radio-wave frequencies.

radioactive waste Radioactive waste is produced in increasing amounts as byproducts of nuclear weaponry, nuclear-power generation, and research. Because much of this waste remains radioactive for long periods of time, it is particularly hazardous and difficult to store.

Fuel from nuclear power generators is considered waste once it cannot be reprocessed anymore. At this point the waste is usually prepared for storage. In the U.S., the Department of Energy has the responsibility of receiving the spent fuel from utilities. The process starts by keeping the waste in temporary storage that then are packaged in corrosion-resistant canisters of steel. Some other types of radioactive waste coming from fission in solution are completely evaporated, leaving the waste in solid form that is subsequently heated until all the constituent nitrate salts are converted to oxides. The oxides are then put into a glass-forming oven and mixed with materials that form a borosilicate glass. Lastly, the glass melt is poured into steel canisters. After the waste has been prepared in canisters, the planned method for ultimate disposal is called *geologic disposal,* which means that the waste is deposited in underground mined tunnels.

Radioactive waste disposal is a world-wide problem that usually demands large amounts of money and resources to keep from doing harm. The lack of appropriate disposal has led some countries to dump radioactive pollutants in the oceans. There is evidence that links various radioactive pollutants to human health problems such as cancer, birth defects, and genetic changes. *See* radiation exposure.

radio channel A radio channel refers to a frequency band in the electromagnetic spectrum dedicated for radio communication. The bandwidth of any radio channel typically depends on the application.

Radio signals in the medium radio frequency range of 525 to 1700 kHz are coded in amplitude

modulation (AM). Medium frequency AM radio waves may propagate between the transmitter and the receiver as a ground wave, a space wave, and/or a sky wave. *Ground waves* embody surface waves traveling along the earth-atmosphere boundary. Ground waves suffer signal attenuation as the wave penetrates the ground or water surface, losing an amount of energy dependent in part on the electromagnetic wave frequency as well as the conductivity and the permittivity of the ground/water surface. *Space waves* represent that part of the transmitted electromagnetic wave traveling through the atmosphere. Space waves are often blocked or impeded from reaching the receiver due to geographic topology or the earth's curvature. *Sky waves* represent reflections of upward-bound space waves off the ionosphere. Sky waves seldom occur during the day, when solar radiation ionizes the ionosphere's D layer (about 40 miles from the earth's surface) which thus absorbs or scatters the upward-bound sky waves. During the night, when the D layer becomes less ionized, the upward-bound sky waves reach and reflect ground-ward at the E layer (60 miles above ground level) and F layer (130 miles above the earth's surface). Medium-frequency AM radio communication depends primarily on ground waves and sky waves; space waves, due to shadowing, are relatively irrelevant.

Radio signals in the very-high frequency range of 88 to 108 MHz are coded in frequency modulation (FM). Very-high frequency radio communications depend primarily on space waves and relatively little on ground waves or sky waves. Due to their very long wavelength, very-high frequency electromagnetic waves can typically diffract round or reflect off geological and human-made structures that would have blocked a medium-frequency wave. Such refractions and reflections result in multiple time-delayed propagation paths between the transmitter and the receiver. The signal power attenuation and multipath profile resulting from refractions and reflection may in principle be calculated using physics principles, but may vary in a very complicated manner with time as the receiver (such as a radio set installed in an automobile) travels amidst a complex surrounding of reflectors and refractors. Global broadcasting in four shortwave sub-bands lying between

5950 kHz and 26.1 MHz, and FM broadcasting in the very-high frequencies between 88 and 108 MHz.

The radio channel may be further corrupted by thermal noises of the electronic hardware, co-channel and adjacent channel interference, multipath propagation effects, non-stationary reflectors and refractors near the transmitters and/or the receivers, obstructions by human-made structures, or by the geographical topology. These physical effects translate into additive stationary and non-stationary impulse noises, time-varying multiplicative distortion to signal amplitude, spectral distortion and inter-symbol interference, depolarization, and prolonged deep attenuation of signal power. These channel non-idealities may be mitigated by various diversity techniques at the transmitter and/or the receiver — spatial diversity by deploying multiple displaced antennas, polarization diversity by deploying antennas of different polarizations, time diversity by interleaving the information symbols, or frequency diversity by modulating the transmitted signal with different or time-varying carrier frequencies. Sophisticated high-speed signal processing techniques are also essential and commonly used to mitigate channel non-idealities.

radio communication   Radio communication represents communications between two or more geographical points using as the transmission medium unguided electromagnetic waves in the radio frequencies. Radio communication in the frequency range of 525 to 1700 kHz uses amplitude modulation (AM), with 10 kHz allowed for each AM channel in the Americas and 9 kHz for much of the rest of the world.

Radio communication in the short-wave frequency range between 5950 kHz to 26.1 MHz are primarily for long-distance international broadcasting. Shortwave radio channels have only a 5 kHz bandwidth, sufficient for speech communication but not for music transmission.

Radio communication in the very-high frequency range of 88 to 108 MHz uses frequency modulation (FM), with a 200 kHz allowed for each FM channel in the United States. FM channels typically transmit in stereo, with a left sub-band and a right sub-band on either side of the carrier frequency for each stereo channel.

radio, for navigation   Radio waves propagate at 300,000 km/s and thus permit measurements of distance as a function of time and direction as a function of differential distance to two or more known points. In free space, radio navigation is capable of considerable accuracy. Along the surface of the earth, however, the effect of multiple propagation paths between a transmitter and receiver reduces the accuracy. Since aircraft and ships may move over large areas, systems that involve cooperation between a vehicle and a ground station required a high degree of international standardization.

radiography   The process of producing radiographs using X-rays as a probe. A radiograph is in simple terms a photograph using X-rays. Radiography has found wide acceptance and use in diverse areas like medicine, biology, civil engineering, the aerospace industry, and environmental protection.

The main difference between radiography and other branches of nuclear medicine is that, in radiography, the patient is subjected to an external source of X-rays. Analysis is based on scattered and transmitted radiation. In other branches, the patient is usually the source of the radiation. Also called *skiagraphy*. *See* radiology.

radiography, neutron   Non-destructive imaging technique that involves the detection of the attenuation of a neutron beam by the object being radiographed. The technique gives information about internal structure of materials because not all parts of the object provide the same attenuation, thus giving an image of the internal attenuation regions.

Neutron radiography is similar to X-ray radiography, but complementary in terms of the information obtained through it. While X-ray gives information about the electron cloud surrounding the nucleus of an atom, neutrons give information about the nucleus itself. Neutrons are either scattered or absorbed by the atomic nuclei. There are two typical detection methods. In the *indirect method* of obtaining an image, a screen is put opposite to the neutron beam on the other side of the sample. After neutron bombardment, the screen becomes radioactive and, after getting it into contact with an X-ray film, a radiograph is obtained where the darkening of the X-ray film corresponds to the highest radiation decay regions of the screen. In the *direct method,* the X-ray film and a conversion screen are both put in the neutron beam, along with the sample.

An example of a neutron source is Californium^{252}, in which one gram emits 2.3×10^{12} neutrons per second.

radioimmunoassay   Radioimmunoassay (RIA) is a technique in biochemistry and medicine to measure extremely small amounts of substances (e.g., antigens, hormones, enzymes, steroids, vitamins, immunoglobulins, drugs) in body fluids. In general terms, the technique consists of injecting a subject with a substance or antigen that will cause the body to produce antibodies. Later, serum from the subject is extracted and the present antibodies are obtained and treated with a radioactive antigen and later with a non-radioactive antigen.

The technique provides then a method of determining the amount of the substance present in the body. Radioimmunoassay is used by hospitals to help diagnose diabetes, thyroid gland disorders, and other diseases. It is also used to measure plasma renin activity.

radioiodination   A technique in which biological molecules (peptides, proteins) are radiolabelled with isotopes of iodine — radioiodines like I^{125} or I^{131} — in tracer studies. The success of the technique lies in the fact that radioiodine is very well suited for radiolabelling hydrophilic and lipophilic compounds widely used in biology and medicine. The versatility of these radioisotopes has made possible the present status of diagnostic nuclear medicine.

In this technique, it is critical that after a molecule is radioiodinated, any remaining radioiodine is removed from the sample to prevent radioactive detection other than the desired molecules. Radioiodine is by itself widely used as a tracer in the diagnosis and treatment of the thyroid gland. *See* radioisotope tracers; radiolabelling; radiology.

radioisotope storage   Storage place to put short-lived or other radionuclides used in radiological studies. Depending on the usage, radioisotope storage should be accessible but at the same time provide proper shielding and security to prevent accidents and mishandling.

Types of radioisotope storage range from simple multi-drawer shielded modules to storage safes. Usually, shielding is provided by 0.5 to 1 inch of lead on all sides, depending on the types of radionuclides used. *See* radionuclides.

radioisotope tracers   A radioactive substance that, when introduced into a biological system, can be followed in time and space. Usually, a biological process or structure is being examined and the tracer provides information about the time development or distribution of the tracer that may be assembled to give an understanding of the process or a diagnosis of a disease. Tracers are located in the body by detecting their radioactive decay radiation, including the emitting of alpha or beta particles, or gamma rays.

Typical tracers in medical applications that involve diagnosis and treatment include isotopes of iodine, technetium, astatine, and scandium. *See* radionuclides; radiolabelling; radiology.

radiolabelling   A technique in which a radioactive tracer is tagged to another molecule (e.g., hormone, enzyme, protein) in order to find or follow its location in the body. The level of radioactivity is a direct measurement of the amount and number of the molecules that are labelled. The selectivity with which the tagging molecule has to work requires specific studies in organic-inorganic chemical synthesis, biochemistry, and receptor binding studies.

The technique is used for the imaging of tumors in the diagnosis of cancer with gamma-emitting radionuclides that are attached to biomolecules that preferentially target tumor cells. The same technique is also applied in cancer therapy. It is also used in the assessment of how the body metabolizes drugs in radiopharmaceutical studies.

radiology   The branch of medicine that deals with the use of radioactive substances and radiation which, based on their ionizing radiation, are used to produce images and give treatment to disease. The most common example is the use of X-rays to produce non-invasive images of bones and dense tissue in the body.

Examples of sub-branches are radiologic anatomy, medical imaging, magnetic resonance imaging, computed tomography, cardiovascular radiology, mammography, musculoskeletal radiology, positron emission tomography, fluoroscopy, ultrasound, radiation medicine, nuclear medicine, neuroradiology, quantitative radiology, gastrointestinal radiology, and genitourinary radiology.

radiometers Mechanical devices for measuring sound energy that are large in comparison to their associated wavelength. If sound waves hit a disk or plate hung on a torsion balance at normal incidence, a pressure is exerted on the disk. Some of the energy is reflected and the acoustic pressure causes the disk to rotate. The amount of acoustic energy is indicated by the amount of twisting or motion. *See also* rayleigh disk.

radiometry (**1**) The science of measuring optical radiation at any wavelength. All fundamental radiometric measurements measure optical energy. Since optical energy induces heat into an absorber, it follows that a thermal-sensitive detector, calorimeter, can be used to measure optical energy.

(**2**) Also refers to an instrument for measuring the intensity or force of radiation.

radio, mobile A small mobile radio terminal used for short-distance radio links.

radionuclide generators Apparatus for the production of radionuclides. Radiation therapy, radionuclides for diagnostic and investigative medicine, and positron emission tomography studies have made radionuclide generators an essential component of hospitals. Their importance lies in providing radionuclides that cannot be transported because of their intrinsic short half-lives.

Radionuclide generators can be of the type of fission reactors that produce a broad range of radionuclides or in-hospital medical cyclotron reactors. *See* tomography, positron emission, ventriculography, radionuclide, radionuclides, radiotherapy.

radionuclides Radionuclides are isotopes of atoms that decay producing radioactivity. They emit various ionizing radiations, for example, electrons, positrons, alpha particles, gamma rays, and X-rays. The particular emissions depend on the radionuclide.

Radionuclides are of importance in medical applications where diagnosis and treatment of disease requires the intervention of tracers and/or ionizing particles and radiation. There are three predominant uses for radionuclides in medicine: in radiographic imaging techniques, in studies of metabolism, and in radiation therapy.

In *imaging techniques,* the purpose is to visualize the distribution of an injected radionuclide within a given organ as a means of studying the anatomic structure and pathological conditions of the organ. In *assessments of metabolism,* the goal is to quantitate the absorption and retention of radionuclides in organs, as a function of time and other medical variables.

Examples of radionuclides include: Carbon[11], Nitrogen[13], Oxygen[15], and Fluorine[18], which are important in PET studies. Other gamma emitting radionuclides are Technetium[99] and Iodine[131], used in imaging and metabolic studies, respectively. *See* radionuclide generators; tomography; positron emission; radiotherapy.

radio sonde A miniature radio transmitter that broadcasts meteorological information, such as pressure and temperature, and other scientific data from various levels of the atmosphere to ground. A radio sonde is usually carried by a balloon or kite.

radiotherapy Radiotherapy refers to the treatment and management of malignant disease with radiation. The treatment usually involves irradiation with X-rays or other ionizing radiation, and by ingestion of radioisotopes. The ionizing radiation results in the production of free electrons that react with water in the tissue to form reactive free radicals. The radicals then interact with the surrounding biological molecules, especially DNA, causing strand breaks. Radiotherapy is a local treatment and may be used when cancerous cells are contained in a few areas of lymph nodes.

Radiotherapy is more effective and safest if the radiation is given in multiple small doses (fractions) over a long period of time. In general, the reason is that small doses separated in time allow normal tissue to repair the damage done to healthy tissue, and this repair is usually faster than the repair to tumors. This technique produces an overall reduced toxicity to normal tissue. Also, the tumor will have better oxygenation, allowing for tumor shrinkage as the treatment progresses (poorly oxygenated tumor is less sensitive to irradiation). Regarding the well-being of the patient, there is better treatment tolerance. Another advantage in the planning of the treatment is that the amount of radiation can be reduced as the size of the tumor shrinks. This usually leads to the delivery of radiation in multiple fractions over a time period that typically spans one to two months.

RAM (random access memory)   Memory in a computer used to read and write data. The locations may be accessed in any order.

Raman effect   *See* scattering, Raman.

Raman scattering, coherent anti-Stokes (CARS)   In CARS, a sample is subjected to two collinear strong laser beams at different frequencies, v_1 and v_2. If the frequency difference $v_1 - v_2$ (assuming $v_1 > v_2$) coincides with the frequency of a Raman-active rotational or vibrational mode nu_R, then there is enhancement of the signal at $v_A = 2v_1 - v_2$ (anti-Stokes) and $v_S = 2v_2 - v_1$ (Stokes). In Stokes, a phonon is produced from the scattering while in anti-Stokes a phonon is absorbed.

Either v_A or v_S can be used for analysis, but the anti-Stokes signal offers the advantage that its frequency is well above v_1 and can be readily separated from the incident beams and spontaneous fluorescence signals by optical filtering. In CARS, intensities of up to 10^4 to 10^5 can be achieved over those of conventional Raman spectroscopy, which allows for faster data acquisition. Other advantages over conventional Raman spectroscopy include the fact that smaller samples can be used, better spatial discrimination can be obtained, and different regions within a sample can be examined. The last two reasons are of special interest to CARS compositional studies of flames and plasmas.

Raman scattering, electrophoretic   The use of Raman scattering techniques can be combined with electrophoretic separation in order to study effects of an electric field on the vibrational modes of the molecules in solution. Also, any property that can be related to the vibrational modes can be analyzed in dependence of the applied electric field.

Electrophoresis by itself is a technique based on the mobility of particles suspended in an electrolytic solution. Because different particles (e.g., proteins) move in the same electric field at different speeds, the difference in speed can be utilized to separate the contents of a suspension. When used in conjunction with Raman scattering techniques, conformational changes of polymers in solution (for example) can be determined as a function of the applied electric field. Raman spectroscopy of vibrational spectra from the polymers is very sensitive to slight changes in the local structure (changes in bond lengths and bond angles). Shifts and changes to the Raman spectra are then reflections of conformational changes.

Modern electrophoretic techniques include capillary electrophoresis in which the solute is contained inside a capillary and subjected to an electric field parallel to the length of the capillary. Its advantages over normal techniques are due to its heat dissipation characteristics and sensitivity.

random code   A random, or pseudo-random, code is used as the signature spreading sequence to modulate the data signal of a particular user in code division multiple access (CDMA). This random code is used to modulate (or spread) that particular user's message bits via either an amplitude modulation scheme (i.e., direct sequence CDMA) or a frequency modulation scheme (i.e., frequency hopping CDMA). The same noise spreading code is used at the receiver to correlate the received signal so as to recover the original data signal. Different users in a CDMA system are typically assigned orthogonal or near-orthogonal signature spreading code sequence, such that the transmitted digits from different users have no or little interference against each other at the receiver after de-spreading.

rating   The specified limit to operating conditions of an electrical or electronic component, e.g., the power rating of a transistor, resistor, etc.

ratio detector   An electric circuit that has two input terminals and one output terminal. The output signal is only determined by the ratio of the two input signals.

ray, bound   In an optical fiber, bound rays are those that are confined to the fiber core.

ray, extraordinary   In double refraction (*see* ray, ordinary), the extraordinary ray does not obey Snell's Law on refraction at the crystal surface. The extraordinary ray can be described in terms of components polarized in directions both perpendicular and parallel to the optic axis, propagating with different velocities. The emergent extraordinary ray is polarized in a direction perpendicular to the polarization of the ordinary ray.

rayleigh disk   (1) A thin disk suspended by a wire into a fluid, either liquid or gas. If the disk makes an angle θ with the direction of fluid flow, the hydrodynamic forces on the disk tend to align it perpendicular to the fluid flow (i.e., $\theta \rightarrow 90°$). This occurs regardless of the absolute value of the direction of flow (e.g., leftward or rightward), so it is particularly suited to measuring the presence of small sound fields present in the fluid. In superfluid helium, it is also possible to create thermal waves, denoted as second sound. Early experiments in superfluid helium used Rayleigh disks to sense second sound waves and test the two-fluid model. *See also* helium-4, superfluid.

(2) These are small paper disks, whose sizes are much smaller compared to the wavelength of the sound under investigation. They are suspended on a fine fiber at 45° and can be used to show streamline flow. Instruments based on this design are also used for comparing intensities of sounds of definite pitch.

rayleigh's criterion   An arbitrary, but useful criterion which relates the lens diameter, D, wavelength, λ, and the limit of the resolution, $\Delta\theta$, for just resolvable images. This requires that the centers of the two image patterns be no closer than the angular radius of the Airy disk; i.e., the principal diffraction maximum of one image falls directly over the first minimum of the other.

$$\Delta\theta = \frac{1.22\lambda}{D}.$$

rayleigh waves   Acoustic waves associated with earthquakes that travel in a thin layer close to the surface of the earth along a great circle from the epicenter. The wavelength is a fraction of the size of the plate the wave travels on. Such waves can also be used for exploration of defects in material.

ray of sound   The concept is used in geometric acoustics in which a sound emanating from a source is assumed to travel in rays of unchanging frequency.

ray, ordinary   When light is propagating through a crystal whose optical axis is at an arbitrary angle with respect to the beam direction, the light would experience double refraction — two refracted beams called *ordinary* and *extraordinary* rays emerge. The ordinary ray obeys Snell's Law and would emerge linearly, polarized perpendicular to the optic axis.

ray, principal   A ray that passes through the lens/optical system undeviated is called the *principal ray* (also called the *chief* or *undeviated ray*).

rays, pencil of   Usually, the section of a ray-bundle made by a plane containing the chief ray (any ray from an off-axis object point that passes through the center of the aperture stop). Also, a bundle of rays diverging from or converging to a single point.

reactance   The imaginary part of the impedance associated with energy storage. Differs from resistance, which is the real part of the impedance and is associated with energy dissipation. The unit of reactance is the ohm. For a pure inductance or capacitance, it is given by

$2\pi f L$ and $1/(2\pi f C)$, respectively, where f is the alternating current frequency.

reactive current   In the phasor representation of an AC current, the component of the AC current perpendicular to the AC voltage is called the *reactive current.*

reactive factor   The ratio of the reactive power of a circuit (i.e., the product of reactive voltage and amperes) to the apparent power (equal to the product of the root-mean-square current and voltage).

reactive voltage   That component of the phasor representing the voltage of an AC circuit that is in quadrature (i.e., 90°) with the current.

reading glass   A large-aperture, simple biconvex lens held such that the object to be viewed remains between the focal point of the lens and the lens itself, producing a magnified virtual image.

read only   A signal is stored in an electric circuit. This signal can only be used as an input signal to outside electric circuit. Any change in outside circuit cannot alter this signal.

receiver, radio   A radio receiver captures amplitude modulated or frequency modulated radio waves and converts them to signals that drive an output transducer such as a loudspeaker. A simple radio receiver consists of six stages. The *first stage* is a radio-frequency section that provides a coupling between the antenna and the radio receiver. It also provides any pre-selection or amplification before the frequency of the incoming signal is changed. The *second stage* is a mixer and local oscillator section that converts the incoming signal to a predetermined fixed intermediate frequency, which is usually lower than the signal frequency. The *next stage* is an intermediate frequency amplifier section, which provides most of the radio receiver's amplification and selectivity. The *main stage* is a second detector section that either detects amplitude-modulated signals or frequency-modulated signals. The *next stage* is a modulation frequency section consisting of either an audio or video amplifier that provides the additional amplification

to drive an output transducer. The *last stage* is an output transducer, such as a loudspeaker used to provide an audio signal.

reception, diversity   A communication system that has two or more paths (referred to as *channels*). An example of space-diversity reception is the use of two antennas at different heights to provide a means of compensating for changes in electrical-path differences between direct and reflected rays. The outputs of the two channels are combined to give a single received signal and thus reduce the effects of fading. Fading is the variation in signal strength at a receiver due to variations in the transmission medium.

reception of sound   For acoustic waves to be detected, a receiver must either partake or otherwise influence the motion of the particles of the medium or must respond in some way to the pressure variations on its surface. This forms the basis of sound detectors such as the ear, where the pressure variations are felt on the ear drum that vibrates with the same frequency as the impinging sound.

reciprocal network   A circuit whose output is reversibly proportional to its input.

reciprocity   Deals with the reciprocal relation between transmitters and receivers of sound waves. If there are any obstacles between the point of origin and reception of sound, the sound emanating at the origin is perceived with the same intensity at the point of reception as if an equal sound had originated at the point of reception and recorded at the original point of emanation.

recording of sound   Mechanical vibrations in a medium need to be converted to electrical signals for recording. Several different media for recording exists such as records, tapes and compact disks, as well as film.

recording, quadraphonic   These sound systems possess a frequency range that includes all the audible components of the sounds being reproduced. It is necessary that the intensity range associated with the recording sounds be distortion-free. The spatial sound pattern and the reverberation characteristics of the original sound should also be preserved.

rectification (cell) The rectification behavior of the cell membrane is exhibited more readily in the behavior of the current as a function of the membrane potential. Because the current across the membrane is a direct measure of the ion flux through it, then, as a function of the membrane voltage, the permeability of the membrane to different ions will show a nonlinear behavior.

The ion channels conduct ions more readily in one direction than in another when the direction of the driving force is reversed. This is the behavior of an electrical rectifier. The rectifying characteristics of the membrane are usually depicted by the plot of the current vs. the voltage for specific channels.

In the figure below (a), the I-V linear curve makes the channel an ohmic channel, while in (b), the channel is a rectifying channel. The units are usually measured in mV for the voltage and pA for the current.

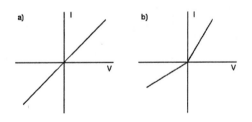

rectifier An electric circuit that changes an alternative input signal into a unipolar output signal. Most of the rectifier makes of the diodes. The figure shows a block diagram of a rectifier.

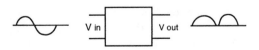

rectifier, bridge This is a specific rectifier. The figure shows this rectifier made of four diodes.

The input signal is alternating and the output signal is only one direction.

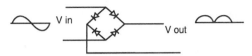

Rectifier, bridge.

rectifier, electrolytic A rectifier that uses electrolytic mechanisms to change an alternative electric signal to a unipolar electric signal. When the alternate signal is applied to the electrolyte, the molecules in this electrolyte become positive and negative ions. These ions produce a unipolar electric signal. Usually this signal is a current or voltage.

rectifier, full-wave A rectifier that has a unipolar output signal during both halves of the input sinusoid.

rectifier, half-wave A rectifier that only has a unipolar output signal during one half of the input sinusoid.

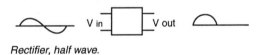

Rectifier, half wave.

rectifier, mercury vapor A rectifier in which the mercury vapor is filled in a tube.

rectifier, metal A rectifier whose anode and cathode terminals are enclosed in a metal chamber.

rectifier, selenium A rectifier in which a selenium layer is deposited on an aluminum plate. Electron flow from a selenium to an aluminum plate is easier than from the opposite direction.

rectilinear propagation In geometric optics, the propagation of light is described in terms of rays. This is stated in terms of the *law of rectilinear propagation of light;* namely, light travels in straight lines. That mode of propagation where the light path can be described by an infinitely thin pencil, the light ray, which propagates perpendicular to the wave front. This limit is commonly known as the *limit of geometrical optics* and can be obtained by setting the wavelength $\lambda \to 0$. Any deviations from this limit are proportional to λ, leading to the existence of diffraction effects for radiation of sufficiently long wavelength.

red light, healing effect   Dating back to Niels Ryberg Finsen in the last part of the 19th century (founder of modern phototherapy: treatment of disease by the influence of light), light has been used to relieve ailments. As part of his discoveries, he found that lengthy exposure to red light by smallpox victims prevents the suppuration of the pustules. Separate studies make use of infrared and ultraviolet light in heat lamps to treat neuritis and arthritis conditions to relieve inflammation.

On UV irradiation treatments, the heating of tissues by IR after the UV tends to suppress the observable manifestation of dilation of the minute blood vessels in the dermis, or erythema (reddening of the skin). If the IR heating occurs before the UV, there is an increases in the degree of erythema.

Biostimulation of biological tissue has been observed during photo-stimulation. Wavelengths of about 660 nm seem to cause an increase in the speed of tissue repair. Different results may be obtained depending upon whether the light source is continuous or pulsating. *See* light, monochromatic, biological action.

red light stimulation and bacteria growth
While the part of the spectrum from 600 nm to 700 nm is important for chlorophyll synthesis and photosynthesis, bacterial photosynthesis takes place close to 900 nm, deeper into the red.

For bacterial photosynthesis, there is a reaction center with the pigment bacteriochlorophyll, which absorbs light of longer wavelengths. These bacteria require some electron donor other than water, and they do not release oxygen. The green bacteria (Chlorobiaceae) and purple sulfur bacteria (Chromatiaceae) use elemental sulfur, sulfide, thiosulfate, or hydrogen gas as the electron donor, whereas the purple nonsulfur bacteria use electrons from hydrogen or organic substrates. All these bacteria require

exposure to red light as well as anaerobic conditions for photosynthetic activity.

reed   A musical box composed of a thin rod clamped at one end, and excited by air pressure from the mouth being blown in through the mouth piece, which in turn makes the free end vibrate.

reflectance   (Also known as *reflectivity*). The ratio of the reflected to the incident power (flux). Depending on the nature of the incident radiation, one talks of specular, diffuse or total reflectance. A large value of reflectance can cause serious loss of light in a multi-component optical system. To reduce the losses due to reflectance, the optical surfaces are coated with film of a transparent substance with a thickness equal to one-quarter wavelength of light in the film.

reflection, acoustic   This occurs when a progressive plane wave in one fluid medium impinges upon the boundary of a second medium causing the acoustic disturbance to bounce back into the first medium.

reflection coefficient   Ratio of the reflected voltage to the incident voltage when a transmission line of characteristic impedance, Z_0, is terminated with an impedance, Z_R. The reflection coefficient is given by

$$\frac{Z_0 - Z_R}{Z_0 + Z_R} .$$

Note that when the characteristic impedance matches the termination impedance (i.e., $Z_0 = Z_R$), the reflection coefficient is zero so there is no reflected voltage.

reflection coefficient, acoustic   This is given by the ratio of the reflected flow of sound energy to the incident flow of sound energy for the transmission of acoustic waves from one medium to another. The reflection coefficient $\alpha_r = I_r / I_I$, where I_r and I_I are the reflected and incident intensities respectively.

reflection density   The negative of natural logarithm of the reflectance. Equivalently, the natural logarithm of the ratio of the luminance of a non-absorbing perfect diffuser to that of the surface under consideration, both surfaces being illuminated at 45° to the normal and viewed normal to the surface.

reflection, laws of For any small region of a surface that can be considered smooth, (a) the incident ray, the normal to the surface at the point of incidence, and the reflected ray, all lie in the plane of incidence, and (b) the angle of incidence (between the incident ray and the normal to the surface) and the angle of reflection (between the normal to the surface and the reflected ray) are equal to each other for any wavelength of light.

reflection, multiple Caused by multiple passages of a light ray between two reflecting surfaces (e.g., in a thin film). The interference patterns provided by transparent material like a film of oil on water, soap bubbles, etc. result in colorful effects.

reflection, selective Reflection by a body showing selective (wavelength-dependent) absorption. The color of specularly reflected light from such materials is complementary to the color of a thin film of the material viewed by transmission.

reflection, specular Reflection of a well-defined narrow beam from a surface, when the beam can be approximated with a ray and the laws of reflection are followed.

reflection, total internal When light from a medium of higher index is incident on an interface at an angle greater than the critical angle, it will be totally internally reflected. No light will emerge into the lower index medium.

reflective power The ratio of the energy flux of the beam reflected from a surface to the energy flux of the beam incident on the surface.

refracting edge The edge along which two refracting plane surfaces of a prism intersect (on extension, if necessary).

refracting face The two non-parallel planes constituting the boundaries of a prism where refraction takes place.

refraction Continuation of the part of the flux associated with a wave (of light) in a medium different from the one in which the wave is traveling currently. For specular incidence, and a smooth surface of separation of the two media, the laws of reflection and refraction are followed.

refraction, acoustic The bending of sound waves due to a change in velocity. When a traveling wave strikes a second medium at an angle in which its velocity is different from that which it is traveling in, the wave continues at a different angle. This effect is significant in the ocean and depends on changes in temperature, salinity and depth in sea. Wind also causes refraction of sound waves.

refraction, atmospheric Changes in direction of light rays due to a gradual increase in air density (and hence index of refraction) on coming closer to the surface of the earth. The apparent direction of a star as seen through a telescope is thus different from its true direction. The continuous thermal and density fluctuations in the atmosphere cause fluctuations in the apparent direction of a star. Some other manifestations of atmospheric refraction are that stars are visible a short time after they have actually set below the horizon, and the disks of the sun and moon near the horizon appear oval instead of circular.

refraction, conical The situation in which refracted light spreads out in a hollow cone inside a biaxial crystal, resulting from the incidence of a beam of unpolarized light such that it is refracted along one of the optic axes (*internal conical refraction*). *External conical refraction* is the emergence of a hollow cone of polarized light when a hollow cone of incident unpolarized light is refracted within a biaxial crystal into a narrow pencil or ray of light.

refraction, double *See* ray, ordinary.

refractivity The name given to some quantitative measure of refraction, usually $(n-1)$, where n is the refractive index of the medium. Its dependence on wavelength causes the phenomenon of dispersion.

refractometer An instrument to determine a refractive index for various purposes like identification of gems and stones, suitability of materials for lenses, and for analysis of gaseous samples. Physical properties — like total internal reflection, angle of minimum deviation from a prism, shift of fringes in an interferometer, etc. — are used to help determine the refractive index. In the case of solid samples, it is convenient to have at least one polished surface. Otherwise, a solid sample with irregular shape is immersed in a mixture of liquids that is adjusted to have the same refractive index as the solid (till the outline of the solid disappears).

refractometer, Pulfrich A device used to measure the refractive indices of solids and liquids, which uses right angle prism and the *principle of internal reflection.*

refrigerants, nuclear Relies on the principles of magnetic cooling or adiabatic demagnetization of the nuclear spins in metals. It is a single-cycle refrigeration process and is effective for producing sub-mK temperatures. Due to the small nuclear moment in metals, the dipole-dipole interactions are reduced and in principle spin temp of less than 1 mK can be reached by nuclear cooling. Copper in wire form or powder form is most commonly used as the nuclear refrigerant. The required polarization is produced by precooling to as low a temperature as possible while applying a large external field.

refrigeration This involves the process of drawing heat from substances to lower their temperature. Mechanical refrigeration systems are based on the principle that absorption of heat by a fluid known as the refrigerant as it changes from a liquid to a gas lowers the temperature of the surrounding air. In the compression system, a compressor exerts pressure on a refrigerant gas causing it to pass through a condenser. It thus loses heat and liquefies. On circulation of the liquid through the refrigeration coils, it vaporizes, drawing heat from the air surrounding the coils. The refrigerant gas then returns to the compressor, and the cycle is repeated.

refrigeration cycle, Stirling This is a reversible cycle that gives rise to one of the most useful type of refrigerators. It is a four cycle process: $1 \rightarrow 2$: the gas or refrigerant is compressed isothermally at temperature θ_H and rejects heat Q_H to a hot reservoir. $2 \rightarrow 3$: Gas is forced through the regenerator at constant volume, giving up some heat Q_R to the regenerator. $3 \rightarrow 4$: An isothermal expansion occurs at the low temperature θ_c during which heat Q_c is absorbed by the gas from the cold reservoir. $4 \rightarrow 1$: Gas is forced at constant volume from the cold to the hot end through the regenerator.

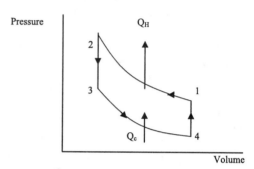

Stirling's refrigeration cycle.

refrigeration, nuclear A nuclear cooling stage is employed in a dilution refrigerator. It provides a high cooling power at a low base temperature. *See* refrigerants, nuclear.

refrigerator, dilution A technique that employs solutions of ^{3}He and ^{4}He which is capable of maintaining continuously low temperatures as low as 2 mK. This is an improvement over the single refrigerant cryostat which tends to be limited to base temperature of 1 K for ^{4}He and 0.3K for ^{3}He. The method depends on the phase separation of ^{3}He-^{4}He mixtures below 0.9 K. The ^{3}He rich phase floats on top of the more dense phase. When the ^{3}He atoms move across the boundary from the concentrated to the dilute phase, a heat of the solution is taken from the liquids. The dilution process can be operated as a single cycle.

refrigerator, helium This system uses helium for cooling below temperatures of 1 K. The process involves evaporation cooling of liquid, Pomeranchuk cooling by adiabatic solidification of the liquid, and dilution cooling by mixing ^{3}He and ^{4}He. *See* refrigerator, dilution.

refrigerator, Joule-Kelvin   These are microminiature refrigerators based on the Joule-Kelvin effect. These devices have no moving parts except for the remote compressors. The fine capillaries are about 63 μm wide and are produced by photo-lithography. Relatively high temperatures are achieved using nitrogen gas as the working fluid. Other units attempt to employ a three-stage cooling process involving nitrogen, hydrogen and helium to achieve helium temperatures.

refrigerator, Leiden   This employs the concept of circulating ^{4}He rather than ^{3}He in a dilution refrigerator. With a circulation rate of $\sim 10^{-3}$ mol/s, temperatures as low as 10 mK have been achieved.

refrigerator, nitrogen   System that uses liquid nitrogen for cooling or as the refrigerant. This type is often used in hospitals where it is necessary to cool to temperatures near 77 K for the preservation of human tissues, blood and bone marrow.

refrigerator, Pomeranchuk   This involves using a method for obtaining temperatures below 1 K, and close to 2 mK using the by-adiabatic solidification of the liquid. The cooling power of this type of refrigerator is proportional to the absolute temperature over the temperature range. By suitable precooling, using a dilution refrigerator or an adiabatic demagnetization stage and by the application of pressure, liquid ^{3}He is brought to a certain point on its melting curve. Increasing the pressure in the cell produces solidification. The cooling power of this method is simply the amount of heat that can be absorbed when n moles /second of liquid are converted into a solid at constant temperature T. One of the main problems with this method is devising a method for compressing the helium without frictional heating.

regeneration   The input signal is amplified through the circuit block (*see* figure). The output signal feedback increases the input signal. This increased signal is further amplified through the circuit if the combined gain is greater than one.

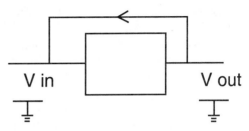

Regeneration.

regenerative depolarization (cell)   (1) In the synaptic junction between two neuron cells, the electrical signal that comes down the membrane of one end is transmitted to the neighboring cell via neuro-transmitters. The quantity of the neuro-transmitters is in proportion to the strength of the signal. The actual signal received at post-synapse can be either hyperpolarizing or depolarizing. In the latter case there is a decrease of the negative charge inside the cell membrane causing an excitatory stimulation.

The flow of neuro-transmitters from one membrane to the membrane of the other cell prevents the signal from dying out and regenerates the action potential or membrane depolarization at post-synapse.

(2) An example of a regenerative depolarization is in the generation of spike trains in the central nervous system (CNS) neurons. Sustained depolarizing input waveforms to neurons, which hold the membrane potential above threshold, provoke rhythmic firing. Less likely, but possible, are extra spikes that come after a depolarizing *after-potential* that crosses threshold. The extra spikes may subsequently cross the firing potential threshold themselves, provoking another spike. The cycle may continue, thus obtaining regenerative or *delayed* depolarizations. *See* potential, graded (membrane); repolarization (cell).

reinforcement   Superposition of waves that are in phase undergo vector addition of the displacement of the particles leading to an augmentation in the amplitude of the acoustic waves. Superposition no longer holds if the intensity of the sound becomes very high, such as in an explosion.

relative permeability   Measure of magnetic softness. Defined by $\mu_r = \mu/\mu_o$, where μ and μ_o are the permeability and the permeability of free space, respectively. *See* permeability, magnetic.

relative permittivity   Measure of electric polarizability. Ratio of the permittivity of a material to ϵ_o, the permittivity of free space. *See* permittivity.

relaxation, acoustic   Molecules are thought to have a certain stiffness that causes a delay in response to a vibration. The time lag between a change in pressure and the corresponding displacement results in some absorption of acoustic waves.

Reproduction of sound involves the preliminary process of reception, then recording, transmission and amplification. Electronics permits the reproduction of sound with fidelity. *See also* recording of sound.

relay   An electrically activated switch. Generally, it consists of an electromagnet that, when energized with a current, attracts a pivoted spring loaded lever. On moving toward the electromagnet, the tip of the lever touches a fixed contact. The pivot of the lever is one terminal of the switch while the fixed contact is the other. As a result, the switch is now closed. When the current in the electromagnet is turned off, the lever will spring back to its original position and break the electrical connection between the two terminals. The relay may consist of several levers and can therefore open and close several switches simultaneously.

relay (in communication)   A relay system uses active satellites to retransmit information to earth. The signals are usually amplified and the frequency changed by a transponder. A transponder is a combined transmitter and receiver system that automatically transmits a signal when a predetermined trigger is received by it.

reluctance   The opposition presented by a magnetic circuit to the establishment of magnetic flux in it. It is defined by the ratio of magnetomotive force to magnetic flux in a magnetic circuit. The reluctance, R, depends on the length, L, cross-sectional area, A, and magnetic permeability, μ, of a magnetic circuit by

$$R = \frac{L}{A \times \mu}.$$

Reluctance is the reciprocal of the permeance of a magnetic circuit. It is analogous to the concept of resistance in an electric circuit.

reluctance, magnetic   Proportionality constant R between magnetomotive force E_m and magnetic flux ϕ in a magnetic circuit. Defined by

$$E_m = R\Phi.$$

Measures the opposition presented by a magnetic circuit to magnetic flux. *See* magnetomotive force. *See also* flux, magnetic.

remanence   Magnetic moment of a material in zero applied field. Usually measured after applying a magnetic field to saturate the material and then reducing the applied field to zero. If all the magnetic moments remain completely aligned then the remanence is close to the saturation magnetization. This is an important parameter for permanent magnets where a large remanence is required.

repeater   A repeater is a device at the physical level that receives signals in one circuit and automatically delivers corresponding signals to one or more other circuits. A repeater is most often used with telephonic or telegraphic circuits to amplify the signal strength. In pulse telegraphy, repeaters perform pulse regeneration of the transmitted pulses. Repeaters can operate either on signals in one direction only or on two-way signals. Telephone repeaters operate on four-wire circuits or two-wire circuits. Repeaters allow a link to span a longer distance.

repolarization (cell)   At the beginning of an action potential, the sodium channels open very quickly when the membrane potential reaches the threshold potential. Once the action potential is at its peak, the sodium channels close with equal speed. This creates the sodium inactivation; it is caused by gates within the channel that are sensitive to the depolarization of the membrane. Following the sodium inactivation is the opening of the potassium channels, which allow the diffusion of K+ out of the cell. The combined effect of sodium inactivation, which blocks the influx of cations, and potassium activation, which causes the efflux of other cations, is the immediate return of the cell membrane to a polarized state, with the inside negative in relation to the outside, i.e., repolarization. *See* nerve impulses, propagation of; potential, membrane; potential, resting.

repulsion, electric   The mechanical force that moves apart charges of the same polarity or objects carrying charges of the same polarity.

residual charge   The charge remaining on the plates or dielectric of a capacitor after an initial discharge of the capacitor.

resistance   A property of an electrical conductor that determines the size of a direct current that can flow on the application of a certain potential difference across it. The relationship between resistance, current and potential difference is given by Ohm's law. The unit of resistance is the ohm (Ω), which can be any value but in practice is between wide limits (10^{-6} to 10^{8} Ω) depending on the material and geometry of the body. Although it is generally a constant for a particular body, a change in temperature may change its value. On the atomic level, resistance is due to collisions of electrons with the atomic lattice of the material, thus dissipating heat energy in the process. *See also* Ohm's law.

resistance, contact   Resistance across the interface of two conductors in contact. This resistance is generally higher than that of the bulk material because the contact area is usually limited to a number of contact points on the two surfaces. Pushing the two surfaces closer together increases the number of contact points and consequently decreases the contact resistance.

resistance, internal   A small resistance inherent in any AC or DC voltage source. Consequently, not all of the EMF of a source is avail-

able for use in its external circuit, and the loss is referred to as the drop in potential due to internal resistance. It is given by

$$r = \frac{E - V}{I},$$

where r is the internal resistance, E is the emf, V is the potential difference across the terminals, and I is the current.

resistance, minimum at low temperatures
Observations have shown that many metals have a negative temperature coefficient at low temperatures and sometimes show a minimum in their resistance curve, for example, metallic oxides and sulfides and semiconductors. Existence of localized moments in dilute alloys that couple to the conduction electrons affects electrical conductivity. The magnetic impurities act as scattering centers, and if they are responsible for lattice imperfection, then at low temperatures the scattering caused will be the primary source of electrical resistance. Nonmagnetic scatterers lead to resistivity dropping monotically with decreasing temperatures. In magnetic alloys, the resistivity has a rather shallow minimum occurring at low temperatures around 10 K that depends weakly on the concentration of magnetic impurities. The minimum arises when the scattering of conduction electrons shows unexpected features when the scattering center has a magnetic moment according to Kondo theory.

resistance, Umklapp   This owes its origin to phonon collisions, in contrast to normal scattering. The Umklapp process presents a direct resistance to the flow of heat. Umklapp scattering is negligible at very low temperatures since there are no phonons of a sufficiently large wave vector in the crystal. An Umklapp process dealing with three-phonon scattering is defined as one in which the total crystal momentum is not conserved — a process more likely to occur at high temperatures than at low temperatures. At the higher temperatures, the mean free path of the phonons is ultimately limited by interatomic spacing thus reducing the spread in thermal conductivity of different crystalline solids.

resistivity   A property of the material from which an electrical conductor is made that does not depend on the physical shape of the conductor. It has the units ohm-meters (Ω-m). It can be used to determine the resistance, R, of a conductor from the following:

$$R = \rho\frac{L}{A},$$

where ρ is the resistivity, L is the length of the conductor, and A is the cross-sectional area.

resistivity, cytoplasm   Because resistance is inversely proportional to the cross-sectional area of the material through which current flows, cytoplasmic resistance is rather high, especially in dendrites that are long and thin processes of the neuron. In addition, the length also contributes directly to the resistance because the greater the length, the higher the number of collisions between the ions (that transport the current) and the cytoplasm components. Values are usually an order of magnitude smaller than the membrane resistance. *See* resistivity, membrane.

resistivity, membrane   The conductance across the membrane is directly proportional to the number of open ion channels. Then, the total conductance is given by the combination of the ion-gated K^+, Na^+, and Cl^- channels. The membrane resistivity is readily obtained from the inverse of the conductance. This resistance is from the ionic (rather than the capacitive) membrane current. Typical values for the membrane resistance are in the proximity of $200\,k\Omega$.

resistor   A component used in circuits to provide a resistance of known value. Electrical energy is converted into heat when current flows through a resistor. Thus resistors are sometimes used as heating elements. They are also used to control current and voltage in circuits. Types are wire wound, metal film, etc. Many resistors used in electronic applications have color bar codes on the outside that indicate the value of the resistance.

resistor, ballast   (**1**) A resistor that mantains a constant current in an electric circuit. It does this by having a resistance that is proportional to the current which flows through it, and that can vary rapidly with the current. It usually consists of a resistive element inside a gas filled tube.

(**2**) A resistor used in electrical circuits to regulate the current. It is made from a material with a positive temperature coefficient; i.e., an increase in temperature will result in an increase in the resistance. It is usually connected in series with a circuit, especially power supplies, so that a sudden increase in the current leads to an increase in temperature which in turn increases the resistance thereby reducing the current.

resistors in parallel   Network of resistances in which the two ends of each resistor are connected to the corresponding ends of every other resistor.

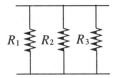

Resistors in parallel.

The network can be replaced by an equivalent resistance, R, given by

$$\frac{1}{R} = \frac{1}{R_1} + \frac{1}{R_2} + \frac{1}{R_3} + \cdots,$$

where R_1, R_2, R_3, etc. are the values of the individual resistances making up the network. The potential difference across each resistance is the same, which enables the current in each resistance to be determined using Ohm's law.

resistors in series   Resistances connected to each other in a similar way to the links in a chain.

Circuit diagram symbol for a resistor.

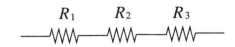

Resistors in series.

These can be replaced by a single equivalent resistance R such that

$$R = R_1 + R_2 + R_3 + \cdots$$

where R_1, R_2, R_3, etc. are the values of the individual resistances. The current through each resistance is the same. This enables the potential difference across each resistance to be determined using Ohm's law.

resolution of an optical system   The distance between two object point sources such that the two points can be distinguished in the image. The resolution is limited by diffraction effects, so that for any form of radiation it will be determined by the wavelength and the aperture of the system.

resolution, principle of   Refers to the use of a quantitative criterion to determine the resolution of an optical system. That most commonly used is the *Rayleigh criterion*, which states that two object points are just resolved when the diffraction pattern of the image of one falls on the first minimum of the diffraction pattern of the other. This gives an angular separation of the object points

$$(\Delta\varphi)_{\min} = 1.22\lambda/D ,$$

where λ = wavelength and D = aperture of the system.

resolving power   For two just resolved points in the image of an optical system of focal length f the center-to-center separation of the images is

$$(\Delta\ell)_{\min} = 1.22 f\lambda/D .$$

The resolving power is then defined as $1/(\Delta\ell)_{\min}$. *See* resolution, principle of.

resonance   This phenomenon occurs in the case where a mass is maintained in a state of vibration by a periodic force. It is a case of forced vibrations, and, when the frequency of the driving force equals the natural frequency of the system, the amplitude of oscillations increases significantly.

resonance in AC circuits   Occurs when the frequency of the applied periodic voltage is at or near the natural frequency of the circuit. The resonant circuit consists of an inductor and capacitor in parallel or in series. The resonant frequency, f, is given by

$$f = \frac{1}{2\pi\sqrt{LC}} ,$$

where L is the inductance, and C is the capacitance. At the resonant frequency, the series circuit has a minimum in the impedance, while the parallel circuit has a maximum.

resonance, optical   When incident monochromatic light has a frequency equal to the resonant frequency of a medium, there will be optical resonance and strong absorption. This energy is then re-emitted as resonance radiation. This gives rise to the close connection of emittance and absorptance of materials, known as *Kirchoff's law.*

resonance scattering, uses in biological studies   Resonance scattering studies provide a wide range of possibilities in the analysis and study of diverse biological processes. Usually, resonance scattering studies the effect on a sample of an incoming probe that could be in the form of radiation or particle bombardment. The radiation "tunes in" with intrinsic energy modes of the substance in consideration, in contrast with elastic and inelastic scattering experiments. The incident frequency of the probe is usually tuned to a particular substance in the sample such that measurements of concentration, abundance, real-time tracking, diffusion rate, reaction rate, or other properties can be realized. The main results come from the analysis of the absorbance of the incident radiation, due to resonance, and of the scattering characteristics.

See particular uses in Raman scattering, coherent anti-Stokes (CARS), Raman scattering, electrophoretic, NMR nuclei in biological materials, nuclear angiography. Other uses are in coherent nuclear resonance scattering studies.

resonance, sharpness of   This depends on the amount of damping in the system. The less damping there is, the sharper the resonance peak. *See also* radiation damping, acoustic.

resonators   That which can vibrate to produce acoustic waves. Examples are pipes and

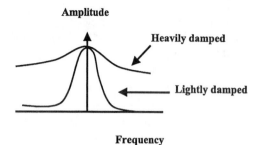

Amplitude

Heavily damped

Lightly damped

Frequency

Sharpness of resonance.

air cavities. They can act as sound producers or detectors.

restoring force   Simple harmonic motion will result whenever this quantity, which is a property of the medium, is directly proportional to the displacement caused. In acoustics the restoring force is provided by the medium. This force always acts to accelerate the particle in the direction of its equilibrium position.

retarded potential   Electric potential at a point which is delayed in reaching its maximum value with respect to another point due to the finite speed of propagation of electromagnetic waves in the medium.

retentivity   Describes the retention of a magnetic moment by a ferromagnet once a magnetic field is removed. Can also refer to the reverse applied magnetic field required to restore the zero overall magnetization state.

retina   The inner transparent near layer of the eye. This is the neurosensory layer of the eye. The light entering the retina encounters about ten layers consisting of ganglion cells, bipolar cells, horizontal cells, and amacrine cell layers, which are different types of neural cells before encountering the photoreceptors (rods and cones) — the site of photon absorption. The retina represents the terminal expansion of the optic nerve.

reverberation   The echo or reflection of sound. Of great importance in architectural acoustics.

reverberation chamber   Small live rooms that can be used to determine the absorption co-

efficient of a material by investigation of the reverberation time. The wall surfaces of such chambers should be highly reflecting so as to produce a large reverberation time in absence of testing material. It should also have a large volume to produce a large number of modes of vibrations and should have irregular wall surfaces to aid diffusion of sound waves.

reverberation time   Relates the empirical relation between echo characteristics of an enclosure, its size, and the amount of absorbing material present and gives a measure of the acoustical properties of a room. It is defined by Sabine as the time in seconds that is required to reduce the intensity from a level of 60 db above the threshold of audibility to the threshold.

Sabine equation for reverberation time T

$$T = 0.049V/a \, ,$$

where V is the volume of the room and a its total absorption.

reverse leakage current   A current that is caused by a reverse bias voltage in an electric device. For example, for a diode that has a positive and negative terminal — if the positive terminal is connected to negative side of the voltage and the negative terminal is connected to the positive side of the voltage — the current flowing from the negative terminal to positive terminal is called *reverse leakage current.*

rheostat   An adjustable resistor used to control the current in a circuit by varying the resistance. This results in the dissipation of unused electrical energy as heat. Rheostats may be constructed using coils or grids from a variety of metals and alloys; they may also be made of carbon, either pulverized and held in tubes or in the form of solid rods or disks. The resistance is adjusted by a linear or rotary sliding contact.

right-handed rotation   For a linearly polarized light beam passing through an optically active crystal, a right-handed rotation of the plane of polarization is that which is in a clockwise direction when looking back toward the source.

right-hand screw rule   When a rotating or circulating quantity is related to a vector, the right-hand screw rule relates the direction of circulation to the vector direction. In this rule, the extended thumb of the right hand gives the vector direction and the direction of the curled fingers of the right hand give the direction of circulation. Examples are:

1. the direction of the magnetic moment of a current loop and the sense of circulation of the current in the current loop,

2. the direction of the electric current in a straight wire and the direction (rotation sense) of the circles of magnetic field lines centered on the wire, and

3. the direction of the magnetic field inside a solenoid and the sense of circulation of the current in the solenoid windings.

rings, vibration of   A set of frequency standards used for producing sounds of definite pitches.

ripple tank   Liquid in a wooden trough that can be used to demonstrate wave phenomena of reflection, refraction, interference, and diffraction.

ripple voltage   The AC voltage superimposed onto the DC voltage from a rectifier, such as the output from a half- or full-wave rectifier. Expressed as a percentage of the average voltage. It may be reduced by a series choke and shunting capacitor, or zener diode. In practice, it is eliminated by including an active circuit (regulator) after the shunt capacitor.

rise time   A time during which a signal is increased from a specific low value (usually 10%) to a high value (usually 90%). The figure shows a resister and capacitor circuit. The input signal is a DC voltage. The output voltage increases from zero to the input voltage value. Rise time is for the output voltage value changing from 10% to 90% of the input voltage value.

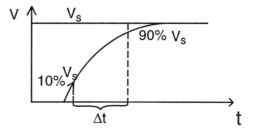

Rise time.

rods, vibration in   Propagation occurs by longitudinal waves, i.e., as the wave disturbance moves along the bar, the displacement of the particles is parallel to its axis. The propagation is similar to acoustic waves. Several acoustic devices utilize longitudinal vibrations in a bar; for example, the sound of definite pitches can be produced by bars of different lengths. Vibrations in nickel tubes are used to drive the vibrating diaphragm of a transducer.

ROM   Read-only memory, an electric chip that permits reading but restricts the writing operation. In computer systems, ROM can be used to store an operating system program.

roton   Quanta of vortex motion or microscopic vortex rings. They can be characterized by the parameters *energy, momentum,* and *effective mass.*

routing   The process of finding a path from a source to every destination in a network.

Rowland ring   Named after H.A. Roland (1848–1901). A magnetic material, usually a ferromagnet, formed into a ring. Magnetic flux is contained entirely within the solid material of the ring so that no demagnetization field is present.

S

sagitta The segment of a line between the vertex of a small curve and the chord joining the two symmetric ends of the curve, which is perpendicular to the chord. Kepler gave this name "sagitta" because of the resemblance of this line with an arrow. Its length measures the maximum elevation (depression) of the convex (concave) surface containing the arc.

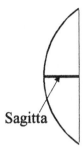

Sagitta.

sagittal focus The point of intersection of the line of secondary image and the sagittal plane. For a conical bundle of rays starting from a non-axial point, the cross-section of the beam after leaving the lens is initially circular, then it becomes gradually elliptical with the major axis in the sagittal plane, until the tangential or the meridional focus where the ellipse degenerates into a straight line. Beyond this point, the beam's cross-section first opens out into a circle of least confusion, and then deforms into a line called *secondary image*. This line image lies in the meridional plane and has a greater image distance than the primary focus.

sagittal plane The plane (also called the *equatorial plane*) containing the principal ray (or the *chief ray*) and perpendicular to the tangential or the meridional plane.

sagittal rays A bundle of rays cut by the sagittal plane to form the pencil of sagittal rays.

sagittal section A section cut by the sagittal plane.

sampling (1) A technique in which only some portions of an electrical signal are measured and used to represent the information contained in the original signal. The sample measurements are usually a set of discrete values of the original continuous signal. The rate of sampling of a periodic quantity must be at least twice the frequency of the signal if the output values are to represent the input signal without significant loss of information.

(2) A technique used in radio navigation systems where information from the navigation signal is extracted only when a sampling gate is activated by a selector pulse.

sampling function The one-dimensional sampling function can be expressed mathematically as

$$S(x) = \sum_{n=-\infty}^{\infty} \delta(x - n\Delta x) \,,$$

where δ is the Dirac-delta function. A two-dimensional version of the sampling function can be written as

$$S(u, v) = \sum \delta(u - u_n, v - v_n) \,,$$

which can be weighted to

$$S(u, v) = \sum R_n T_n D_n \delta(u - u_n, v - v_n) \,,$$

where R_n is a reliability weight, D_n is a density weight and T_n is a taper.

sampling gate A circuit that takes signals from the input signal only when an external pulse is active. This external pulse is applied to the transistor gate.

sampling period A time interval during which the input signal is picked up.

sampling window A time interval during which the input signal is picked up and the output signal is the sum of the input signal and the external triggering signal.

satellite (1) An artificial body that is projected from the earth and may orbit either the earth or another body of the solar system.

(2) Information satellites transmit signals containing many different types of information to the earth. These satellites are typically used to provide atmospheric and meteorological data, infrared, gamma-ray and X-ray studies of celestial objects, surveys of the earth's shape, surface and resources, and as navigational aids.

(3) Communications satellites are used to receive radio-frequency signals from the earth by means of highly directional aerials and return them to another earth location for purposes of long-distance telephony, TV broadcasting, etc. These satellites provide links that traverse very long distances at high bandwidth. They use the 4 to 6 GHz bands, and their potential link bandwidth is in the order of Gbps.

satellite beacon   A beacon satellite acts as a radio-navigation station. Its emissions are intended to enable a mobile-radio service to determine its bearing or direction in relation to the satellite beacon. The primary use is to facilitate a search and rescue operation.

satellite, geostationary   (Also known as *geosynchronous satellite.*) A satellite orbiting above the earth's equator at an altitude of approximately 35,000 km, such that its period of rotation is equal to the earth's rotational period. Thus, the satellite continuously views the same portion of the earth's surface.

Low-earth and geostationary satellites typically rebroadcast messages that they receive from any of the earth stations. Data rates of up to 500 Mbps are possible. The up-and-down propagation delay for geosynchronous satellites is about 250 ms, which is unacceptable for most interactive services (though acceptable for television broadcast).

satellite, synchronous   A synchronous satellite orbits the earth approximately 35,900 km above the equator and moves in a west-to-east direction. At this altitude it makes one revolution in 24 hours and thus is synchronous with the earth's rotation.

saturation   The condition where the output signal of a physical measurement increases to a constant maximum value and cannot increase any further due to subsequent increases in the input signal. The phenomenon is generic in virtually all physical systems and corresponds to complete breakdown of the linear regime.

saturation, current/voltage/resistance   A current/voltage/resistance value that cannot increase further when the outside signal (voltage/current) increases. In a transistor, the source or drain current cannot increase even though the applied voltage increases further. This current is called *saturation current.*

sawtooth wave   A periodic wave whose amplitude linearly changes with time between certain time periods. The shape of wave is like that of a saw.

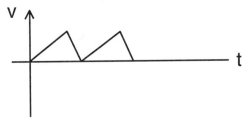

Sawtooth wave.

scale, diatonic   The natural musical scale in Western music in the key of C major, ascending from middle C, which has a frequency of 256 Hz, although it is represented musically as 261.2 Hz.

scale-of-ten   This is similar as scale-of-two, but it has ten inputs. Every input signal from ten inputs is one and the output signal will be one.

scale-of-two   An operator in a circuit in which every two input pulses produce one output pulse. This kind of circuit is normally used in computers for mathematic operations.

scales, equally tempered   On a keyboard instrument like a piano or organ, the notes A to G are sounded by depressing white keys and additional black keys are situated between each pair of consecutive white keys that differ from each other by a tone. Each black key sounds a note that is intermediate in pitch between the white ones on either side of it.

scales, musical   A definite series or succession of tones ascending or descending according to fixed intervals. The range between any note and the octave above it is divided into seven intervals by the insertion of six intermediate notes, the various pitches of which give a number of consonant intervals with each other and with the notes at each extreme of the octave.

scanner   An instrument that can convert the information on materials into an electric signal. Normally, the scanner is the picture converter. It converts the picture on the paper into the electric signal that can be stored in computers.

scattering, acoustic   Incident waves on reflection as secondary waves undergo wavelength shift in increasing proportion toward higher frequencies. It has been shown that the amplitude of the secondary waves varies directly as the volume of the scatterer, and inversely as the square of the wavelength of the incident sound. The intensity I of the reflected sound varies inversely as the fourth power of the wavelength λ

The law of scattering $I = 1/\lambda^4$.

scattering, Brillouin   Light of frequency v_i incident on a solid is scattered at a frequency $v_s = v_i \pm v_e$, due to inelastic scattering by the medium. When the scattering is due to the inelastic phonons of the medium, the effect is known as *Brillouin scattering*. In this case the scattering conditions limit v_e to $v_e < 100\text{GHz}$ or $\leq 1\text{cm}^{-1}$.

The effect can be explained classically by the fact that the operative acoustic phonons set up an effective diffraction grating in the medium so that the incident light is Bragg diffracted and Doppler shifted by the acoustic phonon traveling at the velocity of sound. The quantum description is more exact and is based on conservation of energy and momentum for the scattered photons and the emitted or absorbed phonon. For incident photon frequency v_i, wave vector $\vec{k}_i$, scattered photon $(v_s, \vec{k}_s)$ and phonon $(\Omega, \vec{K})$, this gives $\vec{k}_i - \vec{k}_s = \vec{K}$ and $v_i - v_s = \pm\Omega$.

Brillouin scattering is mostly done on thermal phonons; the phonon wave vector $\vec{K}$ can be determined by fixing $\vec{k}_i$ and $\vec{k}_s$. The experiment is done with a laser source. Since the phonon lines are very close to the elastic (Rayleigh) peak, a very high resolution Fabry-Perot interferometer is used to resolve them. The observed sidebands with decreased (increased) frequency are called Stokes (anti-Stokes) and each contains three peaks corresponding to a longitudinal phonon and two transverse phonons.

Brillouin scattering is a useful technique for the study of the elastic properties of materials such as layer compounds, amorphous materials, gels, phase transitions etc.

scattering, Coulomb   The scattering of an incident charged particle by a stationary charged target due to Coulomb repulsion between them. The classic example is that of Rutherford scattering of alpha particles by gold foils. The experiments showed that the average scattering angle per atom was very small, of the order of 10^{-4} radians. This led to the Rutherford model of the atom where most of the positive charge and essentially all of the mass is assumed to be concentrated in the nucleus at the center of the atom. This model was subsequently quantitatively verified by experiment and led to a good estimate of the size of the nucleus, for copper atoms less than 2×10^{-12}cm.

scattering, cross section   The total cross section for scattering is the ratio of the scattered power to the incident intensity. The cross section has the dimensions of area, so that it gives an intuitive indication of the relative importance of various scattering mechanisms. In general, the cross section is made up of the sum of that for absorption by the target and that for scattering out of the beam.

scattering, incoherent   Scattering in which there is no phase relationship with the incident radiation. Hence there are no associated interference effects and the intensities of such scattered beams can be added to obtain the total scattered intensity.

scattering, inelastic   That in which the incident radiation is absorbed by the scatterer and re-emitted at a different frequency.

scattering, Mie   The scattering of light by a single sphere or incoherently by a group of spheres separated from each other by distances much greater than the wavelength. The theory (G. Mie, 1908) is based on the diffraction of a plane electromagnetic wave by a homogeneous sphere in a homogeneous medium; it is valid for all types of spheres and mediums. The theory, which is very complex, involves the solution of Maxwell's equations in spherical polar coordinates. The polar diagrams for scattering depend strongly on the ratio of sphere diameter to wavelength. The theory is of great practical importance and can be used in the study of light scattering by colloidal suspensions, atmospheric dusts, clouds, fogs, etc. as well as to explain natural phenomena such as rainbows and solar corona. When the wavelength is of the order of the diameter of the sphere, the effect can be used to monitor variations of the latter.

scattering of light in tissue   Light transmitted by tissue is in a large proportion of long wavelengths (red), whereas backscattered and otherwise scattered light is of short wavelengths (blue). This corresponds to the observation that skin looks red when transilluminated by white light. In contrast, blood in veins and melanin deep in the dermis preferentially absorb red and reflect blue light.

Because red and near IR light has a typical absorption length in biological tissue of 1 to 10 cm, as opposed to 10 to 100 mm of scattering length, elastic scattering of light in the optical wavelengths dominates the scattering phenomena, giving rise to turbidity. In general, turbidity destroys phase coherence in the incident light, blurring and making it very difficult to make a sensible image out of the scattered light.

With recent advances in laser technology, the understanding of turbidity has been brought into new importance in the development of spectral diagnosis of disease and optical tissue imaging. The importance of light imaging as compared to X-rays, for example, is that in addition to providing real time imaging, it is a non-invasive procedure that may provide information about biochemical constituents.

Current research in imaging using optical light has been able to image inside tissue that is up to 1 cm in thickness. The imaging is achieved by collecting specific light (transmitted) that goes through the tissue and discarding the scattered light that reaches the other side with a delay relative to the unscattered light.

scattering, Raman   An incident monochromatic beam of light of frequency v_i is scattered inelastically to a frequency $v_s = v_i \pm v_e$. When the inelastic scattering is by electronic excitations or optical phonons, the scattering is called Raman scattering. The sidebands at $\pm v_e$ are typically in the range 10 to 1000 cm^{-1}. The lower frequency band is called Stokes and the upper anti-Stokes.

Raman scattering can be described classically by the induction of a dipole moment in the crystal by the electric field of the laser beam, hence by the polarizability tensor. Symmetry arguments show that a crystal with a center of inversion cannot be both infrared and Raman active; it follows that both are allowed for piezoelectric crystals. Quantum mechanically, the Raman process, as for Brillouin scattering, can be described by the conservation of momentum and energy for incident and scattered photon and the participating phonon.

The experiment is carried out with a laser source and a double monochromator with photomultiplier tube detection. In chemistry, the technique is useful for measurement of rotational and vibrational molecular spectra, while in physics it is generally used for phonon spectroscopy in solids for such studies as structural phase changes and soft modes, magnetic and amorphous materials, doped semiconductors etc. *See* scattering, Brillouin.

scattering, Rayleigh   Scattering of electromagnetic waves by small particles, of diameters less than the wavelength. The result for a classical model, calculated by Lord Rayleigh in 1871, gives for the scattered intensity $I_s = I_0 \lambda^{-4}$, where λ is the incident wavelength. This well-known result has been verified in innumerable contexts and accounts; it has been shown, for example, for blue sky and red sunset, attenuation in optical fibers due to impurities and the central elastic peak in Rayleigh and Raman scattering.

scattering, Thomson   The classical scattering of electromagnetic waves by electrons. It was applied by Thomson to the scattering of X-rays by a metal foil. Under suitable assumptions (the X-ray wavelength is small compared to the atomic diameter and the energy of an X-ray quantum large compared to the binding energy of the atomic electron but small compared to its rest mass energy), the interaction can be considered as being between the oscillating electric field of the X-ray and the charge of the electron. The theory provides the scattering cross section and the angular distribution of scattered X-rays.

Schottky anomalies   Heat capacity of solids decreases with decreasing temperature; however, at extremely low temperatures other contributions to the heat capacity may become significant. Impurities in a crystal give rise to this type of anamoly.

scintillation   Emission of light by a solid bombarded with radiation. The process is traditionally used to detect high energy particles such as α, β and γ radiation in scattering experiments. For the case of γ rays, the scintillation counter may consist of a scintillation detector ($T\ell$ doped, NaI crystal and photomultiplier), amplifiers, and a discriminator. The γ rays interact with the crystal by photoelectric effect, Compton effect, and pair production. For each of these interactions the primary energy goes into the kinetic energy of the electrons and the remainder into secondary photons that are detected.

Scott connection   A connection between transformers that can convert two-phase power to three-phase power or vice versa.

screened cable   A cable with a flexible protective screen of conductive material surrounding it.

search coil   A coil used to measure magnetic field properties. A stationary coil may measure properties of a time-varying magnetic field. A rotating coil may measure properties of a constant magnetic field. In both cases an induced EMF due to the change in magnetic flux in the coil is measured. Can be part of a ballistic galvanometer or a fluxmeter. *See* flux, magnetic. *See also* Lenz's law of induction, fluxmeter.

searchlight   A powerful beam of light produced and projected long distances, usually with the help of an intense light source (e.g., carbon arc) positioned at the focus of a paraboloidal reflector.

secondary cell   An electrochemical cell that can be electrically recharged many times is called a *secondary cell*. They are also called rechargeable batteries. Typical examples are lead-acid cells in automobiles and nickel-cadmium batteries.

secondary focus   *See* sagittal focus.

secondary maxima   These occur between the principal maxima of interference fringes for Fraunhofer diffraction by N identical apertures. The intensity $I(\theta)$ at a point on the screen making angle θ with the center of the grating is given by

$$I(\theta) = I_0 \left(\frac{\sin \beta}{\beta} \right)^2 \left(\frac{\sin N\alpha}{\sin \alpha} \right)^2 ,$$

where

I_0 = flux density in the $\theta = 0$ direction emitted by one of the slits.

$\beta = \dfrac{kb}{2} \sin \theta$

k = wave number of incident radiation = $2\pi/\lambda$

b = slit width

$\alpha = \dfrac{ka}{2} \sin \theta$

a = distance between slit centers .

Principal maxima occur when $\sin N\alpha / \sin \alpha = N$ or $\alpha = 0, \pm\pi, \pm2\pi, \ldots$, minima occur for $(\sin N\alpha / \sin \alpha)^2 = 0$ or $\alpha = \pm\frac{\pi}{N}, \pm\frac{2\pi}{N}, \pm\frac{3\pi}{N} \ldots$ and secondary maxima appear between each pair of neighboring minima.

Seebeck effect   This is the net EMF established in a circuit made from two junctions of dissimilar metals connected in series while the two junctions are maintained at two different temperatures. The magnitude of the potential difference between the two junctions determines the magnitude of the EMF. The Seebeck effect is the basis of a temperature measurement device known as a *thermocouple. See also* thermocouple.

seismic waves   There are two basic types of waves: *P*-waves or primary waves are longitudinal waves that can propagate equally well through a solid and through a fluid medium. *S*-waves or secondary waves are transverse waves that can propagate through solids but not through liquids. These are body waves and travel within the body of the earth. Also, there are *L*-waves or large waves that travel in the upper surface of the earth. *See also* rayleigh waves, surface waves.

seismograph   Very sensitive instrument used for recording earth tremors that relies on the principle of a pendulum driven by the earth tremors.

selectivity   Quality or state of being selective.

selenium cell   Can be one of two types: *photovoltaic* or *photoconductive*. The photovoltaic selenium cell requires no external EMF. An EMF is generated across the terminals of the cell when light shines through an optically transparent window onto a thin film of gold overlaying the selenium film. The photoconductive selenium cell requires an external EMF to operate. The cell also has a transparent window to admit light. Changes in light intensity creates changes in the conductivity of the selenium which changes the current through the cell. Both types of cells are used to measure light intensity.

self-capacitance   Self-capacitance is the ratio of the charge to the potential of a conductor, $C = q/V$, where q is the charge on the conductor and V is the potential of the conductor. Self-

capacitance is the measure of how much charge a conductor can hold per unit potential; it depends on geometry of the conductor and the medium that the conductor is embedded in.

self-diffusion coefficient   The self-diffusion coefficient is defined as the coefficient of proportionality that quantifies the strength of the flux of particles into one another. In a sense, it tells how good the particles in a substance mix with one another.

More precisely: consider a substance composed of two species of particles whose only distinction is that one species is labeled (e.g., radioactive) but are otherwise equal. Assume that initially the two species are separated into two domains along the x axis. If the concentration of labeled particles is described by a function $n(x)$, then after some time the two species will mix, tending to make $n(x)$ more uniform and thus increasing the entropy of the fluid.

If we define J_x as the flux of labeled particles along the x direction, then one expects that

$$J_x = -D\frac{\partial n(x)}{\partial x} .$$

The coefficient D ($D > 0$) is the coefficient of self-diffusion of the substance. Because $n(x)$ changes as a function of time, from particle conservation, it follows that

$$\frac{\partial n}{\partial t} = -\frac{\partial J_x}{\partial x} .$$

Combining the two equations we get:

$$\frac{\partial n}{\partial t} = D\frac{\partial^2 n}{\partial x^2} ,$$

which is the diffusion equation.

Self-diffusion should not be confused with mutual diffusion in which the labeled particles are different from the others.

self-focusing   Self-focusing of a laser beam may occur in a non-linear optical medium. If the refractive index increases with the incident intensity and if one assumes a Gaussian transverse intensity distribution, then the center of the beam will travel slower than the edges. The net effect is that the medium acts as a positive lens and self-focusing occurs, which can lead to focal spots a few microns in diameter. If self-focusing is arranged to exactly cancel diffraction then self-trapping of the light beam will occur. If the intensity is high enough, self-focusing may lead to damage of the crystal.

semiconductors   Materials in which the conductivity is between conductive metals and insulators. Their conductivity can be changed after impurity doping, such as crystal silicon, germanium, and three-five compound crystal materials.

semitone   The interval between any two consecutive notes of the musical scale such as C and $C\#$.

sensitive flame   A tall, steady, high pressure gas flame that ducks and roars in the presence of sound waves.

sensitivity   A minimum input signal that causes a distinguishable output signal. This signal could be light, electric, stretch, etc.

sensitivity of ear   The response of the ear varies with frequency. The hearing threshold for a person with acute hearing is 0 dB and the threshold of pain at about 130 dB. The threshold for hearing, audibility and that of pain also depends on frequency.

sentence (logical)   Logical words that are constructed by the computer software and can be executed in the computer.

sequential logic   Logic in which the output signal is dependent on the previous input signal by a delay time.

server   An entity that controls access to a shared resource.

servomechanism   A feedback system whose output signal represents mechanical motion.

shadow   The shade of finite extent cast upon a screen or another body by an opaque object intercepting the incident rays from a light source. Depending on the size of the source, the shadow is complete (umbra) for a point source and of

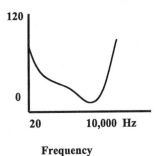

Sound pressure level, dB

Sensitivity of ear.

variable intensity (umbra and penumbra) for an extended source. For macroscopic objects, the shadow can be constructed using the ray approach. For objects with sizes comparable to the wavelength of light, the wave nature of light has to be invoked for a correct description.

shadow, acoustic   Region in which sound intensity is theoretically zero since no rays reach here.

shadow, penetration of sound into   The diffraction of acoustic waves into regions obscured by obstacles allows sound to be propagated in such areas.

shear waves   Type of wave, also known as *transverse,* in which displacement of a particle is perpendicular to the direction of motion.

shielding (low temperature)   Shielding is necessary to reduce heat leaks. Radiation shields help inhibit connective oscillations in the helium gas above the liquid that may be driven by the temperature gradient down the cryostat. This can transfer large quantities of heat to the cold part of the cryostat. As such, built-in infrared shields and traps are used. The cryostat must also be protected from longer wavelength radiation and from emission caused by other equipment in the vicinity. An entire dilution refrigerator can be enclosed in a large Faraday cage — an enclosed metal box known as a *shielded room.* Thermal conduction down support tubes and connecting leads are also major sources of leaks.

shift register   A digital circuit that can store a set of information in the form of pulses. The register has the capability of serially shifting the data from each stage of the register to the adjacent one. It may be a unidirectional or a bidirectional shift register, depending on whether it can shift only from left to right, only from right to left, or both.

shock wave   A wave formed by compression in a medium when an object moves violently through the medium at a speed in excess of that sound.

short-circuit   When the load resistance of a circuit is zero we call the circuit *short-circuited.* It can occur, for example, when a copper wire is connected to the two terminals of a battery.

shortsighted   *See* eye, near-sighted.

short wave   A radio wave that has a wavelength in the range 10 to 100 m or a frequency greater than 3 MHz, i.e., in the high-frequency band (HF).

shunt   A component placed in parallel across the terminals of a circuit or device in order to divert a known current from it. For example, if a current to be measured by a galvanometer is too large for that instrument, a large fraction of the current may be bypassed through a shunt resistor to reduce the sensitivity of the galvanometer in a known way and thus enable the current to be determined.

sidebands   The waveform for simple sine-wave amplitude modulation is given by

$$e(t) = E(t)(1 + m) \sin \omega_s t \, ,$$

where $E(t) = E_0 \sin \omega_c t$ is the unmodulated carrier wave with amplitude E_0 and frequency ω_c. The frequency at which the unmodulated amplitude is varied is ω_s and m is referred to as the *degree of modulation.* The equation of a wave with simple sine-wave amplitude modulation can thus be written as

$$e(t) = E_0 \sin \omega_c t + \frac{m E_0}{2} \cos (\omega_c - \omega_s) \, t$$
$$- \frac{m E_0}{2} \cos (\omega_c + \omega_s) \, t \, .$$

The last two terms in the above equation are called the *lower* and *upper sidebands* with sideband frequencies of $(\omega_c - \omega_s)$ and $(\omega_c + \omega_s)$, respectively. In general, there can be a range of frequencies over which the average amplitude is varied. All the information content of the signal is contained in the sidebands.

sidebands, double (DSB)   Double-sideband modulation is a method of encoding a signal, $g(t)$, with no DC component. The product

$$e(t) = A_c g(t) \cos \omega_c t$$

represents a double sideband suppressed carrier signal, where A_c is the amplitude of the unmodulated carrier. The radio-frequency envelope follows the wave-form of the modulating signal $g(t)$. The spectral components of the DSB signal, $e(t)$, are given by its Fourier transform

$$E(j\omega) = \frac{1}{2} G \left[j (\omega - \omega_c) \right]$$
$$+ \frac{1}{2} G \left[j (\omega + \omega_c) \right] \, ,$$

where $G(j\omega)$ is the Fourier transform of $g(t)$. Note that the upper and lower sidebands are translated symmetrically $\pm \omega_c$ about the origin.

sidebands, independent (ISB)   Independent sideband transmission is double-sideband transmission in which the information carried by each sideband is different. The carrier may or may not be suppressed.

sidebands, single (SSB)   It is possible to transmit all the information represented in a modulated signal by transmitting only a single sideband. The carrier component of an amplitude-modulated wave contains no information since its frequency, amplitude, and phase are not affected by the modulation. Thus the carrier need not be transmitted. Moreover, each sideband contains the same information about the modulation and only a single sideband need be transmitted. Single-sideband transmission requires a frequency band only half as wide as that required to transmit the modulated wave consisting of two sidebands and the carrier. Single-sideband transmission also requires one-third the power of transmitting all the components of the amplitude-modulated wave.

signal   A varying electrical parameter, such as current or voltage, that is used to convey information through an electronic circuit or system. A signal is usually transmitted as electrical impulses or radio waves.

signal, asynchronous   Asynchronous signals are transmitted by prefixing start and postfixing stop information to the original signal. On the other hand, the start and stop information need only appear for each group or block of synchronous signals. Thus asynchronous signals can usually not be transmitted as fast as synchronous signals.

signal, binary   A signal in the form of a binary code, i.e., digital rather than analog.

signaling   (1) A method of controlling communications. Signaling is used to send a signal from the transmitting end of a circuit to inform a user at the receiving end that a message is to be sent.

(2) In a telecommunications network, signaling is the information exchange that involves the establishment and control of a connection and the management of the network. This is in contrast to the user information that is transferred.

signaling, common battery   A common battery is a single electrical power source used to energize more than one circuit, component, equipment, or system. In many telecommunications applications, the common battery is at a nominal −48 VDC.

signaling, common carrier   Common-carrier signaling is a term used in telecommunications that applies to a method employed by a telecommunications company that holds itself out to the public for hire to provide communications transmission service. *See also* carrier.

signaling, composite   Signaling in which an arrangement is made to provide direct current signaling and dial pulsing beyond the range of conventional loop signaling. Composite signaling permits duplex operation; i.e., it permits simultaneous two-way signal synchronous signaling.

signaling, double current   (Also known as *polar direct-current telegraphy transmission.*) Double current signaling is a form of binary telegraph transmission in which positive and negative direct currents denote the significant conditions.

signaling, flashing light   Flashing light signal systems as designed for high visual impact and maximum operating efficiency.

signaling, frequency exchange   (Also known as *two-source frequency keying.*) Frequency-changing signaling is when the change from one significant condition to another is accompanied by the decay in amplitude of one or more frequencies and by the build-up in amplitude of one or more frequencies. Frequency-exchange signaling applies to supervisory signaling and user-information transmission.

signaling, multifrequency   Multifrequency signaling is often applied to trunk circuits for the transmission of switching information. It thus increases the speed of setting up inter-office connections. Digital information is transmitted by combinations of two of the following five audio frequencies: 700, 900, 1100, 1300, and 1500 Hz. A sixth frequency of 1700 Hz is used in combination with the 1100 Hz frequency as a "priming" signal and in combination with the 1500 Hz frequency as a "start" signal.

signaling path   The signaling path in a transmission system is a path used for system control, synchronization, checking, signaling, and service signals used in system management and operation. It is not the path for the data, messages, or calls of the user.

signaling underwater   Communication is primarily done by transmitting ultrasonic waves. High efficiency transducers are used to allow signals to be beamed from source to receiver.

signal, inhibiting   A signal that prevents the occurrence of an event. An inhibiting signal may by used, for example, to disable an AND gate. It thus prevents any signals from passing through the gate as long as the inhibiting signal is present.

signal, mark A mark signal is a term in telegraphy, which represents one of the two significant conditions of encoding. The other, complementary significant condition is called a *space signal.*

signals, AC, whole-organ (**1**) It has been established that heart or cell activities have variations in their electrical properties that can be measured by galvanometers. For example, if an isolated frog heart (still beating) is placed in between two condenser plates, the rhythmic movements of the heart cause variations in the electrical capacitance and also in the impedance of the condenser to the circuit.

(**2**) Low-frequency alternating current measurements indicate that activity from cells contribute to changes in the impedance of tissue. From these measurements it has been determined that the frequency that causes the maximum change in the impedance is in the audiofrequency range, and it is related to changes in the permeability of the cell.

signal sampling Signal sampling is when an initial continuous waveform is replaced by a finite discrete set of signal points that represent samples of the continuous signal. The most common application is when a continuous timevarying analog signal is transferred to a digital system.

signals of specific frequencies The three (ideal) types are *low pass filters,* which only allow frequencies below a particular value to be passed, *band pass filters,* which allow signals between a lower and upper value to be passed, and *high pass filters,* which only allow frequencies higher than a particular value to be passed. Real filter circuits, which may allow some unwanted frequencies to pass, approximate these ideals.

signal-to-noise ratio A measure of the noise of a system, link, channel, etc. The signal-to-noise ratio (SNR) is usually expressed in decibel

(dB) units and defined as

$$\text{SNR (in dB)} = 10 \log \left(\frac{s}{n}\right)^2,$$

where s is the peak signal level and n is the root-mean-squared noise level.

sign conventions A sign convention is used in geometric optics to measure distances in an optical system. This convention is essential for internal consistency in solving problems. A commonly used sign convention is the *Cartesian convention.* If Cartesian axes are drawn at the refracting (or reflecting) surface such that the origin coincides with the vertex of the surface, then assuming that light travels from left to right always:

1. Object and image distances are negative to the left of the origin and positive to the right.

2. The radius of curvature is positive if the center of curvature is to the right of the vertex and negative if to the left.

3. Vertical dimensions above the horizontal axis are positive and negative below.

silencers Device used for deadening the sound of gas escaping from internal combustion engines or for deadening the sound of firearm.

silent zones Phenomenon common to large explosions. Waves of tremendous intensity reveal abnormalities not evident with low intensity waves. Large explosions are accompanied by a succession of pressure pulses arriving in different paths after reflection and refraction in the atmosphere. There are regions where the sound is not audible, while at greater distances the sound is rather evident.

silsbee effect The superconductivity of a wire- or film-carrying current can be quenched or destroyed at a critical value. In thick specimens where the surface effects can be ignored, the critical current is that which creates at the surface of the specimen a field H_c. Smaller samples remain superconducting with much higher currents than those calculated in this manner. The size of the critical current depends on the nature and geometry of the specimen.

simple harmonic motion Deals with the study of motion of vibrating bodies and is any motion that repeats itself in equal intervals of time. The force acting on the particle is proportional to the displacement but is opposite to it in direction. The particle is attracted toward a fixed point in the line of motion.

simplex A simplex communication channel or operation allows transmission of signals in one direction only. *Compare with* duplexing.

simplex code A set of K simplex codewords may be formed from a set of K orthogonal equal-energy codewords by subtracting from each of these K orthogonal codewords their geometric mean. The resulting simplex codeword set would have pairwise cross-correlation equal to $\frac{1}{1-K}$, a codeword energy equal to $\frac{M-1}{M}$ of the energy of each of the original orthogonal codewords, and a pairwise distance identical to that in original orthogonal code.

singing arc If a continuous current arc is shunted by an inductance L and a capacitance C in series, for certain values, a musical note is emitted of frequency $N = 1/2\Pi\sqrt{LC}$. The superposition of the oscillating current on the continuous current in an arc causes heating effects in the gases around the arc.

siren An instrument consisting of a small metal wind chest with a flat top in which a number of equally spaced holes are drilled around the circumference of a circle. A metal wheel, which can rotate freely, is parallel and close to the top of the wind chest and has a similar set of holes drilled in it. The wheel rotates by the slanting jets of air, which issue from the wind chest. As the wheel spins, a puff of air passes through the holes every time two sets of holes are opposite to each other and this gives rise to a sound of frequency equal to the number of times the holes coincide per second.

skin effect Effect by which alternating electric current in a conductor flows near the surface of the conductor. This effect becomes important at high frequencies and for good conductors and is most pronounced for superconductors. It results in an increase in effective resistance since less of the cross-sectional area of the conductor is used to transport current.

skin effect, anomalous This situation occurs when the conducting slab is thick compared to the skin depth — a thin layer on the surface in the material — but not compared to the mean free path of the electrons. An electric field can influence the electrons only when they are within a skin depth of surface, and conversely, electrons can radiate energy back out of the metal only when they are within a skin depth of the surface. If electrons have orbits in the magnetic field that carry them from within a skin depth of the top of a slab to within a skin depth of the bottom, then electrons in such orbits can reproduce on the far side of the slab the current induced by the driving electric field on the near side. Therefore there is resonant increase in the transmission of electromagnetic energy through the slab whenever the thickness and magnetic field are such that orbits can be so matched with the surfaces.

slew rate The change rate of the maximum absolute output voltage in a circuit.

slip ring Provides electrical connection between a continuously rotating object such as a coil and a stationary metal brush. Used in electric motors and electric generators to make electrical connection to a coil that rotates in a magnetic field.

small-signal approximation The signal processing in electric circuits is divided into two ranges: one is a small signal, which is usually for AC signals, and the other is a large signal, which is usually for DC signals. For different processing, the parameters for calculations are different. The small signal processing is called *small-signal approximation*. For small-signal processing, the change of the signal magnitude must be smaller than the total signal magnitude.

Smoluchowski equation The Smoluchowski equation relates the probability for a random walker to go from an initial point to a final point, using a determined number of steps, to intermediate probabilities. Smoluchowski's formulation provides an alternate approach to the problem of Brownian motion as provided by Einstein's formulation.

Consider a one-dimensional Brownian particle moving in a discrete system. Define the probability $p_n(x_0|x)$ as the probability that, in a series of n steps, the Brownian particle initially at point x_0 will reach point x. Smoluchowski's equation is given by

$$p_n(x_0|x) = \sum_{z=-\infty}^{\infty} p_{n-1}(x_0|z)p_1(z|x) ,$$

for $n >= 1$.

Snell's law   When light goes from the medium of index n_1 to the medium of index n_2, it undergoes refraction. This is governed by Snell's Law:

$$n_1 \sin\theta_1 = n_2 \sin\theta_2 ,$$

where θ_1 is the angle of incidence and θ_2 is the angle of refraction. The angles are measured with respect to the surface normals.

sodar   An acoustical system located on the ground or on a ship that emits a sound impulse and receives its *echoes* scattered by atmospheric turbulence. Sodars are used for acoustic remote sensing of the atmospheric boundary layer — in particular, different dynamic regimes of this layer — vertical profiles of wind velocity, and intensities of temperature and wind velocity fluctuations. A frequency band of sodars is from one to a few kHz. The diameter of an acoustical antenna that is used to emit a sound impulse and receive its echoes can reach a value of several meters.

software   Software consists of logical sentences in computer science and can execute certain operations.

solar battery   An electric device that can convert solar energy or light energy to electrical energy.

soldering fluxes (low temperature)   The solder junction is produced by dipping a piece of oxidized wire, usually niobium, into molten solder. When the solder freezes tight mechanical contact is established and the junction is ready.

The flux is seen to penetrate enclosing a region around the niobium oxide. The magnetic field in the area between the junctions is created by passing a current through the niobium wire.

solenoid   A coil of wire, usually uniformly wound with a length greater than its diameter, used to create a uniform magnetic field at its center. The magnetic field within the solenoid is given by

$$\mu_o n i ,$$

where μ_o is the permittivity of free space, n is the number of turns per meter length of the solenoid, and i is the current in the solenoid. When the solenoid is made long compared to its diameter, the field inside is uniform at positions not too close to the ends.

sonar   Acronym for *sound navigation and ranging*. Refers to underwater acoustics involved in the detection and tracking of submarines and surface vessels in naval warfare.

sonic barrier   Also known as *sound barrier,* when a bullet or shell travels with a velocity greater than that of sound, a sound like a crack of an explosion is heard by the shock wave that is created. When the ratio of velocity of aircraft/velocity of sound (Mach number) exceeds 1, a shock wave develops.

sonic boom   A loud explosive sound caused by the sudden dissipation of a pressure field built up and around an airplane as it reaches the speed of sound.

sonobuoy   An acoustic receiver and radio transmitter mounted in a buoy, used to detect and transmit underwater sounds.

sonoluminescence   The transformation of sound energy into light energy by a process using ultrasonic waves. These waves are aimed at an air bubble in a water cylinder, causing the bubble to oscillate vigorously thereby expanding and contracting to a maximum size of about 50 microns. A near-vacuum situation at expansion causes catastrophic collapse of the bubble which is accompanied by a flash of light. Energy is concentrated by a factor of more than a trillion. *See also* triboluminescence.

sonometer   Also known as a *monochord*. An instrument in which a wire, usually metal, is fixed to a steel peg at one end. The wire passes over a freely running pulley and tension is produced by attaching weights to the free end. Usually two fixed bridges and one movable bridge are also provided to vary the length of the vibrating segment. It can be used for the demonstration of standing waves and resonant frequencies.

sonorous   Deep, resonant or rich sound.

sonovox   An electronic device used by a person whose larynx has been removed, for transmitting recorded sounds to the laryngeal area to be emitted in turn as words through the mouth.

sound in gas   The motion of a vibrating body is communicated to a gas by the production of longitudinal waves that travel in the same direction as the vibration of the gas particles. The velocity with which these waves travel in the gas are dependent on its elasticity and density. The velocity of sound in gases is given by $C = \sqrt{\kappa/\rho_o}$ where κ is the bulk modulus of elasticity and ρ_o is the density of the medium. The compressions and rarefactions in sound waves take place very rapidly so that the associated heating and cooling does not have time to be transferred to the surrounding medium and, therefore, the changes are adiabatic. The velocity in gases is independent of pressure and varies as the square root of the absolute temperature.

sounding board   Stringed instruments are generally connected by rigid supports to some form of base. The string itself transfers little sound energy into the surrounding medium. The vibrations are actually transferred to base or board, which is more suitable for transmitting energy into the surrounding medium. Therefore, a much larger vibrating area is in contact with the surrounding medium, and the rate at which energy is transmitted is greatly increased.

sound in solid/liquid   This scenario is more complex than acoustic waves in a gas and involves both shear and longitudinal stresses and strains in the medium. The velocity of longitudinal waves V, in an isotropic solid is

$$V = \sqrt{(B + 4)/3(G)/\rho} \, .$$

Where B is the bulk modulus, G, the shear modulus, and ρ, the density of the solid.

sound intensity   The rate of flow of energy across a unit area perpendicular to the direction of propagation. Also known as *energy flux*.

sound intensity level   Measure based on comparing two intensities. It is a logarithmic function, and the units are in decibels:

$$SL(\text{dB}) = 10 \log I / I_0 \, ,$$

where I is the intensity at the threshold of hearing 10^{-12} W/ m^2 at 1000 Hz.

sound level meters   Instruments used for measuring loudness. They generally consist of a sensitive microphone of good stability, a linear amplifier, one or more attenuators, a set of frequency-weighting networks and an indicating meter. Electrical voltages corresponding to sounds picked up by a microphone are first amplified and then passed through a suitable frequency-weighting network, which ensures the readings of sound level on meter correspond to observed loudness levels, to operate the indicating meter.

sound, range of audibility   The human ear is sensitive to a range from about 30 Hz to 10,000 Hz. *See* audibility, limits of.

sound spectra   The results of sound analysis are often represented with the frequencies of the harmonics represented on the horizontal axis and the amplitude of each particular harmonic represented on the vertical axis. Certain instruments will have specific spectrum.

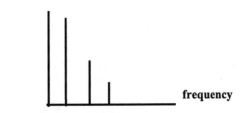

Sound spectra.

sound synthesis   Refers to the fact that a sound of any quality can be regarded as a mixture of harmonics of appropriate relative amplitudes.

source code   Source coding represents an intentional condensation of a data record's symbol rate to maximize coding compaction (i.e., minimize data rate) with minimum distortion in the information content. The source code represents a compact (and thus more efficient) representation of the data (e.g., speech, images, videos) provided by the information source by removing the inherent redundancy in the original data stream. This source coding compression is performed prior to any additional processing by the communication system and is independent of the particular communications system to be used for transmission or storage. Source coding stands in contrast with channel coding, which refers to the additional processing on the source code performed by a communication system to render the transmission more robust against channel distortions. At the receiver, the channel decoder will undo the channel coding, followed by a source decoder to undo the source coding to recover the original transmitted information.

source resistor   An equivalent resistor, which is contained in a power source.

space biology   As part of its scope, space biology considers the study of the biological effects of space travel on living organisms. As the studies involve longer and longer periods of space flight, space biology is more and more concerned on how well and for how long humans and other forms of life can withstand conditions in space. Space biology is especially keen in detecting and studying re-adaptation on earth once space flight has ended.

Specific aspects of space travel under study include weightlessness, inertial forces experienced during lift-off, radiation exposure, the absence of day-night cycle (biorhythm), and heat produced within the spacecraft.

The absence of the night-day division of time has effects on the cyclic patterns of changes in physiological activities that are synchronized with daily, monthly, or yearly environmental changes. These include circadian rhythms that respond to light and dark stimulus (opening and closing of flowers, nighttime increase in activity of nocturnal animals). Circadian rhythms also include changes in blood pressure and urine production during the 24-hour period. In general, internal rhythms that are synchronized to external stimuli (light, temperature, etc.) may be affected and gradually drift out of phase from the original earth environment.

space biology, cardiovascular effects   Because gravity plays a major role in determining the distribution of ventilation, blood flow, gas exchange, alveolar size, intrapleural pressures, and mechanical stresses within the human lung, the absence of gravity affects the cardiovascular system.

The removal of the force of gravity in the lungs during space flight has been linked to increase in diffusing capacity, pulmonary capillary blood volume increase, and increase in membrane diffusing. This means that the lungs are much more uniformly soaked with blood.

Lack of gravity also has important effects on cardiac filling pressure and intravascular fluid distribution. Without gravity, a major central fluid shift occurs. Current data is contradictory regarding the behavior of the central venous pressure, where there are discrepancies between ground-based models and in-flight measurements.

space biology, effects on vision   Although there are no strong effects of space flights on vision, there are changes regarding the ability to orient and position the body relative to the environment. The knowledge of the relative position of the body is a function that involves input from the eyes and inner ear, as well as receptors in the muscles and joints. In space, the body is left with input only from vision and the inner ear.

In the inner ear, the vestibular system composed of the semicircular canals and the otoliths provides information about the sense of rotation and tilting movements. The information coming from these systems are used to control eye movements as a function of postural reflexes. For example, if the head is suddenly rotated, the vestibular system sends information to the brain to help stabilize the position of the eyes while the head is moving. Also, information about the final position of the body is analyzed. All of these functions are affected in space since the inner ear needs gravity to properly measure the different orientational changes. This may translate to symptoms of dizziness and nausea.

space biology, mechanical effects   Long-term space flights negatively affect the structure of muscles and muscular function. Impaired musculoskeletal functions as well as poor muscular coordination have been reported by astronauts after returning to earth. This effect may be due in part to muscle atrophy, that is, the lack of the stimuli provided by the constant pull of gravity. The decrease in muscle strength may also be due to other factors that relate to neuromuscular control and changes in the contractile force of individual muscle fibers.

There are also changes in the bone structure. Because the human skeleton is constantly under a renewing process, disruptions may cause imbalance in the growth-resorption process of the bone. Reduced mechanical workloads on the body in space may induce the skeleton to discard bone it no longer needs. Also, hormonal changes may be linked since some hormones (e.g., parathyroid hormone) are strong stimulators of bone resorption.

space biology, weightlessness   Space studies have revealed that space flight participants initially suffer from symptoms such as nausea, sensory disorientation, and poor muscular coordination. Prolonged flight and exposure to weightlessness affects physical functions that are a direct result of gravity-driven adaptations. These include and affect weight-bearing muscles, gravity-sensing portions of the inner ear, and blood pressure. *See* space biology, mechanical effects; space biology, cardiovascular effects.

space charge   Electric charges that distribute in a material or an electronic device.

space craft   A vehicle used for traveling in space.

space-division multiplexing   A form of multiplexing where the data paths are separated in space. The multiplexing is usually done with switching where data samples in the switch are on separate data circuits.

spark   The ionization and rapid discharge in air and other insulators placed between two conductors that produce a sufficiently high electric field. A sharp snapping sound or a loud explosive noise is emitted depending on the length of the gap between the conductors. For example, thunder is a result of a very large spark discharge between the earth and clouds. Under sufficiently intense fields, such discharges may also take place in liquids and solid dielectrics, in which case the insulating property of the material will be impaired. *See also* lightning flash.

spark gap   Generally, two electrodes separated by a dielectric that breaks down into a spark discharge at a voltage determined by the type of dielectric and the distance between the electrodes. It is mainly used as a voltage limiting safety device (e.g., lightning arrester), generator, of electromagnetic waves, and as a means of depositing concentrated energy, e.g., spark erosion. *See also* lightning arrester.

spark sound   A powerful electric spark can be the source of a single intense pressure pulse, or, if the electric circuit is tuned, the spark may be oscillatory with accompanying pulses of alternating pressure. The pressure effects produce rarefactions and compressions (acoustic waves), which may be sonic or supersonic according to the frequency of the electrical oscillations in spark discharge.

speaking arc   *See* singing arc.

specific charge   The charge to mass ratio of an elementary particle is called the *specific charge* of the particle. For example, the specific charge of an electron is 1.759×10^{11} coulomb/kg while the specific charge of a proton is 9.578×10^7 coulomb/kg.

specific conductance   The ratio of the electrical current to the electric field in a given material, also known as *conductivity*.

specific heat at absolute zero   The third law of thermodynamics states that *the entropy of all systems and of all states of a system is zero at absolute zero*. This implies that the specific heat capacity at constant volume and at constant pressure goes to zero as temperature goes to absolute zero.

specific heat at low temperatures   The specific heat capacity of all solids decreases with decreasing temperature and at 4.2 K for a typical solid it will be between 10^{-3} and 10^{-4} of the value at room temperature. The very low heat capacity at low temperatures also means that only small cooling powers are needed in order to further decrease the temperature of a sample, for example, in a dilution refrigerator. However, at these low temperatures, other contributions may become significant from impurities. *See* Schottky anamolies.

specific heat, cooperative, anomaly   Spin disorders in the paramagnetic state give an inverse relationship with temperature for specific heat at high temperatures. This spin disorder should vanish at 0 K, which can be due to Schottky peak or a cooperative singularity. The former occurs in dilute systems. The latter is more likely if the exchange interactions are strong, thus causing transition to the ferromagnetic and antiferromagnetic states. *See* superconductors; specific heat.

specific heat, Debye's theory   Model for lattice heat capacity at low temperatures that corrects for inadequacies of the Einstein's model where observation shows a T^3 dependence rather than a T^{-1} dependence. Debye assumed that a crystal having N lattice points could be excited by $3N$ acoustic vibrational modes. The factor of three here refers to the three polarizations associated with each wavevector. Debye's theory predicts that the lattice heat capacity is a universal function scaling for all solids through the parameter known as the *Debye temperature*.

specific heat, electronic   Heat capacity is of particular significance at low temperatures when the electrons at the Fermi surface are involved. The electronic-specific heat capacity gives a measure of the electrons at the Fermi surface.

The electronic-specific heat has a shallower temperature variation than thermodynamic-specific heat and only begins to dominate for most metals below a few Kelvins.

specific heat, Schottky   The heat capacity of all solids decreases with decreasing temperature. However, impurities in a crystal at these low temperatures contribute to the heat capacity significantly. This is the *Schottky anomaly* and can increase the heat capacity between 10 to 1000 times the lattice value.

specific impedance, acoustic   The ratio of the acoustic pressure in a medium to the associated particle velocity. It has a magnitude ρc, where ρ is the constant mean density of medium at a point and c is the velocity of propagation of the wave.

spectacles   A pair of ophthalmic lenses of suitable power mounted in a frame to correct for ametropia.

spectra, acoustic   The variation of intensity with frequency which can be represented graphically. *See* sound spectra.

spectra, band   Bands of semi-continuous spectral lines separated by gaps and associated with transitions between molecular energy levels. Each band is actually composed of many fine lines, corresponding not only to electronic transitions but also to vibrational and rotational energy levels.

spectra, continuous   Composed of an infinite number of spectral lines corresponding to the infinite number of wavelengths present, e.g., in a true white light source. One of the most common sources of continuous spectra is the emission spectrum for a heated solid. The standard for such spectra is the black body which absorbs and emits all wavelengths completely.

spectra, ghost   Artifacts of periodic errors in the spacing of the lines of a grating, that give rise to ghost lines accompanying the principal maxima. For *Rowland ghosts,* the error has a single period and the ghosts are symmetrical about the principal maxima. *Lyman ghosts,* involving two incommensurate periods or a single error of very short period, are much harder to identify.

spectra, line   Correspond to images of a spectrometer slit illuminated by one or several distinguishable wavelengths. Since only individual atoms give true line spectra, emission spectral sources are gaseous in nature and may be excited by arcs, sparking, flames, etc. For some elements, spectral lines are arranged in series, being related to the distribution of the atomic energy levels. The simplest and most famous is the Balmer series for hydrogen.

spectral luminous efficacy   The ratio of the radiant flux at a given wavelength in the visible spectrum divided by the radiant power at that wavelength. It rises to a maximum value for a standard observer of 683 lumens per watt near the middle of the visible spectrum. The parameter gives a measure of the luminous flux that is produced by unit radiant power.

spectra, vibrational   Spectra lines or bands corresponding to transitions between molecular vibrational levels. For diatomic molecules, the vibrational levels can be treated in the harmonic oscillator approximation

$$E_n = h v \left(n + \frac{1}{2} \right),$$

with vibrational quantum number $n = 0, 1, 2$. The frequency v is a measure of the spring constant for interatomic forces. Transitions between vibrational energy levels can be observed by infrared absorption or by Raman scattering, with selection rule $\Delta n = \pm 1$. These selection rules are modified by anharmonicity, which changes the form of the potential function and allows overtone transitions $\Delta n = \pm 2, \pm 3, \ldots$ Stacks of rotational levels are associated with each vibrational level and transitions between these levels are governed by $\Delta J = \pm 1$ where J is the rotational quantum number. Such transitions give rise to vibration rotation band spectra.

For polyatomic molecules, one must consider the normal modes of vibration and stretching. The selection rules have additional requirements that there be a change in dipole moment for infrared and a change in the induced dipole moment for Raman.

The selection rules are determined by the symmetry of the molecule, which can in turn be inferred from infrared and Raman spectra. The different vibrations can be represented by potential energy surfaces and anharmonic effects in the potential mix the different modes. Use of model potential functions allows interpretation of the vibrational spectra and hence a complete description of the vibrational modes.

spectrometer   An instrument to measure the wavelengths or radiant intensities of the spectrum of a light source, the design depending on the wavelength of radiation. For example, one uses glass prisms and transmission gratings in the visible region, rock salt prisms for the infrared region, and reflection gratings for ultraviolet and X-ray regions.

spectroscope   An instrument or system that enables us to see the spectrum of a light source, usually consisting of a slit, a collimating lens, a prism or grating and a telescope.

spectroscopy   The study of the absorption, emission or scattering of electromagnetic radiation by matter.

spectroscopy, electron   The collection of techniques involving the spectral analysis of electrons from a sample taken from a beam of incident photons or electrons. There are many different types, principally photoelectron spectroscopy (PES, UPS or MPS for ultraviolet excitation, and ESCA or XPS for X-ray excitation), Auger electron spectroscopy (AES), and others.

The applications in chemistry and surface studies are numerous, where the techniques have become standard analytical laboratory tools. PES is a powerful technique for determining electronic binding energies in various species leading to determination of electronic structure of molecules and solids. XPS and ESCA are complementary techniques to PES for studying inner core energy levels, and all are important techniques for studying surfaces and surface reactions. AES is an extremely useful technique for element detection on surfaces at extremely low coverage.

spectroscopy, interference   Spectroscopy based on examination of light by an interferometer. A light beam with a given wavelength is split into two or more components which are then recombined coherently such that their phase is preserved, giving rise to a set of interference fringes. Since each wavelength has its own set of fringes, spectroscopy can be carried out.

In a two-beam interferometer the incident beam is split so that the light follows two separate paths. The most well known two-beam instrument is the *Michelson* which can also be used for accurate measurement of displacement. In a similar instrument, the *Mach-Zender interferometer,* a sample to be studied is put into one of the light paths and changes in the interferogram are observed.

The most important interference spectrometer based on multiple beam interference is the *Fabry-Perot,* where interference takes place in a thin etalon composed of two glass plates. For high reflectance plates, the fringes are extremely sharp so that the instrument is useful for studying the fine structure of spectral lines.

spectroscopy, laser   The use of a laser source in a spectrometer to replace traditional sources such as the mercury arc lamp. The main advantage of the laser is that it is an intense, almost perfectly monochromatic source. The high intensity allows studying weak lines that are unobservable with other sources. It also permits development of special techniques such as non-linear optical spectroscopy and numerous variations of Raman spectroscopy. The high monochromaticity of the laser permits higher resolution in the spectrometer.

spectroscopy, Mossbauer   Based on the Mossbauer effect, which is the resonant absorption of nuclear gamma radiation in a condensed medium. The Mossbauer spectrometer is based on the measurement of the transmission of resonant gamma rays through the sample as a function of energy. This is accomplished by sweeping the Doppler velocity of the sample with respect to the source. Since the effect is characterized by the absence of recoil or thermal broadening due to transfer of recoil energy directly to the lattice, the spectrum is best measured for strongly bound nuclei at low temperatures. Mossbauer spectra may be measured in transmission and also by scattering, the latter allowing study of surface effects.

There are many applications of Mossbauer spectroscopy in physics, technology, chemistry and biology. These include study of surface films, metal layer interfaces, amorphous ribbons and wires, diffusion in intermetallic alloys, industrial glasses, high pressure effects, chemical isomer shift, chemical bonds, proteins and many more.

spectroscopy, NMR   The phenomenon of absorption of electromagnetic energy by magnetic nuclei of the nucleus, particularly protons, in the presence of an external applied field. Transitions between the proton levels at a given field can be obtained by applying an appropriate RF field at the proton resonant frequency. The experiment is generally done at a fixed RF frequency and by sweeping the magnetic field.

Applications include: (a) determination of the "chemical shift" in the resonance due to the chemical environment; (b) determination of fine structure, due to spin-spin coupling, used for the identification of unknown molecules; and (c) line width determination, which gives information on the molecular motion.

Magnetic resonance imaging in magnetic field gradients for medical diagnostics is one of the most important commercial applications.

spectrum, absorption   Obtained by passing light from a source with a continuous emission spectrum through a target, absorbing material and thence into a spectrometer.

spectrum analyzer   A device used to determine the spectral components of radiation present in a designated range of the electromagnetic spectrum. The technology used depends on the specified frequency range; for radio frequencies, an electronic device is used while for the visible range, a diffraction grating spectrometer is appropriate.

An optical spectrum analyzer uses the acousto-optic effect to analyze the frequency content of an electrical signal. The latter is converted into an acoustic signal by a piezoelectric transducer and the acoustic wave diffracts a monochromatic beam of light incident upon it.

The form of the resulting diffraction pattern depends on the frequency content of the acoustic wave from which the latter can be deduced.

spectrum, angular   The energy spectrum of a given phenomenon as a function of angle with respect to a reference direction.

spectrum, color   *See* spectrum, visible.

spectrum, electromagnetic   The entire range of electromagnetic radiation. At low frequencies (long wavelengths), there is no limit to the wavelength; values up to 50 million km have been observed in radio astronomy. Increasing frequency covers radio waves, microwaves, visible spectrum, ultraviolet, X-rays and γ-rays.

spectrum, emission   The spectrum obtained by observing the radiation emitted from an optical source.

spectrum, normal   In the development of the classification of atomic spectra by Kayser and Runge in the nineteenth century, the spectrum of iron was chosen as a standard or normal spectrum that could be used as a reference for the spectral series of other elements. This work led to the establishment of a series of empirical rules for the positions of the spectral lines of the elements.

spectrum, secondary   The focal length of an optical imaging system can be achromatized in a certain region of the visible spectrum such that the rate of change of focal length with wavelength is zero for that part of the spectrum. Not all wavelengths in the spectrum will have this focal point, leading to a colored zone around the image point, known as the *secondary spectrum.*

spectrum, spark   The emission spectrum obtained by an electrical arc or spark in a gas. When a condenser is connected across the gap, a very rich spectrum characteristic of the metal of the electrodes is obtained. Most emission spectra are currently produced by an inductively coupled plasma source.

spectrum, visible   Corresponds to those wavelengths to which the eye is sensitive. This corresponds roughly to the wavelength range 400 to 700 nm, with violet (shorter than 450 nm), blue (450 to 490 nm), green (490 to 560 nm), yellow (560 to 590 nm), orange (590 to 630 nm) and red (longer than 630 nm).

spectrum, X-ray   Produced when a metal is bombarded with high energy electrons. It consists of a series of sharp lines superposed on a continuum. The continuum, called *Bremstrahlung,* is classical in origin, and is caused by radiation from decelerating electrons in the metal. The discrete spectral lines are due to ejection of electrons from the inner atomic shells and the falling of another electron into the hole left behind, accompanied by the emission of an X-ray. Hard X-rays are associated with transitions to the K atomic shell. Softer X-rays of longer wavelength are associated with transitions to the L atomic shell.

speech, analysis of   Microphones connected to an amplifier can be used to determine the distribution of frequencies in speech.

spontaneous ordering   This refers to the ordering of magnetic spins at the Curie temperature T_c as a result of the exchange interaction being able to overcome the thermal randomization. The figure shows examples of disordered and ordered states respectively.

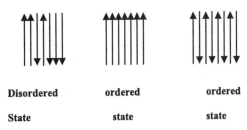

Disordered   **ordered**   **ordered**

State   **state**   **state**

Disordered and ordered states.

sputtering   Charged radicals split out. This process is usually employed for depositing material on films. The charged radicals bombard the target material and produce certain radicals that deposit on the substrate. A film is formed on this substrate.

square wave A periodic wave whose amplitude has two values. The shape of the wave is square.

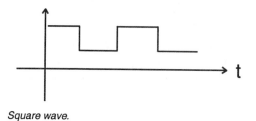

Square wave.

squid Refers to *superconducting quantum interference device.* It is made up of a ring of superconducting material with a weak link. The magnetic flux through this ring is quantized in units of $h/2e$ where h is Planck's constant and e is the electron charge. Can be used to measure magnetic flux very accurately and forms the basis of a magnetometer that is able to measure magnetic moments extremely accurately.

squid, medical use of Superconducting quantum interferometric devices (SQUID) work by detecting minute magnetic fields. Their use is widely spread since magnetic fields appear anywhere there is an electrical current.

Squids are usually circular or square-shaped devices of less than a millimeter in height. A squid consists of a superconducting ring or square interrupted in two places by Josephson junctions (insulating links connecting two superconductors). When sufficient electrical current is applied to the squid, a voltage is generated across its body. In the presence of a magnetic field, this voltage will rise or fall in proportion to the magnetic field.

Magnetometer squids are coupled to a "flux transformer" that functions to amplify the magnetic field input. Gradiometer squids consist of two loops that give signals from two different points in space. In the latter, a flux in the squid only appears if the field is not the same at the two points. This apparatus is thus sensitive to non-uniform fields.

Modern squids can be approximated as close as 20 μm to samples. In this way, squids can be used as "microscopes" on living organisms and other specimens that cannot be frozen. It has uses in the study of biomagnetic phenomena. Namely, migration patterns of bacteria and microbes that possess magnetite particles in their bodies can be tracked down through the environment. These studies play a role in bioremediation, which proposes using bacteria to convert hazardous waste into inert byproducts. In addition, coupling of low-T_c squids to NMR help understand the chemical environments of atoms and molecules.

Squids are widely used to detect magnetic fields from the beating heart (magnetocardiography). In practical terms, they can be used in the measurement of irregularities in the heartbeat of an unborn child. Squid-based apparatuses serve as diagnostic tools in the early detection, diagnosis and follow-up of ischemic heart disease. Also, they serve as clinical tools for arrhythmia localization. They also serve to detect and analyze brain signals in magneto-encephalography.

stability The quality of being stable.

stabilization Making stable.

stabilized power supply A power source that outputs a constant power.

stable states Firm states that do not change with time under certain conditions.

standard cell A primary cell used for instrument calibration because of its very accurately known and constant potential difference. A standard cell differs from other secondary cells in that the electrochemical reactions are not reversible for a standard cell.

standard illuminant A blackbody radiator working at the freezing point of platinum with the luminous intensity per square centimeter defined as 60 candelas. To measure colors of non self-luminous samples, three standard luminants A, B and C are used. The source A is a tungsten filament lamp of specified characteristics operating at a color temperature of 2575° C. The standard illuminants B and C are obtained by using suitable filters with A so as to mimic noon sunlight and normal overcast daylight, respectively.

standing waves   Boundary conditions in any scenario, e.g., for a string with two fixed ends, points of zero displacement cause the string to vibrate with nodes at the ends. The fundamental or the first harmonic consists of two nodes and an antinode. The points of maximum amplitude of vibration are called *antinodes* and spaced evenly between the nodes. The distance between nodal points for the nth harmonic mode of vibration is seen to be l/n.

star connection   Another commonly used three-phase circuit for an AC motor. Here, four wires are employed. The star-connection is also commonly used in a two-phase, four-wire motor as shown in the figure below. It has the same advantages as in the delta connection.

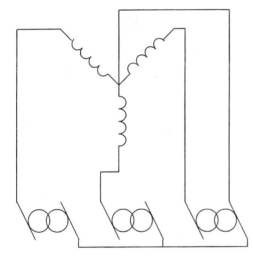

Star connection.

stark effect   The splitting of spectral lines in the presence of a strong electric field. It is analogous to the *Zeeman effect* where lines are split by a magnetic field. The strong electric field can be produced by an electrical discharge in a gas at low pressures.

state attribute   The state attribute refers to the characteristics of a particular state of matter. For example, simple solids possess crystalline symmetry that indicates short- as well as long-range correlations in their positions; i.e., given the position of a point, subsequent positions of the rest of the crystal can be generated starting from the lattice constant. This is due to the regularity of the crystal. In addition, the solid has discrete rotational symmetries, depending on the geometrical configuration of the unit cells. On the other hand, simple liquids and gases have symmetries that correspond to fluids in the sense that they posses only short-range positional correlations that do not continue to long distances.

The thermodynamic attributes are given by the value of the state variables of the substance, and the different phases can be separated into domains, as described by their state diagram. *See* state variables; state diagram.

state diagram   The state diagram of a thermodynamic system is a diagram that denotes the different equilibrium states of matter as a function of a thermodynamic state variable. Two-dimensional representations of state diagrams have to be drawn by having one dependent and one independent state variable, with all the rest fixed at particular values.

The different states (solid, liquid, gas) of the substance whose state diagram is represented are delimited by lines that indicate where the transition from one state to the other occurs. The lines are called *lines of first order transition* because they involve non-analytic behavior in the state variable, as opposed to *lines of second order* that involve non-analytic behavior in the first or higher derivatives of the state variables.

As an example, the P-V state diagram for a simple liquid is given in the figure, where the lines separate the different equilibrium states of the system. The diagram is drawn for a particular fixed volume of the system. The gas and liquid transition line terminates at a point called a *critical point*. The tricritical point is the point at which all of the phases can coexist. Note that at sufficiently high temperature, a continuous path can be drawn from the gas to the liquid (and vice versa) because there are no fundamental differences in the symmetry between each other, only in their density.

state variables   In the study of thermodynamics, the systems that are considered are in equilibrium. The equilibrium state is characterized by being a state stable against internal fluctuations in temperature, pressure, chemical po-

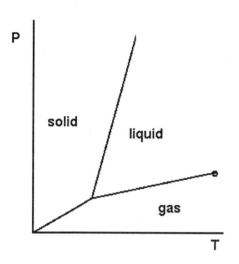

P

solid

liquid

gas

T

State diagrams.

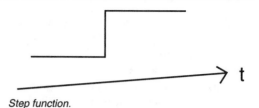

t

Step function.

tential, and composition among other variables. The equilibrium requisite implies that all of the variables have definite values at the equilibrium and thus they are called state variables. Thermodynamics provides a series of relations between these state variables that make it possible to calculate the value of any other variable at equilibrium.

State variables are classified as either *extensive* or *intensive*. The value of an extensive variable depends on the size of the system (e.g., volume). In contrast, intensive variables are not dependent on the system size or number of particles (e.g., temperature). Each state variable has a complementary or conjugate variable of the other type. The variable complementing the volume is the pressure, while the variable complementing the composition of a component is its chemical potential.

stationary channel   A stationary channel refers to a communication channel whose frequency response does not vary with time. Such a channel may be represented as a time-invariant filter.

step function   A function that only has two values. The change from a low value to a high value is abrupt.

stereophony   Necessitates the coinciding of acoustic image with visual image regarding the direction of sound and the production of stereo sound. Systems should possess a frequency range that includes all the audible components of the sounds being reproduced, a distortion-free intensity range that embraces the intensity range associated with the recording sounds, and should be able to preserve the spatial sound pattern of the original sound as well. *See also* quadraphonic sound.

stereopsis   *See* acuity, stereoscopic.

stereoscope   An instrument that separates the field of view of the two eyes (usually by optical methods) such that only certain portions of the stereogram targets viewed through it are seen by one eye and other portions by the other eye. Together, they give rise to a binocular precept of depth. Common stereoscopes include the *Wheatstone* and *Brewster stereoscopes*.

stereoscopic effect   The principle of establishing a three-dimensional image by an optical system using a binocular eyepiece for both eyes. The effect is of great practical importance in several devices, such as prism binoculars and binocular microscopes. The effect is based on two slightly differing images seen by the two eyes; based on normal experience, the brain is able to construct a mental three-dimensional image and obtain depth information.

Stern-Volmer kinetics   *Stern–Volmer kinetics* describe how the fluorescence lifetime of photochemical systems decay in the presence of an interacting environment. The fluorescence described here comes from a chemical reaction whose driving energy is light. Lifetimes of the decay determine the reaction mechanism as well as the electronic structure of the excited states of organic and inorganic molecules.

The simplest way to determine the fluorescent lifetime of a compound is to excite it with a short pulse of light and measure how the fluorescence intensity decays with time. For simple molecules, the fluorescent intensity decays as $I(t) = I_o \exp(-t/\tau)$, where $I(t)$ is the intensity, I_o, the intensity at time 0, t is time, and τ is the lifetime.

When the initial sample is analyzed in the presence of other molecules, they might interact. In typical photochemical experiments, the sample is "quenched" by the surrounding molecules that decrease the lifetime of the fluorescent molecule. If the interaction between the two types of molecules is of collisional nature, the lifetime dependence on the quencher concentration is given by the Stern–Volmer equation as $\tau_o/\tau = 1 + k_q \tau_o [Q]$ where τ is the new lifetime in the presence of the quencher, k_q is the bimolecular quenching rate constant, and $[Q]$ is the quencher concentration.

Applications of fluorescence lifetimes find their way into areas of biomedical research where fluorescent probes are used to study biomembranes, enzymes, photosynthetic systems, nucleic acids, and malignant tissues. It has also found applications in the study of semiconductors, laser dyes, polymers, and other materials.

Stokes parameters   Used to give a complete, quantitative representation of polarized light. For a monochromatic wave propagating in the z direction

$$\vec{E}(t) = \vec{E}x(t) + \vec{E}y(t)$$

$$\vec{E}x(t) = \vec{i}\, E_{ox}(t) \cos\left[(kz - wt) + \delta_x(t)\right]$$

$$Ey(t) = \vec{j}\, E_{oy}(t) \cos\left[(kz - wt) + \delta_y(t)\right] ,$$

then the Stokes parameters are defined as:

$$S_0 = \left\langle E_{ox}^2 \right\rangle + \left\langle E_{oy}^2 \right\rangle$$

$$S_1 = \left\langle E_{ox}^2 \right\rangle - \left\langle E_{oy}^2 \right\rangle$$

$$S_2 = \left\langle 2 E_{ox} E_{oy} \cos \delta \right\rangle$$

$$S_3 = \left\langle 2 E_{ox} E_{oy} \sin \delta \right\rangle ,$$

where $\delta = \delta_x - \delta_y$.

It follows that $S_0^2 = S_1^2 + S_2^2 + S_3^2$ so that S_0 is proportional to the intensity of the wave.

A convenient geometrical representation is that S_1, S_2 and S_3 can be regarded as Cartesian coordinates on the surface of a sphere P of radius S_0, called the *Poincaré Sphere*. Then any possible polarization of the wave can be represented by a point on the surface of P.

stop   Any kind of diaphragm to restrict the amount or angle of light passing through any part of the optical system.

stop, aperture   Any diaphragm or rim of the lens or anything else that restricts the amount of light reaching the image. In addition to limiting the light gathering power, it also limits the resolving power of the system. The aperture stop could be located anywhere inside or outside the optical system. Its image in the object space is called the "entrance pupil". For an optical system with many components, the effective aperture stop for the whole system is the stop whose entrance pupil subtends the smallest angle as seen from the object point.

stop, field   Any device that limits the lateral size or the angular breadth of the object that can be imaged by the system. One usually employs a diaphragm in the image plane to act as a field stop (e.g., the opening in the film holder of a camera). Thus, the field stop is conjugate to the object. Often, to retain only the good quality part of the image (by avoiding the contributions from the far off-axis rays), an aperture is inserted as a field stop in an optical system.

storage vessels, liquefied gas   The efficiency of storage of any liquefied gas is related to its latent heat of vaporization, which controls the evaporation rate as a function of extraneous heat inflow and the normal boiling point which governs the extraneous heat inflow from the surroundings. Glass, usually pyrex, dewar flasks are used for liquid air. Metal dewars have the difficulty of outgassing them completely and therefore require an internal adsorbent trap. Powder and foam insulation are useful. It should also be noted that metal dewars have a larger heat leak with liquid hydrogen in them than with helium. It should be noted that helium gas diffuses through glass.

stored energy in capacitor   When a battery is connected to a capacitor, the positive charges flow into one terminal of the capacitor while the negative charges flow into the other, until the potential difference between the plates reaches the EMF of the battery. In the process, energy is converted from the chemical energy in the battery to the electric energy stored inside the capacitor, $E = 1/2QV = 1/2CV^2$.

stray capacitance   Stray capacitance is the undesirable capacitance that exists between conductors carrying currents or charges. In general stray capacitance is quite small and can be ignored in most cases, but for critical applications, it should be minimized.

stress waves   *See* shear waves.

striking note   *See* subharmonics.

stringed instruments   Examples of plucked string instruments are the harp, guitar, mandolin, and banjo. A plucked string has the full harmonic series of overtones. The stiffness of an actual string and damping due to internal friction causes a slight departure from the harmonic series.

strings, vibrations in   A string can be set into vibration by plucking, striking, or bowing. Nodes are set up at the fixed ends, separated by length L, and the antinodes in between. There is the fundamental, or the first harmonic, which satisfies wavelength $\lambda = 2L$, then second harmonic $\lambda = L$, third harmonic $\lambda = 2/3L$, and so forth.

stroboscope   An instrument for studying periodic or varying motion by illuminating moving bodies with a rapidly flashing electric light, thus making the rotating or vibrating bodies look almost stationary.

subharmonics   The presence of low frequency harmonics in the sound of bells, tuning forks and vibrating piezoelectric crystals, for example, that are not produced by a resonator or from resonances. It is also known as a *striking note* and is possibly due to intermittent contact between the bell and clapper or the tuning fork and surface, respectively.

subscriber   A subscriber station is a telephone station that has access to the public telephone network. A telephone station is a telephone set connected to a telephone system.

subsonics   Speeds less than the speed of sound in air or other medium. Frequencies lower than the auditory capacity of human ear.

subsonic whistle   Whistles that blow at a frequency that is inaudible to human ear but can be picked up by animals such as dogs.

subtracter   A device that has two inputs and one output, the output being equal to the difference of the two inputs.

sufficient (logical)   A condition needed to finish a logic operation.

SUM   An operation that adds all input signals.

sunburn   Reddening of the skin caused by the exposure of the skin to the ultraviolet (UV) light (wavelengths between 290 and 320 nm) coming from the sun. Its effects range from a mild redness to more severe reactions with additional tenderness, pain, swelling and occasional blistering formations. The effects are felt usually within the first 6 to 12 hours, and peak in intensity within 24 hours. After a period of 3 to 5 days, a tan develops that may last a few more days. If the sunburned area is extensive, symptoms such as nausea, headaches, fever, chills, and delirium may occur.

UV light can also reach the skin by reflection from snow, sand, water, sidewalks, and grass. Even on a bright cloudy day with a thin cloud cover, it is possible to receive 60 to 80% of the UV light from a clear day.

Creams to protect against sunburn and the UV rays are called *sunscreens.* They act as barriers, filtering out the transmission of particular wavelength ranges of light (e.g., zinc oxide ointment). Sunscreens are classified according to their sun protective factor (SPF). An SPF rating of 4 provides only limited protection, while a rating of 8 provides maximum sunburn protection.

A rating of 15 gives ultra protection (absorbs burning as well as tanning rays). A disadvantage of most sunscreens is that they need to be applied frequently during the day for protection to be constant. *See* sun exposure and skin cancer.

sundial An instrument that indicates time of day by the position of the shadow of a pointer or gnomon cast by the sun on the face of a calibrated dial.

sun exposure and pigmentation Pigmentation, coming from the darkening of the skin after sun exposure, is due to an increase in the production of melanin in the skin. The melanin (which absorbs UV light) is produced by the skin as a reactive and protective response to protect against further UV rays.

There are three major ranges in wavelength for the classification of UV rays: ultraviolet-A (UVA), ultraviolet-B (UVB), and ultraviolet-C (UVC). UVA has the longest wavelength with UVC the shortest. Because UVC gets absorbed by the ozone layer, only UVA and UVB are of concern to sun exposure. UVA stimulates in greater degree the production of melanin due to its ability to penetrate into deep layers of the skin. Therefore, it is predominantly responsible for tanning, although in sufficient exposure quantity may result in sunburn. UVB probably causes most cases of sunburn and most skin cancers, because it gets absorbed in the outer layers of the skin, burning without stimulating much tanning.

sun exposure and skin cancer There is evidence that links excessive sun exposure to cancer. This evidence even relates sunburns and cancer occurrences separated by years, meaning that a sunburn may have repercussions later in life. Not all of the types of skin cancer turn out to be melanoma that can spread through the body, and may be surgically removed. Another connection that may lead to tumor formation comes from studies of the exposure to ultraviolet-B (UVB, shorter wavelengths than UVA) light that shows interference with the immune system of the skin.

There is also evidence that a risk factor for some skin cancers is light skin color. Cumu-lative, long-term exposure, is associated with higher rates of skin cancer, particularly in light skinned people. Similarly, whites have a higher proportion of skin cancer in the body areas that are routinely exposed to higher levels of sun (e.g., face, shoulders, noses, arms). People with a large number of moles, freckling, or a family history of melanoma have a higher risk of skin cancers. People with dark skin color are somewhat protected by the amount of melanin in their skin because their skin filters twice as much UVB as the skin of whites, although the amount of UV that gets transmitted to the dermal layers of the skin is still significant.

It is estimated that one person in five will develop skin cancer during their lifetimes, in which 90% of all skin cancers are due to sun exposure. *See* sunburn.

superconducting circuits (1) Electrical connections to samples that one wishes to isolate thermally use niobium wires. Such wires are superconducting with a high critical field throughout the entire liquid helium temperature range. They combine low thermal transport with perfect electrical conductivity. Lead wires as heat switches at temperatures below 0.1 K are also commonly used in circuits. Superconductors are used as switching device and memory storage elements in electronic computers.

(2) If a current is flowing in a circuit containing a resistance R and an inductance L and an EMF is suddenly removed, the current falls to $1/e$ of the original value with normal conductors in about 10^{-5}s. However, due to the almost negligible resistance in a superconductor, the current persists long after the EMF has been removed, and in some cases a persistent current has been observed for a period of several years.

superconducting electrons This refers to the electrons in a superconductor that are in the ground state. In a superconductor at any finite temperature, a dynamic equilibrium exists, that is, normal electrons recombine to create pairs — the Cooper pair — continuously created by the break up of pairs. Electron tunneling is a common phenomenon when at least one of the electrodes is a superconductor at a junction.

superconducting thin films These materials offer virtually zero resistance to electrical current and are extremely useful in microelectronics. They have the capacity to carry large currents at significantly higher temperatures than the superconducting material. Thin films of normal metals and superconductors in contact can form superconductive electronic devices, which replace transistors in some applications.

superconducting tunneling It is possible for electrons to tunnel between a superconducting film and a normal one across a thin insulating barrier. Quantum mechanically, an electron has a finite probability of tunneling through the barrier if there is an allowed state of equal or smaller energy available for it on the other side. In this manner, a direct measurement of the energy gap can be made.

superconduction, Heisenberg's microscopic theory of Explains how the influence of the strong intermolecular magnetic field causes spontaneous magnetization of ferromagnetic substances. Heisenberg showed that the field originates in the quantum mechanical exchange integral. There is no classical counterpart of this and it is associated with the difference in the Coulomb interaction energy of electrons when the spins are parallel or antiparallel. This model has also been successful at explaining the spin waves at low temperatures.

superconductive alloys Examples of such alloys include two parts gold and one part bismuth, and rhenium with molybdenum which, as alloys, display superconducting properties.

superconductive compounds *See* superconductors.

superconductive cylinder For the case of a long thin cylindrical sample in a longitudinal magnetic field, the magnetization curve for an ideal superconductor is as follows: at the critical field, the flux penetrates the superconductor and the normal state is restored.

superconductive elements Examples of superconducting elements are aluminum, cadmium, gallium, indium, lead. Superconductiv-

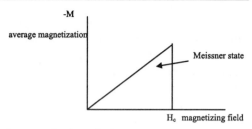

Superconducting cylinder.

ity has been observed in metallic substances for which the number of valence electrons Z lies between about 2 and 8 and the critical temperature T_c shows a sharp maxima for transition metals with $Z = 3, 5,$ and 7.

superconductive sphere The magnetization curve for a sphere is as seen below. Between the states H_p, the penetration field, and H_c, the critical field, the superconductor is in the intermediate state. This is a geometric effect. Since the sample has broken into alternately normal and superconducting states, the magnetic flux is able to pass through the normal state.

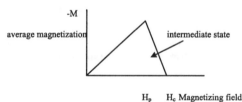

Superconductive sphere.

superconductivity A phenomenon shown by certain metals, alloys, and other compounds of having negligible resistance to the flow of electric current at temperatures approaching absolute zero. Each material has a critical temperature T_c, which generally is under 10 K, above which it is a normal conductor and below which, a superconductor. Recently, some materials have been shown to be superconductive at temperatures hundreds of degrees above absolute zero.

superconductivity, BCS theory A theory of superconductivity proposed by the American physicists John Bardeen, Leon N. Cooper and John R. Schrieffer in 1957 for which the three were awarded the 1972 Nobel Prize in physics. The theory describes superconductivity as a quantum phenomenon, in which the conduction electrons move in pairs and thus show no electrical resistance.

superconductivity, destruction by currents
High current densities cause a superconductor to be restored to its normal or resistive state. When a transport current flows in a superconductor that is in the mixed state, a Lorentz force acts on the flux lines so as to move them at right angles to their axes and to the direction of current flow. Movement of magnetic flux causes a voltage to be induced across the sample in the same direction as the current flow. This causes the superconductor to become resistive, and heat is dissipated. The motion can be reduced if the flux lines are pinned by defects such as dislocations, grain boundaries, vacancies and clusters of impurities in the crystal structure. This can be achieved by work hardening. *See* work hardening.

superconductivity, electrodynamics of   For a mathematical description of the response of a superconductor to an applied DC magnetic field, one needs to take into account perfect conductivity, the Meissner effect. The equations used in conjunction with Maxwell's equations predict that the magnetic flux is excluded from all except a surface region of the bulk superconductor. The decay of the field at the surface has a simple exponential form with a characteristic length. In a normal metal the eddy currents produced by an induced current are quickly reduced by the scattering process. However, these currents on an atomic scale persist and give rise to a weak diamagnetic susceptibility in all materials.

superconductivity, transition of   The transition to superconductivity occurs in several metals, alloys and compounds when they are cooled to below the transition temperature T_c. There is a drastic change in electrical and thermal properties. The DC electrical resistance disappears in the new phase. The transition to superconductivity is a function of temperature and applied magnetic field.

superconductivity, transition, pressure effect
It has been found that the transition temperature T_c below which a substance will behave as a superconductor is a function of pressure and stress generally. However, pressure is not a dominating factor, and the density of the electron states at the Fermi surface is more important as well as the interaction between electrons which arise from coupling with the lattice vibrations.

superconductivity transition, resistive   In the transition to superconductivity, the most significant feature is the disappearance of resistance to DC current.

superconductivity, two-fluid model   This model is based on the postulate that a superconductor possesses an ordered ground state that is characterized by an order parameter, and that the entire entropy of the system resides in the excited states at energies above that of the ground state. Superconducting properties are associated with the superfluid fraction f of the conduction electrons in the ground state, while the remaining "normal" fraction $(1 - f)$ retains the properties of electrons in the normal state at $T > T_c$. The two components are totally interpenetrating and non-interacting. This model is useful in that it provides a simple physical picture providing a semi-quantitative understanding of, and the interrelation between, thermal and magnetic properties.

superconductor, Gibb's free energy in   The transition to a superconducting state is a function of temperature and applied magnetic field. In pure samples, the transition is reversible and can be described by equilibrium thermodynamics. The condition for equilibrium is found by minimizing the magnetic Gibb's free energy.

superconductors   Materials that may be metals, alloys or compounds that display the phenomenon of no resistance to the flow of an electric current below critical temperatures T_c. Superconductors also exhibit strong diamagnetism, meaning that they are repelled by magnetic fields. Superconductivity is manifested only below a certain critical temperature T_c and a critical magnetic field H_c, which vary with the material used.

superconductors, critical field of   Below the critical temperature T_c, the superconducting behavior can be quenched and normal conductivity restored by the application of an external magnetic field — the critical field H_c.

$$H_c \sim H_o \left(1 - (T/T_c)^2\right) \text{ where}$$
$$H_o = H_c \text{ at } T = 0K .$$

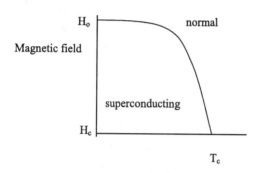

Critical field of superconductors.

superconductors, energy gaps in   The thermal and electrical properties of the superconducting state differ from those of metals in the normal state, and there exists an energy gap between the superfluid ground state and the states of the normal electrons. The energy gap is actually temperature-dependent, falling from its zero temperature value to zero at the critical temperature.

superconductors, high field   A high magnetic field applied to a superconductor will restore a superconductor to its normal or resistive state. *See* superconductors.

superconductors, infrared absorption and transmission   The transmission and absorption effects in superconductors are generally the same in normal and superconducting states for different frequencies (e.g., optical frequencies) except in the microwave and infrared frequencies.   This is determined by observing frequency-dependent conductivity. The absorption becomes very small as the temperature approaches 0 K. This indicates a quantum effect of the excitation of electrons across an energy gap.

superconductors, intermediate state   *See* superconductive sphere.

superconductors, London   Deals with the theoretical treatment of the skin effect in superconductors. It is defined as those superconductors that deal with a penetration depth greater than the intrinsic coherence length. Local electrodynamics is used to treat this type of superconductor. *See* superconductors, Pippard.

superconductors, penetration depth   Currents in a superconductor are not strictly superficial but occupy a layer of finite depth below the surface. The magnetic field also penetrates to a certain depth. This causes the susceptibility for small objects to be reduced, and the observed critical magnetic field for a superconductor becomes greater for one of small dimensions than a larger one. *See* skin effect, anomalous.

superconductors, phase diagram   The transition to superconductivity by substances is akin to a second order phase transition. This means that there is no latent heat involved and a sharp finite discontinuity in the specific heat is seen.

superconductors, Pippard   Deals with the theoretical treatment of the skin effect in superconductors. It is defined as those superconductors that deal with penetration depths less than the intrinsic coherence length but greater than the London penetration depth. Non-local electrodynamics is used to treat this type of superconductor. *See* superconductor.

superconductors, resistance of, high frequency effect   Resistance disappears in superconductivity for the DC case. In the anomalous skin effect, when the skin layer involved in electrical conduction at high frequencies is very thin, if this becomes less than the electron mean free path, the classical theory of electrical conduction breaks down and resistance is proportional to the cube root of the frequency. High frequency resistance occurs in superconductors below the transition point by several factors than that observed at higher temperatures. This has been observed for frequencies of the order of 1200 Hz to 23,000 Hz.

superconductors, specific heat The specific heat displays a strong discontinuity at the transition point for superconducting substances. As the temperature drops below the transition temperature in a zero magnetic field, the specific heat increases substantially and then decreases slowly.

superconductors, transition temperatures in This refers to the temperature at which an element or alloy will display superconducting properties. The values for elements generally range from about 0.105 K for iridium to higher values such as 7.199K for lead and 9.25K for niobium.

superconductors, type I The magnetic flux is expelled thereby producing magnetization that increases with the magnetic field until a critical value is reached, at which it falls to zero as for a normal conductor. These types exhibit Meissner diamagnetism.

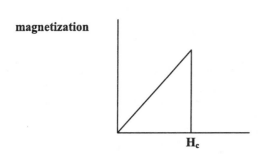

Superconductors, type I.

superconductors, type II In this type the magnetic field begins to penetrate the specimen at a lower critical field but the superconductivity is completely destroyed only at a higher field. This type does not exhibit the full Meissner effect.

supercooling This refers to the cooling of a substance below the temperature at which a phase change is expected to take place without the phase change occurring, causing it to become metastable; i.e., it is possible to cool a liquid below its freezing point without its freezing.

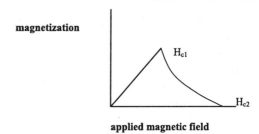

Superconductors, type II.

superfluidity This is shown by particles that obey Bose–Einstein statistics and are in the lowest allowed energy state. The particles therefore have zero resistance to motion and zero entropy. Substances that exhibit superfluidity are Helium II, which can flow through tiny holes impervious to any other liquid, and a pair of electrons in superconductors.

superleak This is used, for example, in a refrigerator. Stainless steel tubes packed tightly with jeweller's rouge which prevents liquid other than superfluid from circulating.

superposition of light waves The principle of superposition for a linear optical system states that, if there are different individual solutions to the wave equation, then a linear combination of them is also a solution. In practical applications to interference and diffraction, a scalar theory is used: e.g., for superposing two waves of the same frequency traveling in the same direction. The superposition can be carried out by direct algebraic addition. Alternatively, a complex representation can be used and the superposition is then carried out by the addition of phasors which incorporate the amplitude and the phase of each wave. This method forms the basis for analysis of interference and diffraction phenomena in optics.

superposition of waves When two waves traverse the same medium, the waves add. The amplitude of the resultant wave at a particular point is the algebraic sum of the two component amplitudes.

supersonic vibrations Speeds greater than that of sound or frequencies above those that can be heard by the ear.

surface charge   The charge that resides on the surface of an object or at the interface between two substances is called *surface charge*. In a perfect conductor, the surface charge results from the fact that the field inside a perfect conductor is zero. For a dielectric material, the surface charge results from the polarization of the material and equals the product of polarization and a unit vector normal to the surface, $\sigma = \mathbf{P} \cdot \mathbf{n}$.

surface charge density   The amount of electric charge, Q, that is spread over the surface area, A, of an object as a thin layer. The surface charge density, σ, is given by

$$\sigma = Q/A .$$

It has the SI units of coulombs per square meter, Cm^{-2}. The charge distribution may not be uniform on insulators and can therefore have different local surface charge densities.

surface tension, cell membrane   Because the cell membrane is the boundary between two dissimilar liquids, but at the same pressure, any kink or deformation in the topology of the membrane will be met by an equivalent imbalance in pressure from one of the liquids. The direction of the restoring forces will be such that the membrane will minimize its surface area, as a direct result of its surface tension.

   The importance of the surface tension of the cell membrane has been linked to olfactory stimulation. In this process, receptor cells may trigger an olfactory response based on chemical reactions that occur at the surface of the cell and to a solution of odorant molecules that may alter the surface tension at the surface of the cell membrane. *See* turgor pressure.

surface waves   *See* Rayleigh waves.

surge protection   To protect from a abrupt change in signals.

susceptance, electric   Unit: siemens. The imaginary part of the admittance, Y, which is given by
$$Y = G + iS ,$$
where S is the susceptance, G is the conductance, and $i = \sqrt{-1}$. For a circuit containing a resistance R, and reactance X, the susceptance is given by

$$S = \frac{-X}{R^2 + X^2} .$$

susceptibility, diamagnetic   A negative magnetic susceptibility due to an induced magnetic moment in a system of electrons by an applied magnetic field. The susceptibility is usually small with a magnitude 10^{-5} to 10^{-6}. The negative value indicates that the induced magnetization is opposite the applied field. Important in materials that have no permanent magnetic moment on their atoms, which include Cu, Bi, B, Si, and many inorganic and organic molecules. Superconductors have the largest diamagnetic susceptibility, with type I superconductors having a value of -1. They are often referred to as *perfect diamagnets* since the induced magnetic moment is large enough that it exactly cancels the applied magnetic flux within the superconductor and so there is no penetration of magnetic flux into the interior of a superconductor. *See* susceptibility, magnetic.

susceptibility, electric   A dimensionless quantity, χ_e, which relates the electric polarization $\mathbf{P}$ of a material to the applied electric field, $\mathbf{E}$, by
$$\chi_e = \frac{\mathbf{P}}{\epsilon_0 \mathbf{E}} ,$$
where ϵ_0 is the permitivity of free space. Electric susceptibility is a measure of the ease with which a dielectric can be polarized. The electric susceptibility is also given by

$$\chi_e = \epsilon_r - 1 ,$$

where ϵ_r is the relative permittivity of a material.

susceptibility, magnetic   Measures how easily a material is magnetized by a (usually) small applied magnetic field. The susceptibility χ of a material is the proportionality constant between the applied magnetic field $\mathbf{H}$ and magnetization $\mathbf{M}$ of a material and is defined by

$$\mathbf{M} = \chi \mathbf{H} .$$

If $\mathbf{H}$ and $\mathbf{M}$ are parallel to each other, then χ is a scalar quantity. χ is related to the relative permeability of the material by $\mu_r = 1 + \chi$. *See* permeability, magnetic.

susceptibility, measurement by Sucksmith's method   Method of measuring magnetic susceptibility in which a sample is suspended from a deformable ring in a non-uniform magnetic field. The force on the sample in the non-uniform field is proportional to the magnetic moment of the sample, which in turn depends on the susceptibility of the sample. The force itself is measured by measuring the deformation of the ring and leads to a determination of the susceptibility of the sample.

sweep generator   An electric circuit that produces a signal to examine other periodic signals.

switch   A mechanical or solid state device for opening and closing a circuit, such as a circuit breaker (mechanical) or a transistor (solid state). Switches are also used to select one of several components to be included in a circuit for a desired mode of operation. Other switches, such as flip-flops, cause the operating condition of a circuit to change between two discrete levels.

switching, time division   A time division switch separates data paths in time; a crossbar switch separates data paths in space. In time division switching, the n inputs are stored in a temporary buffer. The switch reads from the buffer n times faster than the input and writes the data to the outputs in the proper order.

switch, timer   A switch operated by clockwork, electric motor or resistor-capacitor circuit to open or close a circuit at a predetermined time.

synchrocyclotron   A cyclotron in which the magnetic field is held constant, and the frequency of reversal of voltage used to accelerate charged particles is adjusted so that the particles stay in step as they speed up. This allows for such effects as the relativistic increase in mass that occurs at speeds close to the speed of light for the accelerating particles. *See* cyclotron.

synchronization   Any circuit or device that is operated by means of clock pulses is synchronous.

synchronous capacitor   A synchronous motor running without mechanical load and drawing a large leading current, like a capacitor; used to improve the power factor and the voltage regulation of an alternating-current power system.

synchronous motors   Such a motor is one in which the rotor normally rotates at the same speed as the revolving field in the machine. The stator is similar to that of an induction machine consisting of a cylindrical iron frame with windings located in slots around the inner periphery. A revolving field can be produced in synchronous motors by use of the same method as for induction motors. With the main stator winding connected directly to the supply, an auxiliary winding may be connected through a capacitor.

synchroscope   An instrument used to compare both the phase and frequency of two different AC sources. Used mainly to determine when a synchronous motor has been brought to synchronous speed; this occurs when the EMF induced in its armature windings have a phase difference of about 180° with the AC power supply line potential. The motor can then be connected directly to the line. *See also* motor, synchronous.

synchrotron   Device used to accelerate charged particles to high speeds so that experiments to determine their structure and properties may be performed. Uses a magnetic field to make particles move in a circle and an electric field whose polarity changes periodically to accelerate the particles. The strength of the magnetic field and the frequency of the electric field are changed as the particles speed up. Very high speeds are possible with protons reaching speeds of more than 99% of the speed of light. Also used to generate intense electromagnetic radiation (synchrotron radiation) in the infrared to hard X-ray range.

synthetic sound   This is produced by sinusoidal alternating currents of various definite frequencies that are generated continuously and can imitate the quality of various musical instruments.

system (in communication)   A communications system is the complete assembly of apparatus and circuits required to effect a desired transfer of information. In telecommunications, a system is a set of equipment or apparatuses that is combined to perform a function within a carrier's telecommunications network. An example is a switching system.

T

Talbot's Law   A slowly flickering light source can be detected by the eye. However, due to the finite response time of the eye, this flicker cannot be detected above a critical frequency, This principle is known as *Talbot's Law* and can be demonstrated quantitatively by a rotating sector wheel. Above the critical frequency of rotation of the wheel, the intensity of transmitted light varies as the proportion of sectors cut from the wheel. Talbot's Law finds application in motion pictures, where for practical purposes blurring is prevented by projection at 24 frames per second.

tangent law   Assuming that there is no spherical aberration of the principal rays from the object points, the condition to be satisfied for elimination of distortion for a Gaussian image is

$$ny \tan u = n'y' \tan u' \, ,$$

where u and u' are the angles made by the principal rays with the optical axis, y and y' are the sizes of the object and image, and n and n' are the refractive indices of the object and the image spaces, respectively.

tangent plane   The plane containing the optical axis and the off-axis object point. The principal ray always lies in the tangent (or meridional) plane.

telecentricity   A lens is said to be telecentric if the chief rays are parallel to one another. Usually, they are also parallel to the lens axis and perpendicular to the object and/or image plane that are also perpendicular to the axis. A lens is said to be telecentric in object space and/or telecentric in image space. A focal lens can be non-telecentric or telecentric on either side, but cannot be doubly telecentric. An afocal lens can be non-telecentric or doubly telecentric but not telecentric on one side.

telecommunication   Telecommunication is the study or practice of the transfer of information over a distance by electromagnetic means, such as wire in cable telegraph and telephone, or radio waves in broadcasting.

telegram   A message sent by telegraph and then delivered in written form.

telegraph   A system or device for transmitting messages or signals to a distant receiver, usually by making and breaking an electrical connection.

telegraph, polarential   A direct-current telegraph system employing polar transmission in one direction and a form of differential duplex transmission in the other. There are two types of polarential systems, known as types A and B. In half-duplex operation of a type A polarential system, the direct-current balance is independent of line resistance. In half-duplex operation of a type B polarential system, the direct current is substantially independent of the line leakage. Type A is better for cable loops where leakage is negligible but resistance varies with temperature. Type B is better for open wire where variable line leakage is frequent.

telegraphy, radio   Communication by means of a telecommunication system that transmits documentary matter, such as written or printed matter or fixed images, and reproduces it at a distance. The matter is transmitted as a suitable signal code, such as international Morse code, either by means of wire or by radio (radio telegraphy).

telemetry   (1) Measurement at a distance.

(2) A measuring instrument that measures a quantity and transmits the measured data as an electrical signal to a distant recording point is known as a *telemeter*. Space exploration and physiological monitoring in hospitals both require the use of telemetry.

telephone   An apparatus for transmitting sound (especially speech) over a distance by wire, cord or radio. It is an assembly of apparatus that includes a suitable handset containing the transmitter and receiver, and usually a switch hook and the immediate associated wiring.

telephone, analog A plain old telephone service connection with no advanced features.

telephone, radio Normally, communication between two points takes place along suitable cables (telephone lines) except where this is inappropriate, such as ship-to-shore telephony. A radio telephone is used when a radio link is needed to connect a particular access point to the main system.

teleprinter A device for transmitting telegraph messages as they are keyed, and for printing messages received. It is a form of a start-stop typewriter that comprises a keyboard transmitter, which converts keyboard information into electrical signals, and a printer receiver, which reverses the process. Teleprinters are used in telex systems and in some older computing systems.

telescope, astronomical Any telescope used to study astronomical objects. The telescope first used by Galileo for this purpose consisted of two convex lenses, separated by the sum of their focal lengths, with the focal length of the objective larger than that of the eyepiece. In an astronomical telescope, the rays that enter it parallel leave it parallel; the ratio of their angles with the telescope axis give the angular magnification. Instead of a lens system, mirrors can be used to collect light from the stars. (*See* telescope, reflecting). Outside the visible region (for example for radiotelescopes), one uses special radio-dishes and antennas.

telescope, electron Telescopes that are unlike optical telescopes in that the radiation falls on the photocathode surfaces or on charge-coupled devices (CCDs). By accelerating the resulting electrons and making them incident on a fluorescent screen, direct image of objects in the ultraviolet and infrared regions can be observed. In the optical region also, the intensity of faint star images can be enhanced.

telescope, reflecting Astronomical telescopes using a (front polished paraboloidal) mirror to collect light from the stars. Depending on various popular arrangements, the star is observed at the primary focus of the mirror (for a large mirror), or the collected light is brought out at right angles to the beam by use of a plane mirror or prism (Newtonian), or through a hole in the primary by reflection from a concave ellipsoidal secondary mirror (Gregorian), or by the use of a convex hyperboloidal secondary mirror (cassegranian). The reflecting-type telescopes can be made very large because, unlike the difficulties involved in making a large bubble- and strain-free lens for a refracting telescope (with chromatic aberration problems), one can polish a very large geometrically well-defined surface. The mirror can be supported on the back also, in contrast to the lens, which can be supported from the rim alone. The chromatic aberration is absent because refraction is not involved.

teletype (**1**) A kind of teleprinter.
(**2**) To operate a teleprinter or send by means of a teleprinter.

teletypewriter (Also known as *teleprinter.*) A device for transmitting telegraph messages as they are keyed and for printing messages received.

television A display instrument that converts a received electromagnetic signal into a visible image.

telex An international system of telegraphy in which printed messages are transmitted and received by teleprinters using the public telecommunications network. The word *telex* comes from a combination of the words *teleprinter* and *exchange*.

temperament Certain adjustment of tones or intervals of the scale of fixed tone instruments like organs or pipes, so that a larger variety of melodious combinations are possible.

temperature, Bloch The electrical conductivity or resistivity is extremely sample-dependent at low temperature. The Bloch T^5 temperature law is observed in many metals at low temperatures. The residual resistivity is very sensitive to the presence of impurities and structural defects and, for a dirty specimen, may not be much smaller than that at room temperature.

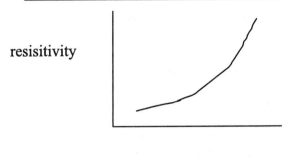

resisitivity

temperature

Bloch temperature.

the isenthalp and reduce on the other side of it. Inversion temperatures are along the isenthalp.

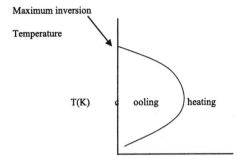

Pressure (Mpa)

Inversion temperature.

temperature changes during passage of sound Compressions and rarefactions take place so rapidly that the gas does not have sufficient time to lose or absorb heat from the surroundings. The process is adiabatic. In general the speed of sound is proportional to the square root of the absolute temperature in general.

temperature, critical This refers to the transition temperature in superconductors below which drastically altered thermodynamic and electrical properties are observed in the sample.

temperature, Debye Dealing with the heat capacity that results from lattice vibrations, the Debye T^3 law predicts that the lattice heat capacity is a universal function scaling for all solids through the parameter θ_d, the Debye temperature. Materials with strong interatomic forces and light atoms such as diamond and sapphire have relatively high Debye temperature, and soft materials with low acoustic velocities have smaller values.

temperature, degeneracy The number of electrons per unit volume in the conduction band and the number of holes per unit volume in the valence band are a function of temperature. However, these depend strongly on the presence of impurities in the semiconductor. Introduction of additional levels of energies due to the impurities can introduce degeneracy.

temperature, inversion According to the Joule-Kelvin effect, when gas is allowed to expand through a porous plug or orifice, it may become warmer or cooler. The temperature change can be determined from isenthalps for the gas. Temperatures increase on one side of

temperature, lowest helium Experimental parameters for helium have been found for temperatures down to about 0.1 K.

temperature, magnetic For paramagnetic salts there is an observable magnetic property, the susceptibility of which is temperature sensitive and provides a thermometric parameter. For most magnetic cooling salts at temperature $T \sim$ 1K the magnetic susceptibility varies inversely with the absolute temperature; at higher temperatures, there is a complicating influence due to magnetic interactions. There may be discrepancies between the thermodynamic temperature and the magnetic temperature below 1 K.

temperature measurements, clinical The temperature of the body is a direct result of the balance between heat production and heat loss. In humans, the metabolism maintains the body temperature within a narrow range (36.5° C to 37.5° C) despite wide variations in heat production or environmental temperature. The part of the human brain in charge of thermo-regulating the body is the hypothalamus, where elevation in the temperature of blood going to that part of the brain initiates heat loss by causing dilation of blood vessels and sweating.

Because of the narrow range of values, temperature measurements are key in prevention and diagnosis of diseases. In particular, fever can easily be detected, being a result of a disturbance in the regulation of temperature. Because the hypothalamus tries to balance the body temperature, during fever, the temperature that the hypothalamus tries to achieve has been shifted from its normal position. This may be due to chemicals called *endogenous pyrogens,* which are derived from white blood cells.

Clinical measurements of temperature have to be taken with care. For example, pressure application to the skin whose temperature is being measured is undesirable due to possible changes in the vascular tissue around it. Also, local sweating may influence the final reading.

temperature, Neel   This is with respect to magnetic behavior at different temperatures. The spins become ordered below the Curie temperature T_c; however, the ordered state is not always ferromagnetic. In some cases the adjacent spins may be antiparallel. In this antiferromagnetic state, there is no net spontaneous magnetization but hysteresis is present and the susceptibility shows a sharp maximum at this transition temperature. Known also as Neel point. *See* spontaneous ordering.

temperature, negative absolute   The absolute temperature scale refers to the Kelvin scale which extends upwards from 0 K representing $-273°$ C. There are no negative values on this scale.

Tesla   The SI unit of magnetic flux density, one tesla (1 T) is equal to 1 Vs/A (or equivalently 1 J/A^2). The average magnetic flux density from the geomagnetic field on the surface of the earth is around 4×10^{-5} T.

Named after Nikola Tesla (1856–1943) whose investigations into time varying electromagnetic fields led to the utilization of AC electric power distribution and the development of radio.

test charge   A minute amount of charge used to evaluate the electric field at a particular position is called a *test charge.* The electric field at a point in space is defined as the force experienced by the test charge divided by the charge of the test charge. The test charge needs to be as small as possible to avoid disturbing the field to be measured.

thaw rigor   Upon death, muscles of animals and humans undergo what is usually called *rigor mortis.* This means that muscles contract after death due to Ca^{2+} leaks from the sarcoplasmic reticulum, artificially initiating the muscle contraction cycle without neuronal signal. After the contraction consumes all of the remnant ATP, the muscles remain contracted because there is no ATP to pump back the calcium or at least to release the myosin from the actin. *See* muscle, mechanics.

A similar process occurs during thaw rigor in which meat shortens considerably after thawing. This happens when the meat has been frozen before the completion of rigor mortis and the ice crystals have slashed open the sarcoplasmic reticulum. Then on thawing, the meat is finally warm enough to respond to the accumulations of Ca^{2+} that causes extreme contraction.

This concept is of importance to the meat industry since the final tenderness of their product depends on the technical steps in freezing the meat.

theorem, the relationship becomes in differential form $\nabla.E = \rho/\epsilon_0$.

thermal drift   A fluctuation caused by the temperature change.

thermionic emission   (**1**) Electrons liberated from the surface of a metal as a result of thermal energy supplied to them by heating the metal to high temperatures. The energy supplied per electron must be greater than the potential barrier at the surface that ordinarily keeps it in.

(**2**) Term describing the emission of electrons or positively charged ions from the surface of a conducting material when it is held at high temperatures. It is caused by some of the electrons/ions gaining sufficient kinetic energy through random fluctuations to overcome the binding energy of the conductor's surface. This effect is the basis for the operation of vacuum tubes.

thermistor   A thermal meter whose readout is responsive to the temperature, such as a thermometer.

thermocouple   A device that converts thermal energy to electrical energy in order to measure the temperature of an object. It consists of two wires of dissimilar metals connected together in a circuit. A current will flow if two such junctions are connected together and maintained at different temperatures. A temperature measurement is usually made by maintaining one of the junctions at a known temperature (e.g., 0° C) while the other junction is brought into contact with the object. The magnitude of the potential difference between the two junctions is a measure of the temperature. Tabulated data of potential differences and their corresponding temperatures are readily available for the different types of thermocouples. *See also* Seebeck effect.

thermodynamics, third law of   The entropy of all systems and of all states of a system is zero at absolute zero. In other words, it is impossible to reach the absolute zero of temperature by any finite number of processes.

thermoelectricity   Conversion of heat energy directly to electrical energy and vice versa. *See* Peltier effect, Seebeck effect, Thomson effect.

thermography   (**1**) Technique in which an apparatus records temperatures in sequential measurements for diagnostic purposes. The thermometer used for the temperature readings is called a *thermograph.*

(**2**) Diagnostic technique in which an infrared camera is used to measure temperature variations on the surface of the body. Images are then produced that may reveal sites of abnormal tissue growth.

thermometer, acoustic   Thermometers that operate by sensing sound waves. This is done by measuring the velocity of sound in a gas at very small pressures since this is directly related to temperature. The velocity of sound is measured by observing acoustic resonances.

thermometer, carbon radio resistor   The most commonly used and cheapest thermometer to work below 4 K. Its characteristics vary widely from manufacturer to manufacturer and different ones are required for different temperature ranges. It is usually made from graphite composite inside a ceramic coating and does not exhibit simple semiconducting behavior. Carbon resistors undergo a change in calibration over a period of time but the change between successive runs gradually decreases after the resistor has been "trained" by repeated cyclings between room temperature and 4 K. The carbon resistor is less sensitive to magnetic fields than germanium.

thermometer, germanium   Semiconductor material thermometer that gives very reproducible results even after many cycles from 300 K to less than 1 K. Recalibration is not usually needed. The resistivity of a semiconductor increases as the temperature is reduced. The semiconductor needs to be doped with suitable impurities to prevent the low temperature resistances from becoming very high. Self heating can sometimes be a problem and the power dissipation during measurement must therefore be kept low.

thermometer, hyperfine   This is a thermometer employed in measurements in the microdegree region by means of nuclear cooling, i.e., hyperfine enhanced nuclear cooling. This relies on the temperature-dependent paramagnetic susceptibility that results from nuclear spins. It involves adiabatic demagnetization to quench the hyperfine fields.

thermometer, Matsushita resistors   Type of thermomenter that employs the resistance property of metals. *See* thermometer, resistance.

thermometer, resistance   Pure metals have a linear resistivity-temperature relationship at room temperature, but the resistivity generally does not follow a simple relationship at low temperatures when it may also be a function of the magnetic field. The cryogenic behavior is a function of sample purity, since residual resistivity arises from impurity scattering of conduction electrons. Generally, the usefulness of metallic resistance thermometers below 10 K is limited. The resistance of certain alloys like constantan is nearly independent at high temperatures, but decreases with decreasing temperature below 5 K. These are used as resistance thermometers but their sensitivity is not high. Sensitivity can be increased by adding lead to bronze or brass.

thermometers, clinical   A clinical thermometer is a thermometer used to measure body temperature. Typically, it is made of glass in the shape of a tube of uniform bore with a narrowing above the mercury bulb in its lower part. The mercury is in a vacuum inside the tube, and a temperature scale is usually etched on its front. When measuring temperatures, the mercury expands or contracts, thus changing the height of the mercury column inside the tube. The narrowing above the mercury bulb permits the mercury column to remain in position when the instrument is removed from the body or positioned in any way other than vertical. *See* temperature measurements, clinical.

thermometers, semiconducting   These type of resistance thermometers maintain their high sensitivity down to the lowest attainable temperatures, where metallic elements become too insensitive although it is difficult to calibrate beyond the lowest point on the ^{3}He vapor pressure scale. *See* thermometer, resistance.

thermometers, vapor pressure   Provides quite accurate measurements of temperature over an extended range, however they usually are not commercially made devices. A thermostat is usually used to achieve and maintain steady temperatures throughout the desired range. The thermometers are based on the principle that the saturation vapor pressure of a pure substance is a monotic and sensitive function of temperature. The following components are needed: supply of working substance in highly pure form, a sensitive, accurate and wide-range pressure measuring device, and a cell to be connected to measuring and supply systems.

thermometry, gamma ray anisotropy   *See* thermometry, nuclear orientation.

thermometry, NMR   Thermometers that employ the principle of nuclear magnetic resonance (NMR) for increased sensitivity. They are suitable for temperature measurements in the mK range. The pulsed NMR thermometer has the advantage of being self-calibrating. It is subjected to an applied field for a short time. The decay of the induced voltage signal is proportional to the nuclear magnetization and hence inversely proportional to temperature. Very pure platinum is the material used in such thermometers.

thermometry, noise   This deals with the fact that a measurable but variable voltage arises in an n electrical conductor of resistance R as a result of random thermal excitation of the conduction electrons. In 1928, it was shown that the noise voltage detected across the resistor varies as the square root of the product of R and T, the thermodynamic temperature.

thermometry, nuclear orientation   Application of quantum level population differences in thermometry. It is applicable in very low temperature thermometry. The basic idea this operates on is that the spatial distribution of nuclear radiation arising in the decay of the 60-Co nuclei emitted from a radioactive source can be related to an identifiable laboratory axis, i.e., the symmetry of a single crystal of cobalt metal and to the temperature of the source atoms. It can obtain values of temperature below 2 K. Nuclear orientation thermometry depends on the anisotropic emission of gamma rays from polarized radioactive nuclei. The scale of the anisotropy in the intensity pattern is determined by the degree of polarization, which in turn depends on the absolute temperature. In practice the polarization is achieved by substituting the radioactive nuclei in a ferromagnetic host lattice so that the nuclear dipoles are aligned by the very large $\sim 10T$ internal field.

thermometry, osmotic pressure   This uses the principle of osmosis as a measure of temperature. In osmosis, certain molecules are able to be transmitted through a semi-permeable membrane connected to a tube. A hydrostatic pressure is therefore built up in the tube. This causes diffusion of water downward until equilibrium is reached. The pressure balancing osmosis obeys the ideal gas law and proportional to the temperature of the gas and is equal to the pressure the solute molecules would exert if they were a gas at that volume. Osmotic pressure can therefore be used as a measure of temperature.

thermophone   Device used for the production of sound by heating a thin conducting wire by an alternating current; the temperature changes cause changes in length. This sets the wire into resonant vibration, which produces sound.

thermopile   Thermocouples connected in series, where every alternate junction is exposed to radiant heat or brought into thermal contact with an object. This arrangement results in the adding together of the EMFs due to pairs of junctions, thus providing greater sensitivity than a single thermocouple to temperature measurement.

Thevenin equivalent   A theory suggested by Thevenin to use an equivalent circuit instead of a real circuit to simplify physical analysis.

Thomson coefficient   Used in measurement of Thomson heat which is evolved or absorbed when a current flows through a conductor across the ends of which a temperature difference is maintained. The rate at which Thomson heat is transferred into a small region of a wire carrying current I with temperature difference dT is equal $\sigma I dT$ where σ is the Thomson coefficient. The coefficient depends on the material of the wire and on the temperature of the small region under consideration.

Thomson effect   The EMF generated in a single electrical conductor by maintaining a thermal gradient in it; a heating and cooling effect in the conductor is then produced by current flow along the thermal gradient. This effect is closely related to the *Peltier* and *Seebeck effects*. *See also* Peltier effect; Seebeck effect.

three-phonon process   An Umklapp process dealing with three phonon scattering, it is defined as one in which the total crystal momentum is not conserved, a process more likely to oc-

cur at high temperature than at low temperature. At higher temperatures, the mean free path of the phonons is ultimately limited by interatomic spacing thus reducing the spread in thermal conductivity of different crystalline solids.

threshold of hearing   The amplitude of the weakest sound wave that can be detected by the ear varies with the frequency of the wave with the normal ear being most sensitive at 3500 Hz. The minimum audible amplitude is of the order of 10^{-9} cm. The lowest frequency that the normal ear can distinguish is about 30 Hz. The highest audible frequency diminishes with age from about 30,000 Hz, down to 10,000 Hz or even lower.

threshold voltage   A voltage at which an electronic device begins to conduct a current.

thyristor   A semiconductor switch device that changes the current direction in an electric circuit.

timbre   The subjective characteristics that make it possible for us to distinguish between two tones having the same intensity level and fundamental frequency but different wave forms; i.e., it expresses our ability to recognize sound of a violin as different from that of a trumpet even when the instruments are sounding the same note with equal loudness. It is primarily dependent on the waveform of the tone being heard and to a lesser extent on the intensity and frequency.

time base   A line produced by sweep circuit operation on a display screen.

time constant   The length of time required for the amplitude of an exponentially changing quantity (current or voltage) to change by 63.2%. For example, the decreasing current $i(t)$ through a series resistance, R, and capacitance, C, is given by

$$i(t) = \frac{V}{R}e^{-t/RC} ,$$

where V is the applied potential difference. The time constant is the value of time that reduces the exponential factor to e^{-1}; i.e., the time constant is RC. Similarly, the time constant for a series inductance, L, and resistance, R, is L/R.

timer circuits   Electric circuits that periodically supply time pulse.

tomography, emission   Any of several techniques for making detailed imaging of only a predetermined plane section of a solid object (planar imaging) while blurring out the images of other planes. The signal usually comes from high energy photons (X-rays, gamma rays). A computer is usually used to assist in forming a composite image.

The computerized axial tomography (CAT) is an example where a CAT scanner produces cross-sectional views of previously inaccessible internal body structure. Another example is neutron tomography where neutrons are used to visualize a sample and detect substances containing water, organic substances, plastics, and lubricants. *See* tomography, positron emission.

tomography, positron emission   Technique for measuring the concentrations of positron-emitting radioisotopes within the tissue of living patients. In *positron emission tomography* (PET), the distribution of positron-emitting radionuclides are imaged in the patient who has been administered with the tracer after the introduction of the compound usually either by injection or inhalation. Like other imaging techniques (CT, MRI, SPECT), PET relies on computerized reconstruction procedures to produce tomographic images.

Radionuclides used in PET include Carbon[11], Nitrogen[13], Oxygen[15], and Fluorine[18], with (relatively short) half-lives of 20 min, 10 min, 2 min, and 110 min, respectively. Because of their short half-lives, facilities equipped for PET that use these radionuclides are usually equipped with a particle accelerator (cyclotron).

In PET, the tracer decays by emitting a positron from the nucleus that, after combining with an outer electron, annihilate each other emitting two high-energy photons (gamma rays) at 180° from each other. The emitted radiation is then detected inside the detector as coming from the tracer. Both photons have to be detected simultaneously on opposite sides of the detector, otherwise the signal is discarded as coming

from the outside. The detection places the signal along the line connecting the two sections of the detector that recorded the signals.

PET is widely used in areas of neurological diseases, including cerebrovascular disease, epilepsy, and cerebral tumors. In general, PET has the capability to image parts of the body where abnormal biochemical changes are occurring. For that same reason, PET is useful in drug research, pharmacokinetics, and pharmacodynamics.

tone arm   The pivoted pickup bar of a record player with a head consisting of a needle set into a cartridge, that follows the grooves of the record, converting the oscillation into the electrical impulses.

tone control   Output at different frequencies can be controlled by variable amounts depending on whether the frequency is high or low. Tone control is easily obtainable for electric guitars by the use of capacitors and resistors in circuits; however for pianos, whether tone control is possible at all is still controversial.

tone deaf   The inability to express or to discriminate distinctions in musical pitch.

tones   Sounds that impress the ear with their individual character, especially pitch, quality of sound or timbre.

tones, combination   When two or more notes of different frequencies are played simultaneously, the sound produced is a composite of the tones. There is the *difference tone,* the *summation tones,* and other combinations too.

tone, warble   A continuous trilling tone as opposed to steady tones particularly used in alarms and sirens.

tooth rigidity   The nature of the rigidity of a tooth comes from its composition. The tooth is a bonelike calcified structure, composed of a core of soft pulp-like tissue containing blood vessels and nerves that is surrounded by a layer of hard dentin. The dentin is coated in turn with cementum or enamel at the crown (the visible part in the mouth). The enamel constitutes the hardest substance in the body.

The dentin in humans is composed of heterogenous material of a solid (circumpulpal) phase surrounding a network of tubules. These tubules, measuring about 1 to 3 μm in diameter, contain elongated cell bodies that radiate from the dental pulp organ throughout the entire dentin.

toroid   A coil of wire wrapped as a solenoid. The solenoid is curved so that the ends join together forming a donut or toroid shape. For a current-carrying toroid with a small width and a large radius, the magnetic field is uniform within the toroid and no magnetic field lines emerge from the toroid. *See* solenoid.

traffic   The message, signal, etc. that is transmitted through a communications system. It is the flow or volume of such information.

transducers   Electrical device that picks up sound vibrations, e.g., a microphone. A device that transforms a signal of one type into a signal of another type. An acoustic signal is a mechanical signal; on being incident on a microphone, it is transformed into an electrical signal.

transducers, medical applications   In general, a transducer is a device that converts a form of input energy into another form of output energy. The relationship between the input and output is usually fixed and well known.

Common examples of transducers are the ones that convert sound and other mechanical inputs into electrical signals, like microphones, cassette recorders, piezoelectric crystals, ultrasound; or light into electricity, like photoelectric cells; or take electricity as input and output in several forms, like loudspeakers, light bulbs, and solenoids. Medical applications of transducers include all sorts of general probes that convert temperature, sound waves, and mechanical force into electrical and electromagnetic output suitable for imaging, and surgical probes.

There are two classifications for transducers: those that require a source of energy in addition to the input signal (active), and those that do not (passive).

transfer function   The ratio of the measured output divided by the applied input of an electronic device or circuit. The transfer function, $\mathbf{H}(j\omega)$, is usually expressed as the ratio of the output to input voltage as a function of frequency, ω.

transformer   An electrical device for increasing or decreasing the voltage of an alternating current source by electromagnetic induction between two or more coils that are not connected electrically. The coils are usually arranged so that the magnetic flux associated with one winding also threads the others either by placing them in close proximity or by winding them on the same ferromagnetic core. The simplest transformer consists of two sets of windings: the primary and secondary windings. The primary is the winding that receives the AC voltage from the supply circuit, while an AC voltage is induced in the secondary by the primary.

The primary voltage, V_1, and secondary voltage, V_2, are related by the following:

$$\frac{V_1}{N_1} = \frac{V_2}{N_2},$$

where N_1 and N_2 are the number of turns in the primary and secondary, respectively. If there were no losses in the transformer, its power output would be the same as its power input so that

$$I_1 \times V_1 = I_2 \times V_2,$$

where I_1 and I_2 are the currents in the primary and secondary, respectively.

The secondary may have several electrical connections (known as taps) on different turns to enable the selection of different output voltages. Transformers are used in power supplies of many electronic devices that plug into the main power supply. They are also widely used in transmitting electrical energy over long distances by raising the secondary voltage to high values in order to reduce line losses.

transformer, acoustic   A device or material that couples two different media of differing characteristics, e.g., acoustic impedance, so as to allow continuity of transmission of acoustic waves across both medium.

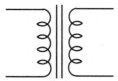

Circuit diagram symbol of a transformer.

transformer, auto   Transformer that consists of a single winding so that the primary is formed by the whole winding while the secondary is formed by a part of this winding that is tapped to give the desired voltage.

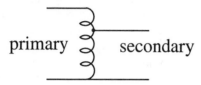

Auto-transformer.

This arrangement acts as a step-down transformer. Such transformers are used in starting devices for induction motors or as a means for varying the AC voltage to be applied to a device. *See also* Variac.

transformer, auto-   A transformer typified by having only one common coil onto which both the input and output terminals are connected. The output terminal has an adjustable brush which can slide along the common coil, thus adjusting the output voltage to values less than or greater than the input voltage.

transformer, current   (**1**) A device that enables the alternating current in a circuit to be measured. There are basically two types:

1. An *auto-transformer configuration* where the primary is connected in series with the circuit in which the current is to be measured, and the secondary is connected directly to the terminals of an ammeter. The current through the ammeter is proportional to, but much less than, the main current and the ammeter will not be subjected to the high potentials of the main circuit.

2. A toroidally wound coil, usually on a ferromagnetic core. The conductor carrying the current to be measured is passed through the center of the toroid, thus forming the primary winding and consequently, the toroidal wind-

ings form the secondary which steps-down the current in a known ratio. An ammeter can then be connected directly to the secondary.

(**2**) A transformer device commonly used to measure currents in a circuit. The primary coil is placed in series with the circuit in which current is to be measured, while the secondary coil, which has more turns than the primary, is connected to a current measuring device.

transformer, input   One used to match the impedance of an AC signal source to the input impedance of a circuit for maximum power transfer. They are also used to exclude DC voltage from the input of the circuit. *See also* transformer.

transformer, output   One that matches the output impedance of a circuit to the impedance of a load connected to the output. For example, the output impedance of an audio amplifier has to match the impedance of a speaker for maximum power transfer and therefore maximum sound output.

transformer, step-down   One in which the output voltage from the secondary is lower than the input voltage to the primary. This is achieved by having a secondary with a smaller number of turns than the primary. *See* transformer.

transformer, step-up   One in which the output voltage from the secondary is higher than the input voltage from the primary. This is the opposite to a step-down transformer and is achieved by having a secondary with a larger number of turns than the primary. *See* transformer.

transformer, voltage   One used to connect high tension lines to an instrument in order to measure voltage. The primary winding is connected in parallel to the main circuit and the secondary winding is connected to a suitable instrument such as a voltmeter.

transient, behavior   Opposite to steady behavior, an instant state.

transistor, *n-p-n*   A transistor consists of *n*-type, *p*-type, and *n*-type semiconductors that form a sandwich configuration.

transistor, *p-n-p* A transistor consists of *p*-type, *n*-type, and *p*-type semiconductors that form a sandwich configuration.

transistor, complementary Two transistors whose geometry is the same, but one is *n*-channel in which the electrons are charge carriers and the other is *p*-channel in which the holes are charge carriers.

transistor, drift A transistor in which the base region has a variable conductivity to reduce the carrier transition time.

transistor, field effect A transistor whose conductance in the current path is controlled by applying an electric field perpendicular to the current.

transistor, IMPATT A transistor that employs an IMPATT (impact avalanche transit time) and is used in high frequency regions.

transistor, junction An *n*-type and a *p*-type semiconductor contact each other and form a junction. A transistor contains two junctions.

transistor, mesa A transistor in which two terminals are positioned at the platform and the emitter is made by the vapor-deposition metal.

transistor, photo A transistor whose output is controlled by external light.

transistor, power A transistor that can be used in high electric power conditions. Its design is different from normal transistors.

transistor, surface barrier A transistor is a semiconductor device. The semiconductor contacts the other materials, which forms a contacting barrier between the surface of the semiconductor and the material.

transistor, unijunction A transistor that consists of a semiconductor bar with two ohmic contacts and a single, small area emitter *p-n* junction positioned between them.

transistor, unipolar A transistor whose current flow is due to one type of carrier, electron or hole.

transition, order-disorder Deals with the situation of spins in magnetic ordering in alloys. There is a disordered phase and a sharp transition temperature above which the ordering disappears. For example, in an alloy β-brass made up of copper and zinc, copper at a site corresponds to spin-up and zinc at a site to spin-down. *See* spontaneous ordering.

transit time A time for a carrier transporting from one terminal to the other.

transmission, analog Analog transmission in telephony is a method of conveying voice, data, image, or video information by a signal that varies continuously in amplitude or frequency with the information being transmitted.

transmission, asynchronous Asynchronous transmission is the transfer of data in which the intervals between the transmitted data packets are of unequal length. The transmission is usually controlled by start and stop signals.

transmission, biternary Biternary transmission is the transfer of two binary pulse trains over a single channel by combining the pulse trains and using a single communications channel in which the available bandwidth is sufficient for the transmission of only one of the two pulse trains at a time.

transmission, blind A form of data transfer that does not require an acknowledge signal from the receiver. Blind transmission may occur or be necessary when security constraints, such as radio silence, are imposed, when technical difficulties with a sender's receiver or a receiver's transmitter occur, or when lack of time precludes the delay caused by waiting for receipts.

transmission, bursty (1) The operation of a data network in which the data transmission is interrupted at intervals. It is a form of transmission that combines a very high data signaling rate with very short transmission times.

Burst transmission allows communication between data terminal equipment and a data network operating at dissimilar data signal rates.

(**2**) A burst in data communication is a sequence of signals, noise, or interference counted as a unit in accordance with some specific criterion or measure.

transmission coefficient, acoustic   This is the ratio of the transmitted sound energy to the incident flow of sound energy when a progressive plane wave in one medium impinges upon the boundary of a second medium and is transmitted. The sound transmission coefficient is independent of the direction of the wave motion; i.e., it is the same from water into air as vice versa.

transmission control protocol (TCP)   The protocol used in the Internet to provide connections. Both ends of a TCP connection can simultaneously read and write packets. The source adjusts its transmission rate to the rate currently supportable in the network. TCP uses timeouts and retransmission to ensure that a destination receives a transmitted packet.

transmission, diversity   A diversity transmission system is a communication system that has two or more signal paths or channels. The outputs of these channels are combined to give a single received signal and thus reduce the effects of fading.

transmission, full-carrier   Full-carrier transmission is a telecommunication system that amplitude modulates a carrier signal and transmits the modulated signal along with the carrier. The other approach, called *carrier-suppression,* does not transmit the carrier signal.

transmission, intercellular   (**1**) Transmission of a signal from one cell to another. An example is in the transmission of an action potential from one cell to another either through synapse or through gap junctions.   In the synapse, the signal is propagated by the release of chemicals (neurotransmitters) from the signaling cell to the other end of the synapse, where the signal is received. In a gap junction the signal is transmitted by an ionic current flow that

flows from one cell's intracellular space directly to the other cell via ion channels that connect both cells directly.

(**2**) Transmission or transport of molecules across a tissue boundary, where the molecules are transported through a passage between cells (paracellular transport). *See* transport, transcellular.

transmission line   Generally any conductor used to transmit electric or electromagnetic energy. In particular it refers to:

1. the power lines that carry electrical power to residential and industrial areas,

2. the electrical cables and waveguides used in telecommunication to transmit electrical signals,

3. the cable that connects an arial to a transmitter or receiver. Some of the common types used are coaxial cable and two-wire line.

All transmission lines can be described by a network of discrete parameters such as inductors, capacitors, and resistors, distributed uniformly along its length. The transmission line is referred to as being balanced if the conductors have identical properties (e.g., resistance and impedance).

transmission loss   The decrease of power in the signal being transmitted from one point to another of a telecommunication system. It is given by the ratio P_1/P_2 where P_1 is the measured power closer to the signal source than the measured power P_2. This is usually expressed in decibels or nepers.

transmission, multipath   Multipath transmission is when radio signals reach the receiving station by more than one path. Some causes for the signals propagating over more than one path are atmospheric ducting, ionospheric reflection and refraction, and reflection from terrestrial objects, such as mountains and buildings. Multipath transmission causes constructive and destructive interference, and phase shifting of the signal.

transmission, serial   Serial transmission is a communication system in which the bits in a word are transmitted sequentially along a single line. This is in contrast to parallel transmission in which the bits in a word are transmitted at the same time over several lines.

transmission, synchronous   Synchronous transmission is a form of digital transmission in which the time interval between any two similar significant instants in the overall bit stream is always an integral number of unit intervals.

transmittance   We consider a beam of light incident at angle θ_i to the normal of the boundary between two media of different refractive indices. The radiant flux density or irradiance of the incident beam is $I = \frac{CE_o}{2}E_o^2 w/m^2$. This beam will be partially reflected at angle θ_i and partially transmitted at angle θ_t.

The transmittance of the interface is

$$T = \frac{I_t \cos\theta_t}{I_i \cos\theta_i} = \frac{n_t \cos\theta_t}{n_i \cos\theta_i}\left(\frac{E_{ot}}{E_{oi}}\right)^2 .$$

Similarly, the reflectance is

$$T = \frac{I_r \cos\theta_r}{I_i \cos\theta_i} = \frac{I_r}{I_i} = \left(\frac{E_{or}}{E_{oi}}\right)^2 .$$

Application of the principle of conservation of energy yields the important result

$$R + T = 1 .$$

Transmittance varies from zero (no transmission) to one (transparent interface). This parameter is of fundamental importance in describing the performance of optical components and devices.

transmitter   A device, circuit or apparatus used in a telecommunication system to generate and transmit an electrical signal to the receiving part of the system.

transport, coupled solute and solvent (cell)
In transport of molecules across the cell membrane sometimes the problem of forcing a molecule (substrate) opposite to its concentration gradient is solved by coupling the movement to the downward flow of another substrate. In this way, the favorable energy balance that comes from the diffusion of the substrate down its concentration gradient is used to drive the other in the energy-absorbing motion from low to high concentration. This transport is sometimes called "secondary" active transport because the energy used to achieve it does not come directly from the energy released from cell metabolism.

Two kinds of coupling can occur in this transport, one in which the two crossing species cross the membrane in different directions (antiport), and the other in which they cross it in the same direction (symport). Because osmotic pressure would drive a solvent to cross in favor of its concentration gradient, coupled solute and solvent would occur in symport, where both species cross in the same direction. In both types of the coupled transport the two species of molecules have to be available either simultaneously or sequentially on the corresponding sides from which they are transported.

Many essential nutrients are transported by symport systems coupled to Na^+ or proton gradients. The system of uptake of neurotransmitters is a similarly coupled symporter process.

transport, electrical (cell)   Refers to the ability of cells to transport electricity along their cell membrane. An example is nerve cells when transporting an electrical signal across the nervous system. This is accomplished by sequential opening and closing of ion channels along the membrane that permit the propagation of the non-equilibrium membrane potential (action potential) for long distances. The strength of the electrical signal is proportional to the permeability of the membrane to the ions that cross, and this is dependent on the action of the ion channels across the membranes.

Another example of electrical transport is seen in the heart, where an electrical transport cycle is established between the Sinoatrial node, the Atrioventricular node, the Bundle of His, and the Purkinjie fibers that are responsible for keeping the heart beating in rhythm. These cells work like conductive wiring in the heart.

transport, ion current (cell)   Changes in the cell membrane polarization may cause flow of ions across the membrane. This is because each ion has an "equilibrium potential" that keeps both sides of the concentration of the ion stable and without flowing; once perturbed, an ion current is established that will stop when the equilibrium potential for that particular ion is obtained.

Ion current across the cell membrane is conducted via "ion channels" or ion pumps (*see* osmotic equilibrium (cell)). Ion pumps require the expenditure of energy in their functioning and form the basis for active transport that regulates the osmotic balance in the cell. *See* Nernst equilibrium potential; nerve impulses, propagation of; potential, membrane.

transport, transcellular   Transport of molecules from one region to another separated by cells. The transport is carried out by transporting the molecules across the plasma membrane of the blocking cells.

An example of this transport is the transport of molecules across the intestinal epithelium. Transport of molecules via a transcellular route happens when they cross the plasma membrane of the epithelial cells. If the transport occurs across the junctions between epithelial cells, it is called *paracellular transport.*

In the intestinal epithelium, water, for example, can be transported both trans- and paracellularly. Large organic molecules (e.g., amino acids and glucose) cannot pass in between cells, so they are transported via the transcellular route with the help of transporter molecules. *See* transport, uncoupled.

transport, uncoupled   Transport that occurs when particular molecules or ions are transported selectively through membranes without any coupling to any other substrate. It is also called *facilitative transport.*

Molecules that fall into this category are many water-soluble molecules, like sugars and amino acids, that cannot penetrate the membrane because they are too large to fit through open channels. Included in this group are also some ions that do not diffuse through channels.

The molecules and ions that undergo uncoupled transport penetrate or leave the cell through the action of membrane transporters. These transmembrane transporters offer a highly specific binding site to particular molecules that, once attached to the binding site, the molecule is transported either out or into the cell. The process by which the molecule crosses is not fully understood, but it is known that the transporters do not offer just a hole for the molecule to go through, like typical channels do.

Because the molecules that get transported move down their concentration gradient, uncoupled transport is considered a type of diffusion. Two examples of transporters are glucose and the bicarbonate ion.

transverse waves   The direction of propagation of the wave is at right angles to the direction of propagation of the particles. This type of wave can be propagated on a stretched string.

traveling wave   A wave pulse will move transporting energy as it does by the vibration of neighboring particles in the media. The medium itself does not move, but energy is therefore transported in this manner.

traveling wave tube   A tube in which interaction of electrons produces a wave.

triad   A chord of three tones, one consisting of a given tone with its major or minor augmented or diminished.

triboluminescence   Phenomena involving luminescence when intense ultrasonic waves exist in materials; e.g., discharge occurs in crystal planes separated by a small distance on which a high voltage gradient exists. *See also* sonoluminiscence.

trichromatic coefficient   Gives a quantitative assessment of color in terms of three standard primary colors: red, green and blue violet for example. In the analysis of an unknown color, a spectrometer is used to measure the spectral reflectance (or transmittance if appropriate) of the three components above, and tristimulus values $\bar{x}$, $\bar{y}$ and $\bar{z}$ are determined at the given wavelengths. The trichromatic coefficients are then given by $x = \bar{x}/(\bar{x} + \bar{y} + \bar{z})$, $y = \bar{y}/(\bar{x} + \bar{y} + \bar{z})$, and $z = \bar{z}/(\bar{x} + \bar{y} + \bar{z})$. *See also* color match.

triggering   Releasing, emitting, and creating by an external pulse.

triggering, ramp   An electric circuit that generates a wave when receiving an external pulse.

triplet   A spectral line that can be split into three component lines upon removal of degeneracy by an appropriate applied field.

Trouton's rule   Involves the heat of vaporization at the normal boiling point for liquids. The ratio giving entropy change due to vaporization at the normal boiling point is not constant but increases with temperature. A rough approximation is about 9, which is useful when the critical temperature is not known.

trumpets, sound from   The blown frequencies with the valves open are in general multiples of 115 Hz with the fundamental frequency of 115 Hz missing. It has a frequency range from about 200 Hz to about 1000 Hz, significantly greater than other orchestral brass instruments.

trunk call   A telephone call on a trunk line in which the charges are calculated according to the distance of the call.

trunk, line   A transmission line that is used to interconnect two electric power stations or two electric power distribution networks. It is considered a main telephone line in the system.

truth table   A table that lists the value of one or zero for each input and output terminals.

tune   Succession of notes or chords forming the characteristic music of a song or other piece.

tuned circuit   The circuit has been adjusted to produce a resonant wave.

tuning forks   Instruments of great purity of tone and constancy of frequency. It is used as a means of indicating and preserving standard pitches.

tunnel diode   A diode in which carriers pass through a sharp barrier by a quantum effect.

turbulence, acoustic   Sound levels of sound waves are strongly affected by random and turbulent fluctuations in wind and temperature in particular in shadow zones. It can also cause the direction of a source to be difficult to identify, and a degradation in signal coherence.

turgor pressure   Turgor pressure gives the normal fullness or tension found in animal tissue and plants. The turgor pressure originates from the pressure that comes from the internal fluids of cells from plants and animals, and the fluid content of blood vessels and capillaries in animals, for example.

turmalin   A mineral with electric properties that is also used as a gem.

turns ratio   The number of active turns in the secondary windings of a transformer divided by the number of turns in the primary windings. *See also* transformer.

tweeter   A small loudspeaker that reproduces high frequency sounds in high fidelity audio equipment for high frequencies 3000 to 20,000 Hz.

twin cable   A transmission line that has two parallel conductors separated by insulating material. The line impedance of a twin cable is determined by the diameter and spacing of the conductors.

twinning, crystal   The mode of plastic deformation of a crystal (particularly hexagonal closed pack or body centered cubic crystals) resulting in a partial displacement successively on each of many neighboring crystallographic planes such that the deformed part of the crystal is a mirror image of the undeformed part.

U

ultra high frequencies (UHF) The short wave band containing radio frequencies from about 300 MHz to 3000 MHz. This band is mainly used for the transmission of television signals, radar, and airplane navigation.

ultrasonics Inaudible sound waves in the range greater than 20,000 Hz. The ability to focus and direct these waves in beams of vibration at high powers in a small area makes them very suitable for applications such as the production of heating effects, destruction of bacteria, underwater signaling, depth sounding, testing of materials, and seismic exploration.

ultrasound therapy Ultrasound therapy involves the use of sound waves in healing soft tissue ailments and pain, and speeding up recovery. The apparatus consists of a soundhead within which a crystal vibrates generally in the frequency range of 1 MHz to 3 MHz.

In ultrasound therapy, it is important to transmit the signal with the least loss from the soundhead to the treated tissue. Because air is not a good sound conductor, a gel or lotion is usually put between the soundhead and the skin. In this way the gel acts as a "coupling" between the two parts hampering air from getting in between. Underwater therapy is also another option to couple the soundhead and the tissue.

Ultrasound has been used in the treatment of arteries clogged by cholesterol plaque and blood clots. Application of the focused sound to the clot in some cases has led to the reduction of the blocking material, thus proving beneficial in the prevention of heart attack. Also, it has been used in the treatment of breast cancer where the ultrasound is used to produce localized heating.

Umklapp process This is the process by which thermal resistance occurs in non-conducting materials. It means "flop over" in German and refers to the interaction of three or more lattice or electron waves in a solid. The sum of the wave vectors is equal to a vector in the reciprocal lattice. *See* resistance, Umklapp.

unary operation An operation that contains one variable.

undercooling The most important factor for the survival of biological cells during cooling is the rate of cooling. This is of particular interest in areas where cells in suspensions are exposed to low temperatures.

If the cells are cooled too slowly, osmotic effects may cause harm to the cell since the osmotic equilibrium between intracellular and extracellular fluid is temperature dependent. If the cooling rate is too fast, formation of ice crystals in the intracellular space may harm the internal structure.

Optimal values for the rate of cooling yielding a high survival curve are related to the hydraulic membrane permeability that is the limiting factor in the cell volume shrinkage.

underwater acoustics Of use primarily to nautical and naval personnel in detecting other ships. *See also* ultrasonics.

unidirectional current A direct current flowing in one direction only is called a *unidirectional current.*

unison Two notes estimated by ear to have the same pitch, possessing the same frequency. The harmonic frequencies of each tone match the other exactly; the sound is pleasant to the ear and is considered a consonant.

unit gain buffer A circuit whose output signal is the same as the input signal and the output resistor is very large.

unit gain op amp An operation amplifier in which the gain is one.

unit planes/points The conjugate planes for an optical system for which the transverse magnification is +1.

unit pole A magnetic pole whose strength is such that when it is exposed to a magnetic field of 1 Oe a force of 1 dyne is exerted on it.

In nature it is impossible to obtain an isolated pole (magnetic monopole) but in practice, if a bar magnet is made very long, a north or a south pole may be approximately isolated to perform this measurement. *See also* magnetic poles.

unstable   A state that is changed with a slight variation of outside conditions.

up-converter   A transmitter that changes a low state to a high state.

up-down waveform generators   An electric device that produces a step wave.

V

van't Hoff's law   This refers either to the law relating to chemical kinetics or the law governing the osmotic pressure of solutions.

Regarding osmotic pressure, *van't Hoff's law* dictates the fundamental law that relates the osmotic pressure in dilute solutions to other parameters as

$$PV = iRT ,$$

where P is the osmotic pressure, V is the volume, T the absolute temperature, R the universal gas constant, and i a measure of the "abnormality" of the substance. Large values of i are due to the dissociation of the dissolved substance into ions.

In chemical kinetics, *van't Hoff's law* relates the equilibrium constant of the reaction to the particular concentration of the reacting species. *See* osmometer, and law of mass action.

varactor   A device whose reactance can be changed by external bias voltage.

variable-length code   A variable-length code represents an efficient source coding method when the source symbols are not equally probable. A variable-length code allows more frequently occurring data symbols to be represented by shorter code words and more rarely occurring data symbols to be represented by longer code words, thereby increasing information transmission rate. A variable-length code maps k input symbols into n output symbols, where either k or n (or both k and n) may vary, depending on the value of the particular symbol in question. The Huffman code and the Morse code exemplify variable-length coding.

Variac   The tradename of a variable autotransformer with windings on a toroidal core. The output voltage is varied by a rotating brush contact on the windings. *See also* transformer, auto.

varistor   A non-linearly variable resistor.

vector potential   A vector quantity, $\mathbf{A}$, used in electromagnetic field theory to deduce the magnetic induction, $\mathbf{B}$, at position (x, y, z) and time t by applying the following:

$$\mathbf{B} = \nabla \times \mathbf{A} .$$

The vector potential, $\mathbf{A}$, is expressed in webers/meter. This is in contrast to the potential V used to determine the electric field strength $\mathbf{E}$ by

$$\mathbf{E} = -\nabla V ,$$

where V is a scalar potential.

vector potential, magnetic   A potential that can be used to describe the magnetic field. The magnetic field $\mathbf{B}$ may be determined from the vector potential $\mathbf{A}$ by

$$\mathbf{B} = \text{ curl } \mathbf{A} .$$

velocity of sound, Laplace equation   This varies in different media, and is 330 m/s in air. Pressure, temperature, density and humidity can affect the velocity of sound in a gas. Laplace introduced a correction to the original formula for velocity of sound V, in a medium given by $\sqrt{E/\rho}$, where E is modulus of elasticity and ρ the density of the medium. The correction involved recognized that the compression and rarefaction took place so rapidly that the gas did not have sufficient time to lose or take heat from surrounding air; i.e., it took place adiabatically rather than isothermally as thought earlier.

velocity selector   A device that selects particles of a certain velocity from a stream of otherwise identical particles (same mass and charge). Operates on the principle that, in an applied magnetic field, the magnetic deflection force on a moving particle depends on the particle's velocity. The principle of operation is similar to that of a mass spectrometer. *See* magnetic force on moving charge, mass spectrometer.

ventriculography, radionuclide   A technique using an intravascular radioactive tracer that gives images of the heart ventricles that ultimately are interpreted for examination of the structure and functioning of the ventricles. Data from several hundred cardiac cycles may be collected to present a single composite cardiac cycle.

When compared with other competing techniques:

1. It assesses the functioning of the ventricles (systolic and diastolic functions) in a more reliable and precise way, such as with echocardiograms.

2. It gives information about regional and global wall motion.

3. It gives the heart chamber size and morphology.

4. It gives left and right ventricular ejection fractions (proportion of blood present in the left ventricle at the end of ventricular diastole that is pumped through the aortic valve during systole).

On the down side, besides being more costly, it requires the puncture of veins and radiation exposure, and it yields less accurate left ventricle data, as with echocardiograms.

ventriloquism   The act of speaking or uttering sounds with barely visible lip movement.

vertical scanning   Movements of the scanning beam with the change of a vertical variable.

very high frequencies (VHF)   The band of radio signals in the frequency range from 30 MHz to 300 MHz. This band is used for FM and amateur radio broadcasting as well as for television transmission.

very low frequencies (VLF)   The band of radio frequencies lower than 30 KHz, sometimes employed in underwater communication between submarines.

vibration of magnet   Refers to the oscillation of a bar magnet when suspended in a magnetic field. The frequency of oscillation f is given by $f = (1/2\pi)(MH/I)^{1/2}$, where M is the magnetic moment of the bar magnet, H is the magnetic field intensity, and I is the moment of inertia of the bar magnet. The measurement of f allows determination of the magnetic moment M of the bar magnet.

vibrations   The to and fro motion of some particles, either freely or acted upon by external periodic forces and friction in a medium. When a vibrating body is immersed in a solid, liquid or gas, sound waves are set up since the vibratory displacements are in the same direction as the propagation of the wave.

vibrators, bone-conduction   Conventional hearing aids make use of amplification of the sound waves and require that the ear still has some perception (to some degree) of the air conduction hearing mechanisms. There are people, however, for whom this poses a problem, and other methods have to be used.

By attaching a special vibrator directly to the cranial bones, sound can be transferred directly to the inner ear, bypassing the air conduction parts (ear canal, ear drum, and ossicles) of the middle ear. This allows pure tones to still be transmitted by bone conduction, as in air conduction hearing. Depending on the apparatus used, the attachment is generally performed as a minor surgical procedure.

vibrometer   Instruments used for monitoring vibrations in systems often using optical techniques.

video frequencies   Frequencies in the band of GHz that allow for the operation of equipment such as video systems, transmitters, receivers, antennas, and power supplies.

videophone   A telephone device that transmits a visual image and sound. The receiver can receive and display visible images simultaneously with the telephone signals.

vidicon   A camera tube that shows an image after the emitting electron beam hits the inside surface of the tube.

viewing distance   The distance between the film or plate and the second nodal point (the point where the extended outgoing ray meets the optical axis). For distant objects, it is simply the focal length of the camera lens, which is usually less than the least distance of vision. Therefore, to get proper perspective, one can use a convex lens with the same focal length as the camera lens and well-accommodated eye. Alternately, one can enlarge the photograph until the viewing distance is equal to or greater than the least distance of distinct vision. In the case of a tele-photo lens, to get proper perspective, one has to use a viewing distance equal to the focal length, which is quite large, thus losing the advantage of the large image size.

vignetting The practice of cutting off or ob-structing the rays in the outer fringes of the image-forming pencil, thus avoiding the asso-ciated aberrations and improving the quality of the image. In a camera, the finite size of the photographic plate effectively performs the vi-gnetting of the image. The term is also used to describe the reduction in the effective beam area with increasing obliquity due to mechan-ical obstructions in the optical system such as apertures, lens holders, etc.

violins, sound from The origin of the sound from the vibrating strings. The vibrations of the string are transmitted to the instrument via the bridge; the sound post and the entire body of the violin is involved in the sound production process.

virtual spaces The virtual extension of the (real) space on the left side of a refracting sur-face to the right of the surface, and vice versa. In detail, a refracting surface can be considered to divide the space into two parts, known as the left- and right-hand spaces. Both spaces can be conceptually extended to infinity in both direc-tions. The left-hand space is considered real on the left side and virtual on the right-hand side of the refracting surface. Similarly, the right-hand space is considered to be real on the right-hand side and virtual on the left-hand side of the re-fracting surface.

viscoelasticity, surface A viscous substance is such that when stress is applied the substance will flow continuously opposing the stress with a constant opposing force. The flow is the result of the movement of molecules past one another, and the resistance is due to intermolecular fric-tion. The deformation does not involve chem-ical bond stretching or bond rotations, and ide-ally all of the mechanical energy is dissipated as heat. At the moment that the stress is removed, the substance will remain in the deformed state because there has been no internal energy stored for the substance to return to its initial (or any other) state. In this sense, viscous deformation is irreversible.

Elasticity describes a substance that, upon deformation, returns quickly and completely to its initial conformation. Contrary to viscosity, the deformation is entirely by chemical bond stretching and rotating. The deformation is re-versible.

Viscoelasticity is a fundamental property of polymers, in which both the viscous and the elas-tic properties are exhibited in some degree that depends on the chemical composition, temper-ature, and time and strength over which the de-formation stress exists. Viscoelastic response is found typically when a polymer undergoes deformation and, upon release, reverses the de-formation in a process that requires substantial periods of time. The same polymer, under dif-ferent conditions of stress, may exhibit either viscous or elastic behavior. For a typical poly-mer, at higher temperatures, the response shifts from elastic towards viscous behavior. At lower temperatures, the response is reversed, resulting in elastic rather than viscous behavior.

Polymers deposited on surfaces or the surface of bulk polymers may exhibit a slightly modi-fied behavior than in bulk since the layers at the surface only interact with the bulk at one side; thus the elastic behavior could be modified.

viscoelastometer A *viscoelastometer* is an instrument for the measurement of viscoelastic properties. The materials under study are usu-ally polymers, prototypes of viscoelastic sub-stances. Results from trying to determine the material's deformational response to mechani-cal force and flow under stress, give information regarding the rheology of these materials.

Techniques vary, but the general method is to measure the resilience (elasticity) of the material and flow properties when subjected to mechan-ical stress. This has application, for example, in measuring the change in flow of a resin as a function of time during the hardening process. *See* viscoelasticity, surface.

vision, binocular Vision in which both eyes are used to give rise to a single, fused precept. This occurs as a coordinated function of both eyes. The fusion of two distinct images from the two eyes by the brain into a single image whereby one can judge depth or portion of objects accurately in a three-dimensional field can be defined as binocular vision. To have good binocular vision without diplopia (double vision), the images of the object must fall on what are called corresponding points on the two retinas of the two eyes.

vision, color   Vision in which color sense is present. *See* vision, photopic.

vision, defects   Any ametropia of the eye (i.e., myopia, hyperopia, hypermetropia).

vision, photopic   Vision due to cone photoreceptor function. This is normally at high levels of illumination (>10 candelas/meter^2) and is characterized by the ability to discriminate colors and small details. Color is perceived because of the trichromatic nature of the three cone types.

vision, scotopic   Vision due to rod photoreceptor function. Rods are active at low levels of illumination (<0.001 candelas/meter^2). This is characterized by a lack of ability to discriminate colors and small details. Scotopic vision (night vision) is characterized by high sensitivity at low light levels and for detection of movement.

vocoder   A synthesizer that produces sounds from an analysis of speech input. The word *vocoder* comes from the combination of the words *voice* and *code*.

voice coil   That part of a loudspeaker that connects to the vibrating diaphragm. It is capable of moving to and fro in a radial magnetic field whose direction is perpendicular to the coil winding. The driving force applied to the diaphragm is directly proportional to the current flowing through the driving coil.

voice, human   The sounds of the human voice are produced when a current of air from the lungs is forced through the *glottis* or narrow slit between the vocal chords. These two membranous reeds are situated just above the junction of the windpipe with the larynx and are coupled to a series of air cavities formed by the larynx, the front and back parts of the mouth, and the nose and its associated cavities.

Voigt effect   Discovered by Voigt in 1902. When a strong magnetic field is applied to a vapor through which light is passing, the medium becomes birefringent; i.e., double refraction takes place. The effect is analogous to the Faraday effect of optical activity induced by a magnetic field.

If a vapor has a response frequency v_o, suppose that a normal Zeeman triplet is formed by application of a magnetic field:

$$v = v_o \pm \Delta v .$$

When white light is incident that spectral range at v_o will be absorbed and components at $\pm \Delta v$ will have 50% absorption compared to the central peak. When plane polarized light is incident, it will be split into components parallel and perpendicular to the magnetic field, which have different refractive indices. Induced birefringence and a change to elliptical polarization result.

volt   Symbol: V. The SI unit of electric potential, potential difference, and electromotive force. One volt potential difference is defined as the ratio of 1 watt of power dissipated by 1 ampere current between two points in a circuit. Alternatively, 1 volt is 1 joule of energy required to transfer 1 coulomb from one point in a circuit to another. An electric potential of 1 volt at a point is 1 joule of energy used to transfer 1 coulomb from infinity to that point. The standard for potential difference is obtained from a special form of electrolytic cell, e.g., a Weston standard cell. The volt is named after count Alessandro Volta (1745–1827) who developed the first rudimentary battery. *See also* potential, electric; potential difference.

voltage   The value of potential difference between two points and electromotive force. *See also* reactive voltage; volt; potential difference; potential, electric.

voltage clamp, ionic current in cell   In the voltage clamp technique it is possible to control the voltage potential across a cell membrane while measuring the current that is directly related to the ionic movement across the membrane. When the voltage clamp establishes a potential across the membrane, the induced current will be a result of the ionic current as well as the capacitive contribution, given by

$$I_c = C(dv/dt) \, .$$

Once the charge distribution on each side of the membrane has been established, the capacitive contribution goes to zero, and for long periods the current will be the result only of the ionic movement across the membrane.

If the clamped voltage is that of a particular equilibrium potential of a particular ion, then the current will be the result of the flow of the other ions in solution. *See* potential, resting.

voltage drop   The decrease in potential along a conductor or across the terminals of a resistive electrical component as a result of the flow of current through them. It is given by the potential difference between two points of the conductor or the two sides of the component. *See* potential difference.

voltmeter   A device that measures potential difference. Generally it can be used to measure both AC and DC voltage. The input impedance of a voltmeter is very high so that minimal current is drawn from the circuit being measured. *Moving coil analog voltmeter, digital voltmeter,* and *cathode ray oscilloscopes* are some commonly used devices for measuring voltages.

vortex sound   If a steady flow of fluid passes an obstacle, eddies are usually formed behind the obstacle. At the edges of the stream, swirling whirlpools are formed. The rotational motion of the vortices causes the superposition of many frequencies and sound is heard. Examples are a person whistling, or sound of wind through a crack.

vowel sound   Characteristic frequencies as a result of the natural vibrations of the oral cavities excited impulsively by the more or less periodic puffs of air from the glottis.

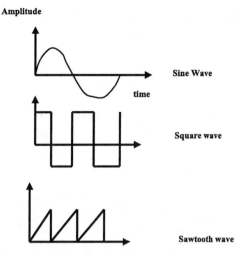

Sine wave, square wave, and sawtooth wave.

water, undercooled   Also known as *sub-cooled water,* this refers to water that continues to exist in a liquid state at a temperature below the freezing point of 0° C. It is in a metastable phase. *See* supercooling.

water, unfreezable   *See* water, undercooled; supercooling.

watt   Symbol: W. The unit of power. It is defined as the dissipation of one joule of energy in one second. The power, P, dissipated by a resistor is given by

$$P = IV ,$$

where I is the current through the resistor in amperes and V is the potential difference across it in volts. *See also* Ohm's law.

wattmeter   An instrument that indicates the instantaneous value of the power expended in a circuit to which it is connected. Generally, it can be used in both AC and DC. Its operating principle is based on the product of current through a circuit with the potential difference across it. Consequently, there are four terminals on the instrument, i.e., two for the current to be connected in series with the circuit, and two for the voltage to be connected in parallel.

wave analyzer   Spectrometers that display simultaneously the frequency and amplitude of the more important component of complex sound.

waveform   Refers to the different types of waves that can be produced, e.g., the sine wave, square wave, and sawtooth waveforms.

wavefront   For a three-dimensional wave pulse, the surface in its path of propagation all of whose points are in the same phase of motion.

wave front or wave surface   This is the locus of points of equal phase of a wave. For a harmonic wave

$$\psi \left(\vec{r} , t \right) = A \left(\vec{r} \right) e^{i (\vec{k} , \vec{r} - ox)} ,$$

where $\vec{k}$ is the propagation vector.

The equation of the wave front is

$$\vec{k} \bullet \vec{r} = \text{constant} .$$

If the amplitude $A(\vec{r})$ is not constant over the wave front, the wave is called *inhomogenous.*

The concept of wave front is important for wave propagation in general in that it allows definition of the phase velocity $v_p \equiv \omega / |\vec{k}|$. It is the basis for the description of all wave phenomena.

wave function   Of a system of particles, this is the solution to the Schrodinger equation. It contains all of the dynamic information of the system. Its main physical interpretation is as a probability density amplitude used to determine the spatial probability function for the particles. From the wave function and the Schrodinger equation one can determine the quantum numbers and energy levels of the system — of prime importance in statistical physics.

waveguide   A tube along which electromagnetic waves are transported.

wavelength   For a harmonic plane wave propagating in a direction of unit vector $\vec{s}$.

$$\psi\left(\vec{r},t\right) = \left(A/\vec{r}\right)\exp i\left[\omega\left(t - \frac{\vec{r}\cdot\vec{s}}{v}\right)\right].$$

The wavelength is the spatial period of the motion such that $\psi(\vec{r},t)$ is the same when $\vec{r}\bullet\vec{s}$ is replaced by $\vec{r}\bullet\vec{s}+\lambda$.

This gives $\lambda = v_p\frac{2\pi}{\omega}$.

wavelets, secondary   The laws of propagation of light waves are based on *Huygen's principle*, which states that every point on a wave front acts as a source of secondary wavelets that have a velocity and frequency corresponding to the primary wave. The envelope of the secondary wavelets corresponds to the new primary wavefront a short time later. This intuitive notion was later put on a firm theoretical basis by Fresnel and Kirchoff.

wave number   This is defined as $2\pi/\lambda$ where λ is the wavelength of the wave in question. It is a convenient way of expressing the wavelength in the wave equation.

wave propagation   A wave is transmitted through a medium by the vibrations of particles. It moves forward by the vibrations being transmitted to adjacent particles. Regions of compressions and rarefactions are set up as it propagates. *See also* traveling wave.

wave pulse   This can be produced in a stretched string by giving it a sideways tug. This will produce a pulse that will travel down the string, each particle in string remaining at rest until the pulse reaches it; then it moves for short time and then returns to rest. *See also* wave train.

waves, intensity from point source   In a three-dimensional wave, such as a sound wave, spherical waves travel outward from the point source, and energy may be absorbed as they travel through space or a medium. The intensity of the space wave is defined as the power transmitted across a unit area normal to the direction in which the wave is traveling and is proportional to the square of the amplitude.

wave speed   Mechanical waves need to travel in a medium and the properties of the medium, such as inertia and elasticity, determine the speed of the wave through it. The speed V is also related to the wavelength λ and frequency f by the following relation.

$$V = f\lambda.$$

wavetrain   A wavetrain is a monochromatic wave of finite length that contains a certain number of cycles. Light emitted from atoms is emitted as wavetrains approximately 10^{-8} to 10^{-10} secs, as a series of random bursts. The width of the wavetrains is a measure of the coherence time; from this, one can define a coherence length over which the phase is well defined. Several wave pulses in succession will produce a train of waves.

weber   International System unit of magnetic flux, equal to Tesla.meter^2. One weber is defined as the flux linking a one turn circuit that, when reduced uniformly to zero in one second, induces an electromotive potential of one volt in the circuit. Named after German physicist Wilhelm E. Weber (1804–1891). Equal to 10^8 maxwells (or gauss.cm^2) in CGS units.

Wheatstone bridge   A very useful electric circuit developed and advocated by Charles Wheatstone in 1843. It is widely used to determine the unknown resistance of a resistor. It consists of two known resistors, a variable resistor, an unknown resistor R_x, as in the following figure. A voltage source is connected to points A and B. Adjust the variable resistor until the current i of the galvanometer is zero. Under this balance condition, the unknown resistance is given by

$$R_x = \frac{R_1}{R_2}R_v.$$

whistle   To make a clear musical sound by the expulsion of breath; an instrument for producing a whistling sound. Sounds may be generated by passing a gas or liquid through an orifice or over an edge. The passage generates vortices, spaced periodically, which propagate as a sound wave.

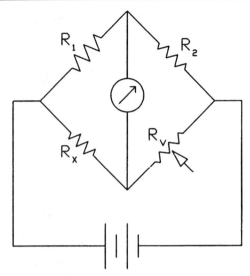

Wheatstone bridge.

whistle, Galton One of the first type of whistles, the Galton whistle consists of a jet that sends out a stream of gas against a small cavity. A miniature organ pipe is used to determine the upper limit of audibility, generally for control and signaling purposes. The small, adjustable closed pipe is about 1 in. in diameter; the resonant length of the pipe is altered by means of a piston controlled by a screw.

whistler waves A type of wave that seeks application in many branches of physics, e.g., plasma physics, the study of magnetic fields, and astrophysics.

whole tone The interval on the musical scale between any two notes such as *C* and *D*.

Wiedemann-Franz law This gives the relationship between the electrical and the thermal conduction in a metal, since the flow of electrons is responsible for both these quantities. According to the law, for all metals at temperatures not too far below 300 K, the ratio of thermal to electrical conductivity is approximately constant and, when divided by the temperature, is known as the *Lorenz number.* However, the law is not obeyed at all temperatures, and *L* eventually starts to fall as the temperature is decreased; at the lowest temperatures, it starts to rise again to its original value. This implies that the mean free path of the electrons in electrical conduction is not always the same as that for thermal conduction.

Wimshurst machine An electrostatic high voltage generator that consists of two counter-rotating glass disks that have a large number of metal plates at their perimeters. The disks usually face each other since their axes coincide, and the metal plates are situated on the outer faces of the disks. Metal combs, placed on the outer sides of the disks, collect charge from the metal plates of opposite polarities. Metal grounding brushes are placed at the top and bottom of each disk. Wimshurst machines can vary in size and method of rotating the disks. Some demonstration models are spun by hand while others used for generating X-rays are motor-driven.

wind effect, on sound Sound travels better with the wind than against it. The velocity of wind increases from the earth's surface upwards. If the wave is traveling against the wind then its upper portion will be retarded more than the lower; the opposite occurs when the sound travels *with* the wind. In this case the upper portion moves with a greater velocity than the lower, the direction of motion of the wavefront is gradually brought down toward the ground, and the observer may experience a concentration of sound.

winding, primary The coil winding in a transformer that receives the energy from the supply circuit. *See* transformer.

winding, secondary The coil winding in a transformer that receives energy from the primary winding. *See* transformer.

wind instruments Any instrument that can be played by breath or air, e.g., flute, horn or organ.

wolf note On stringed instruments, a very difficult note to find, which on being produced causes the whole body to vibrate to an unusual degree.

woofer   A loudspeaker in a sound-reproducing apparatus that picks up low sound frequencies up to 500 Hz.

work hardening   Repeated bendings of a bar of soft metal, until the bar eventually refuses to be bent and breaks. With every bending, more and more dislocations flow into the metal until there are so many dislocations that they impede each other's flow. The crystal is then incapable of further plastic deformation and breaks under subsequent stress.

work of electrical force   The energy expended by an electric field, **E**, in moving a charge, q, a displacement, **l**. For a uniform electric field this is given by

$$W = q\mathbf{E} \cdot \mathbf{l} ,$$

where W is the work done and $q\mathbf{E}$ is the electric force on the charge. Equivalently, it is given by

$$W = qV ,$$

where V is the potential difference between the initial and final position of the charge. In a non-uniform electric field, the work is defined as

$$W = q \int_A^B \mathbf{E} \cdot d\mathbf{l} ,$$

where $d\mathbf{l}$ is an infinitesimally small displacement, and A and B are the initial and final positions respectively. Generally, one deals with the work done by an external force in moving the charge from A to B in the electric field. This is the negative of the work done by the field and is thus given by

$$W = -q \int_A^B \mathbf{E} \cdot d\mathbf{l} .$$

X

xerography   An electrostatic process for reproducing documents.   Many technical approaches have been used but the most common is to use a photoconductive layer. The optical image to be copied is formed on the surface of the layer which is then electrostatically charged by corona discharge. The charge leaks off at the optically exposed regions, forming an electrostatic image; colored particles (toner) are then deposited at the charged regions and the final image is formed.

xeroradiography   Technique in radiology similar to X-rays where instead of an X-ray film, a positively charged selenium plate is used. Exposure to radiation reduces the positive charge of the plate in different regions, depending on the level of exposure received by each region. Developing of the impressed image follows by applying toner powder to the plate.

The xeroradiography imaging process produces an image with higher resolution on the edges of bones and better visualization between different soft tissue structures in the patients' body. It is widely used in mammography.

X-ray fluorescence   Absorption of an incident X-ray and subsequent re-emission of an X-ray by fluorescence from a higher atomic core energy level. The selection rules for the process in terms of atomic quantum numbers are $\Delta n \geq 1, \Delta \ell = \pm 1, \Delta j = 0, \pm 1$.

X-ray fluorescence is carried out with an X-ray source (Coolidge tube) used to irradiate a solid or liquid sample. The emitted X-rays are analyzed by an X-ray spectrometer.

X-rays   Electromagnetic waves in the range of about 0.1 nm wavelength. They are generated by electronic transitions due to the bombardment of materials of high atomic weight by high energy electrons.

X-rays have many applications in science and technology. Since their discovery, elastic X-ray diffraction has been the principal means for the determination of crystal structure and orientation. Diverse spectroscopic tools (*see* spectroscopy, electron) involving X-rays are used for the characterization of ultra-clean surfaces. In both medicine and industry, X-rays have long been used for the non-destructive detection of defects in the interior of opaque objects.

Y

yoke   Completes a magnetic circuit. Usually a soft ferromagnetic material. Can be in contact with a core that is also made of a soft ferromagnetic material and is surrounded by a current-carrying coil. The yoke transmits the magnetic flux from the core through the yoke to another part of the magnetic circuit.

Young-Helmholtz law   When any point of a string is plucked, struck or bowed, all the overtones requiring that point for a node will be absent from the vibration. Therefore, the point of plucking determines the quality of the note emitted.

Y-parameters   Variables that are in response to other parameters and plotted in the y-axis.

Z

Zeeman effect   Discovered by P. Zeeman in 1896 who observed the broadening of spectral lines in the presence of a magnetic field. It was later shown that the lines were split into doublets, triplets and higher order. The effect was explained by Lorentz in his classical theory of the electron: If v_o is the electronic frequency at zero field, then in a magnetic field H, it is split into components $v_o \pm \Delta v$, where $\Delta v = \frac{eB}{m}$, where e = electronic charge, m = electronic mass and c = velocity of light.

The Lorentz theory predicts the following behavior:

(**a**) *Transverse Zeeman effect* (light beam perpendicular to $\vec{H}$): splitting of the line into a triplet with plane polarized components. This is called the *normal triplet*.

(**b**) *Longitudinal Zeeman effect* (light beam parallel to $\vec{H}$): splitting of the line into a doublet with components circularly polarized in opposing senses.

In fact, only the normal triplet can be explained by the classical theory. More complex observed behavior such as multi-component splitting higher than three (*anomalous Zeeman effect*), asymmetric splitting, etc. can all be explained by a detailed quantum mechanical treatment.

Zeeman effect, inverse   The Zeeman effect as observed in absorption. It is produced by sending white light through a vapor subject to a magnetic field. Since the light is not completely absorbed, the transmitted components correspond to those observed in emission but they are circularly polarized in the opposite direction.

zener breakdown   Avalanche breakdown in which electrons in a diode rapidly increase through ionization collisions with atoms.

zener diode   A *pn* junction that forms a diode whose reverse current rapidly increases at some particular reverse bias voltage.

zero sound   The effect that occurs at the characteristic temperature in a Fermi gas when the frequency between collisions becomes equal to the applied frequency. The propagation of sound is impossible since the collisions cannot occur fast enough. For the interacting Fermi liquid, a collisionless sound propagation may be excited. In this situation an increase in propagation velocity and a maximum in attenuation is observed as the sample is cooled through ordinary region where ordinary sound gives way to zero sound.

zone of silence   With highly intense sounds such as explosions, these are regions where the sound is not audible.

zone plates   A circular grating that acts as a condensing lens causing the intensity of sound at the center of the grating to be intensified.

zone refining   A melting region in a crystal growth apparatus. The melting region is smaller than the ingot length. Using this technique a more purifying crystal can be made.

zoom lens   A telephoto lens whose focal length can be varied from about 80 mm to 1000 mm or more without changing the sharpness of the image. This is achieved by employing a system of converging and diverging elements, one or more of which can be moved. To keep the f-number unchanged, one usually has a basic imaging system and a variable focus arrangement.